WITHDRAWN BY THE
UNIVERSITY OF MICHIGAN

Misbehaving Proteins
Protein (Mis)Folding, Aggregation, and Stability

Misbehaving Proteins
Protein (Mis)Folding, Aggregation, and Stability

Edited by

Regina M. Murphy
University of Wisconsin
Madison, Wisconsin

Amos M. Tsai
Human Genome Sciences
Rockville, Maryland

 Springer

Regina M. Murphy
Department of Chemical and Biological
  Engineering
University of Wisconsin
15 Engineering Drive
Madison, Wisconsin 53706–1691
USA
regina@engr.wisc.edu

Amos M. Tsai
Human Genome Sciences
14200 Shady Grove Road
Rockville, Maryland 20850
USA
amos_tsai@hgsi.com

Library of Congress Control Number: 2005938663

ISBN-10: 0-387-30508-4
ISBN-13: 978-0387-30508-0

Printed on acid-free paper.

© 2006 Springer Science+Business Media, LLC.
All rights reserved. This work may not be translated or copied in whole or in part without the written permission of the publisher (Springer Science+Business Media, LLC, 233 Spring Street, New York, NY 10013, USA), except for brief excerpts in connection with reviews or scholarly analysis. Use in connection with any form of information storage and retrieval, electronic adaptation, computer software, or by similar or dissimilar methodology now known or hereafter developed is forbidden.
The use in this publication of trade names, trademarks, service marks, and similar terms, even if they are not identified as such, is not to be taken as an expression of opinion as to whether or not they are subject to proprietary rights.

Printed in the United States of America.   (TB/MV)

10 9 8 7 6 5 4 3 2 1

springer.com

Preface

Misfolded aggregated protein once was considered as interesting as yesterday's trash—a bothersome by-product of important and productive activities, to be disposed of and forgotten as quickly as possible. Yesterday's trash has become today's focus of considerable scientific interest for at least two reasons: (1) protein aggregates are at the core of a number of chronic degenerative diseases such as Alzheimer's disease, and (2) aggregation poses significant obstacles to the manufacture of safe, efficacious, and stable protein products.

As interest in protein misfolding, aggregation, and stability has soared beyond the core group of traditional protein-folding scientists, and as substantial scientific progress in understanding and controlling protein misfolding has been achieved, the need to summarize the state of the art became manifest. Although there are many excellent texts and edited collections on protein structure and folding, these volumes tend to relegate protein misfolding and aggregation to a minor role. Review articles and books focused on the biological role of protein aggregates in diseases have been published recently. *Misbehaving Proteins: Protein (Mis)folding, Aggregation, and Stability* differs from these other recent efforts in its emphasis on fundamental computational and experimental studies and in its linkage of disparate *consequences* of protein misfolding (e.g., from clinical manifestations to manufacturing headaches) to their common *causes*.

The volume begins with a brief review of background information on the nature of folded and misfolded proteins and on the forces that drive protein folding, misfolding, and aggregation. The remainder of the text is organized into four themes: theory and computation (Chapters 2 and 3), experimental techniques (Chapters 4–9), mechanisms and model systems (Chapters 10–13), and industrial applications (Chapters 14–15). Our intent is to demonstrate that problems in protein misfolding and aggregation are best tackled by examining the underlying fundamental physicochemical principles. We hope to reveal the many connections between the molecular and the macroscopic, the monomeric and multimeric, the theoretical and the practical. The experimental and computational techniques discussed should assist both novice and experienced investigators in designing studies and in expanding beyond traditional approaches. The case studies from both academic and industrial laboratories demonstrate that the difficult problems of protein misfolding and aggregation will be conquered by application of sound scientific principles.

This volume is of interest to both academic and industrial researchers interested in the fundamental issues of protein aggregation and in strategies for controlling protein aggregation in disease processes or in manufacturing settings. Industrial researchers at biotechnology companies struggle with protein folding and aggregation problems and will be interested in the experimental techniques and practical approaches described. Medicinal chemists at pharmaceutical companies are searching for compounds that interfere with protein aggregation as therapeutic leads for drugs to treat diseases such as Alzheimer's or amyloidoses. The insights gleaned from this volume will make the search more fruitful. Academic researchers interested in protein stability and aggregation are spread throughout chemistry, biochemistry, chemical engineering, and pharmaceutics departments; all may broaden their base knowledge and find some novel connections in this collection. This volume could serve as a supplemental text for advanced undergraduates and graduate-level courses on topics such as protein structure and function, protein design, or protein pharmaceutical processing.

Contents

Part I. Introduction

Protein Folding, Misfolding, Stability, and Aggregation: An Overview 3
   Regina M. Murphy and Amos M. Tsai

Part II. Mathematical Models and Computational Methods

Nonnative Protein Aggregation: Pathways, Kinetics, and Stability
   Prediction 17
   Christopher J. Roberts

Simulations of Protein Aggregation: A Review 47
   *Carol K. Hall, Hung D. Nguyen, Alexander J. Marchut,
   and Victoria Wagoner*

Part III. Experimental Methods

Elucidating Structure, Stability, and Conformational Distributions during
   Protein Aggregation with Hydrogen Exchange and Mass Spectrometry 81
   Erik J. Fernandez and Scott A. Tobler

Application of Spectroscopic and Calorimetric Techniques in Protein
   Formulation Development... 99
   Angela Wilcox and Rajesh Krishnamurthy

Small-Angle Neutron Scattering as a Probe for Protein Aggregation
   at Many Length Scales.. 125
   Susan Krueger, Derek Ho, and Amos Tsai

Laser Light Scattering as an Indispensable Tool for
   Probing Protein Aggregation 147
   Regina M. Murphy and Christine C. Lee

X-Ray Diffraction for Characterizing Structure in Protein Aggregates......... 167
   Hideyo Inouye, Deepak Sharma, and Daniel A. Kirschner

Glass Dynamics and the Preservation of Proteins 193
   Christopher L. Soles, Amos M. Tsai, and Marcus T. Cicerone

Part IV. Fundamental Studies in Model Systems

Folding and Misfolding as a Function of Polypeptide Chain Elongation:
   Conformational Trends and Implications for Intracellular Events 217
   Silvia Cavagnero and Nese Kurt

Determinants of Protein Folding and Aggregation in P22 Tailspike Protein 247
   Matthew J. Gage, Brian G. Lefebvre, and Anne S. Robinson

Factors Affecting the Fibrillation of α-Synuclein, a Natively
   Unfolded Protein 265
   Anthony L. Fink

Molten Globule–Lipid Bilayer Interactions and Their Implications
   for Protein Transport and Aggregation 287
   Lisa A. Kueltzo and C. Russell Middaugh

Part V. Protein Product Development

Self-Association of Therapeutic Proteins: Implications for
   Product Development 313
   Mary E. M. Cromwell, Chantel Felten, Heather Flores, Jun Liu,
   and Steven J. Shire

Mutational Approach to Improve Physical Stability of Protein
   Therapeutics Susceptible to Aggregation: Role of Altered
   Conformation in Irreversible Precipitation 331
   Margaret Speed Ricci, Monica M. Pallitto, Linda Owens Narhi,
   Thomas Boone, and David N. Brems

Index 351

Part I
Introduction

Protein Folding, Misfolding, Stability, and Aggregation

An Overview

Regina M. Murphy[1,3] and Amos M. Tsai[2]

Protein misfolding and aggregation problems arise in diverse arenas. In the manufacture of commercial protein products, correctly folded proteins in stable formulations are critical for safety and efficacy. In the clinic, there is increasing awareness that protein aggregation is an underlying cause of several severe and chronic diseases. This chapter provides a brief background on the nature of folded and misfolded proteins and on the forces that drive protein folding, misfolding, and aggregation. We then give an overview describing the organization and contents of the remainder of this volume.

1. IMPORTANCE OF PROTEIN MISFOLDING AND AGGREGATION

1.1. Manufacture of Protein Products

In a typical biopharmaceutical production process there are many points at which protein aggregation can occur. A protein expressed in *Escherichia coli* often aggregates shortly after it is synthesized. With varying degrees of success, correctly folded active protein can be recovered by solubilization and refolding. When a recombinant protein is made in mammalian cells, the cellular machinery can process the protein such that solubility is often maintained at the time of harvest. However, successful synthesis in the bioreactor is far from a guarantee of a soluble final drug substance. Unwanted aggregation is a common by-product of rigorous purification processes. For instance, monoclonal

---

[1] Department of Chemical and Biological Engineering, University of Wisconsin, Madison, WI 53706
[2] Human Genome Sciences, 14200 Shady Grove Road, Rockville, MD 20850
[3] To whom correspondance should be addressed: Regina M. Murphy, University of Wisconsin, 1415 Engineering Drive, Madison, WI 53706, regina@engr.wisc.edu

antibodies are often purified using protein A chromatography that requires acid pH for elution. Viral inactivation required for mammalian cell-derived products is carried out in even harsher acidic conditions. Acid denaturation of the protein product as well as other host cell biomolecules leads to aggregation and loss of product. Depth filtration, diafiltration, and other similar processes can cause aggregation due to shear-induced protein denaturation. A protein product that survives purification processes then must be formulated for a useful shelf life, to maintain a drug's stability until the point of administration. Cosolutes and other excipients often are recruited to improve a product's stability. As a greater diversity of proteins reach the market, the industry must incorporate new understanding of mechanisms of protein misfolding and aggregation in order to develop robust manufacturing and formulation processes that result in stable, correctly folded, and active products.

1.2. Protein Misfolding Diseases

The problem of protein structural instability, misfolding, and aggregation is not limited to manufacturing. The "protein misfolding" diseases constitute a newly recognized group of diseases with a diverse array of symptoms. Some of the most common protein misfolding diseases are neurodegenerative, including Alzheimer's disease, Parkinson's disease, Huntington's disease, and the prion diseases. These diseases share a common feature: the deposition of insoluble, usually fibrillar, β-sheet-rich protein aggregates. The source and nature of the aggregating protein, the location of the deposit, and the biological consequences differ from disease to disease (Table 1). The factors that trigger formation of aggregates and the mechanisms by which aggregation leads to disease are poorly understood. Development of effective treatment and prevention therapies for these diseases requires elucidation of the molecular basis for protein misfolding and aggregation.

2. PROTEIN STRUCTURE AND PROTEIN FOLDING

2.1. Amino Acids

Twenty amino acids make up the library from which natural proteins are synthesized (Table 2). Within their side chains is contained a wide diversity of chemical function. Table 3 summarizes important physicochemical properties of these amino acids. At a fundamental level, these properties direct the folding, misfolding, and aggregation of proteins. It should be noted that many different scales have been proposed to measure the relative hydrophobicity of the side chains.

Covalent modifications of these side chains by reactions such as phosphorylation, glycosylation, oxidation, or deamidation introduced by design or by accident further increase chemical diversity and affect protein structure and stability. In a manufacturing environment, the ability to control, minimize, or completely eliminate these modifications directly contributes to the quality attributes of a product. For instance, the glycosylation pattern of a monoclonal antibody can be part of the release specifications and one of

TABLE 1. Protein misfolding diseases[1,2]

Disease	Aggregating protein	Characteristics of protein	Nature of deposits	Major affected brain regions
Alzheimer	Beta-amyloid	4-kDa peptide cleaved from the membrane-bound precursor protein APP	Extracellular amyloid fibrils	Hippocampus, cortex
Huntington	Huntingtin	350-kDa protein, with an expanded (>35) polyglutamine domain in the disease state; number of glutamines correlates inversely with age of onset	Intranuclear inclusions, cytoplasmic aggregates	Striatum, basal ganglia
Parkinson	Alpha-synuclein	14-kDa natively unfolded protein	Lewy bodies, a cytoplasmic inclusion body often localized near the nucleus	Substantia nigra
Prion diseases (kuru, CJD[a], others)	Prion protein	~34-kDa glycoprotein that is normal cell-surface component of neurons	Extracellular and intracellular amyloid deposits	Cortex, thalamus, brain stem, cerebellum

[a] Creutzfeldt-Jakob disease

the quality attributes that defines lot-to-lot consistency. Oxidation or deamidation are considered product-related impurities raising regulatory concerns because such product variants could affect a drug's potency, mode of action, and immunogenicity. Control of the level of covalent modifications consistently is required for product release. In protein-folding diseases, covalent modifications are known to affect the aggregation propensity and aggregate morphology of relevant proteins.

Unnatural amino acids are readily incorporated into peptides and short proteins by solid-phase chemical synthesis. Biosynthetic routes for incorporation of a few analogs of natural amino acids recently have been developed.[6] These methods provide a technology base that could lead to development of a great variety of designer proteins with highly tunable physicochemical properties. Controlling the folding of these designer proteins and preventing aggregation or facilitating self-assembly (depending on the specific application) present formidable challenges.

2.2. Forces Driving Folding, Misfolding, and Aggregation

Correct folding of a polypeptide chain, containing perhaps ~100 amino acids, into a compact structure is truly a remarkable accomplishment of nature. The possible configurations that a polypeptide can sample during folding are enormous. Yet a typical folding process is extremely rapid, taking place in milliseconds to seconds, and a folded structure unique to that particular chain is reliably generated. This suggests that folding of a denatured chain proceeds through multiple pathways and still moves toward the same

TABLE 2. Chemical structures of common 20 amino acids

Glycine, Gly, G	Alanine, Ala, A	Valine, Val, V	Leucine, Leu, L	Isoleucine, Ile, I	
Aspartic acid, Asp, D	Glutamic acid, Glu, E	Asparagine, Asn, N	Glutamine, Gln, Q	Lysine, Lys, K	Arginine, Arg, R
Serine, Ser, S	Threonine, Thr, T	Cysteine, Cys, C	Methionine, Met, M	Proline, Pro, P	
Histidine, His, H	Phenylalanine, Phe, F	Tyrosine, Tyr, Y	Tryptophan, Trp, W		

final intended structure determined by the amino acid sequence. This view assumes that the proper folded structure has the lowest free energy and that the free energy of folding acts to guide the chain along different pathways that lead to the final structure.[7] It is important to recognize that a folded protein is a collection of closely related structures in equilibrium with each other. The conformational distribution of all the states in the ensemble allows a protein to perform its function, but at the same time makes it prone to denaturation. The forces that contribute toward the overall folding free energy include hydrogen bonding, hydrophobic interactions, electrostatics, and conformational or entropic forces. Correct folding of a protein requires a delicate balance of forces between different parts of the polypeptide chain and between the polypeptide and surrounding water molecules. Interestingly enough, the same forces that drive protein folding also

TABLE 3. Key physicochemical properties of amino acid side chains[3-5]

Amino acid residue	Molecular weight (Da)	pKa of side chain (in polypeptides)	Accessible surface area (Å^2)	Hydrophobicity of side chain analogs (kJ/mole)
Alanine	71		113	−3.65
Arginine	156	~12	241	66.61
Asparagine	114		158	21.83
Aspartic acid	115	3.9−4.8	151	40.57
Cysteine	103	8.8−9.5	140	−1.42
Glutamine	128		189	27.21
Glutamic acid	129	3.9−4.8	183	32.55
Glycine	57		85	0
Histidine	137	6−7.5	194	23.52
Isoleucine	113		182	−16.71
Leucine	113		180	−16.71
Lysine	128	9.8−11.1	211	27.25
Methionine	131		204	−5.92
Phenylalanine	147		218	−8.57
Proline	97		143	
Serine	87		122	18.23
Threonine	101		146	14.74
Tryptophan	186		259	−5.84
Tyrosine	163	9.4−10.8	229	4.54
Valine	99		160	−13.02

drive protein misfolding and aggregation. Indeed, perhaps the relevant question is not why some proteins misfold and aggregate, but why most do not!

2.2.1. Hydrogen bonding Hydrogen bonds between carbonyl and amide groups along the polypeptide backbone stabilize the basic structural elements of folded proteins. Although formation of a hydrogen bond between two moieties on the backbone means the loss of hydrogen bonds between peptide amides and water, the multivalency possible in long helices and β-sheets stabilizes these structures relative to the unfolded protein. Urea (H_2N-CO-NH_2) and the guanidinium ion (H_2N-CNH$_2^+$-NH_2) denature proteins in part by competing for hydrogen bonds with the polypeptide backbone.

Hydrogen bonding, although it can explain the stability of folded proteins, cannot explain by itself why one folded structure is native and another folded structure is not. Intermolecular hydrogen bonding is important for stabilizing aggregated proteins with defined structural elements such as β-sheet-rich amyloid fibrils.

Several amino acid side chains also are capable of participating in hydrogen bonds. Of particular note are the amide side chains glutamine and asparagine. Expanded polyglutamine domains are involved in the abnormal protein aggregates occurring in Huntington's disease as well as several less common diseases. It is believed that hydrogen bonding between glutamine side chain and backbone amides stabilize these aggregates.

2.2.2. The hydrophobic effect Burial of hydrophobic side chains in the protein core (hydrophobic collapse) is essential for the development of tertiary structure; indeed, collapse to a molten globule state may sometimes precede formation of secondary structural elements such as helices. The driving force for burial can be considered most simply as the difference between the weak attractive van der Waals interactions between nonpolar groups and the weak attractive van der Waals interactions of nonpolar groups with water. The hydrophobic effect is unusual in that it increases with temperature, whereas the other forces favoring protein structure decrease with temperature. This change in the relative importance of forces often results in misfolding and aggregation at higher temperatures.

2.2.3. Coulombic interactions The basic amino acids lysine and arginine and the acidic amino acids glutamate and aspartate carry charge at neutral pH. Histidine also may be charged at neutral or slightly acidic pH. Charged residues most often are found on the exterior of a correctly folded protein, but ionic pairing between a positively and negatively charged residue may occur in the interior of a protein; such an interaction may be unusually strong because of the very low dielectric constant in the interior of a folded protein.

Perhaps of more importance, although less easily understood, is the role of dipole-ion and dipole-dipole interactions in stabilizing protein structure. The peptide bond itself as well as many side chains function as dipoles because of the different electronegativities of the atoms. For example, interactions between the tyrosine dipole and charged amino acids (e.g., glutamate, aspartate) stabilize some folded protein structures.[8] Protein aggregation and insolubility with pH adjusted to at or near the isoelectric point is a well-known phenomenon, caused by the loss of repulsive electrostatic interactions.

2.2.4. Disulfide bond formation Disulfide bonds are formed during folding when two cysteine side chains are brought into close contact under oxidizing conditions. These covalent bonds cross-link tether two sections of a polypeptide chain and stabilize folded structure. For polypeptides with multiple cysteines, incorrect disulfide bond formation may lock the protein in a misfolded conformation and could lead to further aggregation.

2.3. Role of Cosolutes

Proteins and peptides operate in a complex environment of salts, other proteins, carbohydrates, lipids, and other solutes. Each of these cosolutes influences the folding and aggregation properties of a protein. Cosolutes may be classified as kosmotropes (structure stabilizers) or chaotropes (structure destablizers). Both correctly folded proteins and misfolded structured protein aggregates are more structured than unfolded polypeptide chains; thus, addition of kosmotropes may enhance aggregation as much, if not more, than it enhances correct folding.

The Hofmeister series is a useful tool for correlating the chaotropic or kosmotropic nature of cosolutes (Figure 1). In general, kosmotropes act by a preferential exclusion (also called preferential hydration) mechanism. Essentially, kosmotropes are preferentially excluded from the protein-solvent interface. As a result, the water concentration near the interface is higher than in the bulk solvent and there is an increased driving force for burial of hydrophobic residues. This stabilizes folded (or misfolded) protein structures. Stabilization of protein structure correlates well with the kosmotrope's ability to increase the surface tension of water.[9] Chaotropic action generally is believed to be mediated by preferential binding. The chaotrope preferentially associates with specific chemical moieties on the side chain or backbone, thus favoring increased protein-solvent interfacial area and therefore unfolding. The behavior of some cosolutes such as arginine

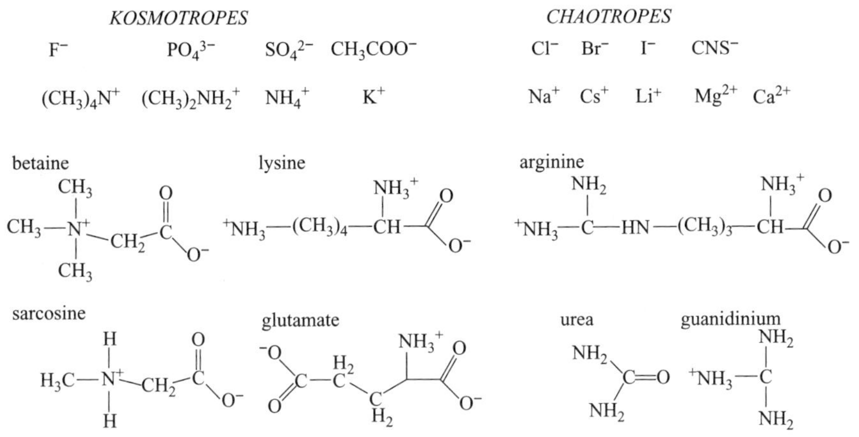

KOSMOTROPES

F⁻   PO₄³⁻   SO₄²⁻   CH₃COO⁻

(CH₃)₄N⁺   (CH₃)₂NH₂⁺   NH₄⁺   K⁺

CHAOTROPES

Cl⁻   Br⁻   I⁻   CNS⁻

Na⁺   Cs⁺   Li⁺   Mg²⁺   Ca²⁺

betaine, lysine, arginine, sarcosine, glutamate, urea, guanidinium [structural formulas]

FIGURE 1. Common kosmotropes and chaotropes of interest in protein-folding studies.

is complex: arginine specifically interacts with and therefore destabilizes many proteins; however, arginine increases water surface tension and can alternatively behave as a kosmotrope. Furthermore, the protein stabilizing or destabilizing activity of a particular cosolute may be a function of the cosolute's concentration.

2.4. Predicting Folded Structure and Aggregation Propensity

Several programs, available free of charge to academic users, are of interest. The Biology Workbench, operated under the San Diego Supercomputer Center (http://workbench.sdsc.edu), provides a slate of tools that, for example, allow the user to search databases for known sequences, predict secondary structure, plot the hydrophobic profile of a sequence, estimate isoelectric points, or align multiple sequences. The European Molecular Biology Laboratory (EMBL) has developed two programs of particular interest. FoldX (http://foldx.embl.de) provides a quantitative analysis of the effects of mutations on protein stability. Tango (http://tango.embl.de) predicts regions of unfolded polypeptides that are most likely to initiate aggregation.

3. KINETICS AND THERMODYNAMICS OF PROTEIN FOLDING AND AGGREGATION

Protein folding from the denatured state often is modeled as a simple two-state transition between the unfolded U and the native N states:

$$U \leftrightarrow N. \tag{1}$$

The folded protein is only marginally thermodynamically stable relative to the unfolded protein. The Gibbs energy of stabilization (net Gibbs energy difference between

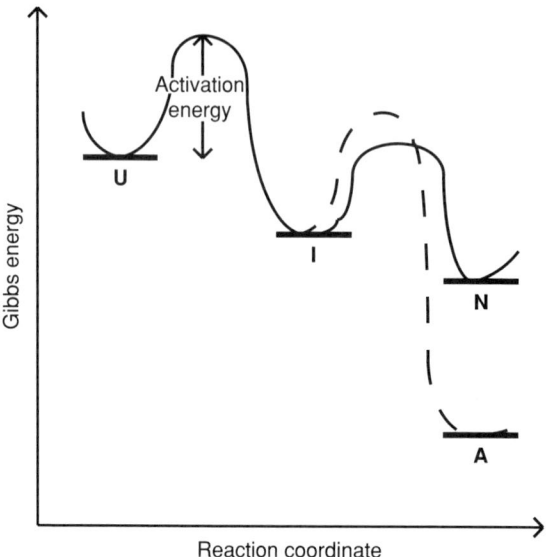

FIGURE 2. Energetic relationship between unfolded protein U, natively folded protein N, intermediate I, and aggregate A. In this example, the partially refolded I can follow two pathways. One path, with a lower activation energy, produces the natively folded protein N. The other path, shown as a dashed line, has a higher activation energy but produces the lower-energy aggregate A. Changes in pH, temperature, or cosolute concentration may affect both the activation energies and the energy of the metastable states.

folded protein and unfolded polypeptide chain) is generally -20 to -60 kJ/mol.[10] Since

$$K = \frac{[N]}{[U]} = \exp\left(\frac{-\Delta G}{RT}\right), \qquad (2)$$

at physiological temperatures, $\sim$0.00001% to 0.01% of polypeptide chains are unfolded.

The relationship between protein folding, misfolding, and aggregation can be understood through a reaction kinetic framework. This framework treats the folding process as a unimolecular equilibrium between the folded and the unfolded states (Eq. 1). In the forward and reverse reactions, the folding energetics can be incorporated in the appropriate form in the reaction constants. The two-state model has been used successfully to explain protein-folding data captured from thermal melting, chemical denaturation, hydrogen exchange, fluorescence spectroscopy, and similar experimental studies.

Proteins that tend to aggregate, however, may not be accurately modeled using a simple two-state model. Rather, formation of one (or more) metastable, partially folded aggregation-prone intermediate I often is postulated (Figure 2). Several examples of these multistate transitions are given in later chapters in this volume. Another plausible model for aggregating proteins postulates the formation of two alternate conformers (I_1 and I_2) from the unfolded polypeptide chain, one of which leads to aggregates and the other to correctly folded protein. Perhaps proteins with regions of "conformational confusion," where alternate folded structures have similar thermodynamic stability, are most likely to be aggregation-prone.

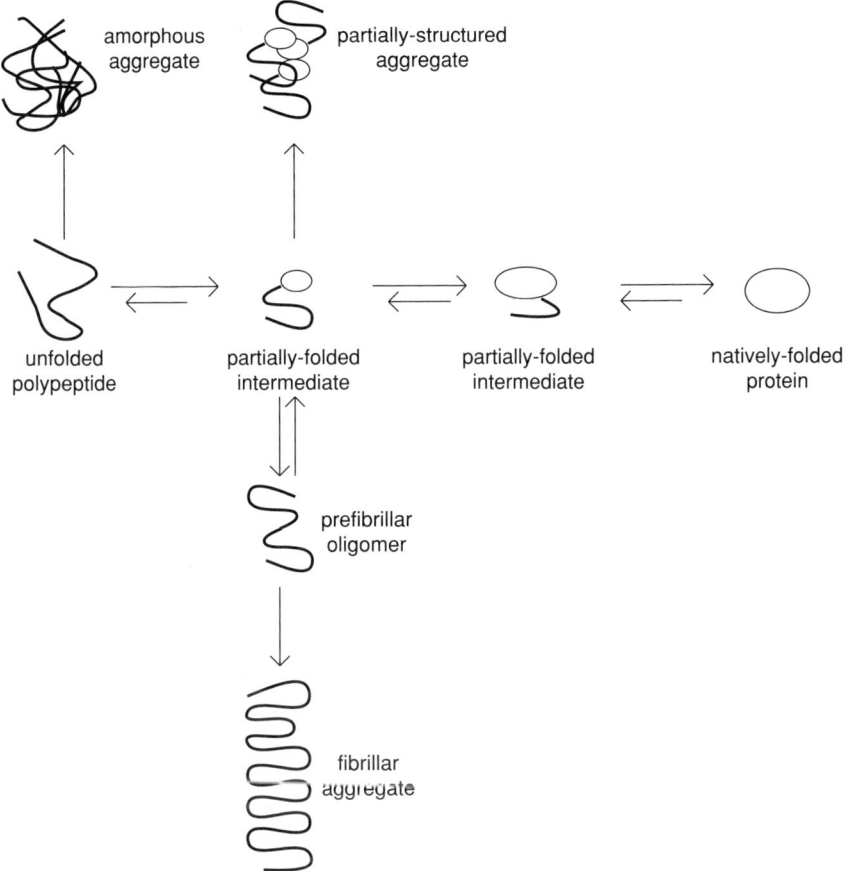

FIGURE 3. Kinetic pathways leading from unfolded protein to aggregates with various morphologies. In this sketch formation of large aggregated species is shown as irreversible; in some cases these steps may be partially reversible. Adapted from Foguel and Silva.[11]

In favorable cases, small proteins refold rapidly and correctly within seconds. In these cases, protein-refolding kinetics often can be modeled using a single forward and reverse rate constant. Slow steps include *cis-trans* isomeration of peptide bonds preceding prolines and disulfide bond formation. If metastable intermediates form, however, complex kinetic expressions may be required to model protein refolding. Refolding kinetics are generally first-order, while aggregation kinetics may be first-order or higher-order, depending on the exact mechanism and rate-limiting steps (Figure 3).

At one point it was thought that misfolded aggregated protein was simply an amorphous mass, lacking any kind of organized structure. This turns out to be untrue; many aggregates that appear grossly amorphous retain some evidence (by FTIR or other means) of secondary structural elements and some aggregates such as amyloid fibrils are highly structured, although not crystalline. Micelles, globular assemblies, fibrils, ribbons, sheets, and other morphologies have been reported. There are examples of polypeptides forming alternatively folded conformational states, one of which leads to amorphous aggregates and the other leading to fibrils.[12] There is continuing discussion as to whether these various aggregate morphologies are stuck in kinetic traps (local minima in the energetic landscape) or are globally equilibrated.

4. OUTLINE OF CHAPTERS

This book is organized into four themes: theory (Chapters 2 and 3), experimental techniques (Chapters 4–9), mechanisms and model systems (Chapters 10–13) and industrial applications (Chapter 14–15).

In Chapter 2, Chris Roberts extensively discusses a reaction kinetics framework to model the aggregation pathways and provides guidelines for experimental design and data interpretation. Carol Hall and colleagues, in Chapter 3, give us a glimpse of the future in computer simulation. While full atomic scale modeling of protein aggregation processes is still far into the future, the approaches presented in this chapter provide detailed and novel insights into aggregation phenomena.

The second section of this volume introduces us to techniques of special interest to the study of protein aggregation. Fernandez and Tobler discuss the use of hydrogen exchange coupled with mass spectrometry to identify regions within a molecule that are involved in aggregation. Three experimental tools widely used for protein folding studies—circular dichroism, fluorescence spectroscopy, and calorimetry—can be adapted for investigations of aggregating systems, as discussed by Wilcox and Krishnamurthy in Chapter 5. In separate chapters, Krueger and colleagues. and Murphy and Lee, describe how neutron and light scattering, respectively, afford us the ability to probe structure over a wide range of length scales. By using appropriate mathematical models to analyze scattering data, one can gain additional insight into the geometry of the aggregates and the kinetics of growth. Daniel Kirschner and colleagues describe X-ray diffraction techniques and theory and demonstrate how this method is employed to identify the existence of different morphologies along the aggregation pathway. Lastly, Soles and colleagues. explain how the dynamics of a protein at the atomic level can be directly measured and how molecular motion is intimately related to the macroscopic stability of a protein, giving us a glimpse of the earliest events that may eventually lead to aggregation.

Chapters 10–13 describe the mechanisms through which folding, misfolding, and self-assembly or aggregation may occur in model systems. Cavagnero and Kurt demonstrate that cotranslational folding, leading to early formation of nativelike secondary and tertiary structures, often make a protein aggregation-prone. Anne Robinson and co-workers show that aggregation is species-specific, i.e., protein aggregation is a self-assembly process even in the midst of other proteins that can be recruited into the process. Using a natively unfolded protein, Fink uses chemicals identified as amyloid disease risk factors to study the mechanism of fibril formation *in vitro*. Kueltzo and Middaugh discuss how formation of the structurally labile molten globule state facilitates protein transport across membranes. Read together, these chapters offer us a comprehensive insight on how protein structure, dynamics, and stability contribute to the formation of either highly structured or amorphous aggregates.

The last section of the volume contains two case studies from the biotechnology industry. Mary Cromwell's chapter discusses the unique challenges presented during protein product development in trying to monitor and control native protein aggregation. Ricci and co-workers describe in detail an exhaustive search for mutations in a protein that lead to improved stability and reduced aggregation. These two case studies illustrate beautifully the fundamental themes and approaches discussed in earlier chapters, thereby confirming their universal appeal.

In this volume, our intent is to show that protein misfolding and aggregation is best tackled by breaking down the problem into its fundamental components. We hope to reveal the connections between the molecular and the macroscopic, the monomeric and the multimeric, the theoretical and the practical. The many techniques reviewed here should assist both novice and experienced investigators to formulate their studies and to expand beyond traditional approaches in their search for additional insight. The examples included in this volume, from both academic and industrial laboratories, should give hope that the apparently unwieldy problem of protein misfolding and aggregation will succumb to sound scientific approaches.

REFERENCES

1. C. A. Ross and M. A. Poirier, Protein aggregation and neurodegenerative disease, *Nature Med.* **10**, S10–S17 (2004).
2. R. M. Murphy, Peptide aggregation in neurodegenerative disease, *Annu. Rev. Biomed. Eng.* **4**, 155–174 (2002).
3. N. J. Darby and T. E. Creighton, *Protein Structure* (Oxford University Press, New York, 1993).
4. T. E. Creighton, *Proteins: Structures and Molecular Properties* (New York: W. H. Freeman and Co., 1984).
5. J. D. Rawn, *Biochemistry* (New York: Harper and Row, 1983).
6. T. L. Hendrickson, V. de Crecy-Lagard, and P. Schimmel, Incorporation of nonnatural amino acids into proteins, *Annu. Rev. Biochem.* **73**, 147–176 (2004).
7. K. A. Dill, Polymer principles and protein folding, *Protein Sci.* **8**, 1166–1180 (1999).
8. T. Cserhati and M. Szogyi, Role of hydrophobic and hydrophilic forces in peptide-protein interaction: new advances, *Peptides* **16**, 165–173 (1995).
9. M. G. Cacace, E. M. Landau, and J. J Ramsden, The Hofmeister series: salt and solvent effects on interfacial phenomena, *Q. Rev. Biophys.* **30**, 241–277 (1997).
10. R. Jaenicki, Protein stability and protein folding, *Ciba Foundation Symposium* **161**, 206–221 (1991).
11. D. Foguel and J. L. Silva, New insights into the mechanisms of protein misfolding and aggregation in amyloidogenic diseases derived from pressure studies, *Biochemistry* **43**, 11361–11370 (2004).
12. R. Khurana, J. R. Gillespie, A. Talaptra, L. J. Minert, C. Ionescu-Zanetti, I. Millett, and A. L. Fink, Partially folded intermediates as critical precursors of light chain amyloid fibrils and amorphous aggregates, *Biochemistry* **40**, 3525–3535 (2001).

Part II
Mathematical Models and Computational Methods

Nonnative Protein Aggregation

Pathways, Kinetics, and Stability Prediction

Christopher J. Roberts[1,2]

Protein aggregation via nonnative conformational states is effectively irreversible for a variety of systems of scientific and commercial interest, particularly during processing and storage. The formation of irreversible aggregates typically involves one or more reversible conformational changes leading to nonnative, aggregation-prone conformers that subsequently assemble to form soluble or insoluble aggregates. Experimentally observed kinetics of this process are controlled by a combination of the dynamics and thermodynamics of conformational transitions and association or assembly steps. Many of the useful models describing nonnative aggregation kinetics can be understood within a generalized Lumry-Eyring scheme, with each particular model appearing as a limiting case. This chapter presents such a generalized kinetic scheme for nonnative protein aggregation and the model solutions that result for limiting cases that follow from different rate determining step(s) in the overall aggregation process. These limiting cases are presented and discussed from the perspective of their physical and mechanistic bases. Implications for measuring and interpreting experimental kinetics and for prediction of protein aggregation kinetics from accelerated stability studies are detailed. It is stressed that truly *a priori* predictions of protein stability, either quantitative or qualitative, are untenable by these or other presently available methods. Rather, the utility of such kinetic models lies in the guidance they provide for devising experimental studies to predict stability via extrapolation from known conditions, and the constraints they place on mechanistic interpretation of experimental kinetics

---

[1] Department of Chemical Engineering, University of Delaware, Newark, DE 19716
[2] To whom correspondence should be addressed: Christopher J. Roberts, University of Delaware, Newark, DE 19716; email: roberts@che.udel.edu

1. INTRODUCTION

Many proteins of therapeutic and biotechnological interest are susceptible to degradation via a variety of mechanisms: aggregation via nonnative states being one of the most prevalent.[1-3] Commercially, nonnative or irreversible aggregation often is of concern during manufacturing and purification steps and during subsequent product storage, shipment, and administration, as it leads to loss in potency and potentially to increased safety risk.[1-3] Irreversible aggregation is necessarily a kinetic phenomenon, although thermodynamic considerations such as conformational free energy also play significant roles. If one accepts the hypothesis that aggregates are lower in free energy than the (unaggregated) native state,[4,5] it follows that any protein product or sample ultimately will aggregate over sufficiently long storage or operating times, and implies a number of important practical questions: What are the kinetics of (active) protein loss or aggregate formation, and how do they arise mechanistically? How can those kinetics be practically controlled for a given product? How (if at all) can those kinetics be predicted or estimated for conditions for which direct measurements are unavailable or untenable, e.g., long-term (multiyear) storage, or unforeseen deviations in operating or storage conditions? The primary objectives of this chapter are to illustrate the utility and limitations of classical thermodynamic and macroscopic (mass action) kinetic models for addressing questions such as these, and to provide aids in the interpretation and design of experimental stability measurements.

Irreversible protein aggregation, as opposed to reversible flocculation or precipitation, is highly sensitive to protein conformation. Generally, nonnative, denatured, or partially folded protein conformers, collectively termed the denatured state (D) here, are thought to be much more prone to irreversible aggregation than folded or native proteins (N), due to increased solvent exposure of hydrophobic groups believed to result in strong, effectively irreversible, interprotein contacts.[1-3,6-12] The "denatured" state in this context encompasses the ensemble of protein conformers that are partitioned from the relatively small ensemble of native or folded conformers by a cooperative "all-or-none" transition, with N and D states separated by a significant free energy barrier.[5,13,14] Available evidence indicates irreversible aggregation often occurs between protein monomers or relatively small oligomers that have lost most if not all tertiary structure but maintain significant secondary structure in their D state.[1,7-12] Retention of secondary structure is argued to help to provide large, contiguous "patches" of hydrophobic protein-solvent interface that would be less prevalent in fully unfolded proteins and that presumably impart stronger hydrophobic or solvent-mediated attractions. Interestingly, recent evidence suggests a mechanism for some proteins in which D proteins first associate *reversibly*, followed by a structural rearrangement that creates even stronger intermolecular contacts, resulting in those proteins being committed to the aggregation pathway.[9,15-17] In the terminology commonly used to describe protein fibril formation kinetics, this would constitute the "nucleation" or initiation stage.[16] Such a structural rearrangement or initiation step within an otherwise reversible oligomer is one possible explanation for the observation that aggregated proteins may exhibit significantly different secondary structures from those observed in the parent D or N states.[8,9,15,17]

A reaction scheme for nonnative aggregation that captures the qualitative considerations outlined above is given in Figure 1 for a protein that is natively monomeric.

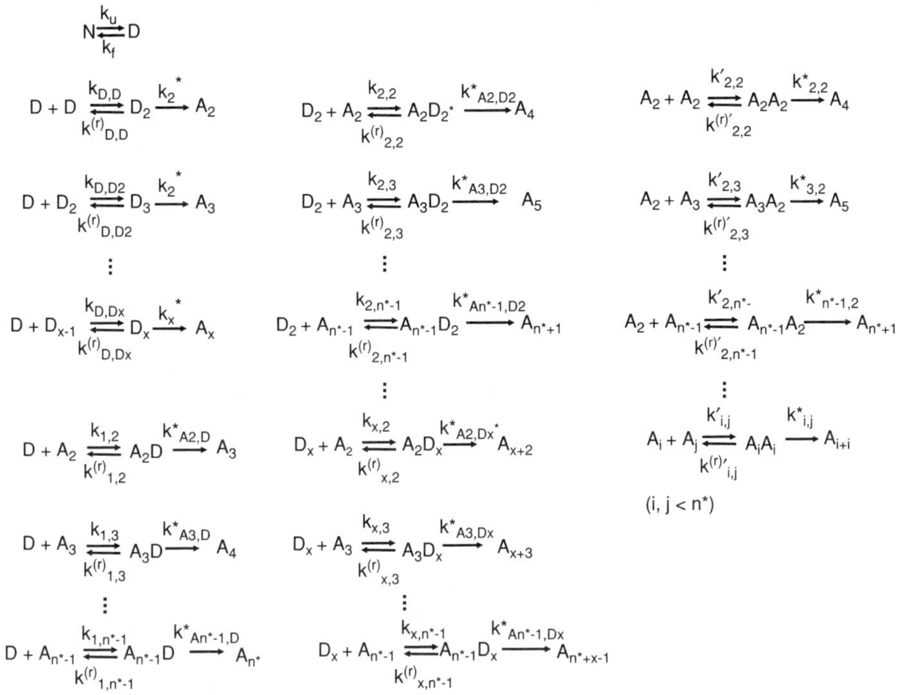

FIGURE 1. Kinetic scheme for protein aggregation proceeding via the nonnative state. N and D denote native and denatured or nonnative monomeric conformational states, respectively. D_j is a reversible oligomer composed of j nonnative monomers. A_j is an irreversible aggregate composed of j monomers. Rate coefficients and equilibrium relations are described in the text.

"Monomer" is not strictly a single protein chain in the context of this chapter and the models considered here. It instead denotes the protein in whatever state of association is dominant for the native state. So long as the native and denatured states incorporate the same number of protein chains, one may define the total monomer content unambiguously as the sum of unaggregated native and denatured proteins.

The folding-unfolding transition is two-state and reversible, though not necessarily in equilibrium. At the level of a classical kinetic description, such as given here, a simple two-state N-D description is a good approximation in a variety of cases, and is used throughout the majority of this chapter. Some simple arguments regarding the sufficiency, or lack thereof, of a two-state approximation in the context extended Lumry-Eyring aggregation kinetics are presented in Section 3.4 for the interested reader. A natural way to extend the analyses to "multistate" proteins also follows from that discussion and is included there as a guide in dealing with proteins exhibiting more complex conformational thermodynamics. Extension of such a description to account for folding-unfolding involving changes in the level of association, e.g., for a multimeric native protein, can be done provided one accounts properly for the differences in stoichiometry of the N and D states, and any concentration dependences of the equilibrium constant(s) for folding-unfolding. Explicit consideration of natively multimeric proteins (i.e, N_2, N_3, etc.) are foregone here, but can be included by extending the analysis in Section 3.4. Proteins in which the D state is multimeric[18] are implicitly included via the reversible association steps involving only D oligomers in Figure 1.

In the scheme in Figure 1, D monomers may self-associate reversibly, and the reversible oligomers so formed may further undergo irreversible conformational transitions to create (irreversible) aggregates (A_j) composed of j monomers that were originally denatured. In general, aggregates also may associate reversibly with D monomers and oligomers, and with other aggregates. It is presently unknown whether association involving previously formed aggregates would also involve a sequence of reversible association(s) and irreversible conformational steps. This is, in principle, possible and is treated as such in Figure 1 to provide a more general scenario. Addition of native monomers to previously formed aggregates is not included in Figure 1 simply because the large majority of published reports implicate only nonnative conformers as the "reactive" species in irreversible aggregate formation. For completeness, we note there is at least one recent report in which both native and nonnative monomers appear to be aggregation prone.[19]

The remainder of this chapter is presented in four sections. Section 2 presents the mathematical description of the nonnative aggregation scheme in Figure 1. In its most general form, the model is intractable for direct application to experimental data or predicting protein stability in practical situations, if only because it is not typically possible to independently measure time- and initial-concentration-dependence of the populations of all molecular species involved, and therefore impossible to reliably estimate all of the relevant kinetic and thermodynamic parameters in a scheme as detailed as Figure 1. As a result, solution of such a model for practical applications requires making at least a (hopefully small) number of physically motivated approximations that can be justified qualitatively or quantitatively by available data. In this spirit, Section 3 considers a number of limiting cases of the general model and derives tractable mathematical expressions describing each case that may then prove more useful in practical applications and data analysis. (Reference 20 may be useful to some readers as a detailed introduction to mathematical modeling of reacting systems, of which protein aggregation can be considered a subset if "reactions" are generalized to include conformational and non-covalent association states.) Section 4 presents a series of straightforward experimental analyses that are suggested by the results of Section 3. These have utility in identifying which kinetic description(s) is(are) most likely to be appropriate for a given experimental system and by inference which steps in the generalized aggregation pathway are most important for control of the observed kinetics of that system. Section 5 addresses potential application of the limiting kinetic models for prediction of protein stability via extrapolation from accelerated (high temperature) conditions.

2. GENERALIZED OR EXTENDED LUMRY-EYRING MODEL OF NONNATIVE AGGREGATION

The reaction scheme in Figure 1 is an extended or generalized version of the original Lumry-Eyring (LE) picture,[21] in which only folding-unfolding events were treated in detail, i.e., the $N \leftrightarrow D$ transition in Figure 1 and variations upon it. Classic Lumry-Eyring models either do not account explicitly for the process that consumes D protein, or treat it empirically by assuming a pseudo-first-order irreversible step.[21–25] With the exception of simple unfolding-limited kinetics (Section 3.1), it is not possible to capture many of the experimentally observed kinetics of protein aggregation within such a classic LE description. An extended or generalized LE description, such as discussed here, encapsulates both the standard model behaviors and more complex situations of practical interest, and so provides a more general, self-consistent approach to modeling nonnative aggregation.

All steps in Figure 1 are taken as *elementary* steps in the reaction pathway. That is, the stoichiometry of the step, as written, also connotes the inherent reaction order with respect to the concentration of each species involved. No higher-than-second-order elementary steps are considered, based on general probability arguments regarding infrequency of true multi-body events.[20] In Figure 1, we first consider reversible conformational transitions between N and D states of single monomers, with forward (unfolding) and reverse (folding) rate coefficients denoted by k_u and k_f, respectively. From the principle of dynamic equilibrium, we also have

$$\frac{k_u}{k_f} = K = \left(\frac{[D]}{[N]}\right)_{eq} = \exp\left(-\Delta G_D/RT\right) \qquad (1)$$

where K is the equilibrium unfolding constant, the subscript eq indicates that the ratio of D and N concentrations are equal to K only when they are in (pseudo-) equilibrium, and ΔG_D is the standard Gibbs energy, or free energy, of D relative to N at a given temperature, pressure, and composition. T is the absolute temperature, and R is the gas constant. Considering next the simplest aggregation step, two D monomers may self-associate to form a reversible dimer D_2. In Figure 1 this process and all association steps are shown as being potentially reversible. This is not a requirement (see, e.g., Section 3.2) but is a more general representation than simply assuming the association step itself is irreversible. In general, any reversible step in a kinetic mechanism that occurs ahead of the rate-determining (i.e., slowest) step(s) in a process will achieve pseudoequilibrium if the rate-determining step(s) is(are) significantly slower than the characteristic equilibration rate of the reversible step(s). The term pseudoequilibrium, rather than simply equilibrium, is used to acknowledge that the macroscopic concentrations of the species involved are equilibrated with respect to one another—i.e., their ratio equals the associated equilibrium constant—at all times, but their concentrations are time dependent because of the irreversible "downstream" process(es) of aggregation that deplete one or more of them with time. Because those "downstream" steps occur much more slowly than the steps in the reversible ("upstream") processes, the species involved in the upstream steps can easily reequilibrate every time their concentrations are perturbed from the equilibrium ratios due to the downstream aggregation steps that consume them. This is the same as approximating steps prior to the rate-determining step in a reaction mechanism to be in "preequilibrium."

D_2 may further associate with D monomers up to some maximum size. Here, x denotes the number of D monomers in the largest oligomer that is significantly populated. For practical purposes, x will be finite so long as the free energy of association is a relatively strongly increasing function of x. Equivalently, we may stipulate that the equilibrium constant (K_x) for the process of assembling D_x from x monomers

$$K_x = \left(\frac{[D_x]}{[D]^x}\right)_{eq} = \exp(-\Delta G_x / RT) \qquad (2)$$

decreases with increasing x, with ΔG_x denoting the standard Gibbs energy for the process. To be consistent with Figure 1 we also define equilibrium constants $K_{1,j}^D$ for the sequential association steps shown in Figure 1, each involving addition of one D monomer at a time,

$$K_{1,j}^D \equiv \exp(-\Delta G_{1,j}^D / RT) = \left(\frac{[D_{j+1}]}{[D][D_j]}\right)_{eq} = \frac{k_{D,D_j}}{k_{D,D_j}^{(r)}} \qquad (3)$$

with $\Delta G_{1,j}^D$ the associated standard Gibbs energy of $D + D_j \leftrightarrow D_{j+1}$, and the right-hand-most equality also providing a relationship between the forward and reverse rate coefficients for the association of a D monomer to a preexisting D_j oligomer. From Eqs. (2) and (3) it follows that

$$K_j \equiv \exp(-\Delta G_j / RT) = \left(\frac{[D_j]}{[D]^j}\right)_{eq} = \prod_{i=1}^{j-1} K_{1,j}^D \qquad (4a)$$

$$\Delta G_j = \sum_{i=1}^{j-1} \Delta G_{1,j}^D \qquad (4b)$$

Recent experiments[9,15,17] and previous theoretical arguments[16,28] suggest some form of conformational change or rearrangement of monomers within the complexes of D monomers (also referred to as micelles or nuclei[15–17,28]) may occur to cause a given D_j complex to convert to the irreversible oligomer or aggregate A_j. Whether this occurs in a single concerted, activated step, as approximated in the scheme in Figure 1, or whether it is really a rapid sequence of steps in which only one or two monomers convert at a time is unclear. What is clear is that such initiation steps are strongly implied by available experimental evidence and structural changes upon transition from D states to aggregate states.[9,15–17] In the scheme in Figure 1, only the simplest case (kinetically) is shown. D_j converts to A_j in a single kinetic step characterized by a rate coefficient k_j^*. This is a valid approximation if it is a concerted or cooperative process, or conversion of the first D monomer in the D_j cluster is the slowest step in the process.

The remaining steps shown in Figure 1 are essentially a generalization of the steps described above to include reversible association of D monomers and oligomers with pre-formed aggregates A_j and a subsequent irreversible rearrangement step for each reversible complex. For completeness, the equilibrium relations for the remaining association steps are defined below.

$$K_{i,j}^{DA} = \frac{[D_i A_j]}{[D_i][A_j]} = \frac{k_{i,j}}{k_{i,j}^{(r)}} \qquad (5)$$

$$K_{i,j}^{AA} = \frac{[A_i A_j]}{[A_i][A_j]} = \frac{k'_{i,j}}{k_{i,j}^{(r)'}} \qquad (6)$$

At this point we can write a general set of differential equations describing the kinetic scheme in Figure 1. These are given in Table 1.

In Eqs. (8)–(16) the parameter n^* is introduced, which represents the size of irreversible aggregate A_{n^*} at which aggregates are no longer appreciably reactive with respect to further growth, either due to a kinetic bottleneck or because of low solubility of $A_{j \geq n^*}$.[26] Thus only aggregates up to a size of $n^* - 1$ are included in the equations listed above.

3. PRACTICAL LIMITING CASES

Enumerated over all possible $n^* - 2$ species of A_j, and x species of D_j, the total number of differential equations represented above is $[n^*(n^* - 3) + (n^* - 1)x + 3]$, which for realistic values of x (≥ 2) and n^* ($\sim 10^2$ or greater) yields an intractable model for practical application to typical experimental data. This poses only a minor difficulty, however, since the above model can be solved analytically in a number of limiting cases that capture a range of experimental behaviors commonly observed. The remainder of this chapter focuses on such limiting solutions of the more general nonnative aggregation model and their application to experimental analysis and stability prediction. The full model is included above in order to provide a detailed background for those workers interested in solving alternative (limiting) cases not covered here. The scenarios described below are not exhaustive, but illustrate the key physics involved in nonnative aggregation as it is currently understood, and are easily adaptable to specific protein systems having, e.g., more complex folding-unfolding pathways.

3.1. Unfolding-Limited Aggregation

If the characteristic time scale of folding-unfolding is significantly longer than those of *both* association and rearrangement involving D proteins (i.e., the "down-stream" processes in Figure 1), then folding-unfolding will be significantly slower than the steps that incorporate D into (irreversible) aggregate. In this case, D monomer is consumed by the first association step, i.e., $2D \rightarrow D_2$, as quickly as it can be replenished by the $N \rightarrow D$ step. If the protein is predominantly in the N state at the beginning of the experiment or process, the rate-determining step for the overall aggregation process will become unfolding of N protein after a (short) lag time in which the D monomers

TABLE 1. Summary of general system of differential equations for extended Lumry-Eyring model in Figure 1

Equation	No.

$$\frac{d[N]}{dt} = -k_u\left([N] - \frac{[D]}{K}\right) \tag{7}$$

$$\frac{d[D]}{dt} = k_u\left([N] - \frac{[D]}{K}\right) - \sum_{j=1}^{x-1} k_{D,D_j}\left([D_j][D] - \frac{[D_{j+1}]}{K_{1,j}^D}\right) - \sum_{j=2}^{n^*-1} k_{1,j}\left([D][A_j] - \frac{[A_jD]}{K_{1,j}^{DA}}\right) \tag{8}$$

$$\frac{d[D_2]}{dt} = -k_2^*[D_2] + k_{D,D}\left([D][D] - \frac{[D_2]}{K_{1,1}^D}\right) - k_{D,D_2}\left([D_2][D] - \frac{[D_3]}{K_{1,2}^D}\right)$$

$$- \sum_{j=2}^{n^*-1} k_{2,j}\left([D_2][A_j] - \frac{[A_jD_2]}{K_{2,j}^{DA}}\right) \tag{9}$$

$$\frac{d[D_x]}{dt} = -k_x^*[D_x] + k_{D,D_{x-1}}\left([D][D_{x-1}] - \frac{[D_x]}{K_{1,x-1}^D}\right) - \sum_{j=2}^{n^*-1} k_{x,j}\left([D_x][A_j] - \frac{[A_jD_x]}{K_{x,j}^{DA}}\right) \tag{10}$$

$$\frac{d[A_2]}{dt} = k_2^*[D_2] - \sum_{j=1}^{x} k_{j,2}\left([D_j][A_2] - \frac{[A_2D_j]}{K_{j,2}^{DA}}\right)$$

$$- \sum_{j=2}^{n^*-1} k'_{2,j}\left([A_2][A_j] - \frac{[A_2A_j]}{K_{2,j}^{AA}}\right) \tag{11}$$

$$\frac{d[A_3]}{dt} = k_3^*[D_3] + k^*{}_{A_2D}[A_2D] - \sum_{j=1}^{x} k_{j,3}\left([D_j][A_3] - \frac{[A_3D_j]}{K_{j,3}^{DA}}\right)$$

$$- \sum_{j=2}^{n^*-1} k'_{3,j}\left([A_3][A_j] - \frac{[A_3A_j]}{K_{3,j}^{AA}}\right) \tag{12}$$

$$\frac{d[A_x]}{dt} = k_x^*[D_x] + \sum_{j=1}^{x/2} k^*{}_{A_{x-j}D_j}[A_{x-j}D_j] - \sum_{j=1}^{x} k_{j,x}\left([D_j][A_x] - \frac{[A_xD_j]}{K_{j,x}^{DA}}\right)$$

$$- \sum_{j=2}^{n^*-1} k'_{3,j}\left([A_x][A_j] - \frac{[A_xA_j]}{K_{x,j}^{AA}}\right) + \sum_{j=2}^{x/2} k^*{}_{j,x-j}[A_j][A_{x-j}] \tag{13}$$

$$\left.\frac{d[A_i]}{dt}\right|_{x<i<n^*} = \sum_{j=1}^{i/2} k^*{}_{A_{i-j}D_j}[A_{i-j}D_j] - \sum_{j=1}^{x} k_{j,i}\left([D_j][A_i] - \frac{[A_iD_j]}{K_{j,i}^{DA}}\right)$$

$$- \sum_{j=2}^{n^*-1} k'_{i,j}\left([A_i][A_j] - \frac{[A_iA_j]}{K_{i,j}^{AA}}\right) + \sum_{j=2}^{i/2} k^*{}_{j,i-j}[A_j][A_{i-j}] \tag{14}$$

$$\left.\frac{d[A_iD_j]}{dt}\right|_{\substack{2 \le i < n^* \\ j \le x}} = k_{j,i}\left([D_j][A_i] - \frac{[A_iD_j]}{K_{j,i}^{DA}}\right) - k^*{}_{A_iD_j}[A_iD_j] \tag{15}$$

$$\left.\frac{d[A_iA_j]}{dt}\right|_{2\le i,j<n^*} = k'_{i,j}\left([A_i][A_j] - \frac{[A_iA_j]}{K_{i,j}^{AA}}\right) - k^*{}_{i,j}[A_iA_j] \tag{16}$$

originally present are consumed to the point that the formation of D via unfolding of N exactly balances the consumption of D via aggregation. That is, the $N \to D$ step becomes rate-determining and $d[D]/dt \approx 0$. The resulting equations governing this process are

$$\frac{d[N]}{dt} = -k_u[N] + k_f[D] \tag{17}$$

$$\frac{d[D]}{dt} = k_u[N] - k_f[D] - k_{eff}[D]^\gamma \approx 0 \tag{18}$$

where we have introduced k_{eff} as the effective rate coefficient for aggregation under these conditions, and γ as the apparent reaction order for aggregation with respect to D concentration. A mechanistic basis of k_{eff} and γ will be given more precisely in Sections 3.2 and 3.3. For present purposes the important feature is that γ is either 1 or 2. This follows because the aggregation steps that immediately follow this unfolding step are necessarily fast and irreversible, thereby constituting essentially the kinetics described in Section 3.2 minus the N-D preequilibrium, which result in either pseudo-first or pseudo-second-order kinetics with respect to $[M]$ (cf. exponents of $[M]$ in Eqs. (30) and (31)). Further, unfolding-limited kinetics require $(k_u + k_f)/k_{eff} \ll 1$, i.e., the time scale of folding-unfolding $\tau_{conf} = (k_u + k_f)^{-1}$ must be much longer than that of aggregation. Typical experimental ranges for τ_{conf} are $\sim 10^{-1}$ to 10^3 seconds even at low temperatures,[31] with the latter value providing an approximate upper bound for the time scale one expects unfolding-limited kinetics to be operable. See Sections 4 and 5 for a discussion of the implications of this last criterion in practical applications.

Using $\gamma = 1$ in Eq. (18) gives

$$[D] = k_u[N]/(k_f + k_{eff}) = K[N]/(1 + k_{eff}/k_f) \tag{19a}$$

and using $\gamma = 2$ gives

$$[D] = \frac{k_f}{2k_{eff}}\left[\sqrt{1 + 4k_u k_{eff}[N]/k_f^2} - 1\right] \tag{19b}$$

Historically, $\gamma = 1$ was assumed,[21-25] which upon substitution of Eq. (19a) into Eq. (17) yields

$$\frac{d[N]}{dt} \cong -k_u[N]\left(\frac{1}{1 + k_f/k_{eff}}\right) \tag{20}$$

Experimentally, $[N]$ and $[D]$ cannot typically be assayed directly, and it is rather the total monomer concentration $[M] = [N] + [D]$ that is measured. Further, $d[M]/dt = d[N]/dt$, since $d[D]/dt = 0$, and $[N] \approx [M]$ because $[D] \ll [N]$ (consider Eq. (19a) with $K \sim 1$ or smaller and $k_f k_{eff}^{-1} \ll 1$). The resulting expression is

$$\frac{d[M]}{dt} \cong -k_u[M] = -k_{obs}[M] \tag{21}$$

which is the form of unfolding-limited kinetics employed most often. Interestingly, $\gamma = 1$ was used historically either for mathematical convenience or because chemical degradation rather than aggregation was being modeled.[21–25] More recent work[26] suggests a mechanistic reason for $\gamma = 1$ (i.e., pseudo-first-order kinetics of the aggregation process itself), which will be reviewed in Section 3.2.

However, the $\gamma = 1$ case may not always hold, and so we also consider the apparently equally plausible case of $\gamma = 2$.[26] In the latter case, Eq. (19b) is substituted into Eq. (17) to give

$$\frac{d[N]}{dt} = -k_u[N]\left(1 - \frac{k_f/k_{eff}}{2K[N]}\left[\sqrt{1 + \frac{4K[N]}{k_f/k_{eff}}} - 1\right]\right) \tag{22}$$

which, for $k_f k_{eff}^{-1} \ll 1$ reduces to

$$\frac{d[N]}{dt} \cong -k_u[N]\left(1 - \sqrt{\frac{(k_f/k_{eff})}{K}}[N]^{-1/2}\right) \tag{23}$$

showing that significant deviations from pseudo-first-order kinetics would be expected if $K \ll 1$, i.e., under native-favoring conditions. It was found elsewhere[26] for association-limited kinetics (see also Section 3.2) that $\gamma = 1$ was likely only for $K \sim 1$ or higher, or equivalently, near and above the midpoint unfolding temperature T_m. Conversely, $\gamma = 2$ was prevalent under native-favoring conditions. Independent of these theoretical considerations, Eq. (21) is empirically the form of unfolding-limited kinetics most commonly chosen for comparison with experiment. The arguments in Section 4 suggest Eq. (23) might not be realized in practice because at lower T most systems switch to association- or rearrangement-limited aggregation, described next.

3.2. Association-Limited Aggregation

We consider next the scenario in which association of D monomers and preformed aggregates is significantly slower than the preceding conformational changes of folding and unfolding, and the subsequent rearrangement and assembly steps. Folding-unfolding is therefore in pseudoequilibrium. The forward association steps in Figure 1 will be the rate-determining kinetic steps and will be effectively irreversible because the later steps of rearrangement and assembly will occur too rapidly for the association steps to reverse. The effective reaction scheme that results is given in Figure 2A. With some minor changes in nomenclature, this is the same kinetic scheme that was explored in

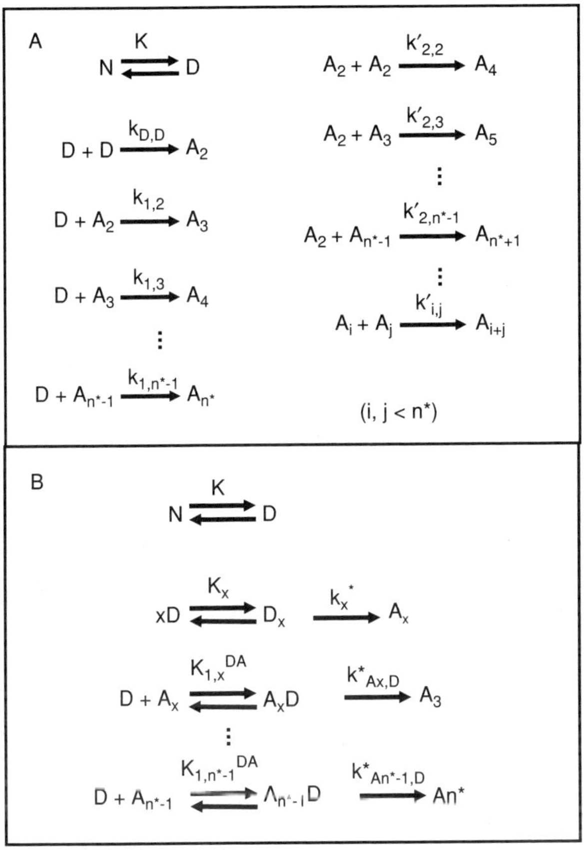

FIGURE 2. Simplified kinetic scheme for (A) association-limited aggregation and (B) rearrangement-limited aggregation.

detail in Ref. 26 and used to interpret and predict the temperature dependence of shelf life for bG-CSF.[11]

Incorporating pseudoequilibrium for the N-D transition, and the material balance $[M] = [N] + [D]$ gives

$$[N] = \frac{1}{1+K}[M] = (1 - f_{D,eq})[M] \quad (24)$$

$$[D] = \frac{K}{1+K}[M] = f_{D,eq}[M] \quad (25)$$

This defines the equilibrium fraction of denatured monomer ($f_{D,eq}$) in the standard manner as $K/(1+K)$.[29,30] Making all (forward) association steps irreversible results in the following expressions.[26]

$$\frac{d[M]}{dt} = -2k_{D,D}f_{D,eq}^2[M]^2 - \left(\sum_{j=2}^{n^*-1} k_{1,j}[A_j]\right) f_{D,eq}[M] \quad (26)$$

$$\frac{d[A_2]}{dt} = k_{D,D}f_{D,eq}^2[M]^2 - k_{1,2}f_{D,eq}[A_2][M] - \sum_{j=3}^{n^*-1} k'_{2,j}[A_2][A_j]$$
$$- 2k'_{2,2}[A_2]^2 \quad (27)$$

$$\left.\frac{d[A_i]}{dt}\right|_{i<n^*} = \left(k_{1,i-1}[A_{i-1}] - k_{1,i}[A_i]\right) f_{D,eq}[M] + \sum_{j=2}^{i/2} k'_{j,i-j}[A_{i-j}][A_j]$$
$$- \sum_{j=2}^{n^*-1} k_{i,j}[A_i][A_j] \quad (28)$$

Eqs. (26)–(28), while incorporating all details in Figure 2A, are not tractable for analyzing experimental data, since the set $\{[A_j]\}$ is not typically known, and therefore the full set of rate coefficients cannot be unambiguously obtained. We consider instead an easily measurable quantity: the monomer concentration $[M]$ as a function of time (t) and experimental conditions. Thus, we are seeking the solutions of Eqs. (26)–(28), but in a mathematical form that describes $[M](t)$ or equivalently $d[M]/dt$. To this end, a number of approximate, analytical solutions for the association-limited scenario are enumerated below. One useful limiting case that is excluded is one in which aggregate solubility limits impose constraints on the achievable aggregate concentrations, and result in mixed-order kinetics. A description of this scenario is foregone as it will not illustrate dramatically different behavior from the pseudo-steady-state cases described below for the purposes of Section 4 and 5, and because its characteristic behaviors are enumerated theoretically and experimentally elsewhere.[11,26]

3.2.1. Individual association steps at pseudo-steady state

The concentrations of the set of aggregates $\{A_j\}_{j>1}$ will reach pseudo-steady-state levels at very low extents of monomer loss in association-limited aggregation if:[26] (1) aggregate-aggregate association rates are comparable to, or more rapid than, D monomer addition to aggregates, or (2) aggregates grow predominantly via D monomer addition, and n^* is reasonably low (~ 50 or less). In either of these cases, the summation in Eq. (26) will be well approximated by its pseudo-steady-state level. If the former situation occurs, we have[26]

$$\frac{d[M]}{dt} = -2k_{D,D}f_{D,eq}^2 \left(1 + \sqrt{\frac{\langle k_{ij}\rangle}{2\alpha_I k_{D,D}}}\right)[M]^2 = -k_{obs}[M]^2 \quad (29)$$

while if the latter occurs we have instead

$$\frac{d[M]}{dt} = -n^* k_{D,D} f_{D,eq}^2 [M]^2 = -k_{obs}[M]^2 \quad (30)$$

where α_I is a constant of order one, $<k_{i,j}>$ denotes the value of the forward rate coefficients for $A_i + A_j \rightarrow A_{i+j}$ and $D + A_j \rightarrow A_{j+1}$ averaged over all sizes of aggregates ($i \geq 1, j \geq 2$), and all other symbols were defined previously. By inspection, Eqs. (29) and (30) are indistinguishable from the perspective of kinetics at an isolated state point (temperature, pressure, and composition). Although beyond what will be covered here, what distinguishes these two scenarios practically is that the physical situation giving rise to Eq. (30) is predicted to transition to that described in Section 3.2.2 as $f_{D,eq}^{-1}$ due to an inability of the system to reach steady state if the "chain" of aggregation events is too long ($n^* \sim 50$ or higher). Such a transition does not occur in the former situation because the much larger number of competing reactions involving each aggregate species effectively maintain the pseudo-steady-state approximation for each $[A_j]$ unless $<k_{ij}>$ becomes unrealistically small.[26]

If one has evidence that a particular set of aggregate species do not achieve pseudo-steady-state levels, e.g., a well-populated dimer or oligomeric species, one may still utilize steady-state approximations for the other aggregate species, but must treat the non-steady-state species explicitly. Alternatively, if essentially all aggregates do not achieve steady state before the majority of monomeric protein is converted to aggregates, one can instead utilize the solution described next.

3.2.2. Individual aggregate levels do not achieve steady state If aggregate-aggregate association steps are intrinsically much slower (i.e., have lower rate coefficients) than association steps involving monomer addition to preexisting aggregates, *and* the maximum size of soluble aggregates n^* is large (~ 100 or higher), steady-state conditions will not be reached by a significant fraction of the soluble aggregates until a majority of the monomeric protein is consumed, i.e., $[M]/C_0 \ll 1$, with C_0 denoting the initial monomer concentration. This occurs because aggregates are formed predominantly by sequential addition of monomer and because the number of steps in the "chain" leading to the largest aggregates is large. Since initially most if not all protein is monomeric, a large fraction of monomers originally in solution must be consumed to form aggregates at sufficiently high concentrations to satisfy the steady-state conditions that formation and depletion rates of each j-mer be approximately equal. However, although the individual aggregate concentrations do not achieve steady state, the summation over all aggregate concentrations *does* achieve a pseudo-steady state. This can be understood by realizing that since aggregate growth is via monomer addition, the depletion of a j-mer necessarily results in the formation of a $(j+1)$-mer. Thus, after a short lag time, the sum across all j-mers is approximately constant at its pseudo-steady-state value (denoted by the subscript *ss*), and this summation term dominates in Eq. (26). The resulting aggregation kinetics therefore will appear to be pseudo-first order with respect to $[M]$ at constant temperature.[26] Numerical solution of the full set of differential equations (unpublished results and Figure 8 in Ref. 26) shows the value of $(\Sigma_j[A_j])_{ss}$ scales as $C_0\{<k_{1,j}>^{-1}f_{D,eq}k_{D,D}\}^{1/2}$, which upon substitution in Eq. (26) gives

$$\frac{d[M]}{dt} = -\sqrt{k_{D,D}\langle k_{1,j}\rangle} f_{D,eq}^{3/2} C_0[M] = -k_{obs}(C_0)[M] \qquad (31)$$

using a scaling prefactor of unity. Eq. (31) shows that first-order behavior is expected if one considers only $[M](t)$. However, the apparent or observed first-order rate coefficient is a function of C_0, and so true first-order behavior will be violated if one measures, e.g., initial rates versus C_0 (see also Section 4).

3.3. Rearrangement-Limited Aggregation

The final limiting cases we will consider are two in which the N-D transition and the bimolecular association steps, together representing all reversible steps in Figure 1, occur much more rapidly than the conformational rearrangement(s) of the associated species that subsequently convert them to irreversible aggregates. In this scenario, both N-D conformational transitions and all association steps are at pseudoequilibrium, and the rearrangement steps constitute the rate-determining steps in the process. Therefore, all terms in Eqs. (7)–(16) that represent the balance between forward and reverse steps must be zero, e.g., $[N]$-$[D]/K = 0$ for folding-unfolding, and $[A_j][D]$-$[A_j \cdot D]/K_{1,j}^{DA} = 0$ for association of D and A_j. The particular limiting cases considered here are those in which the following three conditions are met: (1) the standard Gibbs energy of oligomers composed of D are much greater than the thermal energy of the system, and thus the associated equilibrium constants (K_j) are much less than one; (2) there is a single dominant initiation step for the conversion of the largest appreciable D oligomer, i.e., $D_x \to A_x$ is the dominant "nucleation" process; and (3) aggregate growth beyond the formation of A_x occurs predominantly via addition of D monomers. Conditions (1) and (2) are those commonly used in describing protein aggregation kinetics, and are based in part on experimental data[9,15–17,28] and in part on the mathematical simplifications they permit.[16] The third condition is not strictly necessary, but it enables a simple illustration of rearrangement-limited kinetics in the cases considered here.

Imposing the above constraints, substituting for $[N]$ and $[D]$ with Eqs. (24) and (25), and utilizing the definitions of the equilibrium constants in Eqs. 2–4, the general model equations [Eqs. (7)–(16)] reduce to those given below and, the corresponding kinetic scheme in Figure 2B.

$$\frac{d[M]}{dt} = -xk_x^* K_x f_{D,eq}^x [M]^x - \left(\sum_{j=x}^{n*-1} k_{Aj,D}^* K_{1,j}^{DA}[A_j]\right) f_{D,eq}[M] \quad (32)$$

$$\frac{d[A_x]}{dt} = -k_{Ax,D}^* K_{1,x}^{DA}[A_x] f_{D,eq}[M] + k_x^* K_x f_{D,eq}^x [M]^x \quad (33)$$

$$\left.\frac{d[A_j]}{dt}\right|_{j>x} = -\left(k_{Aj,D}^* K_{1,j}^{DA}[A_j] - k_{Aj-1,D}^* K_{1,j-1}^{DA}[A_{j-1}]\right) f_{D,eq}[M] \quad (34)$$

with

$$K_{1,j}^{DA} = \frac{[D \cdot A_j]}{[D][A_j]} = \frac{k_{1,j}}{k_{1,j}^{(r)}} \quad (35)$$

Note that the summation in Eq. (32) has a lower bound of $j = x$, since irreversible aggregates smaller than x cannot form in this limiting case.

The solution of Eq. (32)–(34) to obtain a practical form for $d[M]/dt$ requires consideration of two different scenarios. The first is analogous to that described in Section 3.2.1, where each aggregate concentration achieves a pseudo-steady state at low extents of monomer loss. The second is analogous to that in Section 3.2.2, in which individual aggregate species do not achieve steady state but the sum over their concentrations does. Direct numerical solution of Eq. (32)–(34) over a range of physically reasonable values for the model parameters indicate that the first scenario is dominant for $x = 2$, while the second one predominates for $x > 2$, because the first term on the right-hand side of Eq. (32) becomes relatively small as x is increased significantly above 2.[27]

Considering the first of these scenarios, the steady-state $[A_j]$ values are given by setting $d[A_j]/dt = 0$ and solving for $[A_j]$.

$$[A_j]_{j \neq x} = \frac{k^*_{Aj-1,D} K^{DA}_{1,j-1}}{k^*_{Aj,D} K^{DA}_{1,j}} [A_{j-1}] \tag{36a}$$

$$[A_x] = \frac{k^*_x K_x f^{x-1}_{D,eq}}{k^*_{Ax,D} K^{DA}_{1,x}} [M]^{x-1} \tag{36b}$$

Substitution of Eq. (36) in the summation of Eq. (32) yields

$$\left(\sum_{j=x}^{n^*-1} k^*_{Aj,D} K^{DA}_{1,j} [A_j] \right)_{ss} = (n^* - x) k^*_x K_x f^{x-1}_{D,eq} [M]^{x-1} \tag{37}$$

and then

$$\frac{d[M]}{dt} = -n^* k^*_x K_x f^x_{D,eq} [M]^x = -k_{obs} [M]^x \tag{38}$$

For $x = 2$, this result is formally identical to the association-limited case described by Eq. (30), except that $k_{D,D}$ has been replaced with $k^*_x K_x$. The second scenario is similar to that in Section 3.2.2, where the solution follows from consideration of Eq. (32) as

$$\frac{d[M]}{dt} = -\left[xk^*_x K_x f^{x-1}_{D,eq} [M]^{x-1} + \left(\sum_{j=x}^{n^*-1} k^*_{Aj,D} K^{DA}_{1,j} [A_j] \right) \right] f_{D,eq} [M]$$

$$\approx -\langle k^*_{Aj,D} K^{DA}_{1,j} \rangle \left(\sum_{j=x}^{n^*-1} [A_j] \right)_{ss} f_{D,eq} [M] \tag{39}$$

In Eq. (39) brackets <...> denote an average over all aggregate sizes j. Direct numerical simulation of Eqs. (32)–(34) with $x \geq 3$ (note: x must be an integer) shows that the steady-state value of the summation over $[A_j]$ follows a scaling law described by

$$C_0^{-1}\left(\sum_{j=x}^{n^*-1}[A_j]\right)_{ss} \cong \varepsilon \left(\frac{k_x^* K_x C_0^{x-2} f_{D,eq}^{x-1}}{\langle k_{Aj,D}^* K_{1,j}^{DA}\rangle}\right)^\Gamma \qquad (40)$$

with $\varepsilon \sim 1$ and $\Gamma \sim 0.35$ for $3 \leq x \leq 6$.[27] Substitution of Eq. (40) into Eq. (39) gives

$$\begin{aligned}\frac{d[M]}{dt} &\cong -f_{D,eq}^{1+\Gamma(x-1)}\left(k_x^* K_x\right)^\Gamma \left(<k_{Aj,D}^* K_{1,j}^{DA}>\right)^{1-\Gamma} C_0^{1+\Gamma(x-2)}[M] \\ &= -k_{obs}(C_0)[M]\end{aligned} \qquad (41)$$

Eq. (41) shows that the apparent or observed rate constant can depend strongly and nonlinearly on C_0 (see also Section 4 discussion regarding Figure 4).

3.4. Multistate Monomeric Folding-Unfolding

At first the two-state approximation may seem overly simplified and incapable of capturing the important features of more complex proteins, e.g., those in which it has been possible to observe signatures of more than one structural motif in spectroscopic measurements under nonnative favoring conditions, or proteins for which one or more "intermediates" on the folding pathway are postulated or observed.[1,7-10] Scenarios such as these are in fact readily incorporated into the preceding description, either with minor changes in the working formulas or without changing the formal results. The following arguments are intended to help to clarify this. At the outset, it should be realized that multistate proteins do not induce any change to the preceding analysis for unfolding-limited aggregation, as nonnative states are never appreciably populated in that case.

In what follows, we maintain the same "reversibility" assumption for conformational states of nonaggregated proteins as used earlier in this chapter. For concreteness, we consider the case of a protein in which three *thermodynamically* distinct conformational states can simultaneously equilibrate among one another: native (N), "fully" unfolded (U), and a partially folded or near-native "intermediate"(I). In practical terms, thermodynamically distinct states are ones that by definition are local free energy minima (only one of which of course may be a global minimum), and therefore are also separated by significant free-energy barriers relative to the thermal energy of the system (i.e., more than a few kT, $k =$ Boltzmann's constant). Simply being structurally distinct is insufficient. Structurally dissimilar states with negligible free-energy barriers between them are effectively members of the same thermodynamic conformational state, as any observable property will unavoidably be an ensemble average over all such structures.[5,7,13,14] If these criteria are met, it is rational to treat those conformational states as each having well-defined thermodynamic properties such as enthalpy, entropy, and free energy. Meaningful equilibrium constants between them can then be written as

$$K_{NI} = \exp\left(-\frac{\Delta G_{NI}}{RT}\right) = \left(\frac{[I]}{[N]}\right)_{eq} \tag{42}$$

$$K_{NU} = \exp\left(-\frac{\Delta G_{NU}}{RT}\right) = \left(\frac{[U]}{[N]}\right)_{eq} \tag{43}$$

$$K_{IU} = \exp\left(-\frac{\Delta G_{IU}}{RT}\right) = \left(\frac{[U]}{[I]}\right)_{eq} \tag{44}$$

with ΔG_{ij} defined as the standard Gibbs energy of conformational state j minus that of state i. The fraction of M existing in each state is then

$$f_{U,eq} = \frac{K_{NU}}{1 + K_{NU} + K_{NI}} \tag{45}$$

$$f_{I,eq} = \frac{K_{NI}}{1 + K_{NU} + K_{NI}} \tag{46}$$

$$f_{N,eq} = \frac{1}{1 + K_{NU} + K_{NI}} \tag{47}$$

Defining D as consisting of both I and U gives

$$f_{D,eq} = f_{I,eq} + f_{U,eq} \tag{48}$$

$$f_{I,eq} = \frac{1}{1 + K_{IU}} f_{D,eq} \tag{49}$$

$$f_{U,eq} = \frac{K_{IU}}{1 + K_{IU}} f_{D,eq} \tag{50}$$

One may now assume that either I or U is the "reactive" state with respect to aggregation, rather than D as a whole. All of the analyses in Sections 3.2 and 3.3 remain essentially unchanged, except $f_{D,eq}$ is replaced with $f_{I,eq}$ or $f_{U,eq}$, respectively, from Eq. (49) or (50). Extending this further to deal with additional substates within the D ensemble is analogous to what is done above and is foregone here. Additionally, notice we recover a simple two-state picture if $\Delta G_{IU}/RT$ is much greater than or less than one. In the former case $f_{D,eq} \to f_{I,eq}$ and $f_{U,eq} \to 0$, while in the latter $f_{D,eq} \to f_{U,eq}$ and $f_{I,eq} \to 0$.

The mathematical forms of the models that will result if I or U is taken as the reactive state with respect to aggregation are then formally identical to those given in Sections 3.2 to 3.3, as the additional factor of $K_{IU}/(1 + K_{IU})$ or $1/(1 + K_{IU})$ preceding $f_{D,eq}$ in each rate expression will be constant and therefore indistinguishable using data at a single-state point (T, pressure, and solvent composition). The discussions in Section 4 remain essentially unchanged in this situation. The discussion in Section 5 remains predominantly the same, except one needs additional information about the temperature dependence of K_{IU} in order to apply all of the methods presented for extrapolating kinetic predictions to lower temperatures. This should not be an issue, however, if one adopts the approach advocated below, as such thermodynamic information should arise naturally in the process of gathering the necessary data to decide whether to apply a two-state or multistate description.

What remains is to determine if it is feasible to experimentally determine whether it is necessary to incorporate a more complex, multistate description for a given protein. One such way relies on the availability of at least nominal data on thermal or chemical unfolding for a given protein. For example, if a three-state folding-unfolding model is appropriate, one would encounter one of two scenarios experimentally in an unfolding experiment: either the protein exhibits two sequential "unfolding" transitions or effectively exhibits a single transition. The former would be evidenced, e.g., by two calorimetric peaks or the convolution thereof upon heating, or by multiple, cooperative sigmoidal stages during a spectroscopically monitored unfolding experiment. The latter scenario would show only one such transition within experimental uncertainty.

If the former occurs, one may conclude not only are ΔG_{NI} and ΔG_{IU} both highly dependent on temperature or solvent composition, but also that both I and U are significantly populated ($f_{I,eq}/f_{U,eq} \sim 1$) over some window of temperature or solvent composition of practical relevance. In this case, it is recommended that one forego the two-state approximation, modify Eqs. (29)–(31), (38), and (41) as described above, and use a three-state thermal unfolding model[29,30] to extract the relevant thermodynamic parameters to be able to predict both $f_{D,eq}$ and $f_{I,eq}$ (or $f_{U,eq}$) as a function of T in the extrapolation procedures in Section 5.

If only a single transition occurs in thermal unfolding experiments, one may be justified in approximating that K_{IU} is only weakly dependent on temperature. In this case, the prefactors before $f_{D,eq}$ in Eqs. (49) and (50) can be taken as constant when performing the analyses and extrapolations in Section 5, and therefore have no practical impact on the resulting predictions.

Finally, although not considered in any depth here, we note that $f_{D,eq}$ is itself a function of $[M]$ for folding-unfolding transitions that involve a change in stoichiometry, e.g., a multimeric native state but single-chain denatured state. Following the type of procedure utilized above, this may lead to rate laws in which $[M]$ is raised to powers other than those in Sections 3.1–3.3 and Table 2, i.e., apparent reaction orders other than 1 or 2, including non-integer powers.[27]

4. INTERPRETATION OF MONOMER LOSS KINETICS

This section focuses on relatively simple analyses of monomer loss kinetics that can be used to elucidate which limiting case or cases, if any, are consistent with the data for a particular protein of interest, and thereby also to infer the rate-controlling step(s) at the conditions of interest. Implications of the kinetic models for subsequent stability prediction are deferred to the next section.

For many commercial applications, only small extents of aggregation are tolerable, i.e., on the order of one to ten percent of the initial monomer content. Quantitative

TABLE 2. Summary of rate laws and experimental signatures of limiting cases of aggregation kinetics as described in Section 3

Category	Rate Law $v_M = -d[M]/dt$	k_{obs} [a]	Eqs.	Experimental Signatures[b,c]
Unfolding-limited	$v_M = k_{obs}[M]$	k_u	(21)	True first-order kinetics Arrhenius behavior Minimum $k_{obs} \sim 10^3$ s^{-1} Rate-determining step likely changes as T lowered Stability predictions via extrapolation from high T unreliable
Association-limited				
$\{[A_j]\}$ at steady state	$v_M = k_{obs}[M]^2$	$n^* k_{D,D} f_D^2$ or $2k_{D,D} f_D^2 \times$ $\left[1 + \sqrt{\langle k_{ij}\rangle / 2\alpha k_{D,D}}\right]$	(30) (29)	True second-order kinetics Non-Arrhenius Possible switch to Eq. (21) or (31) as T increases Stability prediction possible using extrapolation from $T < T_m$
$\{[A_j]\}$ not at steady state	$v_M = k_{obs}[M]$	$\sqrt{k_{D,D}\langle k_{1j}\rangle} f_D^{1.5} C_0$	(31)	Apparent first-order, but true second-order kinetics Non-Arrhenius Possible switch to Eq. (29) or (30) as T decreases Stability predictions via extrapolation from high T unreliable
Rearrangement-limited				
$[A_j]$ at steady state[d]	$v_M = k_{obs}[M]^x$	$n^* k_x^* K_x f_D^x$	(38)	True reaction order ≥ 2 Non-Arrhenius Effect of T on rate-determining step, and stability prediction using T extrapolation, not yet tested
$[A_j]$ not at steady state[e,f]	$v_M = k_{obs}[M]$	$f_D^{\delta+\Gamma}\left(k_x^* K_x\right)^{\Gamma} \times$ $\left(k_{A_j,D}^* K_{1,J}\right)^{1-\Gamma} C_0^{\delta}$	(41)	Apparent first-order, but true reaction order ≥ 2 Non-Arrhenius Effect of T on rate-determining step, and stability prediction using T extrapolation, not yet tested

[a] k_{obs} is the observed rate coefficient if Co and temperature are held fixed; $f_{D,eq}$ abbreviated to f_D.
[b] $\delta = 1 + \Gamma(x-2)$, Γ 0.35; $\delta \geq 2$ [Eq.(41)] and not necessarily an integer.
[c] True reaction order is taken as that based on initial rate kinetics.
[d] Bounds on k_{obs} values based on estimate that minimum $k_u \sim 10^{-3}$ s^{-1} for typical proteins.
[e] x = number of monomers in maximum reversible oligomer or aggregate "nucleus"; x must be at least 2 and an integer.
[f] Assuming monomer addition to preformed aggregates is primary mode of aggregate growth.

detection of, and distinction among, aggregate species or sizes at such low extents of reaction are often not reliable by techniques such as high-pressure liquid chromatography (HPLC) or turbidity that are commonly the primary assays in commercial quality control and analytical R&D laboratories. Conversely, analytical gel permeation HPLC measurements of $[M]$ are capable of achieving routine levels of precision on the order of one to two percent, which are sufficient to distinguish the model behaviors considered here. For the purposes of initial rate kinetics (discussion below), this also is sufficient, as the initial rate is effectively constant over the first five to ten percent loss for any rate law with a simple power-law dependence of $[M]$,[20] such as those in Section 3. The kinetics of monomer loss, rather than aggregate buildup, therefore are the primary focus here.

On a related point, high precision light-scattering techniques also may be used to quantitate aggregation kinetics,[16,28] and provide much greater sensitivity in detection of aggregates than do typical chromatographic methods. From a modeling perspective, both types of data have advantages. Analysis of monomer loss kinetics can be simpler than that for aggregate accumulation kinetics, as one is able to take advantage of the highly mono-disperse nature of monomeric proteins. This removes a need to deal with ambiguity about the identity of different species and determination of higher moments of the (aggregate) size distribution than may be experimentally feasible in order to discriminate among competing models. On the other hand, it is not unheard of for soluble aggregates to be detectable and even quantitated at levels as low as one part in 10^5 (mole basis). This constitutes an extent of reaction ~ 0.1 percent (assuming $n^* \sim 100$), at least an order of magnitude smaller than what is practical with monomer loss detection. Ideally one would benefit greatly from having both monomer loss and aggregate growth kinetics, as they provide powerful self-consistency checks on the models employed. Indeed, aggregate growth kinetics could provide a means to discriminate between the limiting cases represented by Eqs. (29) and (30) via tests of the assumptions inherent in their derivations.[26] The interested reader is directed to earlier references[16,28] for recent examples of kinetic model development based on aggregate formation kinetics.

In what follows, we consider the effects of three independent experimental variables that are easily controlled: incubation time, initial protein concentration, and temperature. When used together they can yield significant insight into the mechanistic step(s) governing the aggregation process, and ultimately help identify the experimental quantities that must be measured, if possible, to properly account for both thermodynamic and kinetic contributions to the observed kinetic rate coefficient(s) and protein stability.

All of the final rate expressions for monomer loss, $v_M = -d[M]/dt$, derived in Section 3, are explicitly of order 1 or 2 with respect to $[M]$, cf., Eqs. (21), (23) (with $K \sim 1$), (29)–(31), (38), and (41). For those kinetics with apparent first-order dependence on $[M]$ [Eqs. (21), (31), and (41)] a plot of $\ln[M]/C_0$ versus time at fixed C_0, temperature, and solution conditions will be linear, with slope $-k_{obs}$. By contrast, a system obeying steady-state association-limited (Eqs. 29 and 30) or rearrangement-limited (Eq. 38, $x = 2$) kinetics will instead be linear with a slope of k_{obs} when plotted as $C_0/[M] - 1$ versus time. Since multiple limiting cases can follow pseudo-first-or pseudo-second-order kinetics when considered as a function of time at fixed C_0 and incubation conditions, it necessarily follows that [M] versus time data *cannot on their own* be used to conclusively identify the rate-determining step, or equivalently to identify what k_{obs} represents in terms of intrinsic rate coefficient(s) and equilibrium constant(s).

Two particularly useful orthogonal variables to help alleviate such ambiguity are the initial (unaggregated) protein concentration C_0 and the temperature T. The former is useful because the time scales of bimolecular events such as association and assembly will generally decrease with increasing C_0, if only because the average molecular separation diminishes and/or the translational free volume of the protein monomers and aggregates decreases upon increasing C_0. For example, true first-order kinetics characteristic of monomer unfolding can be distinguished from apparent first-order kinetics characteristic of two of the limiting cases involving association (Eq. 31) and rearrangement (Eq. 41) as the rate-determining steps. k_{obs} is independent of C_0 for unfolding-limited kinetics but has a first- or possibly higher-order dependence on C_0, respectively, for the other two cases. Experimentally, initial rate data provide a simple means to measure the C_0 dependence of the observed rate coefficient(s). Specifically, the rate of monomer loss at short times ($v_{M,0}$) or more precisely, at low extents of reaction ($[M]/C_0 \geq 0.9$),[11,20] is measured over a range of C_0 at fixed temperature and sample composition. All rate expressions for monomer loss in Section 3 follow a form such as

$$v_{M,0} \equiv - \frac{d[M]}{dt}\bigg|_{t \to 0} = -k_{obs} C_0^{z_t} \qquad (51)$$

with z_t defined as the reaction order on the basis of [M] versus time ($z_t = 1$ or 2 for the models considered here). However, k_{obs} is not necessarily independent of C_0. Inspection of Eqs. (21), (29)–(31), (38), and (41) reveals that a plot of $\ln(v_{M,0})$ versus $\ln(C_0)$ such as shown in Figure 3A will be linear with a slope of 1 only for unfolding-limited

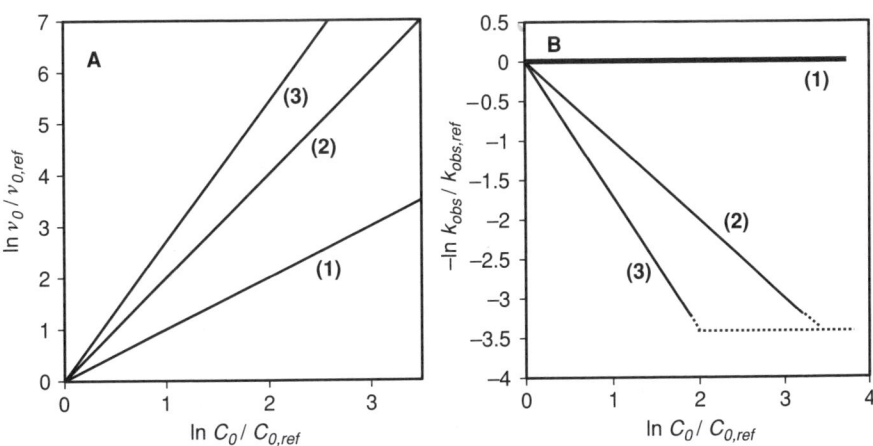

FIGURE 3. Effects of initial protein concentration C_0 on the (A) initial rates and (B) observed rate coefficient of aggregation. In both panels the different limiting cases are: (1) only unfolding-limited, (2) all association-limited, and steady-state rearrangement-limited cases; and (3) unsteady-state rearrangement-limited case, with x = 3.

kinetics. Its slope will be 2 for the remaining limiting scenarios except for unsteady-state rearrangement-limited kinetics, in which case the slope will be greater than 2 and no longer necessarily an integer (in Figure 3, $x = 3$ and $\Gamma = 0.35$, cf. Eq. 41). Equivalently, if one were to plot a measure of the characteristic time scale for aggregation, k_{obs}^{-1}, versus C_0 on a log-log plot, behavior such as depicted in Figure 3B would be expected. In such a plot, unfolding-limited aggregation shows no dependence on C_0, while association-limited and rearrangement-limited scenarios show first- or higher-order dependences. In Figure 3, all quantities are normalized relative to a reference condition with concentration $C_{0,ref}$, and corresponding $v_{M,0,ref}$ and $k_{obs,ref}$, so as to permit the curves to be shown on the same scale.

In Figure 3B we also have extended the range of $C_0, / C_{0,ref}$ to illustrate that beyond some upper limit of C_0 the time scale of aggregation may decrease to below that of folding/unfolding, and therefore two regimes may occur: an "aggregation-limited" regime at lower C_0, and an unfolding-limited regime at higher C_0. The lower and upper plateaus in Figure 3B do not overlay, simply because they are scaled by different values of $k_{obs,ref}$. Knowledge of such a transition concentration separating the two regimes not only helps to confirm or refute that a given system is displaying unfolding-limited or aggregation-limited kinetics, but also identifies a boundary in terms of C_0 at constant temperature and composition, across which the rate-determining step fundamentally changes. Therefore, predictions of protein stability that rely on extrapolation across such a boundary will necessarily be incorrect and potentially can be in error by orders of magnitude.

Such a range of kinetic regimes is observed experimentally, as illustrated in Figure 4, which show time profiles (Figure 4A) and reduced initial rates (Figure 4B) for bovine

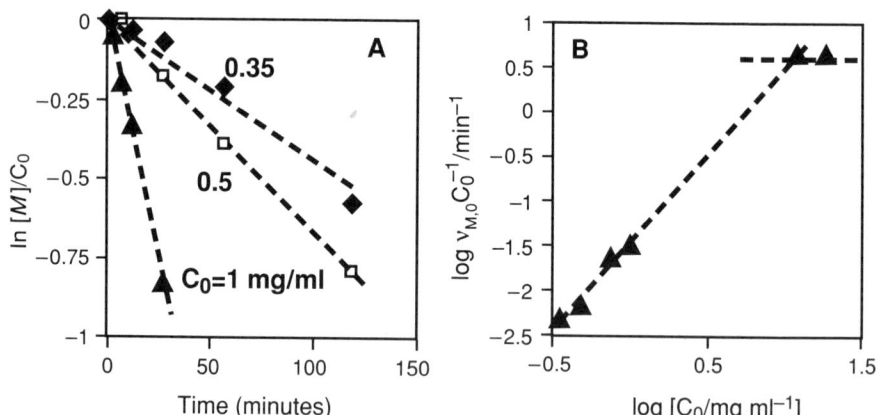

FIGURE 4. (A) Monomer concentration vs. time determined from size-exclusion HPLC for aggregation of α-chymotrypsinogen at pH 3.5, 10 mM sodium citrate buffer, 65°C. The curves are labeled with the values of their initial concentration to illustrate the C_o dependence of k_{obs} despite apparent first-order behavior. (B) Initial aggregation rate as a function of initial monomer concentration for the same protein, temperature, and solvent conditions as in (A). The slope is approximately 2 for data below the plateau, implying via Eq. (51) that the true reaction order is approximately 3. Equation (41), using $\Gamma = 0.35$, gives an estimate of x of almost 6, suggesting an interpretation in which α-chymotrypsinogen forms a reversible hexameric "nucleus" prior to subsequent irreversible aggregation steps at these experimental conditions.[27]

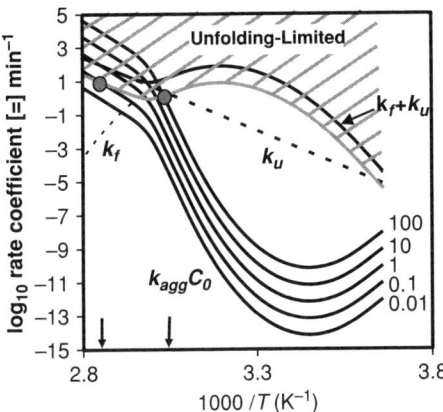

FIGURE 5. Temperature-rate diagram for a model protein exhibiting a switch from unfolding- to association-limited aggregation kinetics as T and/or C_0 is lowered, analogous to that for bG-CSF in Ref. 11, using $k_{agg} = k_{obs}$ for association-limited aggregation (Eq. 30). The numbers next to the $k_{agg}C_0$ curves indicate the value of C_0. The value of T_x at two different C_0 values is indicated by vertical arrows at the x axis; $C_0 = 0.1$ mg/ml and 100 mg/ml for the left- and right-most arrows, respectively.

α-chymotrypsinogen aggregation as a function of C_0 at 65°C.[27] Figure 4A shows apparent first-order behavior (linear $\ln[M]/C_0$ vs. t) but with k_{obs} values, or slopes, that are C_0 dependent. Figure 4B shows the dependence of the initial rate (scaled by C_0) on the value of C_0 for the same temperature and solution conditions as Figure 4A. The slope up to a $C_0 \sim 15$ mg/ml is ~ 2, indicating a true reaction order ~ 3 (cf. Eq. 51). Also shown in Figure 4B is an apparent plateau at high C_0, indicative of a change to unfolding-limited kinetics (or some other true first-order process) at sufficiently high C_0. Thus, under these solution conditions and temperature it appears Eq. (41) is operative at C_0 below ~ 15 mg/ml (~ 60 μM) and Eq. (21) above that concentration.

Finally, we consider the effects of temperature on the observed aggregation kinetics; k_{obs} in all of the models considered here has an intrinsic, or purely kinetic, contribution from the rate coefficient(s) of the rate-determining step(s), and in all cases except unfolding-limited kinetics also has one or more thermodynamic contributions from N-D equilibrium ($f_{D,eq}$) and reversible association (e.g., K_x). Temperature affects both the thermodynamic and intrinsic kinetic contributions, and its relative effects may be difficult to deconvolute. With this in mind, it is useful to construct a temperature-rate diagram to compare the relative time scales and the associated temperature dependence for association and/or rearrangement with that for folding-unfolding. An example of such a diagram for a hypothetical globular protein is shown in Figure 5, which is an adaptation of a similar diagram for bovine G-CSF.[11]

Figure 5 shows the temperature dependence of the first-order rate coefficients for folding (k_f) and unfolding (k_u), as well as that for the effective rate coefficient for aggregation, k_{agg}, multiplied by C_0^{zt-1} to account for the reaction order being greater than one for cases in which $z_t > 1$. For example, for association-limited kinetics $k_{agg} = n^* k_{D,D} f_{D,eq}^2 C_0$ would be used (cf., Eq. 30). The inverse of k_{agg} has units of time and gives the characteristic time scale for aggregation via association or rearrangement steps such as discussed in Section 3, but this time scale is dependent on C_0 because association necessarily includes bi-molecular steps. Therefore, rather than a single curve as a function of temperature, a family of curves with different C_0 are shown. $(k_u + k_f)^{-1}$ provides the characteristic time scale for the reversible first-order process of folding-unfolding.[11,20,26] k_u and k_f exhibit typical behavior for proteins, including an Arrhenius T dependence for k_u, a highly non-Arrhenius dependence for k_f, and a characteristic time scale, i.e., $(k_u + k_f)^{-1}$, on the order of 10^{-3} to 10 sec at temperatures near and above ambient.[31] This also indicates unfolding-limited kinetics should be Arrhenius and have a maximum time scale $\sim$ 10 min, since $d[M]/dt \sim k_u$ (Section 3.1). When k_{agg} lies significantly below $k_u + k_f$ in Figure 5, folding-unfolding can be taken as equilibrated at all times, relative to monomer depletion occurring via aggregation. Thus, the region lying above the $k_u + k_f$ curve in Figure 5 denotes unfolding-limited conditions (Section 3.1). For practical purposes, the region one order of magnitude below $k_u + k_f$ also is included in order to provide a conservative estimate for the unfolding-limited regime for use in Section 5.

In addition to the above discussion, there are a number of key points to take from Figure 5. The first is that in general k_{agg} will decline more quickly with decreasing T than will $k_u + k_f$, because of the compensating effects of temperature on k_u and k_f for typical proteins,[31] and k_{agg} is proportional to $(f_{D,eq})^r$, $r \geq 1$. Because unfolding-limited aggregation requires the characteristic time scale of folding-unfolding to be lower than or approximately equal to the time scale for association, this indicates that decreasing temperature will ultimately result in association- or rearrangement-limited kinetics, even if unfolding-limited kinetics are observed at higher T. With this in mind, one may define a crossover temperature (or temperature range) T_X where the time scale of aggregation (association or rearrangement) crosses that for folding-unfolding; i.e., where $k_{agg} \sim (k_u + k_f)$ is satisfied. Two such crossover points are indicated in Figure 5, one for each of the aggregation curves at the extremes of C_0. At $T \geq T_X$, unfolding-limited kinetics will be observed. At lower temperature, N-D equilibrium will be maintained and association- or rearrangement-limited kinetics will be found. T_X increases with decreasing C_0. Therefore, decreasing (increasing) C_0 will expand (contract) the region of association- or rearrangement-limited kinetics. By decreasing C_0, it is possible therefore to suppress T_X to temperatures significantly above the equilibrium midpoint unfolding temperature T_0 ($\Delta G_D = 0$ at T_0).[29,30] Such a strategy can permit folding-unfolding thermodynamics to be measured even for an aggregation-prone system.[11] In the example used to generate Figure 5, T_X increases by almost 25 deg, C with a decrease in C_0 from 100 to 0.1 mg/ml, and T_X is effectively infinite for $C_0 \sim 0.01$ mg/ml and below. Conversely, thermal unfolding or "melting" temperature (T_m) measurements that are performed at high C_0 have a higher potential to be convoluted with aggregation kinetics that may invalidate any subsequent thermodynamic stability calculations.[21–23,29] Finally, there is significant curvature to the k_{agg} curves in Figure 5. This curvature is due to the non-Arrhenius temperature dependence of $f_{D,eq}$, and the fact that $f_{D,eq}$ occurs implicitly in k_{agg} (cf., k_{obs} in Eqs. 29–31, 38, and 41). The strong, non-Arrhenius temperature dependence of $f_{D,eq}$ follows directly from that of K[11,26,29,30] and the fact that ΔH_{un}

is highly temperature dependent for most proteins.[29,30] This point will be elaborated in Section 5.

To conclude this section, we reiterate a number of key points that follow from the above results and related findings elsewhere.[11,26] A partial summary of these, along with key expressions from Section 3 and conclusions from Section 5 regarding viability of stability prediction via temperature extrapolation, are given in Table 2.

1. It is impossible to conclusively determine whether an aggregation process is unfolding-limited, association-limited, or rearrangement-limited solely by considering kinetic data in the form of concentration versus time. Although not discussed in detail here, it can be shown that this conclusion holds for data on aggregate formation just as it does for monomer loss data.
2. Unfolding-limited kinetics must show all of the following behaviors: first-order kinetics in terms of $[M]$ versus t *and* in terms of initial rates, or equivalently concentration-independent k_{obs}; values of k_{obs} that are consistent with orthogonally measured k_u values (typically $\sim 10^{-3}$ to 10 s^{-1}); and k_{obs} that exhibit Arrhenius kinetics with an activation energy that is comparable to that measured for k_u.
3. Association-limited kinetics may exhibit first- or second-order kinetics in terms of $[M]$ versus time, but still show second-order dependence on C_0 in terms of initial rates. Such behavior can be convoluted if aggregates precipitate during the course of the experiment.[11] Association-limited kinetics are expected to exhibit non-Arrhenius temperature dependence of k_{obs} if a large enough temperature range is considered ($\sim 20°$ to $30°$C) due to the temperature dependence of folding-unfolding equilibrium controlling $[D]$ (i.e., $f_{D,eq}$ vs. T).
4. Steady-state association-limited kinetics are indistinguishable from steady-state rearrangement-limited kinetics (with $x = 2$) without thermodynamic or kinetic measurements beyond the classic macroscopic measures discussed here. Unsteady-state rearrangement-limited kinetics may be distinguishable via initial rate kinetics if a greater-than-second-order dependence on C_0 is observed (i.e., for $x > 2$ in Figures 3 and 4).

5. APPLICATIONS AND LIMITATIONS FOR STABILITY PREDICTION

5.1. Unfolding-Limited Conditions

As the discussion in Section 4 demonstrated, unfolding-limited aggregation is operable under a rather limited set of conditions, i.e., relatively high temperatures or equivalently $K \sim 1$ and higher, and with time scales that are necessarily no slower than that for unfolding. Further, because decreasing temperature typically increases the time scale for association-and rearrangement-limited aggregation to much greater extent than it does the time scale for folding-unfolding, a system exhibiting unfolding-limited kinetics under accelerated (high temperature) conditions will switch to association- or rearrangement-limited kinetics at some point as temperature is lowered. Therefore, accelerated stability data gathered under unfolding-limited conditions cannot, in general, be extrapolated to significantly lower temperatures.

An additional complication arises from the effects of initial protein concentration (cf., Figure 5). Decreasing C_0 also has the potential to shift the rate-determining step from unfolding to association or rearrangement, and thereby invalidate stability data taken at higher C_0. Conversely, such concerns should not be an issue if one chooses to extrapolate to increased C_0, provided of course any effects of C_0 on the unfolding process itself also are accounted for. Overall, great care must be taken if one wishes to attempt to predict "real-time" (low temperature) aggregation kinetics from accelerated (high temperature) data obtained under unfolding-limited conditions. A possible exception to this would be for a protein that is inherently incapable of refolding (k_f is effectively zero), e.g., due to a chemical alteration or a lack of required chaperone.[32] It appears, however, that such an extrapolation has yet to be tested beyond the stage of model calculations.[33]

It is therefore recommended that if possible one adjust the temperature and initial protein concentrations included in a given experimental design so as to operate in the aggregation-limited region of the temperature-(concentration-) rate diagram for a given protein. In such cases, extrapolations of accelerated data may be possible, e.g., via approaches such as discussed below, without the threat of a change in rate-determining step within the temperature range of extrapolation that thereby would invalidate the resulting stability prediction.

5.2. Aggregation-Limited Conditions

The following analysis applies to systems exhibiting association- or rearrangement-limited kinetics, collectively termed aggregation-limited kinetics here, under accelerated (high temperature) conditions. Note that C_0 is fixed in the following analysis. From the preceding section, it is clear that extrapolations of accelerated data obtained under different protein concentrations should be done with care and avoided altogether if detailed information on the *true* reaction order with respect to $[M]$ is not available as a function of temperature.

In general, the equilibrium constant for any process i obeys the following relation at constant pressure and composition

$$\frac{d \ln K_i}{d 1/T} = -\frac{\Delta H_i}{R} \tag{52}$$

with ΔH_i denoting the enthalpy change for process i. To simplify the analysis and because we do not presently have direct experimental data to the contrary, we will assume that all association equilibria (but *not* folding-unfolding) in Figure 1 have constant ΔH_i over the T range of interest and that all intrinsic rate coefficients follow pseudo-Arrhenius temperature dependences with constant activation energy. With these approximations, the temperature dependence of k_{obs} for all the limiting cases other than unfolding-limited will share a common mathematical form

$$\frac{d \ln k_{obs}}{d 1/T} = \phi \frac{d \ln f_{D,eq}}{d 1/T} - \frac{\langle E_{a,int} \rangle}{R} \qquad (53)$$

where $\langle E_{a,int} \rangle$ is a constant including the contributions from the activation energies of the appropriate intrinsic rate coefficient(s) and association equilibria (if applicable), and ϕ is the power to which $f_{D,eq}$ appears in k_{obs} for a given limiting scenario (see also Table 2). From the definition of $f_{D,eq}$ [Eq. (25)] and standard thermodynamic derivatives with respect to temperature, Eq. (53) can be rewritten as

$$\frac{d \ln k_{obs}}{1/T} = -R^{-1} \left[\phi \left(1 - f_{D,eq}\right) \left(\Delta H_0 + \Delta c_p (T - T_0)\right) + \langle E_{a,int} \rangle \right] = -\frac{E_{obs}}{R} \qquad (54)$$

In Eq. (54), T_0 denotes the temperature at which $K = 1$ ($\Delta G_D = 0$) or approximately the midpoint-unfolding temperature (see also discussion associated with Figure 5), ΔH_0 denotes the enthalpy of unfolding at T_0, and Δc_p the heat capacity of unfolding. These are the quantities typically measured in equilibrium differential scanning calorimetry or thermal unfolding experiments.[29,30] The term in brackets in Eq. (54) defines an effective or observed activation energy for aggregation outside of unfolding-limited conditions. Even with the intrinsic activation energy for association or rearrangement ($\langle E_{a,int} \rangle$) and ϕ both being independent of T, the observed activation energy E_{obs} still will be highly temperature dependent. This is due to the temperature dependence of $f_{D,eq}$,[11,26,29,30] and because of the term proportional to Δc_p.[26] Physically, this arises because decreasing T shifts the N-D transition toward the N state, thereby depleting the amount of D and all species in pseudo-equilibrium with it, i.e., the reactive species in the rate-determining step. It is re-emphasized that unfolding-limited kinetics are *necessarily* independent of the thermodynamics of folding-unfolding, as there is essentially no refolding. Further, based on data for the experimental systems in which unfolding rate constants are available[31] the observed activation energy for k_u, and therefore unfolding-limited aggregation, is expected to be independent of temperature, and at most of order 100 kcal/mole. In contrast, as the analysis above shows, observed activation energies in aggregation-limited scenarios can be both highly non-Arrhenius and have much larger absolute values, easily over 100 to 200 kcal/mole.[11,26,34] Based on the discussion of Figure 5 in Section 4 and the potential for a switch from unfolding-limited to association-/rearrangement-limited aggregation as T is lowered from the vicinity of T_0, it is advisable to choose accelerated stability conditions below the unfolding-limited regime. As this will likely also be at temperatures more than a few degrees below the midpoint unfolding temperature, $f_{D,eq}$ will be much less than 1 and Eq. (54) will reduce to

$$\frac{d \ln k_{obs}}{d 1/T} \cong -\frac{\langle E_{a,int} \rangle + \phi \left(\Delta H_0 - \Delta c_p T_0\right)}{R} - \frac{\phi \Delta c_p}{R} T \qquad (55)$$

and therefore

$$E_{obs} = [\langle E_{a,\text{int}} \rangle + \phi(\Delta H_0 - \Delta c_p T_0)] + \phi \Delta c_p T = E_0 + \phi \Delta c_p T \tag{56}$$

where E_0 is a constant and $\phi \Delta c_p T$ gives the temperature dependence of the observed activation energy. The above analysis suggests a simple procedure for extrapolating accelerated kinetic data to lower temperatures while accounting for the nonconstant observed or apparent activation energy.

Specifically, one first measures k_{obs} over a small range of T under accelerated conditions, e.g., a range of temperatures below the unfolding-limited region but over a 5 to 10 degree window above the lowest temperature at which the observed kinetics are fast enough to be sufficiently "accelerated" to make their measurement practical within experimental or product development constraints. We will denote this lower temperature T_L. The $k_{obs}(T)$ data at $T_L \leq T \leq T_L + 5°C$ can be used to calculate $E_{obs}(T_L)$ from a simple Arrhenius plot. From Eq. (56) it follows that E_{obs} at lower T will be given by

$$E_{obs}|_{T<T_L} = E_{obs}|_{T=T_L} + \phi \Delta c_p (T - T_L) \tag{57}$$

Therefore, one need only estimate ϕ and Δc_p to perform a non-Arrhenius extrapolation to lower temperatures. From the results of Section 3, it is apparent that $\phi = 2$ is a reasonable approximation in the absence of experimental evidence to the contrary. Further, the most conservative estimates of stability at lower T will come from choosing the upper bound on $\phi \Delta c_p$, because $T - T_L$ is necessarily negative and because a lower E_{obs} implies aggregation kinetics will be slowed to a lesser degree with decreasing temperature. A conservatively large estimate of Δc_p is that for complete unfolding, which can be calculated solely from knowledge of the primary sequence of a protein chain.[35,36] Finally, incorporating Eq. (57) into an integrated form for extrapolating k_{obs} to $T < T_L$ gives[26] (note: this extrapolation holds at constant C_0)

$$\ln \frac{k_{obs}(T)}{k_{obs}(T_L)} = -R^{-1} \left\{ E_{obs}|_{T=T_L} \left(\frac{1}{T} - \frac{1}{T_L} \right) - \phi \Delta c_p \left[1 - \frac{T_L}{T} + \ln \frac{T_L}{T} \right] \right\} \tag{58}$$

Thus, rational and conservative extrapolation of association-limited and rearrangement-limited aggregation kinetics to lower temperatures is in principle possible using this approach.[11] It requires k_{obs} measured over a rather small temperature window of accelerated conditions (e.g., 5 to 10 deg C), somewhat below the midpoint unfolding temperature (typically 10 deg C or more below T_0), and some estimate of $\phi \Delta c_p$. The apparent reaction order ϕ can be estimated easily in the process of obtaining the accelerated k_{obs} data, or taken as 2 (cf., Table 2) if only preliminary, conservative estimates are needed. The heat capacity change upon unfolding, Δc_p, can be determined from thermodynamic folding-unfolding data[29,30] if available, or in the absence of such data may be well approximated from semi-empirical theoretical calculations.[35,36]

6. NOTE ADDED IN PROOF

Recent calculations in the author's laboratory indicate $\Gamma \approx 1/2$ holds for a larger range of model parameters in the rearrangement-limited scenario associated with Eq. (41). This does not change any of the formal expressions or qualitative conclusions in this chapter, but does slightly increase the slope of lines marked (3) in Fig. 3 and changes the estimate of x from 6 to 5 in the caption of Fig. 4.

REFERENCES

1. A. L. Fink, Protein aggregation: folding aggregates, inclusion bodies, and amyloid, *Folding Des.* **3**, R9–R23 (1998).
2. P. M. Bummer and S. Koppenol, Chemical and physical considerations in protein and peptide stability, *Drugs Pharm. Sci.* **99**, 5–69 (2000).
3. J. L. Cleland, M. F. Powell, and S. J. Shire, The development of stable protein formulations: a close look at protein aggregation, deamidation, and oxidation, *Crit. Rev. Therapeutic Drug Carr. Sys.* **10**, 307–377 (1993).
4. E. Gazit, The "correctly folded" state of proteins: is it a metastable state? *Angewandte Chemie, Int. Ed.* **41**, 257–259 (2002).
5. J. N. Onuchic, Z. Luthey-Schulten, and P. G. Wolynes, Theory of protein folding: the energy landscape perspective, *Annu. Rev. Phys. Chem.* **48**, 545–600 (1997).
6. M. G. Mulkerrin and R. Wetzel, pH dependence of reversible and irreversible thermal denaturation of γ interferons, *Biochemistry* **28**, 6556–6561 (1989).
7. J. M. Finke, M. Roy, B. H. Zimm, and P. A. Jennings, Aggregation events occur prior to stable intermediate formation during refolding of interleukin-1β, *Biochemistry* **39**, 575–583 (2000).
8. A. Dong, T. W. Randolph, and J. F. Carpenter, Entrapping intermediates of thermal aggregation in α-helical proteins with low concentration of guanidine hydrochloride, *J. Biol. Chem.* **276**, 27689–27693 (2000).
9. R. Khurana, J. R. Gillespie, A. Talapatra; L. J. Minert, C. Ionescu-Zanetti, I. Millet I., and A. L. Fink, Partially folded intermediates as critical precursors of light chain amyloid fibrils and amorphous aggregates, *Biochemistry* **40**, 3525–3535 (2001).
10. A. O. Grillo, K.-L. T. Edwards, R. S. Kashi, K. M. Shipley, L. Hu, M. J. Besman, and C. R. Middaugh, Conformational origin of the aggregation of recombinant human factor VIII, *Biochemistry* **40**, 586–595 (2001).
11. C. J. Roberts, R. T. Darrington, and M. B. Whitley, Irreversible aggregation of recombinant bovine granulocyte-colony stimulating factor (bG-CSF) and implications for predicting protein shelf life, *J. Pharm. Sci.* **92**, 1095–1111 (2003).
12. B. S. Kendrick, J. L. Cleland, X. Lam, T. Nguyen, T. W. Randolph, M. C. Manning, and J. F. Carpenter, Aggregation of recombinant human interferon gamma: kinetics and structural transitions, *J. Pharm. Sci.* **87**, 1069–1076 (1998).
13. P. L. Privalov, Intermediate states in protein folding, *J. Mol. Biol.* **258**, 707–725 (1996).
14. A. V. Finkelstein, Proteins: structural, thermodynamic, and kinetic aspects, in *Slow Relaxations and Nonequilibrium Dynamics in Condensed Matter, NATO Advanced Study Institute*, ed. J.-L. Barrat, M. Feigelman, J. Kurchan, J. Dalibard, (Berlin: Springer-Verlag, EDP Sciences, Les Ulis, 2003), 649–704.
15. P. O. Souillac, V. N. Uversky, and A. L. Fink, Structural transformations of oligomeric intermediates in the fibrillation of the immunoglobulin light chain LEN, *Biochemistry* **42**, 8094–8104 (2003).
16. A. Lomakin, D. S. Chung, G. B. Benedek, D. A. Kirschner D. A., and D. B. Teplow, On the nucleation and growth of amyloid β-protein fibrils: detection of nuclei and quantitation of rate constants, *Proc. Natl. Acad. Sci. USA* **93**, 1125–1129 (1996).

17. M. D. Kirkitadze, M. M. Condron, and D. B. Teplow, Identification and characterization of key kinetic intermediates in amyloid β-protein fibrillogenesis, *J. Mol. Biol.* **312**, 1103-1119 (2001); G. Bitan, S. S. Vollers, and D. B. Teplow, Elucidation of primary structure elements controlling early amyloid β-protein oligomerization, *J. Biol. Chem.* **278**, 34882–34889 (2003).
18. C. R. Robinon, D. Rentzeperis, J. L. Silva, and R. T. Sauer, Formation of a denatured dimer limits the thermal stability of Arc repressor, *J. Mol. Biol.* **273**, 692–700 (1997).
19. A. M. Buswell and A. P. J. Middelberg, Critical analysis of lysozyme refolding kinetics, *Biotech. Prog.* **18**, 470–475 (2002).
20. K. J. Laidler, *Chemical Kinetics, 3rd ed.* (New York: HarperCollins Pub., 1987).
21. R. Lumry and H. Eyring, Conformational changes of proteins, *J. Phys. Chem.* **58**, 110–120 (1954).
22. S. E. Zale and A. M. Klibanov, On the role of reversible denaturation (unfolding) in the irreversible thermal inactivation of enzymes, *Biotech. Bioeng.* **25**, 2221–2230 (1983).
23. J. M. Sanchez-Ruiz, Theoretical analysis of Lumry-Eyring models in differential scanning calorimetry, *Biophys. J.* **61**, 921–935 (1992).
24. C. La Rosa, D. Milardi, D. Grasso, R. Guzzi, and L. Sportelli, Thermodynamics of the thermal unfolding of Azurin, *J. Phys. Chem.* **99**, 14864–14870 (1995).
25. S. Tello-Solis and A. Hernandez-Arana, Effect of irreversibility on the thermodynamic characterization of the thermal denaturation of *Aspergillus saitoi* acid proteinase, *Biochem. J.* **31**, 969–974 (1995).
26. C. J. Roberts, Kinetics of irreversible protein aggregation: analysis of extended Lumry-Eyring models and implications for predicting protein shelf life, *J. Phys. Chem. B* **107**, 1194–1207 (2003).
27. D. J. Caravoulias et al., unpublished.
28. M. M. Pallitto and R. M. Murphy, A mathematical model of the kinetics of β-amyloid fibril growth from the denatured state, *Biophys. J.* **81**, 1805–1822 (2001).
29. P. L. Privalov, Stability of proteins, *Adv. Protein Chem.* **33**, 167–241 (1979).
30. P. L. Privalov and S. A. Potekhin, Scanning microcalorimetry in studying temperature-induced changes in proteins, *Methods Enzymol.* **181**, 4–51 (1986).
31. S.-I. Segawa and M. Sugihara, Characterization of the transition state of lysozyme unfolding. I. Effect of protein-solvent interactions on the transition state, *Biopolymers*, **23**, 2473–2488 (1984); M. Oliveberg, Y.-J. Tan, and A. R. Fersht, Negative activation enthalpies in the kinetics of protein folding, *Proc. Natl. Acad. Sci. USA* **92**, 8926–8929 (1995).
32. L. Lopez-Arenas, S. Solis-Mendiola, and A. Hernandez-Arana, Estimating the degree of expansion in the transition state for protein unfolding: analysis of the pH dependence of the rate constant for caricain denaturation, *Biochemistry* **38**, 15936–15943 (1999).
33. I. M. Plaza del Pino, B. Ibarra-Molero, and J. M. Sanchez-Ruiz, Lower kinetic limit to protein thermal stability: a proposal regarding protein stability in vivo and its relation with misfolding diseases, *Proteins: Struct. Funct. Genet.* **40**, 58–70 (2000).
34. A. Fatouros, T. Osterberg, and M. Mikaelsson, Recombinant factor VIII SQ—inactivation kinetics in aqueous solution and the influence of disaccharides and sugar alcohols, *Pharm. Res.* **14**, 1679–1684 (1997).
35. D. Milardi, C. la Rosa, S. Fasone, and D. Grasso, An alternative approach in the structure-based predictions of the thermodynamics of protein unfolding, *Biophys. Chem.* **69**, 43–51 (1997).
36. J. Gomez, V. J. Hilser, D. Xie, and E. Freire, The heat capacity of proteins, *Prot. Struc. Func. Genet.* **22**, 404–412 (1995).

Simulations of Protein Aggregation

A Review

Carol K. Hall,[1,2] Hung D. Nguyen,[1] Alexander J. Marchut,[1] and Victoria Wagoner[1]

Protein aggregation is a cause or associated symptom of a number of neurodegenerative diseases including Alzheimer's, Parkinson's, Huntington's, and the prion diseases. In this chapter we review recent efforts by others and by ourselves to simulate the aggregation of proteins into both amorphous aggregates (inclusion bodies) and ordered aggregates (fibrils). Since protein aggregation occurs on long timescales and involves many proteins, the detailed all-atom approaches that are often used to simulate the folding of isolated proteins are not entirely suitable for studying aggregation because they are too computationally intensive. Instead, a variety of simulation approaches, with concomitant compromises in system size or level of realism, are being used. These approaches range from atomistic-resolution simulations of systems containing two to three small peptides to low-resolution lattice Monte Carlo simulations of systems containing many peptides. We have adopted an intermediate-resolution approach that allows the simulation via discontinuous molecular dynamics (a fast alternative to standard molecular dynamics) of relatively large systems of polyalanine peptides. Polyalanine was chosen for study because it forms β-sheet complexes (fibrils) *in vitro* at high temperatures and high peptide concentrations. In our simulations, a system of 48- to 96-$KA_{14}K$ peptides initially in the random coil state forms alpha-helices at low concentrations and temperatures but assembles spontaneously into fibrillar structures at high concentrations and temperatures. The effect of temperature, peptide concentration and chain length on the kinetics and thermodynamics of fibril formation will be described.

---

[1] Department of Chemical Engineering, North Carolina State University, Raleigh, NC 27695-7905
[2] To whom correspondence should be addressed: Carol K. Hall, Department of Chemical Engineering, North Carolina State University, Raleigh, NC 27695-7905; email: hall@turbo.che.ncsu.edu

1. INTRODUCTION

Protein aggregation [1] can be both a curse and a blessing. It is viewed as a curse because it is a cause or an associated symptom of diseases such as Alzheimer's and Parkinson's,[2-7] can interfere with the recovery of recombinant proteins from inclusion bodies,[4-8] can limit the stability of protein-based drugs during shipping, storage, and administration,[15,16] can lead to the production of off-flavors and unwanted color changes in the foods industry,[17] and often is the unwanted last step in a protein-folding experiment. It is viewed as a blessing by those who seek to take advantage of the built-in ability of biological macromolecules to self-assemble into supramolecular structures,[18,19] thereby creating building blocks for advanced nanomaterials with superior mechanical, sensing, electronic, magnetic, or optical properties.[18-23] Despite the obvious practical importance of protein aggregation, our understanding of its physical basis is far from complete.

While experimental research is invaluable in the current effort to understand the mechanisms of protein aggregation, computer simulation is a complementary tool that can help to elucidate these mechanisms in molecular-level detail. Simulations can yield microscopic information about the system under investigation that cannot be obtained directly from experiments, provided that the model employed is a reasonable representation of the real molecule.

The two principal computer simulation techniques are molecular dynamics and Monte Carlo. In molecular dynamics, the trajectories of studied atoms are computed by solving Newton's equation of motion at regularly space intervals, called time steps. At the beginning of each time step the net force acting on each molecule due to all the other molecules is calculated. Knowledge of this force allows the determination of each molecule's acceleration ($F = ma$), which in turn allows the prediction of the position and velocity of each molecule at the beginning of the next time step. At each time step, the instantaneous values of thermodynamic properties are computed and recorded. Thermodynamic properties are obtained by averaging the instantaneous properties over time.

A variant on standard molecular dynamics that is applicable to systems of molecules interacting via discontinuous potentials (e.g., hard-sphere and square-well potentials) is the discontinuous molecular dynamics (DMD) method. Unlike soft potentials such as the Lennard-Jones potential, discontinuous potentials exert forces only when particles collide, enabling the exact (as opposed to numerical) solution of the collision dynamics. DMD simulations proceed by locating the next collision, advancing the system to that collision, and then calculating the collision dynamics. DMD of chainlike molecules is generally implemented using the "bead string" algorithm introduced by Rapaport[24,25] and later modified by Bellemans *et al.*[26] Briefly, adjacent beads along the chain are bonded together by short, invisible strings whose length ensures that the bond length between beads varies freely between $(1 - \delta)L$ and $(1 + \delta)L$, where L is the bond length and $\delta \ll 1$. This means that bonded beads along the chain are partially decoupled from each other, moving freely along linear trajectories between core collisions and bond stretch collisions. Chains of square-well spheres can be accommodated in this algorithm by introducing well-capture, well-bounce, and well-dissociation "collisions" when a sphere enters, attempts to leave, or leaves the square well of another sphere. In canonical ensemble simulations (constant system size, volume, temperature) the temperature is maintained constant by implementing the Andersen thermostat method[27]; all beads are subjected to random, infrequent collisions with ghost particles whose velocities are chosen randomly from a Maxwell-Boltzmann distribution centered at the system temperature.

In Monte Carlo simulation, molecular configurations are generated randomly on the computer; thermodynamic properties are obtained by averaging over large numbers of these configurations. The justification for this approach lies in the basic postulate of statistical mechanics, which says that the time average of a property is equal to the average over all possible microscopic states, which in classic statistical thermodynamics is equivalent to the average over all possible configurations. The most popular way to do Monte Carlo simulation is the Metropolis sampling algorithm, which works well because it does not require enormous amounts of computer time. The Metropolis sampling algorithm generates a new configuration in which one molecule in the old configuration is moved at random to a trial position. If the energy of the new configuration is less than the energy of the old configuration, the new configuration is automatically accepted; otherwise, it is accepted according to an a priori transition probability designed to generate configurations in proportion to their probability of occurring.

Since computer simulation studies of protein aggregation are built on protein folding models, it is appropriate to briefly review those aspects of the protein folding literature that are most relevant to this review.

2. PROTEIN MODELS

Computer simulation studies of protein folding can be divided roughly into two categories: high-resolution models and low-resolution models.

High-resolution folding models are based on a realistic representation of protein geometry and energetics. They typically account for the motion of every atom on the protein (with the exception of some hydrogen atoms) and every solvent atom. The models are highly popular with experimentalists as well as theorists, due in part to the availability of simulation packages containing well-tested atom-atom force fields such as AMBER,[28,29] CHARMM,[30] ENCAD,[31,32] DISCOVER,[33] and ECEPP.[34-36] Unfortunately, the detail that makes high-resolution models so realistic also makes them extremely computationally intensive, precluding their application to problems involving large conformational changes or long time scales. The closest that high-resolution models have come to being able to simulate the entire folding process is the remarkable 1-microsecond simulation of Duan and Kollman,[37] who monitored the "folding" of a 36-residue protein surrounded by 3000 water molecules from a conformation with some native turns to a partially folded intermediate using 256 dedicated processors for two full months.

Low-resolution models, also called simplified folding models, are based on a coarse-grained representation of protein geometry and energetics. They typically account for the motion of groups of atoms along the chain and ignore the motion of the solvent atoms in order to enhance computational efficiency. The absence of solvent atoms in low-resolution models means that effective potentials, or potentials of mean force, must be used to describe the interactions between residues. There are two types of low resolution models: lattice models, which represent a protein as a linear chain of beads confined to a lattice, and off-lattice models, which represent a protein as a chain of beads (spheres) or groups of beads moving through continuous space. Low-resolution models have provided valuable insights into the basic principles of protein folding due to their ability to monitor large conformational changes and long time scales. Their weakness is their inability to make definitive statements about the folding of specific proteins.

Obviously high-resolution models and low-resolution models represent two ends of a spectrum of models with varying levels of detail. Which type of model to use depends on what questions are being asked. Trade-offs must be made: the larger the number of degrees of freedom to be simulated, the simpler the model must be. Since most computer simulation studies on protein aggregation have employed a simplified representation of a protein, here we review simplified folding models in detail.

Simplified or low-resolution folding models were first introduced in the late 1970s.[38-46] Go and co-workers[38,44-46] modeled proteins as chains confined to two- and three-dimensional lattices with long-range residue-residue interactions assigned so as to favor native tertiary contacts, the so-called "Go potential," and short range interactions assigned to favor native local conformations. Levitt and Warshel[39] modeled bovine pancreatic trypsin inhibitor (BPTI) using an off-lattice representation in which backbone residues are depicted as chains of single spheres with attached single-sphere side chains interacting via empirical potentials derived from correlations of data on dipeptide structures. This work was criticized[43] because its empirically derived local interaction potential contained a built-in bias toward BPTI's native conformation, causing this and other simplified models based on empirical potentials to fall out of favor for a number of years. They nonetheless occupy a significant place in protein-folding model history because they demonstrated for the first time that folding could be studied using simplified models based on reduced coordinate representations.

Low-resolution models were popularized again in the late 1980s by the work of the group of Skolnick and Kolinski and the group of Dill. Skolnick, Kolinski, and co-workers[47-52] developed a series of three-dimensional lattice models designed to fold without assigning tertiary interactions that favor the native conformation. They identified the minimal set of interactions required to successfully fold model chains via Monte Carlo simulation into nativelike protein motifs. In the early 1990s, Skolnick, Kolinski, and collaborators[53-58] began adding more detail to these lattice models in an effort to predict the folding of specific proteins, the first in a new wave of so-called intermediate resolution models. Dill and co-workers[59-68] introduced a new class of lattice models, the HP lattice models, in which a protein is modeled as a chain of hydrophobic (H) and polar (P) residues arranged in a specific sequence on a two- or three-dimensional lattice. Nonbonded H beads attract each other with strength ε to account for the tendency of hydrophobic residues to bury themselves in the protein interior, while nonbonded P-P and H-P interactions are set equal to zero. The success of the HP model in investigations of the general theoretical principles that underlie the connection between a protein's sequence and its native structure, folding pathways, and kinetics has stimulated the creation of a number of related lattice models.[69-79]

Off-lattice low-resolution models became fashionable in the 1990s as many investigators began to explore the myriad possibilities offered by mixing and matching single- or multibead protein representations with various kinds of site-site energy functions. Protein geometry was represented off-lattice using a one-bead per residue approach[80-95] or a two-bead per residue approach.[39,96-99] The types of energy functions used in these calculations can be divided roughly into three categories—Go-type potentials in which the parameters are chosen to favor the protein's known native state, potentials based on the relative hydrophobicity of the side chains, and knowledge-based potentials in which statistical data on residue-residue contacts from the Protein Data Bank are used to infer side-chain/side-chain potentials.

More recently the number of beads used to represent protein geometry has increased, leading to what we would call intermediate-resolution models. The idea here was to incorporate more realistic features into simplified folding models in an effort to predict a priori the native state's three-dimensional structure of specific proteins based solely on their amino acid sequence. Various protein representations were chosen including: three-bead approaches (one bead and one side-chain bead[100-107] or one backbone bead and two side-chain beads[108]), four-bead approaches,[109] five-bead approaches,[110-116] and seven-bead approaches.[117-119] Also falling into the intermediate-resolution model category (but not in the off-lattice category) is the discretized lattice model of Skolnick, Kolinski, and co-workers.[120-131] Energy functions for the intermediate-resolution models include not only the three categories described in the previous paragraph but also hydrogen-bonding potentials, multibody terms, burial terms (in which the strength of the hydrophobic interaction depends on the extent of burial), and special potentials for disulfide bonds and proline. These models are used in conjunction with various highly efficient computer algorithms that constrain and guide the search through conformational space, allowing the attainment of the native state in a reasonable time frame.

3. PROTEIN AGGREGATION STUDIES

3.1. Lattice Models

There are numerous low-resolution lattice model studies aimed at answering some of the basic questions regarding protein aggregation. The methods of choice are Monte Carlo simulation, in which configurations are generated in proportion to their probability of occurring, and exact enumeration studies, in which all possible molecular conformations are generated. Exact enumeration studies were conducted on HP-related chains by Giugliarelli et al. and by Harrison et al. Giugliarelli et al.[132] considered a system containing multiple two-dimensional HP chains and probed the influence of the interresidue interaction strength on the formation of isolated native structures, aggregates composed of chains with native structures, and aggregates composed of chains with nonnative structures. They found that at an interaction potential fine-tuned to give proteinlike hydrophobicities, the number of sequences that have compact, soluble native states and the number of nonsoluble "prionlike" proteins are both large and of the same order. By "prion-like" they meant that the structure of the protein in the aggregate was different than that in the native state. Harrison et al.[133,134] enumerated all possible configurations of pairs of two- and three-dimensional HP chains to examine the thermodynamics underlying the conformational change associated with aggregation of model proteins. They found that model proteins that form marginally stable monomers are highly likely to form alternative conformations as homodimers, which is a key feature of prion aggregation. They also showed that aggregation behavior depends on the specific sequence; sometimes a single mutation resulted in nonpropagating oligomers.

The earliest work that employed the Monte Carlo method is that of Patro and Przybycien[135,136] who simulated a system containing proteins modeled as two-dimensional hexagons with a mix of hydrophobic and polar sides. They monitored the association of the hexagons to ascertain how the structure of the resulting protein aggregates are dependent on protein surface characteristics, protein-protein interaction energies, and the entropic penalty accompanying the immobilization of protein in a solid phase. Although their work provides insight into the aggregation process, their assumption that all protein monomers are in the folded state means that they could not explore how folding and unfolding impacts the conformational changes that occur during aggregation.

Broglia et al.[137] investigated the simultaneous folding of two designed identical 36-amino-acid lattice chains using Monte Carlo simulations. They used a 20-letter potential energy model that was proposed by Miyazawa and Jernigan.[138] In this model, the effective interresidue contact energies are estimated from the numbers of residue-residue interactions observed in protein crystal structures appearing in the protein data bank. These investigators found that protein aggregates are formed from partially folded intermediates whose intramolecular contacts are associated with the most strongly interacting amino acids; these contacts are formed at an early stage in the normal folding process. Their model contains a more detailed description of the specific residue-residue interactions than most other lattice models of protein aggregation.

Istrail et al.[139] performed Monte Carlo simulations on an HP lattice model containing two chains. In addition to attractive hydrophobic-hydrophobic residue interactions, they added a repulsive interaction between solvent sites and hydrophobic residues to account for the penalty of having hydrophobic residues exposed to solvent. The aggregation of a system containing two chains was studied by determining the probability that the two chains associate as a function of the energy matrix, folding time, initial conformation, the number and sequence of the hydrophobic beads along the chain, and the packing fraction. They reported that aggregation propensity is dependent on the ratio of the number of hydrophilic residues and hydrophobic residues; a large ratio can protect chains from aggregation.

Bratko and Blanch[140] performed Monte Carlo simulations on systems containing up to six lattice proteins modeled as chains of 27 beads each interacting through a variation of the Go potential. Recall that in the Go potential the pair interaction strengths are assigned so as to favor the native structure of the protein. They found that the chains in the global free energy minimum dimer state are in the native structure. Clusters of three or more misfolded chains were observed and found to be stable compared to similar-sized aggregates of folded chains, although some native structure was preserved. As the number of chains present in the system increased, the refolding yield of native proteins increased. They also performed conventional and replica-exchange Monte Carlo simulations on a set of sequences of different lengths (16 to 40 residues) containing varying amounts of secondary-structural motifs (α-helices and β-sheets). They found that increasing the proportion of β-sheet structures facilitates the folding of isolated chains but favors the formation of misfolded aggregates in multichain systems.

Dima and Thirumalai[141] performed Monte Carlo simulations on a system containing two three-dimensional HP lattice chains with side chains to explore the mechanism associated with the conformational change from a monomeric compact state to an oligomeric β-sheet state. The idea here was to probe the general principles that govern aggregation by mapping out a phase diagram in the temperature-concentration plane. They found three distinct ordered structures, one of which contained native state proteins. They also found that ordered aggregates formed directly from the unfolded state rather than from partially structured intermediates. They concluded that polypeptide sequence and external conditions (e.g., peptide concentration, temperature, pH, and salt concentration) both play important roles in the aggregation process.

Leonhard et al.[142] conducted Monte Carlo simulations on a system containing two to six cubic-lattice model chains interacting via the 20-letter potential energy model proposed by Miyazawa and Jernigan.[138] Their aim was to understand how the values of the interaction parameters and chain length affect aggregation. They found that aggregation behavior is strongly influenced by small changes in the interaction energies.

All of the above-mentioned studies were limited to only a few chains, which is not enough to fully explore the competition between folding and aggregation in a multichain system. Studying only two or three chains does not allow exploration of situations in which chains interact with more than one other chain at a time such as would occur during fibril formation. In contrast, the systems simulated by Combe and Frenkel,[143] Toma and Toma,[144] and Hall and co-workers[145,146] were truly multichain systems. Combe and Frenkel[143] performed Monte Carlo simulations on systems containing up to 20 cubic-lattice Go-model 8-mer peptides. They calculated the phase diagram of a multiprotein system and found three phases: vapor, liquid, and crystal. Since the liquid-vapor transition lay below freezing temperatures, the gas-liquid transition was metastable. In the work by Hall and co-workers[145,146] and by Toma and Toma,[144] a simple two-dimensional HP lattice model proposed by Lau and Dill[59] was used to represent a protein chain. Toma and Toma[144] conducted Monte Carlo simulations on relatively large systems containing between 16 and 20 molecules. They considered three short two-dimensional lattice HP sequences: (1) PHPPHPPHPPHP, (2) PHPHPHPHPHP, and (3) PHHHPPHHHP. They found that the peptide sequence, concentration, and temperature were key to determining whether or not ordered structures formed.

Hall and co-workers focused their study on a single 20-bead sequence (PHPPHPPHHPPHHPPHPPHP).[145,146] This sequence was chosen for study (1) because of its ability to fold into the β-sheet-motif native state and (2) because of its ability to form multiple, easily recognized, partially folded intermediates. Typical configurations in the unfolded state and in the native state are shown in Figure 1. Gupta and Hall performed exact enumerations of the conformations of this isolated protein chain. At low values of $|\varepsilon^*|$, the model chain was found to be in an unfolded random-coil state characterized by a large radius of gyration and a small number of HH contacts. As $|\varepsilon^*|$

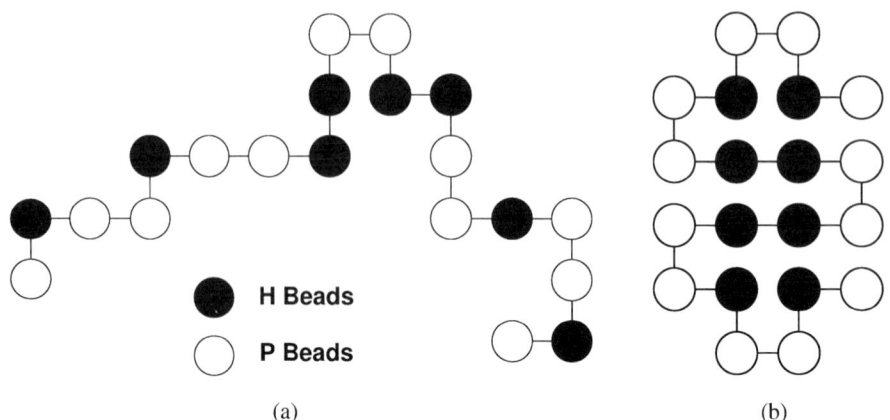

FIGURE 1. The 20-bead polypeptide molecule in (a) an unfolded state and (b) the native state.

increased, the chain folded into a compact native state with eight HH contacts arranged in a conformation with a hydrophobic core surrounded by polar beads.[147] The model protein exhibited proteinlike folding pathways and intermediates that change with the denaturant concentration.[148]

Gupta and Hall then performed dynamic Monte Carlo simulations on multichain systems containing between 20 and 40 of these same HP model protein chains.[145] They found that aggregation arises from association among partially folded intermediates, not random coil states. In addition, they found that there exists an optimum level of denaturant where the refolding yield is highest; moving away from the optimum levels resulted in a reduction of the refolding yield. (The existence of an optimum was later proven to be erroneous, as indicated below.)

Nguyen and Hall[146] conducted simulations on the same model peptide as Gupta and Hall. They investigated how changing the rate of chemical or thermal renaturation affects the folding and aggregation behavior of the model protein molecule. Three methods for the removal of denaturant were modeled: infinitely slow cooling (mimicking dialysis),[149] slow-but-finite cooling (mimicking multistep dilution or diafiltration),[150] and quenching (mimicking instantaneous removal of denaturant or reduction in temperature).[151] In addition, they modeled the pulse renaturation[152] or fed-batch operation method[153] in which denatured proteins are slowly added to a solution of already-folded or almost-folded proteins. Dynamic Monte Carlo simulations were performed on multichain systems containing 50 model proteins at seven values of the packing fractions. The average aggregate size and the average fractions of chains that fold, aggregate, and remain unfolded were monitored as a function of time at varying values of the effective denaturant concentration, ε, and the protein packing fraction. They found that the infinitely slow cooling method provided refolding yields that are relatively high, with only small aggregates (dimers or trimers) formed. The slow-but-finite cooling method provided refolding yields that are almost as high, but in a shorter time frame. The quenching method had low refolding yields and was relatively slow. A maximum in the refolding yield as a function of denaturant concentration was observed for the quenching case (as was seen by Gupta and Hall),[145] but the maximum disappeared after a very long simulation time. The pulse renaturation method provided refolding yields that are substantially higher than those observed in the other methods, including the infinitely slow cooling case, even at high packing fractions. A strategy for rapidly obtaining high refolding yields was suggested that involved instantaneous denaturant removal to intermediate denaturant concentrations followed by dialysis to the final state. Nguyen and Hall's predictions regarding different renaturation protocols are supported by experimental results on the lysozyme system.[154–156]

3.2. Low-Resolution Off-Lattice Models

There have been a few studies of protein aggregation that use low-resolution off-lattice models. Jang et al.[157–160] considered an off-lattice model with each amino acid residue represented by a single bead. Since protein aggregation often includes β-strand association or the conversion of α-helices to β-sheets, they focused part of their efforts on the development of models for β-strand proteins and complexes of β-strand proteins (fibrils) in an effort to understand why multiprotein systems sometimes form fibrils and other times form amorphous aggregates. They introduced three minimalist models of four-strand antiparallel β-strand peptides:[157] the β-sheet, the β-clip, and the β-twist as shown in Figure 2. These β-strand peptides each had 39 connected residues and different native state conformations. Nonbonded beads interacted via a hybrid Go-type potential[38,44–46] modeled as a square-well or square-shoulder potential depending on the value of the bias gap parameter g,[87,90,91] which measures the difference in interaction strength between the native and nonnative contacts. The larger the bias gap, the more the native state is favored over the nonnative state; intermediate values of the bias gap are thought to be most representative of real proteins. For example, $g > 1$ implies that the nonnative contacts are repulsive so that the native state structure is strongly favored over any nonnative state structures; $0 < g < 1$ implies that all nonbonded contact pairs are attractive but native contacts are always more favorable than nonnative contacts; and

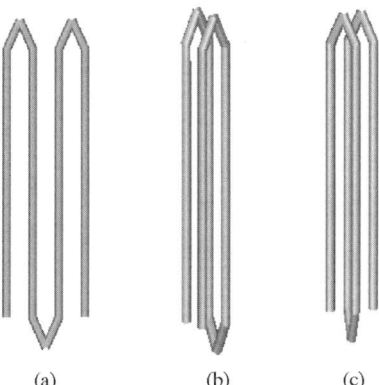

FIGURE 2. Global energy minimum structures for: (a) the β-sheet, (b) the β-clip, and (c) the β-twist model proteins.

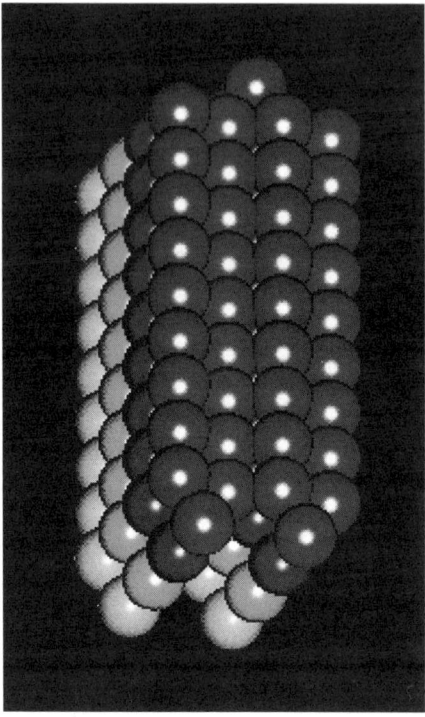

FIGURE 3. Global enegy minimum for the model β-sheet complex. (A color version of this figure appears opposite this page.)

$g = 0$ implies that native and nonnative contacts are equally favorable. The bias gap is an artificial measure of a model protein's preference for its native state; in a real protein this preference for the native state might be measured, for example, by the energy difference between the native and nonnative state.

Jang et al. began by performing discontinuous molecular dynamics (DMD) simulations of the equilibrium properties, folding pathways, and kinetics of all three isolated β-strand peptides at different values of the bias gap parameter and a wide range of temperatures.[157,158] They then investigated the thermodynamics[159] and kinetics[160] of the assembly of four of the model β-sheet peptides into a tetrameric β-sheet complex as shown in Figure 3. To describe the native intermolecular contacts, they introduced an additional parameter: the intermolecular native contact parameter, η, which measures the ratio of the strengths of the intermolecular and intramolecular native attractions. This parameter can be thought of as being a measure of the ratio of the hydrophobic interaction strength and the hydrogen-bonding interaction strength. For $\eta = 1$, the intramolecular and intermolecular native contacts are equally favorable. For $\eta < 1$, the intramolecular native contacts are more favorable than the intermolecular native contacts. DMD simulations were performed on the model complex at an intermediate value of the bias gap, $g = 0.9$, for different intermolecular native contact parameters at different temperatures. The phase diagram for the β-sheet complex in the temperature-intermolecular native contact parameter plane is shown in Figure 4. Seven different phases were observed:

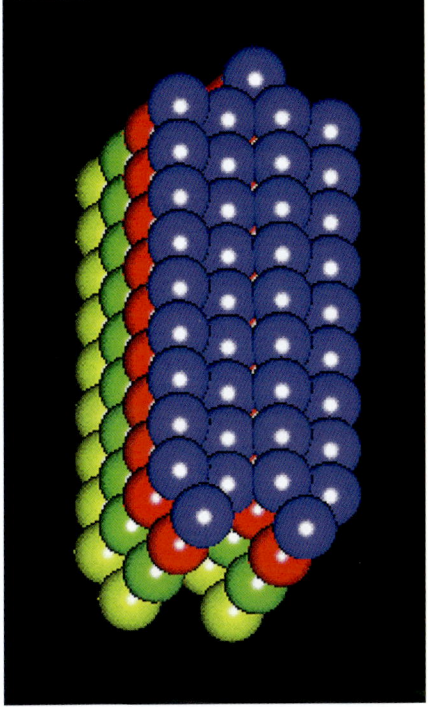

FIGURE 3. Global enegy minimum for the model β-sheet complex.

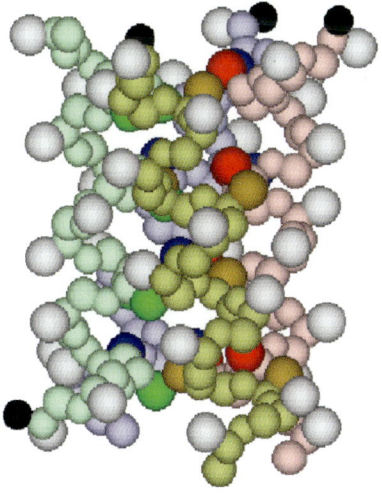

FIGURE 6. A tetrameric α-helical bundle.

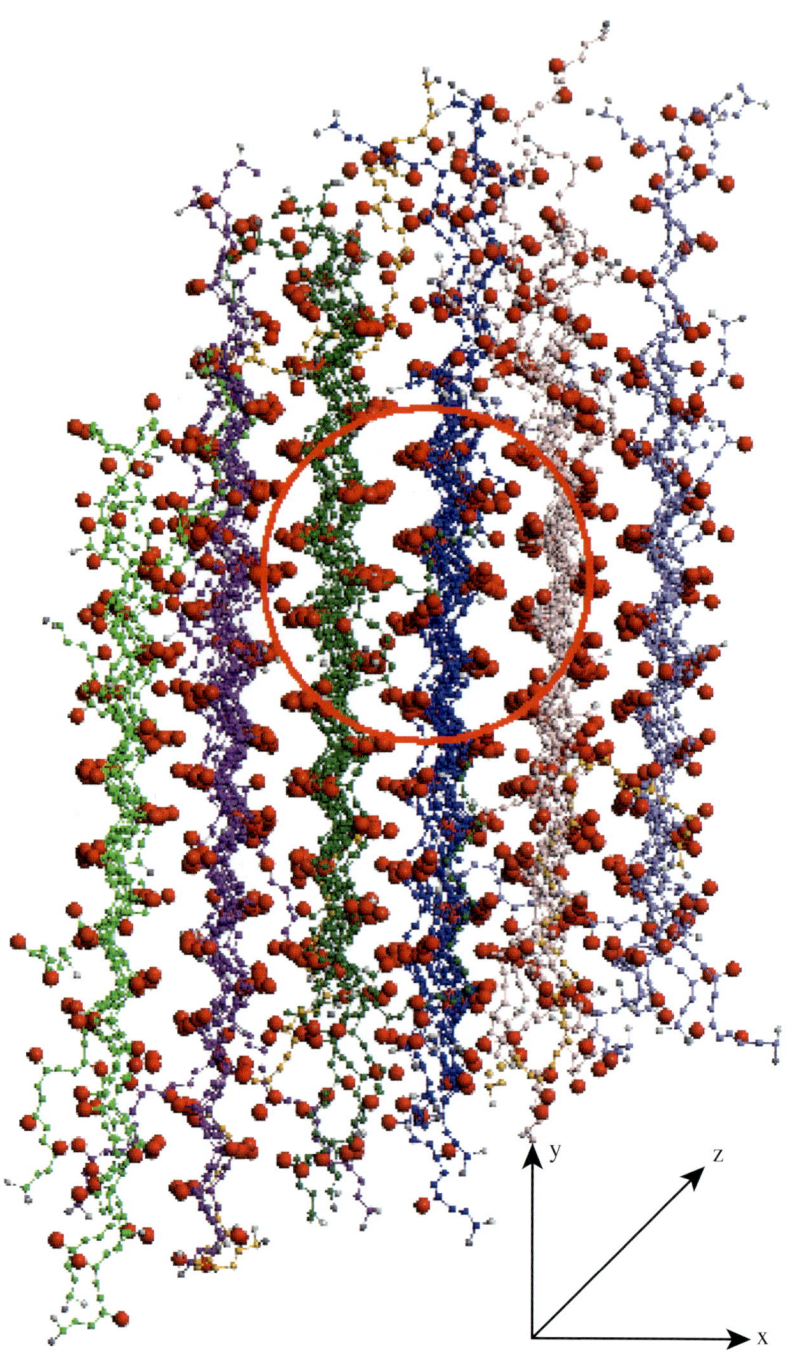

FIGURE 8. A snapshot of the 96-peptide fibrillar structure formed at $c = 5.0$ mM and $T^* = 0.13$ viewed down the fibril (z) axis.

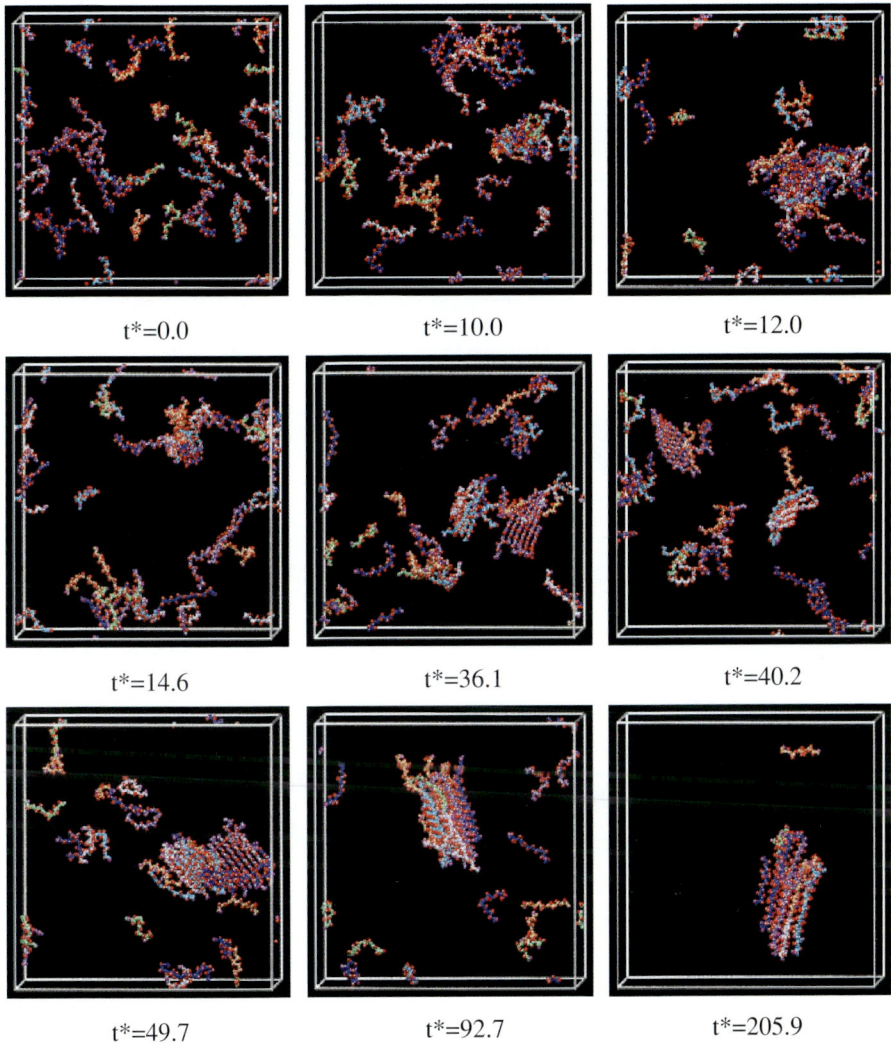

FIGURE 9. Snapshots of the 48-peptide system at various times during a simulation at $c = 10$ mM and $T^* = 0.14$ that results in a fibril.[206] Hydrophobic side chains are red; colors of backbone atoms on different peptides are assigned to make it easy to distinguish the various sheets in the resulting fibril. United atoms are not shown full size for ease of viewing.

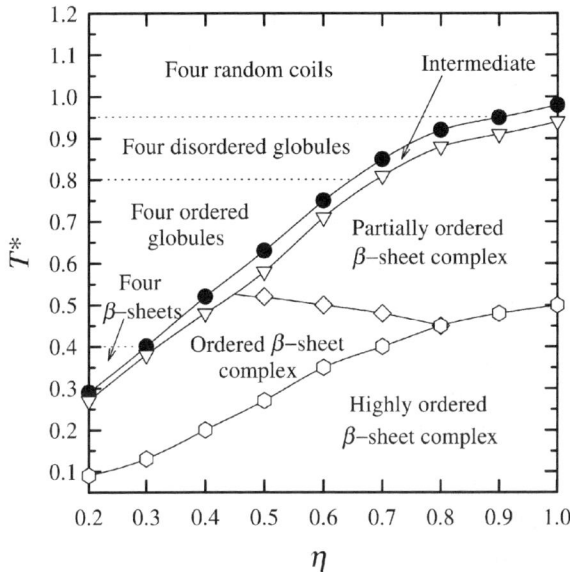

FIGURE 4. Phase diagram for the β-sheet complex as a function of the reduced temperature T^*, and the intermolecular native contact parameter η.

the four monomer states found previously for the isolated β-sheet peptide—random coil, disordered globule, ordered globule, and native β-sheet)—and three oligomer state phases—the partially ordered β sheet complex (four ordered globules), the ordered β-sheet complex (four native β-sheets), and the highly ordered β-sheet complex (global minimum energy state in Figure 3).

Jang et al. also examined the folding (assembly) pathways and kinetics[160] of the β-sheet complex (data not shown). Constant-temperature discontinuous molecular dynamics (DMD) simulations were performed at an intermediate value of the bias gap, $g = 0.9$, for different intermolecular native contact parameters starting from a state containing four separated random coils. The folding yield (fraction of 50 independent simulations that resulted in the β-sheet complex) at small η was low with most trajectories following paths through a trimer or dimer intermediate. Secondary structure began to form before the complex assembled. Many trajectories folded to a misfolded state containing two well-aligned eight-stranded β-sheets. The folding yield at intermediate η was very high with most trajectories following paths through a trimer intermediate. The folding yield at high η was low with most trajectories following paths through a dimer intermediate.

Ding et al.[99] applied an off-lattice model with each amino acid residue represented by one backbone bead and one side chain bead interacting via Go potentials to study the aggregation of a system containing eight model Src SH3 domain proteins. They observed a fibrillar double β-sheet structure, formed by packing the flexible segments (called RT-loops) from different proteins. The structure obtained from their simulations had inter-β-strand spacing distances and inter-β-sheet spacings similar to those observed in experiments. This simulation study and those by Jang et al.[159,160] provide more detail on the mechanisms that govern protein aggregation than the lattice models. However, since the Go potential contains a built-in bias toward the native conformation, they are not suitable for the study of spontaneous fibril formation from random-coil configurations.

3.3. High-Resolution Off-Lattice Models

Most computer simulations using high-resolution protein models have been devoted to the study of systems containing already-formed amyloid fibrils. Li et al.[161] tested the stability of a model fibril structure that was proposed by Sunde et al.[162] based on synchrotron X-ray studies. They constructed an untwisted β-sheet stack containing 16 Aβ(12–42) peptides and a β-sheet stack with a 15° twist containing 48 Aβ(12–42) peptides in the antiparallel β-sheet conformation with a turn at residues 25–28. Molecular dynamics (MD) simulations in explicit aqueous solution were performed using the Amber 4.1 package with the Cornell et al.[163] forcefield on the 16-mer untwisted β-sheet stack for 2.5 nanoseconds, which is long enough for the structure to reach equilibrium, and on the 48-mer protofilament for 600 picoseconds, which is not long enough for the structure to reach equilibrium. Li et al.[161] found that the 16-mer structure remains stable in the original untwisted conformation, while the 48-mer structure lessens its original 15°-twist over time.

George et al.[164] conducted Monte Carlo simulations on Aβ, the β-amyloid peptide, using the CHARMM package to determine the difference between interpeptide interactions in Aβ(1–40) and in Aβ(1–43). They found well-defined enhanced side-chain interactions between residues 41, 42, and 43 and residues on adjacent molecules. This explains the increased fibrillogenic nature of the Aβ(1–43) peptide over the Aβ(1–40) peptide and also the increased fibrillogenic nature of the Dutch family mutation of Aβ (i.e., residue E22 is replaced with Q22). They also examined the interaction between the surface of an Aβ(1–43) fibril and a proposed aggregation-inhibitor, the molecule IDOX.

Nussinov and co-workers conducted MD simulations using the packages Discover and CHARMM to study the stability of fibrils containing between 6 and 24 model peptides arranged in different parallel/antiparallel stacking arrangements. The model peptide systems studied include AGAAAAGA and AAAAAAAA,[165] fragments of the Alzheimer's β-amyloid peptide, Aβ(16–22), Aβ(16–35), and Aβ(10–35),[166] and segments of the human islet amyloid polypeptide 22–27 (NFGAIL) and 22–29 (NFGAILSS).[167,168] For the AGAAAAGA and AAAAAAAA peptide simulations, which lasted from one to four nanoseconds, Nussinov and co-workers found that the most stable oligomers are hexamers and octamers. They also found that the most stable fibril structures containing short poly-A-based peptides, Aβ(16–22) and NFGAIL, are those in which the β-strands are antiparallel within each sheet but parallel across sheets.

Liotta and co-workers[169,170] performed MD simulations using the AMBER package to study the stability of amyloid fibrils containing eight β-sheets each with 14 Aβ(10–35) fragments. They found large spatial and temporal fluctuations about a central core whose "micelle" or "molten-globule-like" properties served to lower the entropic penalty for fibril formation, and thus to enhance the overall stability of the fibril structure.

Hwang et al.[171] examined the supramolecular structure of helical ribbons containing between 40 and 60 β-sheet peptides with the sequence KFE8 by constructing different molecular packing geometries suggested by experimental data. They conducted MD simulations using the CHARMM package to identify the most stable structure. They found that the helical ribbons that were the most stable contained a double β-sheet in which the inner and outer helices had the same hydrogen-bonding pattern. They also found that electrostatic interactions between side chains play an important role in the stability and curvature of the helix.

Kuwata et al.[172] performed MD simulations using Discover 2.98 to study the stability of fibrils containing eight mouse prion protein fragments 106–126 in different parallel/antiparallel stacking arrangements. They found that the most stable structure contains two layers of parallel β-strands, stacked in parallel orientation. They also observed that hydrogen bonds and close-packed side-chain interactions in the fibril's middle region (which contains the central hydrophobic segments) fluctuate less than those in the flanking regions.

High-resolution simulation studies of the formation of fibrils in systems containing random coils have been conducted but they are generally limited to early events in the assembly of a few peptides. Gsponer et al.[173] performed MD simulations on a system containing three heptapeptides, GNNQQNY, from the yeast prion Sup35 using the CHARMM program and observed the formation of a β-sheet. Mager and co-workers[174] conducted MD simulations on a system containing two to four Aβ(1–42) peptides using the AMBER program and observed the formation of a β-sheet. Fernandez and Boland[175] performed MD simulations on systems containing two Aβ(12–24) or two β2-microglobulin(21–31) fragments using the AMBER95 force field to study how crowding affects aggregation. Klimov and Thirumalai[176] conducted long MD simulations on systems containing three Aβ(16–22) peptides using the MOIL program. They observed the formation of an anti-parallel β-sheet through an obligatory α-helical intermediate.

Although the abovementioned studies offer considerable insight into the properties of fibrils and the mechanisms by which they are formed, the systems considered do not contain enough peptides to mimic the nucleus that stabilizes the large fibrils observed in experiments. An alternative approach was taken by Ma et al.[165] who conducted MD simulations using the package Discover 2.98 for up to 4 nanoseconds on a system containing an already-formed fibrillar aggregate of eight AAAAAAA peptides that is surrounded by either a single α-helical AAAAAAA monomer or eight random coil AAAAAAA peptides. They did not observe fibril growth due to the limited simulation time in either case. Given current computational capabilities, simpler models are required to simulate spontaneous fibril formation in multiprotein systems.

3.4. Intermediate-Resolution Off-Lattice Models

Intermediate-resolution protein models, which are essentially a compromise between the simplified chain models described above and the detailed all-atom models like AMBER, CHARMM, and ENCAD, have been used extensively in recent years to simulate the folding of isolated proteins. These models also have been used by our group to study fibril formation. Our approach allows the treatment of large multichain systems while maintaining a fairly realistic description of protein dynamics without built-in bias toward any conformation. In this approach, each amino acid residue is composed of four

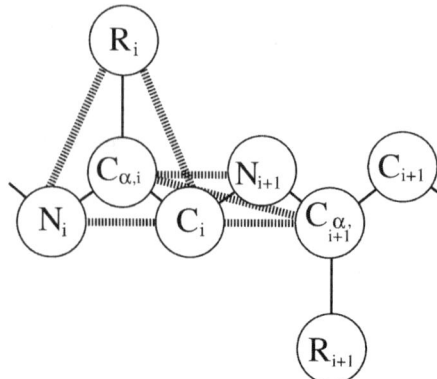

FIGURE 5. Geometry of the intermediate-resolution protein model of Hall and co-workers. Covalent bonds are shown with straight lines connecting united atoms. Pseudobonds, which are shown with disjointed lines, are used to maintain backbone bond angles, consecutive C_α distances, and residue L-isomerization. The united atoms are not shown full size for ease of viewing.

spheres, a three-sphere backbone composed of united atom NH, C_αH, and CO, and a single-sphere side chain (CH$_3$- for alanines). Ideal backbone bond angles, C_α-C_α distances, and residue L-isomerization are achieved by imposing pseudobonds, as shown in Figure 5. Details of the model are given in the literature.[177,178]

All forces are modeled by either hard-sphere or square-well potentials to ensure compatibility with the discontinuous molecular dynamics algorithm (DMD). The solvent is modeled implicitly; its effect is factored into the energy function as a potential of mean force. The excluded volumes of the four united atoms are modeled using hard-sphere potentials with realistic diameters. Interactions between hydrophobic side chains are represented by a square-well potential of depth ε_{HP}; interactions between polar side chains or between polar and hydrophobic side chains are represented by a hard sphere interaction. Hydrogen bonding between amide hydrogen atoms and carbonyl oxygen atoms is represented by a directionally dependent square-well attraction of strength ε_{HB} between NH and CO united atoms.[177] The strength of a hydrophobic contact, ε_{HP}, is proportional to the strength of a hydrogen bond, ε_{HB}.

Simulations are performed in the canonical ensemble with periodic boundary conditions imposed to eliminate artifacts due to box walls. Constant temperature is achieved by implementing the Andersen thermostat method.[27] Simulation temperature is expressed in terms of the reduced temperature, $T^* = k_B T/\varepsilon_{HB}$, where k_B is Boltzmann's constant and T is the temperature. Reduced time is defined to be $t^* = t/\sigma(k_B T/m)^{1/2}$, where t is the simulation time, and σ and m are the average bead diameter and mass, respectively.

Our first aggregation study was on a 16-residue peptide GELEELLKKLKELLKG, a member of the *de novo* designed α family of proteins of Ho and Degrado.[179–182] This peptide folds into an amphipathic α-helix with the polar residues glycine (G), glutamic acid (E), and lysine (K) on one face and the hydrophobic leucine (L) residues on the opposite face. Four of these peptides assemble into a four-helix bundle with the hydrophobic residues in the interior. Following the lead of Guo and Thirumalai,[84] we approximated the

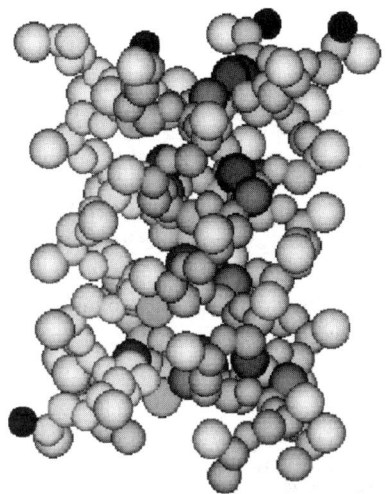

FIGURE 6. A tetrameric α-helical bundle. (A color version of this figure appears between pages 56 and 57.)

real peptide sequence by a model sequence of alanine-sized hydrophobic (H) and polar (P) residues, PPHPPHHPPHPPHHPP. The hydrophobic interaction strength, ε_{HP}, was set at $\varepsilon_{HP} = \varepsilon_{HB}/6$ because this allowed the hydrophobic interactions to exert an influence on native-state (α-helix) stability without constantly trapping the chain in misfolded conformations. We began by performing simulations on isolated PPHPPHHPPHPPHHPP chains starting in the random coil state; this served to establish that α-helices form at reduced temperatures $T^* = k_B T/\varepsilon_{HB}$ below a folding temperature of $T^* = 0.12$. Next we performed simulations on systems containing four of these chains, again starting in a random-coil state. These resulted in the formation of a four-helix bundle (the native state as shown in Figure 6) a fraction of the time depending on the temperature.

The structure of the model bundle was in good agreement with experimental findings in that the helices aligned lengthwise at angles offset by 20° from the bundle axis. Bundle folding followed many different trajectories, with the majority folding via a trimeric intermediate and the minority folding via two dimer intermediates. An optimal temperature range for bundle assembly was observed, defined by the tendency of the peptides to unfold at high temperatures and to aggregate into misfolded structures at low temperatures. Finally we performed simulations on systems containing eight of these peptides,[183] again starting in random coil configurations, at a variety of temperatures to study the competition between the formation of two four-helix bundles and amorphous aggregates. Once again an optimal range for folding (bundle assembly) was observed, but with different boundaries than in the one- and four-peptide systems. The high-temperature boundary appeared to be a function of the complexity of the protein (or oligomer) being assembled. The low-temperature boundary appeared to be a function of the number of surrounding chains; this temperature increased when the number of chains, and hence the probability of aggregation, increased.

We then embarked on a series of studies aimed at simulating the spontaneous formation of fibrils. Since Blondelle and co-workers[184] had shown that systems of polyalanine peptides form fibrillar aggregates at high temperatures, and since our single-bead side chain model (Figure 5) is well-suited to the treatment of polyalanine, we chose to study the competition between folding and aggregation of polyalanine-based peptides of the sequence Ac-KA$_{14}$K-NH$_2$ at a variety of temperatures starting from a random coil configuration.

We began by investigating the folding thermodynamics of isolated polyalanine (Ac-KA$_{14}$K-NH$_2$) chains to see how the value of the energy parameter, $R = \varepsilon_{HP}/\varepsilon_{HB}$, the ratio of the hydrophobic and hydrogen-bonding interactions, affects the nature of the transitions observed.[185] In effect R measures solvent conditions, with $R = 0$ corresponding to the vacuum. The motivation here was to determine which value of R to use in our multipeptide simulations. The replica exchange method, as originally formulated by Sugita and Okamoto,[186] was used for these simulations, since this method is ideally suited for equilibrium studies. In replica exchange, a number of replicas of the system are created and simulated at a spectrum of temperatures usually on a system of parallel computers. At set time intervals, replicas whose temperatures are nearest neighbors along the spectrum are exchanged, provided that a Metropolis criterion is satisfied. At the end of the simulation the thermodynamic properties at a wide variety of temperatures can be deduced using the weighted histogram method.[187,188]

The peptide Ac-KA$_{14}$K-NH$_2$ was found to mimic real polyalanine in that it can form different structures: α-helix, β-structures (including β-hairpins and β-sheetlike conformations), and random coil, depending on the solvent conditions.[189–195] At low values of the hydrophobic interaction strength ($R = 0$, 1/12, and 1/10), there was a relatively sharp transition between an α-helical conformation at low temperatures and a random coil conformation at high temperatures. Increasing the hydrophobic interaction strength to $R = 1/8$, 1/6, and 1/4 induced a second transition to the β-hairpin state, resulting in an α-helices conformation at low temperatures, a β-hairpin at intermediate temperatures, and a random-coil at high temperatures. At very high values of the hydrophobic interaction strength ($R = 1/2$), β-hairpins and β-sheetlike structures were observed at low temperatures and random coils were observed at high temperatures. Since the appearance of the β-hairpin state is controversial (although it has been seen by other investigators[196,197]), we decided to choose $R = 1/10$ for their multipeptide simulations.

We next conducted DMD simulations on systems containing 12, 24, 48, and 96 model Ac-KA$_{14}$K-NH$_2$ peptides at a wide variety of concentrations and temperatures to determine how peptide concentration and temperature affect the formation of β-helices, β-sheets, amorphous aggregates, and fibrils.[198] All simulations were performed in the canonical ensemble starting from a random coil configuration equilibrated at a high temperature and then slowly cooled to the temperature of interest so as to minimize kinetic trapping in local free energy minima. These simulations took approximately 40 hours on an AMD Athlon MP 2200 + single-processor workstation. The percentage of peptides that form α-helices, β-sheets, or fibrils was monitored during the simulation as was the peptide arrangement and packing of fibrils. The stability of the resulting fibrillar structures was evaluated by comparing the abilities of the system to maintain the fibrillar structures at various temperatures that are higher than the fibril formation temperature.

The simulation results showed that the populations of α-helices and fibrils were highly dependent on temperature and peptide concentration as shown in Figure 7, which

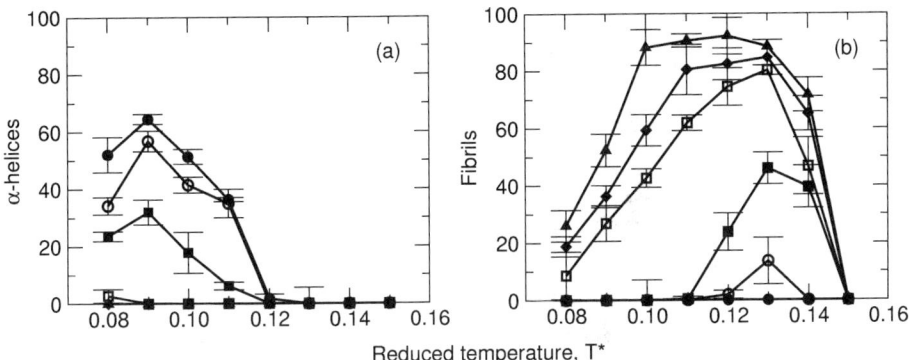

FIGURE 7. The percentage of peptides in (a) alpha-helices and (b) fibrils versus reduced temperature T^* at different peptide concentrations: 0.5 mM (●), 1.0 mM (○), 2.5 mM (■), 5 mM (□), 10 mM (♦), and 20 mM (▲).

plots the percentage of peptides at the end of the 48-peptide slow-cooling simulations that form these structures versus the reduced temperature, T^*, at different peptide concentrations, c. Figure 7A shows that the percentage of peptides that formed α-helices (the Ac-KA$_{14}$K-NH$_2$ native state) decreased with increasing concentration and ceased at concentrations, $c \geq 5$mM. At $c < 5$mM there was an optimal temperature range for forming α-helices centered about a maximum at $T^* = 0.09$. Below this temperature, the system was kinetically trapped and formed either nonfibrillar β-structures or amorphous aggregates that contain numerous β-helices (data not shown). Fibril formation was found primarily at high temperatures and concentrations, as indicated in Figure 7B. At high concentrations ($c = 5$, 10, and 20 mM), fibril formation increased as the temperature increased up to $T^* \sim 0.13$ and then decreased. The maximum in the percentage of peptides in fibrils as a function of temperature broadened and shifted to lower temperatures as the concentration increased from $c = 5$ mM to 20 mM. This means that the critical temperature for forming fibrils decreased with increasing peptide concentration. The results described above agree qualitatively with the experimental results of Blondelle and co-workers on Ac-KA$_{14}$K-NH$_2$ peptides[184,199] showing that the critical fibril formation temperature decreases with increasing peptide concentration. They observed monomeric α-helices structures at low concentrations (100 μM) and temperatures (25°C). As c increased to 1 mM, β-sheet complex formation increased with increasing T, exhibiting an S-shaped dependence on T with a critical temperature of 65°C. As c increased to 1.8 mM, the critical temperature at which β-sheet complexes start to form decreased to 45°C.

The fibrils observed in our slow-cooling simulations mimic the structural characteristics observed in experiment in that most of the fibrillar peptides were arranged in an in-register parallel orientation,[200–202] with intrasheet and intersheet distances similar to those observed in experiments.[162,203,204] The fibrils contained a maximum of six β-sheets, each with many peptides, and were disproportionately long along the fibril axis.[162,205] A snapshot of the 96-peptide fibril formed at $c = 5.0$ mM and $T^* = 0.13$ in Figure 8 revealed that the alanine side chains within a particular β-sheet (e.g.,

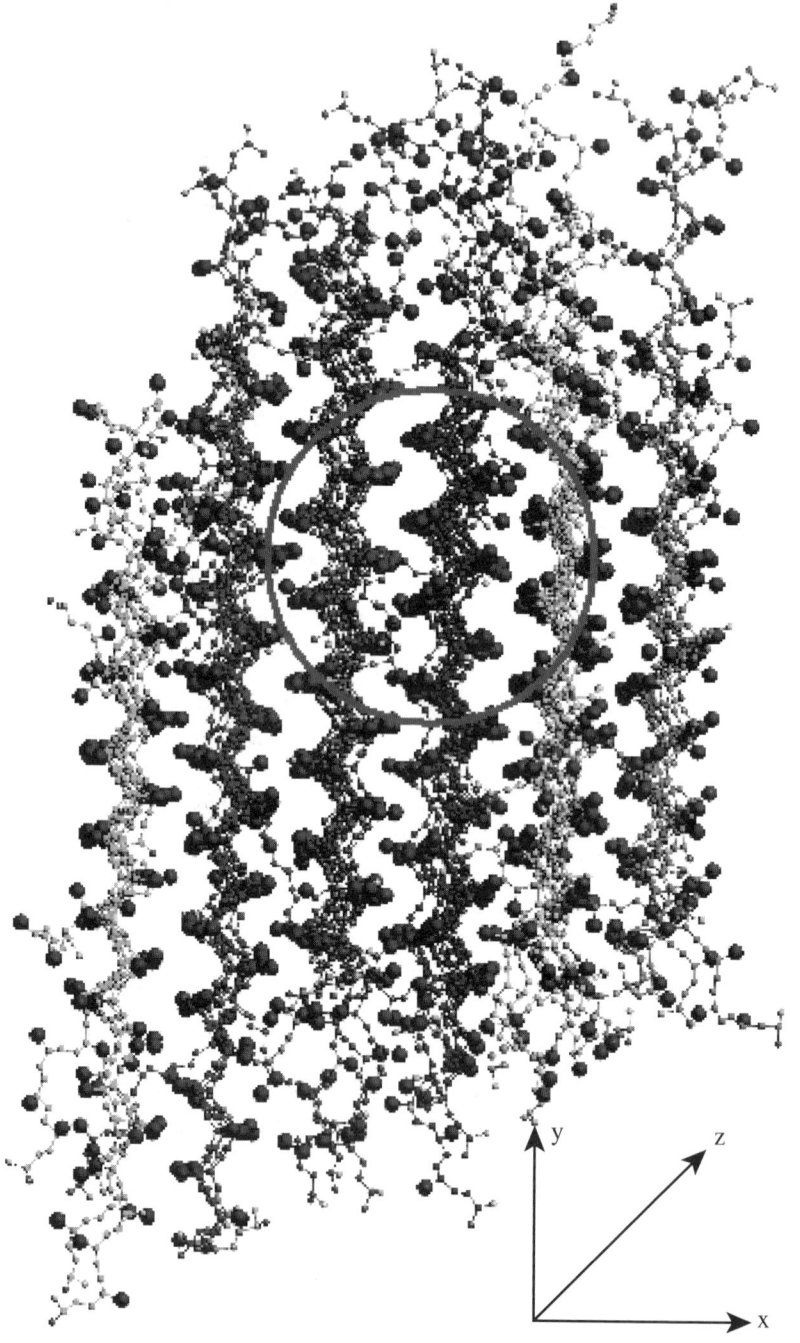

FIGURE 8. A snapshot of the 96-peptide fibrillar structure formed at $c = 5.0$ mM and $T^* = 0.13$ viewed down the fibril (z) axis. (A color version of this figure appears between pages 56 and 57.)

green sheet in red circle) were aligned (stacked on top of one another in the figure) and alternated from one side of the sheet to the other. Alanine side chains on an adjacent sheet (e.g., dark blue sheet) also alternated from side to side but were shifted so that they fit into the "pockets" on the neighboring sheet. Finally, we found that when the strength of the hydrophobic interaction between nonpolar side chains relative to the strength of hydrogen bonding was high, the system formed amorphous rather than fibrillar aggregates.

We then investigated the kinetics of fibril formation of Ac-KA$_{14}$K-NH$_2$ peptides as a function of the peptide concentration and temperature.[206] Constant-temperature simulations were conducted on systems containing 48-model 16-residue peptides at a wide variety of concentrations and temperatures using the same protein model as in the previous studies. During each simulation, the formation of different structures such as α-helices, amorphous aggregates, β-sheets, or fibrils was monitored as a function of time. Key fibril-forming events that are associated with the four proposed fibril-forming mechanisms described in the literature[207–213] were identified. Two types of simulations were conducted: unseeded and seeded. In the unseeded simulations, the peptides, which were initially in a random coil configuration, were equilibrated at a high temperature and then quickly cooled to the temperature of interest. In the seeded simulations, a previously created single fibril was immersed in a sea of denatured chains.

The results showed that fibril formation for polyalanines is nucleation dependent, which is similarly observed in various experimental studies.[203,212,214] The lag time before fibril formation commences decreased with increasing concentration and increased with increasing temperature as is observed in experiments.[211–213] Furthermore, the formation of small fibrils was preceded by the appearance of small amorphous aggregates, then β-sheets, and finally rapid growth of a stable fibril (which is similar to the key features of the nucleated conformational conversion model proposed by Serio et al.[211]). This can be seen in Figure 9, which shows snapshots of the fibrillization process taken at various reduced times, t^*, for the 48-peptide unseeded constant-temperature simulation at $T^* = 0.14$ and a peptide concentration of $c = 10$ mM. Starting in random-coil conformations at reduced time $t^* = 0$, the chains begin to form small amorphous aggregates almost immediately. These amorphous aggregates have grown by $t^* = 10.0$ and have collapsed into one big amorphous aggregate by $t^* = 12.0$. This big amorphous aggregate disperses by $t^* = 14.6$ into smaller aggregates, one of which is a two-peptide β-sheet (purple sheet at the middle right side of the box). By $t^* = 36.1$, the other amorphous aggregates have converted into three three-peptide β-sheets (i.e., light blue sheet in box middle, white sheet, behind light blue sheet, and dark blue sheet in upper right corner); in addition, the purple sheet has grown into a six-peptide β-sheet. By $t^* = 40.2$, the light blue sheet has associated with the white sheet, creating a two-sheet fibrillar structure. By $t^* = 49.7$, this two-sheet fibrillar structure has been joined by the purple sheet and the dark blue sheet, the latter of which is at an oblique angle. By $t^* = 92.7$ the dark blue sheet has realigned itself, docking to create a four-sheet fibril, which itself has grown by adding peptides to the β-sheets ends. Even after a long equilibration time, $t^* = 205.9$, the fibrillar structure remains stable and grows by adding more peptides to the ends of the β-sheet.

Fibril growth in our simulations involved both β-sheet elongation, in which the fibril grows by adding individual peptides to the end of each β-sheet, and lateral addition, in

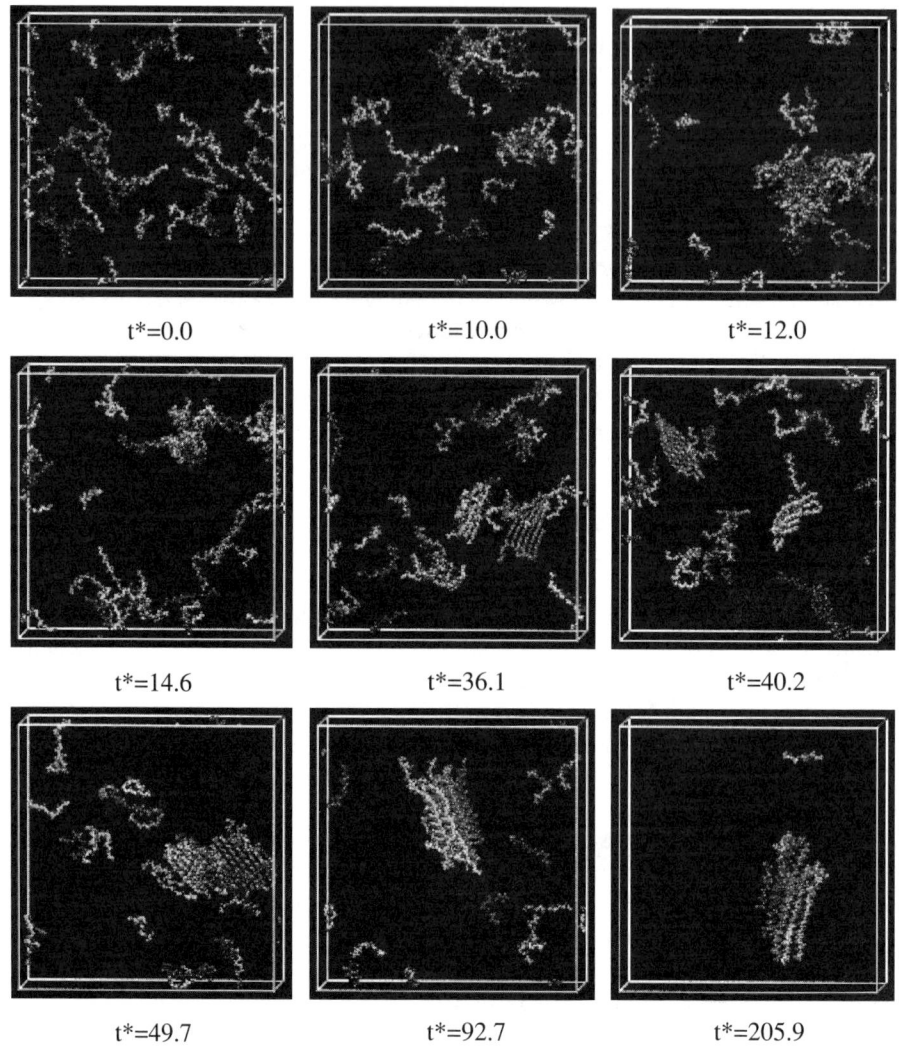

FIGURE 9. Snapshots of the 48-peptide system at various times during a simulation at $c = 10$ mM and $T^* = 0.14$ that results in a fibril.[206] Hydrophobic side chains are red; colors of backbone atoms on different peptides are assigned to make it easy to distinguish the various sheets in the resulting fibril. United atoms are not shown full size for ease of viewing. (A color version of this figure appears between pages 56 and 57.)

which the fibril grows by adding already-formed β-sheets to its side. These two growth mechanisms are similar to those observed in experiments by Green et al.,[214] who found two distinct phases in human amylin (hA) fibrillogenesis in which lateral growth of oligomers is followed by longitudinal growth into mature fibrils.

Finally, we performed replica exchange equilibrium simulations on systems containing 96 Ac-KA$_{14}$K-NH$_2$ over a very wide range of temperatures and peptide concentrations to study the system's thermodynamics.[215] The goal here was to map out a phase

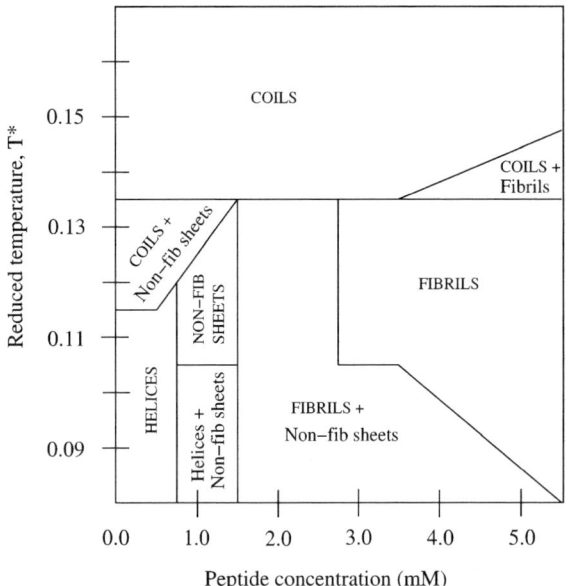

FIGURE 10. Phase diagram for the 96-peptide system as a function of the reduced temperature and peptide concentration. The one-structure phases are α-helices, fibrils, nonfibrillar β-sheets, and random coils. The two-structure phases are random coils/nonfibrillar β-sheets, random coils/ fibrils, fibrils/nonfibrillar β-sheets, and α-helices/nonfibrillar β-sheets.

diagram in the temperature-concentration plane delineating the regions where random coils, α-helices, β-sheets, fibrils, and amorphous aggregates are stable. Based on the system heat capacity and peptide radius of gyration and on data on the percentage of the peptides that form the various structures, a phase diagram in the temperature-concentration plane was constructed as seen in Figure 10. We found four distinctive single-phase regions (α-helices, fibrils, nonfibrillar β-sheets, and random coils) and four two-phase regions (random coils/nonfibrillar β-sheets, random coils/fibrils, fibrils/nonfibrillar β-sheets, and α-helices/nonfibrillar β-sheets). The α-helical region is at low temperature and low concentration. The nonfibrillar β-sheet region is at intermediate temperatures and expands to higher temperatures as concentration is increased. The fibril region is primarily at intermediate temperatures and intermediate concentrations and expands to lower temperatures as the peptide concentration is increased. The random-coil region is at high temperatures and all concentrations and shifts to higher temperatures as the concentration is increased.

The results, which are summarized in a phase diagram in Figure 10, hopefully provide experimentalists some guidance in locating the temperature and concentration at which to conduct *in vitro* fibrillization experiments or to avoid fibrillization. Although this phase diagram is not expected to be quantitatively accurate, especially for an arbitrary protein, we speculate that its shape may be universal. An experimentalist that is aware of this universal shape is less likely to conclude that fibrillization does not occur when instead the wrong region of the phase diagram is being accessed.

4. CONCLUSION

We have described three approaches to the simulation of protein aggregation: atomistic resolution models, lattice models, and intermediate-resolution models. Each approach has its strengths and weaknesses. Atomic resolution models offer insights into the stability of model fibrils and β-sheet structures containing specific proteins but, since they access short timescales (∼ nanoseconds), do not allow the simulation of spontaneous fibril formation. Lattice models can be used to examine the big picture and have enhanced our understanding of the types of sequences and conditions under which aggregation occurs but provide little information on the aggregation of specific proteins. Off-lattice intermediate resolution models stand at the crossroads between atomistic resolution models and lattice models. They are computationally efficient like the lattice models yet represent protein geometry and interactions in a fairly realistic way like the atomistic models.

While it might be tempting for the authors to say that intermediate resolution models offer the most promise for future contributions to general knowledge of amyloid formation, this would be shortsighted (as well as self-serving). In our view the best hope for the future lies in cross-fertilization between the three approaches. We envision a broad multiscale modeling effort across the computational biology community in which insights and parameters obtained from simulations on one scale are used to feed parameters into simulations on another scale; in effect, a cascade of coarse-graining efforts. (Such efforts are now common in the computational polymer-physics community.) For example, CHARMM simulations of very early events in the formation of beta-amyloid or in the disassociation of an already formed small fibril could be used to develop better intra- and intermolecular potentials for input into intermediate resolution protein models. In turn the results from intermediate-resolution protein model simulation of fibril formation could be used to devise lattice models that examine the growth of very large (proto-) fibrils and the subsequent coming together of several protofibrils into a multistranded fibril. Whether or not this vision becomes a reality remains to be seen. Nevertheless, it is clear to the authors that protein aggregation simulations have matured to the point where they are poised on the brink of being useful to experimentalists and members of the medical community seeking to understand the root causes of amyloid formation.

REFERENCES

1. J. King, Deciphering the rules of protein folding, *Chem. Eng. News* **34**, 32–54 (1989).
2. J. W. Kelly, The alternative conformations of amyloidogenic proteins and their multi-step assembly pathways, *Curr. Opin. Struct. Biol.* **8**, 101–106 (1998).
3. J. C. Rochet and P. T. Lansbury, Amyloid fibrilogenesis: Themes and variations, *Curr. Opin. Struct. Biol.* **10**, 60–68 (2000).
4. C. M. Dobson, The structural basis of protein folding and its links with human disease, *Phil. Trans. R. Soc. Lond. B*. **356**, 133–145 (2001).
5. J. W. Kelly, Towards an understanding of amyloidogeneisis, *Nat. Struct. Biol.* **9**, 323–325 (2002).
6. E. Zerovnik, Amyloid-fibril formation, *Eur. J. Biochem.* **269**, 3362–3371 (2002).

7. B. S. Shastry, Neurodegenerative disorders of protein aggregation, *Neurochem. Internatl.* **43**, 1–7 (2003).
8. R. L. Baldwin, Intermediates in protein folding reaction and the mechanism of protein folding, *Annu. Rev. Biochem.* **44**, 453–475 (1975).
9. P. S. Kim and R. L. Baldwin, Specific intermediates in the folding reaction of small proteins and the mechanism of protein foldings, *Annu. Rev. Biochem.* **51**, 459–489 (1982).
10. F. A. O. Marston, The purification of eukariotic polypeptides synthesized in *E. coli*, *Biochemistry* **240**, 1–12 (1986).
11. J. F. Kane and D. L. Hartley, Formation of recombinant protein inclusion-bodies in *E. coli*, *Trends Biotechnol.* **6**, 95–101 (1988).
12. A. Mitraki and J. King, Protein folding intermediates and inclusion body formation, *Biotech.* **7**, 690–697 (1989).
13. G. Georgiou and G. A. Bowden, Inclusion body formation and the recovery of aggregated recombinant protein, in ed. A. Prokop, R. K. Bajpas, and C. Ho, *Recombinant DNA Technology and Applications* (New York: McGraw Hill, 1990), 333–356.
14. E. DeBernardez-Clark and G. Georgiou, Inclusion bodies and recovery of proteins from the aggregated state, *ACS Symp.* **470**, 1–20 (1991).
15. M. Manning, K. Patel, and R. Borchardt, Stability of protein pharmaceuticals, *Pharm. Res.* **6**, 903–918 (1989).
16. H. R. Costantino, R. Langer, and A. M. Klibanov, Aggregation of lyophilized pharmaceutical protein, recombinant human albumnin: Effect of moisture and stabilization by experiments, *Biotech.* **13**, 493–496 (1995).
17. K. M. Persson and V. Gekas, Factors influencing aggregation of macromolecules in solution, *Proc. Biochem.* **29**, 89–98 (1994).
18. A. Aggeli, N. Boden, and S. Zhang, eds. *Self-Assembling Peptide Systems in Biology, Medicine and Engineering* (Dordrecht, Kluwer Academic Publishers: 2001).
19. T. O. Yeates and J. E. Padilla, Design supramolecular protein assemblies, *Curr. Opin. Struct. Biol.* **12**, 464–470 (2002).
20. C. E. MacPhee and C. M. Dobson, Formation of mixed fibrils demonstrates the generic nature and potential utility of amyloid nanostructures, *J. Am. Chem. Soc.* **122**, 12707–12713 (2000).
21. J. D. Hartgerink, E. Beniash, and S. I. Stupp, Self-assembly and mineralization of peptide-amphiphile nanofibers, *Science* **294**, 1685–1688 (2001).
22. Z. Megeed, J. Cappello, and H. Ghandehari, Genetically engineered silk-elastin protein polymers for controlled drug delivery, *Adv. Drug Del. Rev.* **54**, 1075–1091 (2002).
23. T. Scheibel, R. Parthasarathy, G. J. Sawicki, X.-M. Lin, H. Jaeger, and S. L. Lindquist, Conducting nanowires built by controlled self-assembly of amyloid fibers and selective metal deposition, *Proc. Natl. Acad. Sci. USA* **100**, 4527–4532 (2003).
24. D. C. Rapaport, Molecular dynamics simulation of polymer chains with excluded volume, *J. Phys. A* **11**, L213–L217 (1978).
25. D. C. Rapaport, Molecular dynamics study of polymer chains, *J. Chem. Phys.* **71**, 3299–3303 (1979).
26. A. Bellemans, J. Orbans, and D. V. Belle, Molecular dynamics of rigid and non-rigid necklaces of hard disks, *Mol. Phys.* **39**, 781–782 (1980).
27. H. C. Andersen, Molecular dynamics simulation at constant temperature and / or pressure, *J. Chem. Phys.* **72**, 2384–2393 (1980).
28. S. J. Weiner, P. A. Kollman, D. A. Case, U. C. Singh, C. Ghio, G. Alagona, S. Profeta, and P. Weiner, A new force field for molecular mechanical simulation of nucleic acids and proteins, *J. Am. Chem. Soc.* **106**, 765–784 (1984).
29. S. J. Weiner, P. A. Kollman, D. T. Nguyen, and D. A. Case, An all atom force field for simulations of proteins and nucleic acids, *J. Comp. Chem.* **7**, 230–252 (1986).
30. B. R. Brooks, R. E. Bruccoleri, B. D. Olafson, D. J. Stales, Swaminathan, and M. Karplus, CHARMM: A program for macromolecular energy minimization and dynamics calculation, *J. Comp. Chem.* **4**, 187–217 (1983).

31. M. Levitt, Molecular dynamics of native protein: I. Computer simulation trajectories, *J. Mol. Biol.* **168**, 595–620 (1983).
32. M. Levitt, M. Hirshberg, R. Sharon, and V. Daggett, Potential energy function and parameters for the simulations of the molecular dynamics of proteins and nucleic acids in solution, *Comp. Phys. Commun.* **91**, 215–231 (1995).
33. P. Dauber-Osguthorpe, V. A. Roberts, D. J. Osguthorpe, J. Wolff, M. Genest, and A. T. Hagler, Structure and energetics of ligand binding to proteins: *Escherichia coli* dihydrofolate reductase trimethoprim, a drug receptor system, *Proteins: Struct. Funct. Genet.* **4**, 31–47 (1988).
34. F. A. Momany, R. F. McGuire, A. W. Burgess, and H. A. Sheraga, Energy parameters in polypeptides. VII. Geometric parameters, partial atomic charges, nonbonded interactions, hydrogen bond interactions, and intrinsic torsional potentials for the naturally occurring amino acids, *J. Phys. Chem.* **79**, 2361–2381 (1975).
35. G. Nemethy, M. S. Pottle, and H. A. Sheraga, Energy parameters in polypeptides. 9. Updating of geometrical parameters, nonbonded interactions, hydrogen bond interaction for the naturally occurring amino acids, *J. Phys. Chem.* **87**, 1883–1887 (1983).
36. Y. K. Kang, T. K. No, and H. A. Sheraga, Intrinsic torsional potential parameters for conformational analysis of peptides and proteins, *J. Phys. Chem.* **100**, 15588–15598 (1996).
37. Y. Duan and P. A. Kollman, Pathways to a protein folding intermediate observed in a 1-microsecond simulation in aqueous solution, *Science* **282**, 740–744 (1998).
38. H. Taketomi, Y. Ueda, and N. Go, Studies on protein folding, unfolding, and fluctuations by computer simulation. Effect of specific amino-acid sequence represented by specific inter-unit interactions, *Int. J. Pept. Protein Res.* **7**, 445–459 (1975).
39. M. Levitt and A. Warshel, Computer simulation of protein folding, *Nature* **253**, 694–698 (1975).
40. M. Levitt, A simplified representation of protein conformation for rapid simulation of protein folding, *J. Mol. Biol.* **104**, 59–107 (1976).
41. I. D. Kuntz, G. M. Crippen, P. A. Kollman, and D. Kimelman, Calculation of protein tertiary structure, *J. Mol. Biol.* **106**, 983–994 (1976).
42. S. Tanaka and H. A. Scheraga, Medium and long range interaction parameters between amino acids for predicting three dimensional structure of proteins, *Macromolecules* **9**, 945–950 (1976).
43. A. T. Hagler and B. Honig, On the formation of protein tertiary structure on a computer, *Proc. Natl. Acad. Sci. USA* **75**, 554–558 (1978).
44. Y. Ueda, H. Taketomi, and N. Go, Studies on protein folding, unfolding, and fluctuations by computer simulation, *Biopolymers* **17**, 1531–1548 (1978).
45. N. Go and H. Taketomi, Respective roles of short ranged and long ranged interactions in protein folding, *Proc. Natl. Acad. Sci. USA* **75**, 559–563 (1978).
46. N. Go and H. Taketomi, Studies on protein folding, unfolding and fluctuations by computer simulation, *Int. J. Protein Res.* **13**, 235–252 (1979).
47. A. Kolinski, J. Skolnick, and R. Yaris, Monte Carlo simulations on an equilibrium globular protein folding model, *Proc. Natl. Acad. Sci. USA* **83**, 7267–7271 (1986).
48. A. Kolinski, J. Skolnick, and R. Yaris, Monte Carlo studies on equilibrium globular protein folding. 1. Homopolymeric lattice models of β-barrel proteins, *Biopolymers* **26**, 937–962 (1987).
49. J. Skolnick, A. Kolinski, and R. Yaris, Monte Carlo simulations of the folding ofgβ-barrel globular proteins, *Proc. Natl. Acad. Sci. USA* **85**, 5057–5061 (1988).
50. J. Skolnick, A. Kolinski, and R. Yaris, Monte Carlo studies on equilibrium globular proteins. II. β-barrel globular models, *Biopolymers* **28**, 1059–1095 (1989).
51. J. Skolnick and A. Kolinski, Computer simulations of globular protein folding and tertiary structure, *Annu. Rev. Phys. Chem.* **40**, 207–235 (1989).
52. J. Skolnick, A. Kolinski, and R. Yaris, Dynamic Monte Carlo study of the folding of a six-stranded Freek key globular protein, *Proc. Natl. Acad. Sci. USA* **86**, 1229–1233 (1989).
53. J. Skolnick and A. Kolinski, Simulations of the folding of a globular protein, *Science* **250**, 1121–1125 (1990).
54. A. Godzik, J. Skolnick, and A. Kolinski, Simulations of the folding pathway of triose phosphate isomerase-type α/β barrel proteins, *Proc. Natl. Acad. Sci. USA* **89**, 2629–2633 (1992).

55. A. Godzik, J. Skolnick, and A. Kolinski, *De novo* and inverse folding predictions of protein structure and dynamics, *J. Comp.-Aided Molec. Des.* **7**, 397–438 (1993).
56. A. Kolinski and J. Skolnick, Monte Carlo simulations of protein folding. I. Lattice model and interaction scheme, *Proteins: Struct. Funct. Genet.* **18**, 338–352 (1994).
57. A. Kolinski and J. Skolnick, Monte Carlo simulations of protein folding. II. Application to protein A, rop and crambin, *Proteins: Struct. Funct. Genet.* **18**, 353–366 (1994).
58. M. Vieth, M. Kolinski, C. L. Brooks, and J. Skolnick, Prediction of the quaternary structure of coiled coils. Applications to mutants of the GCN4 leucine zipper, *J. Mol. Biol.* **251**, 448–467 (1995).
59. K. F. Lau and K. A. Dill, A lattice statistical mechanics model of the conformational and sequence spaces of proteins, *Macromolecules* **22**, 3986–3997 (1989).
60. H. S. Chan and K. A. Dill, Intrachain loops in polymers: Effects of excluded volume, *J. Chem. Phys.* **90**, 492–509 (1989).
61. H. S. Chan and K. A. Dill, Compact polymers, *Macromolecules* **22**, 4559–4573 (1989).
62. H. S. Chan and K. A. Dill, The effects of internal constraints on the configuration of chain molecules, *J. Chem. Phys.* **92**, 3118–3135 (1990).
63. H. S. Chan and K. A. Dill, Origins of structure in globular proteins, *Proc. Natl. Acad. Sci. USA* **87**, 6388–6392 (1990).
64. K. F. Lau and K. A. Dill, Theory for mutability in biogenesis, *Proc. Natl. Acad. Sci. USA* **87**, 638–642 (1990).
65. K. A. Dill, Dominant forces in protein folding, *Biochemistry* **29**, 7133–7155 (1990).
66. H. S. Chan and K. A. Dill, Sequence space soup of proteins and copolymers, *J. Chem. Phys.* **95**, 3775–3787 (1991).
67. H. S. Chan and K. A. Dill, Transition states and folding dynamics of proteins and heteropolymers, *J. Chem. Phys.* **100**, 9238–9257 (1994).
68. R. Miller, C. A. Danko, M. J. Fasolka, A. Balazs, H. S. Chan, and K. A. Dill, Folding kinetics of proteins and copolymers, *J. Chem. Phys.* **96**, 768–780 (1992).
69. E. I. Shakhnovich and A. M. Gutin, Engineering of stable and fast-folding sequences of model proteins, *Proc. Natl. Acad. Sci. USA* **90**, 7195–7199 (1993).
70. N. D. Socci and J. N. Onuchic, Folding kinetics of proteinlike heteropolymers, *J. Chem. Phys.* **101**, 1519–1528 (1994).
71. H. Li, Helling, T. R., C., and N. Wingreen, Energency of preferred structures in a simple model of protein folding, *Science* **273**, 666–669 (1996).
72. E. M. O'Toole and A. Z. Panagiotopoulos, Monte carlo simulation of folding transitions of simple model proteins using a chain growth algorithm, *J. Chem. Phys.* **97**, 8644–8652 (1992).
73. E. M. O'Toole and A. Z. Panagiotopoulos, Effect of sequence and intermolecular interactions on the number and nature of low-energy states for simple model proteins, *J. Chem. Phys.* **98**, 3185–3190 (1993).
74. E. M. O'Toole, R. Venkataramani, and A. Z. Panagiotopoulos, Simple lattice model of proteins incorporating directional bonding and structured solvent, *AIChE J.* **41**, 954–958 (1995).
75. E. Shakhnovich, G. G. Farztdinov, G. Gutin, and M. Karplus, Protein folding bottlenecks: A lattice Monte Carlo simulation, *Phys. Rev. Lett.* **67**, 1665–1668 (1991).
76. E. Shakhnovich, Proteins with selected sequences fold into unique native conformation, *Phys. Rev. Lett.* **72**, 3907–3910 (1994).
77. A. Sali, E. Shakhnovich, and M. Karplus, How does a protein fold? *Nature* **369**, 248–251 (1994).
78. A. Sali, E. Shakhnovich, and M. Karplus, Kinetics of protein folding—a lattice model study of the requirements for folding to the native state, *J. Mol. Biol.* **235**, 1614–1636 (1994).
79. D. A. Hinds and M. Levitt, Exploring conformational space with a simple lattice model for protein structure, *J. Mol. Biol.* **243**, 668–682 (1994).
80. C. Wilson and S. Doniach, A computer model to dynamically simulate protein folding: Studies with crambin, *Proteins: Struct. Funct. Genet.* **6**, 193–209 (1989).
81. J. D. Honeycutt and D. Thirumalai, Metastability of the folded states of globular proteins, *Proc. Natl. Acad. Sci. USA* **87**, 3526–3529 (1990).

82. Z. Guo and D. Thirumalai, Kinetics of protein folding: Nucleation mechanism, time scales and pathways, *Biopolymers* **36**, 83–102 (1995).
83. T. Veitshans, D. Klimov, and D. Thirumalai, Protein folding kinetics: Timescales, pathways and energy landscapes in terms of sequence-dependent properties, *Folding Des.* **2**, 1–22 (1996).
84. Z. Guo and D. Thirumalai, Kinetics and thermodynamics of folding of a *de novo* designed four-helix bundle, *J. Mol. Biol*, **263**, 323–343 (1996).
85. Z. Guo and D. Thirumalai, The nucleation-collapse mechanism in protein folding: Evidence for the non-uniqueness of the folding nucleus, *Folding Des.* **2**, 377–391 (1997).
86. D. K. Klimov, M. R. Betancourt, and D. Thirumalai, Virtual atom representation of hydrogen bonds in minimal off-lattice models of alpha-helices: Effect on stability, cooperativity, and kinetics, *Folding Des.* **3**, 481–496 (1998).
87. Y. Zhou and M. Karplus, Folding thermodynamics of a model three-helix-bundle protein, *Proc. Natl. Acad. Sci. USA* **94**, 14429–14432 (1997).
88. Y. Zhou and M. Karplus, Folding of a model three-helix bundle protein: A thermodynamic and kinetic analysis, *J. Mol. Biol.* **293**, 917–951 (1999).
89. Y. Zhou and M. Karplus, Interpreting the folding kinetics of helical proteins, *Nature* **401**, 400–403 (1999).
90. Z. Guo and C. L. Brooks, Thermodynamics of protein folding: A statistical mechanical study of a small all-beta protein, *Biopolymers* **42**, 745–757 (1997).
91. J. E. Shea, Y. D. Nochmovitz, Z. Guo, and C. L. Brooks, Exploring the space of protein folding hamiltonians: The balance of forces in a minimalist beta-barrel protein, *J. Chem. Phys.* **109**, 2895–2903 (1998).
92. H. Nymeyer, A. E. Garcia, and J. N. Onuchic, Folding funnels and frustration in off-lattice minimalist protein landscapes, *Proc. Natl. Acad. Sci. USA* **95**, 5921–5928 (1998).
93. C. Clementi, H. Nymeyer, and J. N. Onuchich, Topological and energetic factors: What determines the structural details of the transition state ensemble and 'en-route' intermediates for protein folding? An investigation for small globular proteins, *J. Mol. Biol.* **298**, 937–953 (2000).
94. N. V. Dokholyan, S. V. Buldyrev, H. E. Stanley, and E. I. Shakhnovich, Discrete molecular dynamics studies of the folding of a protein-like model, *Folding Des.* **3**, 577–587 (1998).
95. G. F. Berriz and E. I. Shakhnovich, Characterization of the folding kinetics of a three-helix bundle protein via a minimalist Langevin model, *J. Mol. Biol.* **310**, 673–685 (2001).
96. Z. Miyazawa and R. L. Jernigan, Equilibrium folding and unfolding pathways for a model protein, *Biopolymers* **21**, 1333–1363 (1982).
97. A. Rey and J. Skolnick, Comparison of lattice Monte Carlo dynamics and Brownian dynamics folding pathways of alpha-helical hairpins, *Chem. Phys.* **158**, 199–219 (1991).
98. F. Ding, N. V. Dokholyan, S. V. Buldyrev, H. E. Stanley, and E. Shakhnovich, Direct molecular dynamics observation of protein folding transition state ensemble, *Biophys. J.* **83**, 3525–3532 (2002).
99. F. Ding, N. V. Dokholyan, S. V. Buldyrev, H. E. Stanley, and E. Shakhnovich, Molecular dynamics simulation of the sh3 domain aggregation suggests a generic amyloidogenesis mechanism, *J. Mol. Biol.* **324**, 851–857 (2002).
100. A. Liwo, M. R. Pincus, R. J. Wawak, S. Rackovsky, and H. A. Scheraga, Calculation of protein backbone geometry from α–carbon coordinates based on peptide-group dipole alignment, *Prot. Sci.* **2**, 1697–1714 (1993).
101. A. Liwo, M. R. Pincus, R. J. Wawak, S. Rackovsky, and H. A. Scheraga, Prediction of protein conformation on the basis of a search for compact structures: Test on avian pancreatic polypeptide, *Prot. Sci.* **2**, 1715–1731 (1993).
102. A. Liwo, S. Oldziej, R. Kazmierkiewicz, M. Groth, and C. Czaplewski, Design of a knowledge-based force field for off-lattice simulations of protein structure, *Acta Biochim. Polon.* **44**, 527–547 (1997).
103. A. Liwo, S. Oldziej, M. R. Pincus, R. J. Wawak, S. Rackovsky, and H. A. Scheraga, A united-residue force field for off-lattice protein-structure simulations. I. Functional forms and parameters of long-range side-chain interaction potentials from protein crystal data, *J. Comp. Chem.* **18**, 849–873 (1997).

104. A. Liwo, M. R. Pincus, R. J. Wawak, S. Rackovsky, S. Oldziej, and H. A. Scheraga, A united-residue force field for off-lattice protein-structure simulations. II. Parameterization of short-range interactions and determination of weights of energy terms by z-score optimization, *J. Comp. Chem.* **18**, 874–887 (1997).
105. A. Liwo, R. Kazmierkiewicz, C. Czaplewski, M. Groth, S. Oldziej, R. J. Wawak, S. Rackovsky, M. R. Pincus, and H. A. Scheraga, United-residue force field for off-lattice protein-structure simulations: III. Origins of backbone hydrogen-bonding cooperativity in united-residue potentials, *J. Comp. Chem.* **19**, 259–276 (1998).
106. C. Hardin, Z. Luthey-Schulten, and P. G. Wolynes, Backbone dynamics, fast folding, and secondary structure formation in helical proteins and peptides, *Proteins: Struct. Funct. Genet.* **34**, 281–294 (1999).
107. C. Hardin, M. P. Eastwood, M. Prentiss, Z. Luthey-Schulten, and P. G. Wolynes, Folding funnels: The key to robust protein structure prediction, *J. Comp. Chem.* **23**, 138–146 (2002).
108. A. Wallqvist and M. Ullner, A simplified amino acid potential for use in structure predictions of proteins, *Proteins: Struct. Funct. Genet.* **18**, 267–280 (1994).
109. S. Takada, Z. Luthey-Schulten, and P. G. Wolynes, Folding dynamics with non-additive forces: A simulation study of a designed helical protein and a random heteropolymer, *J. Chem. Phys.* **110**, 11616–11629 (1999).
110. S. Sun, N. Luo, R. L. Ornstein, and R. Rein, Protein structure prediction based on statistical potential, *Biophys. J.* **62**, 104–106 (1992).
111. S. Sun, Reduced representation model of protein structure prediction: Statistical potential and algorithms, *Prot. Sci.* **2**, 762–785 (1993).
112. S. Sun, P. D. Thomas, and K. A. Dill, A simple protein folding algorithm using a binary code and secondary structure constraints, *Prot. Eng.* **8**, 769–778 (1995).
113. S. Sun, Reduced representation approach to protein tertiary structure prediction: Statistical potential and simulated annealing, *J. Theor. Biol.* **172**, 13–32 (1995).
114. A. Irback, F. Sjunnesson, and S. Wallin, Three-helix-bundle protein in a Ramachandran model, *Proc. Natl. Acad. Sci. USA* **97**, 13614–13618 (2000).
115. A. Irback, F. Sjunnesson, and S. Wallin, Hydrogen bonds, hydrophobicity forces and the character of the collapse transition, *J. Biol. Phys.* **27**, 169–179 (2001).
116. G. Favrin, A. Irback, and S. Wallin, Folding of a small helical protein using hydrogen bonds and hydrophobicity forces, *Proteins: Struct. Funct. Genet.*, **47**, 99–105 (2002).
117. P. Derreumaux, From polypeptide sequences to structures using Monte Carlo simulations and an optimized potential, *J. Chem. Phys.* **111**, 2301–2310 (1999).
118. F. Forcellino and P. Derreumaux, Computer simulations aimed at structure prediction of supersecondary motifs in proteins, *Proteins: Struct. Funct. Genet.* **45**, 159–166 (2001).
119. K. K. Koretke, Z. Luthey-Schulten, and P. G. Wolynes, Self-consistently optimized energy functions for protein structure prediction by molecular dynamics, *Proc. Natl. Acad. Sci. USA* **95**, 2932–2937 (1998).
120. A. Kolinski, M. Milik, and J. Skolnick, Static and dynamic properties of a new lattice model of polypeptide chains, *J. Chem. Phys.* **94**, 3978–3985 (1991).
121. A. Kolinski and J. Skolnick, Discretized model of proteins. I. Monte Carlo study of cooperativity in homopolymers, *J. Chem. Phys.* **97**, 9412–9426 (1992).
122. A. Kolinski, M. Milik, J. Rycombel, and J. Skolnick, A reduced model of short range interactions in polypeptide chains, *J. Chem. Phys.* **103**, 4312–4323 (1995).
123. J. Skolnick, A. Kolinski, and A. R. Ortiz, Monsster: A method for folding globular proteins with a small number of distance restraints, *J. Mol. Biol.* **265**, 217–241 (1997).
124. A. Kolinski and J. Skolnick, Assembly of protein structure from sparse experimental data: An efficient Monte Carlo model, *Proteins: Struct. Funct. Genet.* **32**, 475–494 (1998).
125. A. Kolinski, L. Jaroszewski, P. Rotkiewicz, and J. Skolnick, An efficient Monte Carlo model of protein chains. Modeling the short range correlations between side group centers of mass, *J. Phys. Chem. B* **102**, 4628–4637 (1998).
126. J. Skolnick, A. Kolinski, and A. R. Ortiz, Reduced protein models and their application to the protein folding problem, *J. Biomol. Struct. Dynam.* **16**, 381–396 (1998).

127. J. Skolnick, A. Kolinski, and A. R. Ortiz, Derivation of protein-specific pair potentials based on weak sequence fragment similarity, *Proteins: Struct. Funct. Genet.* **38**, 3–16 (2000).
128. H. Lu and J. Skolnick, A distance-dependent atomic knowledge-based potential for improved protein structure selection, *Proteins: Struct. Funct. Genet.* **44**, 223–232 (2001).
129. D. Kihara, H. Lu, A. Kolinski, and J. Skolnick, Touchstone: An *ab initio* protein structure prediction method that uses threading-based tertiary restraints, *Proc. Natl. Acad. Sci. USA* **98**, 10125–10130 (2001).
130. Y. Zhang, D. Kihara, and J. Skolnick, Local energy landscape flattening: Parallel hyperbolic Monte Carlo sampling of protein folding, *Proteins: Struct. Funct. Genet.* **48**, 192–201 (2002).
131. A. Sikorski, A. Kolinski, and J. Skolnick, Computer simulations of protein folding with a small number of distance restraints, *Acta Biochim. Polon.* **49**, 683–692 (2002).
132. G. Giugliarelli, C. Micheletti, J. R. Banavar, and A. Maritan, Compactness, aggregation, and prionlike behavior of protein: A lattice model study, *J. Chem. Phys.* **113**, 5072–5077 (2000).
133. P. M. Harrison, H. S. Chan, S. B. Prusiner, and F. E. Cohen, Thermodynamics of model prions and its implications for the problem of prion protein folding, *J. Mol. Biol.* **286**, 593–606 (1999).
134. P. M. Harrison, H. S. Chan, S. B. Prusiner, and F. E. Cohen, Conformational propagation with prion-like characteristics in a simple model of protein folding, *Prot. Sci.* **10**, 819–835 (2001).
135. S. Y. Patro and T. M. Przybycien, Simulations of kinetically irreversible protein aggregate structure, *Biophys. J.* **66**, 1274–1289 (1994).
136. S. Y. Patro and T. M. Przybycien, Simulations of reversible protein aggregate and crystal structure, *Biophys. J.* **70**, 2888–2902 (1996).
137. R. A. Broglia, G. Tiana, P. S., H. E. Roman, and E. Vigezzi, Folding and aggregation of designed proteins, *Proc. Natl. Acad. Sci. USA* **95**, 12930–12933 (1998).
138. S. Miyazawa and R. Jernigan, Estimation of effective interresidue contact energies from protein crystal structures: Quasichemical approximation, *Macromolecules* **18**, 534–552 (1985).
139. S. Istrail, R. Schwartz, and J. King, Lattice simulation of aggregation funnels for protein folding, *J. Comp. Biol.* **6**, 143–162 (1999).
140. D. Bratko and H. W. Blanch, Competition between protein folding and aggregation: A three-dimensional lattice-model simulation, *J. Chem. Phys.* **114**, 561–569 (2001).
141. R. I. Dima and D. Thirumalai, Exploring protein aggregation and self-propagation using lattice models: Phase diagram and kinetics, *Prot. Sci.* **11**, 1036–1049 (2002).
142. K. Leonhard, J. M. Prausnitz, and C. J. Radke, Solvent-amino acid interaction energies in 3-d-lattice mc simulations of model proteins. Aggregation thermodynamics and kinetics, *Phys. Chem. Chem. Phys.* **5**, 5291–5299 (2003).
143. N. Combe and D. Frenkel, Phase behavior of a lattice protein model, *J. Chem. Phys.* **118**, 9015–9022 (2003).
144. L. Toma and S. Toma, A lattice study of multimolecular ensembles of protein models. Effect of sequence on the final state: Globules, aggregates, dimers, fibrillae, *Biomacromolecules* **1**, 232–238 (2000).
145. P. Gupta, A. Voegler, and C. K. Hall, Effect of denaturant and protein concentrations upon protein refolding and aggregation: A simple lattice model, *Prot. Sci.* **7**, 2642–2652 (1998).
146. H. D. Nguyen and C. K. Hall, Effect of rate of chemical or thermal renaturation on refolding and aggregation of a single lattice protein, *Biotechnol. Bioeng.* **80**, 823–834 (2002).
147. P. Gupta and C. K. Hall, Computer-simulation of protein refolding pathway and intermediates, *AIChE J.* **41**, 985–990 (1995).
148. P. Gupta and C. K. Hall, Effect of solvent conditions upon refolding pathways and intermediates for a simple lattice protein, *Biopolymers* **42**, 399–409 (1997).
149. J. London, C. Skrzynia, and M. E. Goldberg, Renaturation of *Escherichia coli* tryptophanase after exposure to 8 M urea, *Eur. J. Biochem.* **47**, 409–415 (1974).
150. S. Tandon and P. Horowitz, Detergent-assisted refolding of guanidinium chloride-denatured rhodanese, *J. Biol. Chem.* **261**, 15615–15618 (1986).
151. D. Brems, Solubility of different folding conformers of bovine growth hormone, *Biochemistry* 4541–4546 (1988).

152. B. Fischer, B. Perry, I. Summer, and P. Goodenough, A novel sequential procedure to enhance the renaturation of recombinant protein from *E. coli* inclusion bodies, *Protein Eng.* **5**, 593–596 (1992).
153. S. Katoh, Y. Sezai, T. Yamaguchi, Y. Katoh, H. Yagi, and D. Nohara, Refolding of enzymes in a fed-batch operation, *Proc. Biochem.* **35**, 297–300 (1999).
154. H. Yoshii, T. Furuta, T. Yonehara, D. Ito, Y. Y. Linko, and P. Linko, Refolding of denatured/reduced lysozyme at high concentration with diafiltration, *Biosci. Biotechnol. Biochem.* **64**, 1159–1165 (2000).
155. Y. Maeda, H. Koga, H. Yamada, T. Ueda, and T. Imoto, Effective renaturation of reduced lysozyme by gentle removal of urea, *Prot. Eng.* **8**, 201–205 (1995).
156. E. DeBernardez-Clark and D. Hevehan, Oxidative renaturation of lysozyme at high concentrations, *Biotech. Bioeng.* **54**, 221–230 (1997).
157. H. Jang, C. K. Hall, and Y. Zhou, Folding thermodynamics of model four-strand antiparallel β-sheet proteins, *Biophys. J.* **82**, 646–659 (2002).
158. H. Jang, C. K. Hall, and Y. Zhou, Protein folding pathways and kinetics: Molecular dynamics simulations of β-strand motifs, *Biophys. J.* **83**, 819–835 (2002).
159. H. Jang, C. K. Hall, and Y. Zhou, Thermodynamics and stability of β-sheet complex: Molecular dynamics simulations on simplified off-lattice protein models, *Prot. Sci.* **13**, 40–53 (2004).
160. H. Jang, C. K. Hall, and Y. Zhou, Assembly and kinetic folding pathways of a tetrameric β-sheet complex: Molecular dynamics simulations on simplified off-lattice protein models, *Biophys. J.* **86**, 31–49 (2004).
161. L. Li, T. A. Darden, L. Bartolotti, D. Kominos, and L. G. Pedersen, An atomic model for the pleated beta-sheet structure of abeta amyloid protofilaments, *Biophys. J.* **76**, 2871–2878 (1999).
162. M. Sunde, L. C. Serpell, M. Bartlam, P. E. Fraser, M. B. Pepys, and C. C. F. Blake, Common core structure of amyloid fibrils by synchotron x-ray diffraction, *J. Mol. Biol.* **273**, 729–739 (1997).
163. W. D. Cornell, P. Cieplak, C. I. Bayly, I. R. Gould, K. M. Merz, Jr., D. M. Ferguson, D. C. Spellmeyer, T. Fox, J. W. Caldwell, and P. A. Kollman, A second generation force field for the simulation of proteins, nucleic acids, and organic molecules, *J. Am. Chem. Soc.* **117**, 5179–5197 (1995).
164. A. R. George and D. R. Howlett, Computationally derived structural models of the beta-amyloid found in Alzheimer's disease plaques and the interaction with possible aggregation inhibitors, *Biopolymers* **50**, 733–741 (1999).
165. B. Ma and R. Nussinov, Molecular dynamics simulations of alanine rich β-sheet oligomers: Insight into amyloid formation, *Prot. Sci.* **11**, 2335–2350 (2002).
166. B. Ma and R. Nussinov, Stabilities and conformations of Alzheimer's β-amyloid peptide oligomers (Aβ16–22, Aβ16–35, and Aβ10-35):Sequence effects, *Proc. Natl. Acad. Sci USA* **99**, 14126–14131 (2002).
167. D. Zanuy, B. Ma, and R. Nussinov, Short peptide amyloid organization: Stabilities and conformations of the islet amyloid peptide NFGAIL, *Biophys. J.* **84**, 1884–1894 (2003).
168. D. Zanuy and R. Nussinov, The sequence dependence of fiber organization. A comparative molecular dynamics study of the islet amyloid polypeptide segments 22–27 and 22–29, *J. Mol. Biol.* **329**, 565–584 (2003).
169. A. Lakdawala, D. Morgan, D. Liotta, D. Lynn, and J. Snyder, Dynamics and fluidity of amyloid fibrils: A model of fibrous protein aggregates, *J. Am. Chem. Soc.*, **124**, 15150–15151 (2002).
170. D. M. Morgan, D. G. Lynn, A. S. Lakdawala, J. P. Snyder, and D. C. Liotta, Amyloid structure: Models and theoretical considerations in fibrous aggregates, *J. Chin. Chem. Soc.* **49**, 459–466 (2002).
171. W. Hwang, D. M. Marini, R. D. Kamm, and S. Zhang, Supramolecular structure of helical ribbons self-assembled from a β-sheet peptide, *J. Chem. Phys.* **118**, 389–397 (2003).
172. K. Kuwata, T. Matumoto, H. Cheng, K. Nagayama, T. James, and H. Roder, NMR-detected hydrogen exchange and molecular dynamics simulations provide structural insight into fibril formation of prion protein fragment 106–126, *Proc. Natl. Acad. Sci. USA* **100**, 14790–14795 (2003).
173. J. Gsponer, U. Haberthur, and A. Caflisch, The role of side-chain interactions in the early steps of aggregation: Molecular dynamics simulations of an amyloid-forming peptide from the yeast prion sup35, *Proc. Natl. Acad. Sci. USA* **100**, 5154–5159 (2003).

174. P. P. Mager, Molecular simulation of the amyloid β-peptide Aβ(1–42) of Alzheimer's disease, *Mol. Sim.* **20**, 201–222 (1998).
175. A. Fernandez and M. D. L. Boland, Solvent environment conducive to protein aggregation, *FEBS Lett.* **529**, 298–302 (2002).
176. D. K. Klimov and D. Thirumalai, Dissecting the assembly of Aβ(16–22) amyloid peptides into antiparallel beta sheets, *Structure* **11**, 295–307 (2003).
177. A. V. Smith and C. K. Hall, A-helix formation: Discontinuous molecular dynamics on an intermediate resolution model, *Protein: Struct. Funct. Genet.* **44**, 344–360 (2001).
178. A. V. Smith and C. K. Hall, Assembly of a tetrameric α-helical bundle: Computer simulations on an intermediate-resolution protein model, *Proteins: Struct. Funct. Genet.* **44**, 376–391 (2001).
179. S. P. Ho and W. F. DeGrado, Design of a 4-helix bundle protein: Synthesis of peptides which self-associate into a helical protein, *J. Am. Chem. Soc.* **109**, 6751–6758 (1987).
180. L. Regan and W. F. DeGrado, Characterization of a helical protein designed from first principles, *Science* **241**, 976–978 (1988).
181. D. P. Raleigh, S. F. Betz, and W. F. DeGrado, A *de novo* designed protein mimics the native state of natural proteins, *J. Am. Chem. Soc.* **117**, 7558–7559 (1995).
182. S. F. Betz, J. W. Bryson, and W. F. DeGrado, Native-like and structurally characterized designed α-helices bundles, *Curr. Opin. Struct. Biol.* **5**, 457–463 (1995).
183. A. V. Smith and C. K. Hall, Protein refolding versus aggregation: Computer simulations on an intermediate resolution model, *J. Mol. Biol.* **312**, 187–202 (2001).
184. B. Forood, E. Perez-Paya, R. A. Houghten, and S. E. Blondelle, Structural characterization and 5'-mononucleotide binding of polyalanine β-sheet complexes, *J. Mol. Recognit.* **9**, 488–493 (1996).
185. H. D. Nguyen, A. J. Marchut, and C. K. Hall, Effects of the solvent on the conformational transition of polyalanines, *Protein Sci.* **13**, 2909–2924 (2004).
186. Y. Sugita and Y. Okamoto, Replica exchange molecular dynamics method for protein folding, *Chem. Phys. Letts.* **314**, 141–151 (1999).
187. Y. Zhou, C. K. Hall, J. M. Wichert, and M. Karplus, Equilibrium thermodynamics of homopolymers and clusters: Molecular dynamics and monte carlo simulations of systems with square-well interactions, *J. Chem. Phys.* **107**, 10691–10708 (1997).
188. A. M. Ferrenberg and R. H. Swendsen, Optimized Monte Carlo data analysis, *Phys. Rev. Lett.* **63**, 1195–1198 (1989).
189. R. Ingwall, H. Scheraga, N. Lotan, A. Berger, and E. Katchalski, Conformational studies of poly-L-alanine in water, *Biopolymers* **6**, 331–368 (1968).
190. K. Platzer, V. Ananthanarayanan, R. Andreatta, and H. Scheraga, Helix-coil stability constants for the naturally occurring amino acids in water. IV. Alanine parameters from random poly(hydroxypropyl-glutamine-co-L-alanine), *Macromolecules* **5**, 177–187 (1972).
191. A. Shoji, T. Ozaki, T. Fujito, K. Deguchi, S. Ando, and I. Ando, ^{15}N chemical shift tensors and conformation of solid polypeptides containing ^{15}N-labeled L-alanine residue by ^{15}N NMR. 2. Secondary structure reflected in alpha2-2, *J. Am. Chem. Soc.* **112**, 4693–4697 (1990).
192. H. Kimura, T. Ozaki, H. Sugisawa, K. Deguchi, and A. Shoji, Conformational study of solid polypeptides by 1h combined rotation and multiple pulse spectroscopy nmr. 2. Amide proton chemical shift, *Macromolecules* **31**, 7398–7403 (1998).
193. D. Lee and A. Ramamoorthy, Determination of the solid-state conformations of polyalanine using magic-angle spinning NMR spectroscopy, *J. Phys. Chem. B* **103**, 271–275 (1999).
194. S. E. Blondelle, B. Forood, R. A. Houghten, and E. Perez-Paya, Polyalanine-based peptides as models for self-associated β-pleated-sheet complexes, *Biochemistry* **36**, 8393–8400 (1997).
195. R. Warrass, J. Wieruszeski, C. Boutillon, and G. Lippens, High-resolution magic angle spinning NMR study of resin-bound polyalanine peptides, *J. Am. Chem. Soc.* **112**, 1789–1795 (2000).
196. Y. Levy, J. Jortner, and O. M. Becker, Solvent effects on the energy landscapes and folding kinetics of polyalanine, *Proc. Natl. Acad. Sci. USA* **98**, 2188–2193 (2001).
197. F. Ding, J. M. Borreguero, S. V. Buldyrev, H. E. Stanley, and N. V. Dokholyan, Mechanism for the alpha-helix to beta-hairpin transition, *Proteins: Struct. Funct. Genet.* **53**, 220–228 (2003).

198. H. D. Nguyen and C. K. Hall, Molecular dynamics simulations of spontaneous fibril formation by random-coil peptides, *Proc. Natl. Acad. Sci. USA* **101**, 16174–16179 (2004).
199. T. L. Benzinger, D. M. Gregory, T. S. Burkoth, H. Miller-Auer, D. G. Lynn, R. E. Botto, and S. C. Meredith, Propagating structure of Alzheimer's beta-amyloid(10–35) is parallel beta-sheet with residues in exact register, *Proc. Natl. Acad. Sci. USA* **95**, 13407–13412 (1998).
200. O. N. Antzutkin, J. J. Balbach, R. D. Leapman, N. W. Rizzo, J. Reed, and R. Tycko, Multiple quantum solid-state NMR indicates a parallel, not antiparallel, organization of beta-sheets in Alzheimer's beta-amyloid fibrils, *Proc. Natl. Acad. Sci. USA* **97**, 13045–13050 (2000).
201. J. Balbach, A. Petkova, N. Oyler, O. Antzutkin, D. Gordon, S. Meredith, and R. Tycko, Supramolecular structural constraints on Alzheimer's beta-amyloid fibrils from electron microscopy and solid-state nuclear magnetic resonance, *Biophys. J.* **83**, 1205–1216 (2002).
202. J. Harper, C. Lieber, and P. Lansbury Jr., Atomic force microscopic imaging of seeded fibril formation and fibril branching by the Alzheimer's disease amyloid-beta protein, *Chem. Biol.* **4**, 951–959 (1997).
203. J. Jarvis, D. Craik, and M. Wilce, X-ray diffraction studies of fibrils formed from peptide fragments of transthyretin, *Biochem. Biophys. Res. Commun.* **192**, 991–998 (1993).
204. L. Serpell, M. Sunde, M. Benson, G. Tennent, M. Pepys, and P. Fraser, The protofilament substructure of amyloid fibrils, *J. Mol. Biol.* **300**, 1033–1039 (2000).
205. H. D. Nguyen and C. K. Hall, Kinetics of fibril formation by polyalanines, *J. Biol. Chem.* (published online Dec 10, 2004).
206. Y. Uratani, S. Asakura, and K. Imahori, A circular dichroism study of *Salmonella* flagellin: Evidence for conformational change on polymerization, *J. Mol. Biol.* **67**, 85–98 (1972).
207. G. H. Beaven, W. B. Gratzer, and H. G. Davies, Formation and structure of gels and fibrils from glucagon, *Eur. J. Biochem.* **11**, 37–42 (1969).
208. J. Hofrichter, P. D. Ross, and W. A. Eaton, Kinetics and mechanism of deoxyhemoglobin s gelation: A new approach to understanding sickle cell disease, *Proc. Natl. Acad. Sci. USA* **71**, 4864–4868 (1974).
209. S. B. Prusiner, Novel proteinaceous infectious particles cause scrapie, *Science* **216**, 136–144 (1982).
210. J. S. Griffith, Self-replication and scrapie, *Nature* **215**, 1043–1044 (1967).
211. T. R. Serio, A. G. Cashikar, A. S. Kowal, G. J. Sawicki, J. J. Moslehi, L. Serpell, M. F. Arnsdorf, and S. L. Lindquist, Nucleated conformational conversion and the replication of conformational information by a prion determinant, *Science* **289**, 1317–1321 (2000).
212. J. T. Jarrett and P. T. S. Lansbury, Seeding "one-dimensional crystallization" of amyloid: A pathogenic mechanism in alzheimer's disease and scrapie? *Cell* **73**, 1055–1058 (1993).
213. D. Wilkins, C. Dobson, and M. Gross, Biophysical studies of the development of amyloid fibrils from a peptide fragment of cold-shock protein B, *Eur. J. Biochem.* **267**, 2609–2616 (2000).
214. J. Green, C. Goldsbury, J. Kistler, G. Cooper, and U. Aebi, Human amylin oligomer growth and fibril elongation define two distinct phases in amyloid formation, *J. Biol. Chem.* **279**, 12206–12212 (2004).
215. H. D. Nguyen and C. K. Hall, Phase diagrams describing fibrillization by polyalanine peptides, *Biophys. J.* **87**, 4122–4134 (2004).

Part III
Experimental Methods

Elucidating Structure, Stability, and Conformational Distributions during Protein Aggregation with Hydrogen Exchange and Mass Spectrometry

Erik J. Fernandez[1,3] and Scott A. Tobler[2]

> Engineering processes and proteins to control aggregation behavior has been hindered by the lack of detailed information about the mechanisms of protein aggregation. In the studies described here, hydrogen-deuterium isotope exchange detected by mass spectrometry (HX-MS) has revealed kinetic, thermodynamic, and structural aspects of model and pharmaceutical protein unfolding under destabilizing, aggregation conditions. First, hen egg white lysozyme was studied during salt-induced precipitation. Bimodal mass distributions in the HX-MS-labeling experiments for precipitates indicated that lysozyme continues to exhibit two-state unfolding behavior under these conditions, with the denatured state being only partially unfolded, resembling molten globule states observed in other folding studies. The stability and structure of recombinant human interferon-γ (IFN-γ) also were explored. While KSCN and GdnHCl yield aggregates with similar tertiary structure, the pathways by which aggregates form are different. HX analysis also showed that apparent unfolding rates were markedly increased in the presence of benzyl alcohol, a multidose preservative that induces IFN-γ aggregation. Taken together, the observations show that protein folding can remain cooperative, but that the denatured states often are partially unfolded. The patterns of retained native structure in both

---

[1] Department of Chemical Engineering, University of Virginia, 102 Engineers Way, Charlottesville, VA 22904-4741
[2] Drug Substance Development, Wyeth BioPharma, 1 Burtt Road, Andover, MA 01810
[3] To whom correspondence should be addressed: Erik J. Fernandez, University of Virginia, 102 Engineers Way, Charlottesville, VA 22904-4741; email: erik@virginia.edu

systems suggest that native-state HX patterns may presage regions of the protein susceptible to unfolding during aggregation. Such detailed information about tertiary structure changes and local instability could be a valuable aid in engineering stable formulations and proteins themselves for biopharmaceutical applications.

1. INTRODUCTION

Dill and co-workers have pointed out that the parallel nature of funnellike energy landscapes associated with protein folding means that folding "intermediates" and transition states are not well-defined structures, but ensembles of conformations.[1–4] While such intermediates may possess substantial native state structure in certain regions of the molecule, other portions can be substantially disordered in a wide distribution of configurations. Aggregates and aggregation-prone "intermediates" likely also exist in an ensemble of conformations with partial, localized native structure.

Much of the structural analysis of aggregates to date has involved secondary structure measurements of average total secondary structure or measurements of local tertiary structure using intrinsic or extrinsic probes that report on a single or undefined site. Such information must be complemented with higher-resolution tertiary structure information in order to elucidate detailed structural mechanisms for aggregation and to guide potential protein engineering strategies to address the problem. Unfortunately, the conformational heterogeneity and large size of aggregates generally precludes analysis by conventional high-resolution X-ray or solution phase nuclear magnetic resonance (NMR) techniques. Thus, there is a need for alternative experimental methods to determine the regions of aggregates and aggregation-prone intermediates that retain native structure.

It is well established that exchange rates between hydrogen atoms in water and amide hydrogens along the peptide protein backbone are extremely sensitive to the state of folding.[5] Depending on conditions, these exchange rates can be used to determine unfolding rates or equilibrium constants for folding. When these exchange events are recorded by NMR spectroscopy,[6] residue level maps of solvent accessibility can be generated. These approaches have provided a significant new level of detailed information about the structural features and dynamics of protein folding.[7–9] More recently, hydrogen-deuterium exchange (HX) and mass spectrometry (MS) have revealed distributions of conformations and intermediates in protein folding experiments.[10,11] Combined with suitable proteolytic fragmentation procedures, MS can be used to determine exchange rates on the peptide level with high sensitivity.

In recent years, hydrogen exchange has been applied productively to investigations of protein aggregation, including studies of pharmaceutical proteins[12–14] as well as amyloid forming proteins.[15–19] In this chapter, we first summarize the basic theory and experimental issues associated with HX and MS. We then discuss examples that illustrate the structural, kinetic, and thermodynamic insights into pharmaceutical protein aggregation that can be provided by hydrogen exchange.

2. MECHANISMS AND INFORMATION PROVIDED BY HYDROGEN EXCHANGE KINETICS

Hydrogen-deuterium exchange labeling takes advantage of the compact, water-excluding nature of proteins. Amide hydrogen atoms in the polypeptide backbone exchange with hydrogen or deuterium atoms in solvent water at rates that are dramatically affected by solvent accessibility. For example, backbone amides in solvent-accessible residues of lysozyme exchange within a small fraction of a second, while the amides buried in the hydrophobic core of the protein are stable for months.[20] This exquisite sensitivity to protein structure, together with advances in detection methods, has led to extensive application of hydrogen exchange to protein folding as well as protein-protein interactions,[21] including mapping the details of contact surfaces.[22] There are several recent reviews available.[9,11]

2.1. Kinetic Equations for Hydrogen-Deuterium Isotope Exchange

The well-established model for describing hydrogen-deuterium isotope exchange is a two-step model[5,23] in which unfolding is followed by exchange

$$N_H \underset{k_{fold}}{\overset{k_{unfold}}{\rightleftarrows}} U_H \overset{k_{intr}}{\longrightarrow} U_D \underset{k_{unfold}}{\overset{k_{fold}}{\rightleftarrows}} N_D, \tag{1}$$

where N and U refer to native- and solvent-exposed unfolded states, respectively. H and D refer to hydrogen- and deuterium-labeled forms. The folding and unfolding rates constants are denoted by k_{fold} and k_{unfold}. For fully solvent-exposed polypeptides, amide hydrogen atoms in the peptide backbone exchange with deuterium atoms at intrinsic rates that can be catalyzed by H_2O, H_3O^+, and OH^-:

$$k_{intr} = k_{acid}(H_3O^+) + k_{water}(H_2O) + k_{base}(OH^-). \tag{2}$$

The acid- and base-catalyzed terms generally dominate. Given the typical values for the rate constants, the intrinsic exchange rate is dominated by acid catalysis above pH 4, with a minimum exchange rate that occurs at approximately pH 2.5–3.0. Trends in exchange rates can be illustrated by the peptide polyalanine, for which k_{acid} and k_{base} have values of 41.7 and 1.12×10^{10} M^{-1} min^{-1}, respectively, at 20°C and low salt concentrations.[24,25] A rough estimate of average intrinsic exchange rates for fully solvent-exposed residues above pH 5 is given by[26]

$$<k_{intr}> = 10^{pH-5} \text{ min}^{-1}. \tag{3}$$

Detailed estimates of these rates for individual peptide are available, including the effects of amino acid sequence.[25] Such effects can be significant; notably, the residue neighboring the N-terminus exhibits intrinsic rates that are typically an order of magnitude faster than other residues in a fully solvent-exposed polypeptide.

Under conditions where proteins are reasonably stable ($k_{fold} \gg k_{unfold}$), the observed exchange rate for the two-step model can be expressed as[5]

$$k_{obs} = \frac{k_{unfold}\, k_{intr}}{k_{fold} + k_{intr}}. \tag{4}$$

There are two very useful limiting cases for interpreting hydrogen exchange data. First, for more stable conditions under which the refolding step is fast compared with intrinsic exchange rates ($k_{fold} \gg k_{intr}$), the observed exchange rate reduces to

$$k_{obs} = K_{unfold}\, k_{intr}, \tag{5}$$

where K_{unfold} is the unfolding equilibrium constant. In this "EX2" exchange regime, the observed rate can be used together with calculated intrinsic exchange rate constants[25] to estimate the free energy of unfolding. This has been used as a high-throughput screening method for protein stability measurement[26] as well as a means for estimating "local" stability at different locations within a protein.[27] When refolding is slow compared to intrinsic labeling rates (EX1 limit, $k_{fold} \ll k_{intr}$), then the observed exchange rate becomes

$$k_{obs} = k_{unfold}, \tag{6}$$

and observed rates reveal rates of solvent exposure (i.e., unfolding) at the residue(s) interrogated. Of course, because solvent accessibility and dynamics will vary dramatically within the protein, the kinetic regime can vary within the protein. A useful test to determine the exchange regime is to investigate whether the observed exchange rate varies with pH in a manner consistent with intrinsic rates [e.g., plot the degree of exchange measured at different pH values vs. (k_{intr})(labeling time)]. If EX2 kinetics are at play, then the results will fall on a single curve.

In an experiment where deuterium is exchanged into the protein, the fractional deuterium incorporation will be pseudo-first order *at each amide* with a rate constant unique for each amide. Thus, the total increase in deuterium content will be a sum of first-order reactions, one at each residue, with the exception of proline residues and the N-terminus, both of which lack backbone amide hydrogen atoms. Ionizable side-chain groups also can undergo exchange, although their rates are very fast compared to backbone amides, and when polypeptides are resolved and analyzed in deuterium-free solvents, their exchange can generally be neglected. The fractional exchange-in of deuterium thus will be

$$\frac{D}{N} = 1 - \sum_{i=1}^{N} e^{-k_i\, t}, \tag{7}$$

where D is the number of deuterium atoms exchanged in, N is the number of exchangeable residues in the peptide or protein, and k_i is the pseudo-first-order rate constant for exchange of residue i. Because of the wide range of solvent accessibilities in proteins (e.g., lysozyme exchange times can vary from subsecond to months[20]), it generally is not possible to meaningfully represent exchange with an average exchange time. Furthermore, the native state stability often is of particular interest (i.e., through Eq. 5), and this parameter is related not to the average of the exchange distribution, but is determined by the residues in the core of the protein that represent the slowest exchanging tail of the distribution. In practice, it can be quite reasonable to represent the exchange of a large number of amides as a collection into a small number of pools, so that

$$\frac{D}{N} = 1 - A_{fast}\, e^{-k_{fast}\, t} - A_{med}\, e^{-k_{med}\, t} - A_{slow}\, e^{-k_{slow}\, t}, \tag{8}$$

where A_{fast}, A_{med}, and A_{slow} are the fraction of N amides that belong to fast-, medium-, and slow-exchanging collections of amides, respectively. Likewise, k_{fast}, k_{med}, and k_{slow} refer to the exchange rate constants for the fast, medium, and slow groups. Certainly extended data or residue level measurements of exchange (e.g., from NMR) can allow more detailed representations of exchange distributions. However, the approach of Eq. (8) permits the distribution of exchange times to be represented relatively simply, with a reasonable amount of HX-MS data.

2.2. Detection Methods

The most detailed information about hydrogen exchange can be made available from NMR, since the exchange can be followed at individual amides.[28] This approach has been applied to studies of protein conformational changes during precipitation[29] and adsorption to surfaces.[30] However, NMR has limited sensitivity, molecular weight range, and rather extended analysis time. Alternatively, mass spectrometry can detect the small mass changes that occur upon isotope exchange, even in intact proteins. In concert with fragmentation using proteases[31] the degree of hydrogen exchange can be localized at the peptide level. With more complex fragmentation methods using the mass spectrometer itself, even higher resolution conceivably can be attained.[32] In all cases, mass spectrometry can be performed with small (e.g., picomole) levels of material, especially if extra care in sample handling is taken.[33] The rather convenient high-throughput capabilities of MALDI-TOF mass spectrometry also can be exploited.[26,34]

2.2.1. HX-MS protocols and continuous versus pulsed-labeling approaches

Two general hydrogen-deuterium labeling protocols are shown in Figure 1. Figure 1A shows the simplest procedure used for NMR preparation of the sample or for electrospray ionization mass spectrometry (ESI-MS) detection of intact protein mass. Hydrogen-deuterium exchange experiments involve a labeling period under the conditions of interest (e.g., native state, aggregation prone conditions, precipitate, etc.). This step can be conducted under the pH, temperature, and solvent conditions of interest. The isotope exchange reaction must be quenched to preserve the labeling information created in the first step. Exchange between amide hydrogens and those in the solvent is strongly pH dependent with a minimum typically around pH 2.5, although this value varies measurably with primary sequence.[25] Thus, quenching is usually carried out at pH $\sim$ 2.5 and low temperature (near 0°C). Depending on the type of detection performed, some kind of purification may be required. If detection is performed with NMR, relatively high concentrations of soluble protein are required, and molecular weights are limited to less than about 20 kDa. High concentrations of salt also will have to be removed. If detection is by mass spectrometry, then salt concentrations may need to be reduced to millimole

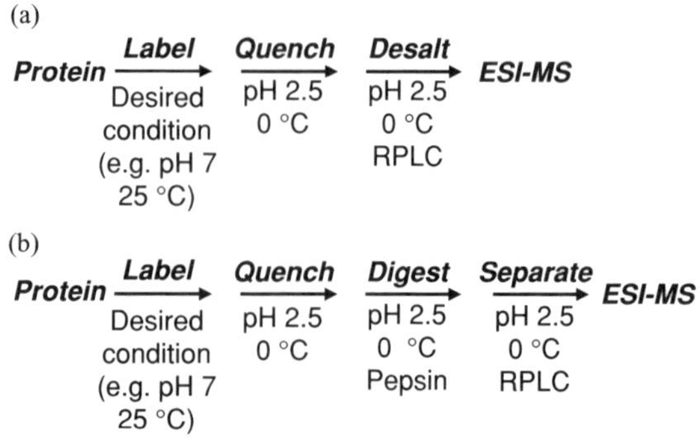

(a)
Protein →[Label; Desired condition (e.g. pH 7 25 °C)] →[Quench; pH 2.5 0 °C] →[Desalt; pH 2.5 0 °C RPLC] ESI-MS

(b)
Protein →[Label; Desired condition (e.g. pH 7 25 °C)] →[Quench; pH 2.5 0 °C] →[Digest; pH 2.5 0 °C Pepsin] →[Separate; pH 2.5 0 °C RPLC] ESI-MS

FIGURE 1. Protocols for hydrogen-deuterium isotope exchange (HX) detected by electrospray ionization mass spectrometry (ESI-MS). (a) HX analysis of intact protein. In this case, mass increases indicate average degree of exchange of entire protein. (b) HX analysis of peptides produced by proteolysis under quenched conditions. Mass changes at individual peptides reveal degree of conformational flexibility at multiple, specific regions within the protein.

levels depending on instrumentation. Thus, salt levels often must be reduced, and reversed phase chromatography often is used to simultaneously concentrate and desalt the sample.

Figure 1A shows a modified procedure used if peptide-level structural information is desired with MS detection. The procedure is similar except that a protease stable to the acid-quench conditions (typically pepsin) is used to digest the protein under quenched exchange conditions. The resulting peptides rapidly must be separated, resolved, and concentrated by reversed-phase HPLC. Regardless of the particular protocol, one key to the success of the approach is minimizing loss of the label by exchange following the quench period. During HPLC with deuterium-free solvents, peptides to be analyzed can lose deuterium atoms to solvents containing exchangeable hydrogens (e.g., water and alcohols). The quenching conditions of low pH and temperature slow, but do not stop the loss of labeling information. Thus, low pH and temperature conditions must be strictly maintained during sample preparation, and the processing times also must be kept to a minimum (i.e., less than 15 minutes). The measured deuterium contents can be corrected for back exchange rather quantitatively, if it is not too great[31]:

$$D_{corr} = \frac{\langle m \rangle - \langle m_{0\%} \rangle}{\langle m_{100\%} \rangle - \langle m_{0\%} \rangle} \cdot N, \qquad (9)$$

where D_{corr} is the corrected deuterium content and $\langle m \rangle$, $\langle m_{0\%} \rangle$, and $\langle m_{100\%} \rangle$ are the measured average molecular masses of a polypeptide obtained by analysis of partially deuterated (i.e., produced from a period of labeling in a hydrogen exchange experiment), nondeuterated, and fully deuterated samples, respectively. Because the intrinsic rates for the residue neighboring the N-terminus are so fast, back exchange will be particularly high for this residue. Consequently, under most conditions, the degree of exchange at this residue generally will not be measurable.

The hydrogen-deuterium isotope exchange reaction can be carried with labeling occurring during the process of interest or in a pulsed (e.g., quenched-flow) mode. The continuous and pulsed labeling methods have been contrasted in a useful review.[35] The former is much simpler and more common, but the latter[7] has provided a rich level of detail about the structural features and dynamics of protein folding in recent years with residue level detection of hydrogen exchange.[9] One exciting application of pulsed labeling has been to the characterization of fleeting folding intermediates.[10]

At this point, there are a number of interesting applications of the HX method[11] including detailed studies of protein folding.[9,36,37] and analysis of protein-protein interactions.[22,38,39] However, for the purposes of this review, we focus on examples relevant to protein aggregation and summarize findings from our own laboratory that illustrate information that HX-MS can provide, including conformational distributions, thermodynamics, kinetics, and molecular structure.

3. APPLICATIONS TO PRECIPITATION AND AGGREGATION

3.1. Studies of Precipitation by Stabilizing and Denaturing Salts

For our initial studies of hydrogen exchange applied to protein precipitation and aggregation, we chose hen egg white lysozyme as a model protein for which extensive folding and stability information is available. In earlier studies of lysozyme precipitation by HX, we used NMR spectroscopy to detect exchange at the residue level for a significant number of residues in the core of the protein.[29] The results showed that ammonium sulfate precipitates were nativelike, as expected, while KSCN-induced precipitates showed enhanced exchanged revealing conformational changes. More specifically, preferential disruption of the β-sheet/loop domain was observed in the presence of KSCN, while the α-helical domain remained largely protected from exchange. The extent of unfolding as measured by exchange increased with KSCN concentration. However, NMR detection cannot readily provide information about the distributions of conformers present. Where partial exchange is observed by NMR, this could arise from either a small fraction of molecules undergoing complete and rapid exchange or a larger fraction of molecules undergoing slower exchange. The ability of MS to detect mass distributions can address such questions.

Consequently, we carried out a study on lysozyme precipitated with three different salts: $(NH_4)_2SO_4$, NaCl, and KSCN, generally following the protocol outlined in Figure 1A. The protein was fully deuterated in all the exchangeable amide and other positions beforehand as previously described, and isotope exchange was carried out by exposing the deuterated protein to H_2O. Figure 2 shows the mass spectrum obtained for lysozyme under native conditions. Lysozyme that was not predeuterated showed a mass consistent with that for the native molecule (14,304 Da; see light curve shown in Figure 2). In contrast, lysozyme that was predeuterated at the exchangeable positions showed significantly higher masses with a slow drop in the measured mass as a function of exchange time (heavy curves in Figure 2). The increased mass indicates that even after many hours, a core of residues in lysozyme is resistant to solvent exposure and exchange, even under these low pH conditions. This is a testimony to the stability of this protein. The slow shift to lower mass is characteristic of the EX2 kinetic regime

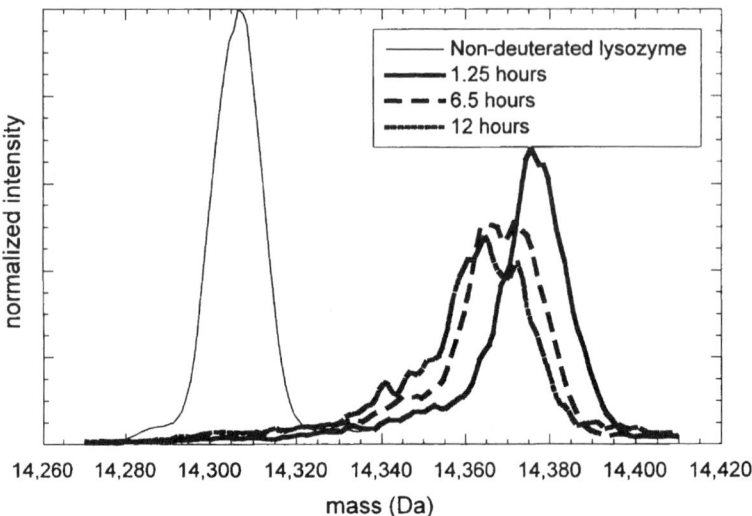

FIGURE 2. Intact protein HX of lysozyme without precipitating agents in 0.1 M phosphate buffer (H_2O, pH 2.16). Predeuterated lysozyme was exposed to H_2O for the periods of time shown (heavy solid and dashed curves). The light curve shows the mass spectrum of lysozyme that was not predeuterated. Reproduced from Tobler et al.,[12] with permission from Wiley Periodicals, Inc.

of exchange, indicating that the refolding rates are much faster than intrinsic exchange rates.

Figure 3 shows the different exchange behavior exhibited by lysozyme when precipitated by kosmotropic salts, $(NH_4)_2SO_4$ and NaCl (Figure 3A), versus a chaotropic salt, KSCN (Figure 3B). Figure 3A shows that mass spectra for the kosmotropic salts yield single peaks similar to each other and to native protein (Figure 2), indicating that the protein has remained native, with stability at least as high as the native under precipitating conditions. In contrast, the protein precipitated with KSCN showed very different spectra, with a significant fraction of protein exhibiting a very different, lower mass. This fraction increased at the higher KSCN concentration (1.0 M).

The appearance of a distinct peak at high KSCN concentrations indicates that refolding rates are slower than exchange rates and the protein is effectively under the EX1 exchange kinetic regime. Also, as far as can be detected, the folding appears to remain two-state. Under these conditions, the fraction of the low mass peak reveals the fraction of protein that has unfolded during the labeling period. Thus, while KSCN was found to reduce solubility very effectively, it has done so at the expense of lower stability. Interestingly, the mass of the exchanged protein (low mass peak) did not reach that of undeuterated protein. This suggested that the nonnative state of the molecule is not fully solvent exposed, but retains a core that protects amide from exchange for at least a few hours. Thus, this "unfolded" state may in fact be a molten-globule-like state with some residual native structure. For the KSCN precipitation experiments, there was a fraction of the precipitate that could not be redissolved in the original buffer. These precipitates were analyzed separately and found to have a much greater fraction of unfolded protein

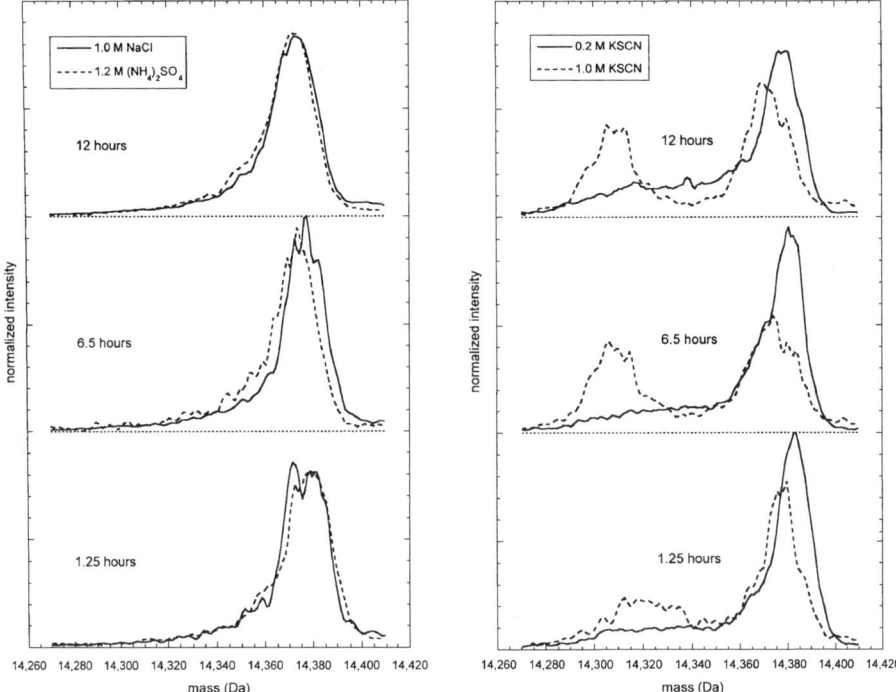

FIGURE 3. Mass spectra for predeuterated lysozyme exchanging in a precipitate slurry containing 0.1 M phosphate buffer (H$_2$O, pH 2.16) and the indicated precipitating salt. (A) NaCl and (NH$_4$)$_2$SO$_4$; (B) KSCN. The spectra shown were obtained from the soluble fractions of the precipitate. Reproduced from Tobler et al.,[12] with permission from Wiley Periodicals, Inc.

(data not shown). From the fraction of the low-mass peak, the kinetics of unfolding during the precipitation process could be followed (data not shown). Investigation of the HX behavior of lysozyme in a KSCN solution at low protein concentrations confirmed the destabilizing effect of the salt on the protein structure, even when there was almost no solid phase present. The HX/ESI-MS results provide insight into the mechanism combining precipitation and denaturation for such a system, both in terms of obtaining quantitative kinetic and stability information and the identification of the conformers present.

3.2. Analysis of Pharmaceutical Protein Aggregation

We applied HX-MS to study stability and structural changes of recombinant human interferon-γ (IFN-γ) during aggregation induced by guanidine hydrochloride (GdnHCl) and potassium thiocyanate. This 32-kDa homodimer consists primarily of α-helical secondary structure with no β-sheet or disulfide bonds. The protein has been shown to aggregate under a variety of denaturing conditions such as high temperature, low pH, and in the presence of chaotropes or the formulation preservative benzyl alcohol.[40–43] These prior studies have shown that aggregates created under different denaturing conditions possess similar and significant amounts of helical structure. Thus, the aggregated protein was thought to retain some degree of nativelike structure and was postulated to form aggregates under a common pathway under the different conditions. Using HX-MS in concert with proteolysis we have probed the structure of aggregates created using GdnHCl and KSCN and found that while the aggregates have common, nativelike structural features, the pathways by which they are formed are different.

In this study, proteolysis was incorporated into the protocol, as shown in Figure 1B, to localize the degree of hydrogen exchange at the peptide level. Aggregates were formed, diluted into D_2O, and at various times aliquots of the protein were quenched and digested using pepsin. The resulting peptides were resolved on reversed-phase chromatography column and mass analyzed to generate a HX curve for each peptide. Because pepsin cleaves after a number of hydrophobic residues, a significant number of peptides spanning almost the entire sequence were generated. Of these, 19 of the most abundant peptides were used as reporters, shown in Figure 4. The variable size of the peptides provided a resolution of exchange ranging from 3 to 20 residues.

Sample HX data from four of the peptides (2, 4, 7, and 12) are shown in Figure 5. As can be seen from Figure 5, peptide 2 was quickly deuterated (mostly fast exchangers), peptides 7 and 12 consisted mainly of medium exchangers, and peptide 4 underwent essentially no exchange during the course of the 24-hour experiment (i.e., consisted of all slow exchangers). For the purposes of quantitative data analysis, we considered the individual amides to fall into three groups: (1) fast-exchanging amides that were deuterated by the time of the first data point, (2) slow-exchanging amides that remained

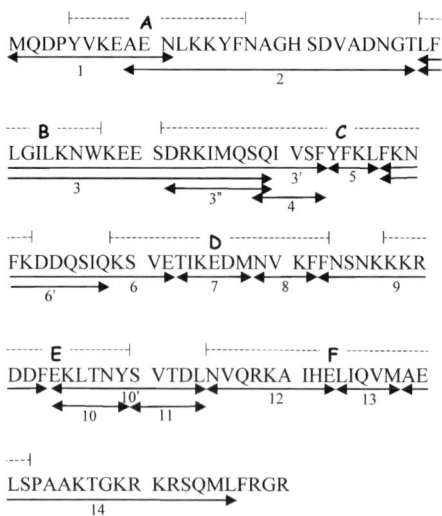

FIGURE 4. Primary sequence of recombinant human IFN-γ monomer. The numbered arrows indicate the peptides identified and analyzed for this study. The dashed lines indicate the locations of the six helices labeled A-F. Reproduced from Tobler and Fernandez,[13] with permission.

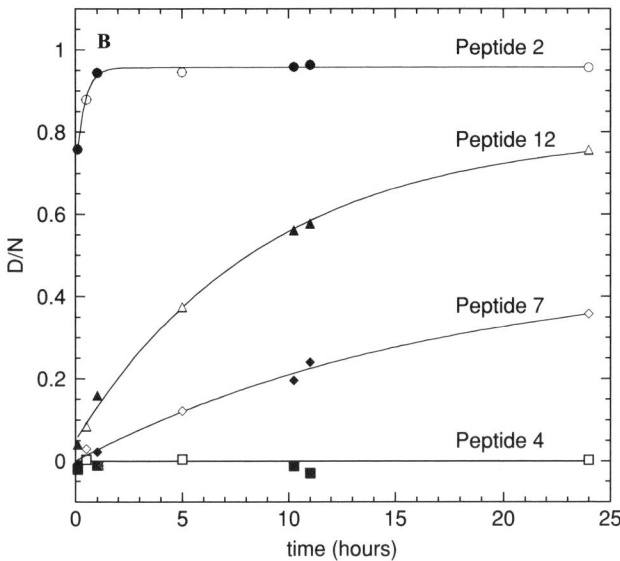

FIGURE 5. Fractional amide incorporation of deuterium, D/N, into four sample peptides from IFN-γ in 5 mM succinate at a measured pH of 5. D/N was calculated and corrected for back exchange as previously described.[13] The curves shown are fits to a simplified form of Eq. 10 as described in the text. Reproduced from Tobler and Fernandez,[13] with permission.

protected at the last time point, and (3) medium-exchanging amides that underwent exchange during the duration of the experiment. The curve fits shown in Figure 5 represent fits to a simplified version of Eq. (8), where $\exp(-k_{fast} t) \ll 1$ and $\exp(-k_{slow}t) \sim 1$:

$$\frac{D}{N} = 1 - A_{med}\, e^{-k_{med}\, t} - A_{slow}. \qquad (10)$$

The curve fits were for the three remaining parameters in Eq. (10), with the size of the fast-exchange group being determined from $A_{fast} = 1 - A_{med} - A_{slow}$.

Figure 6 shows the breakdown of the fraction of fast, medium, and slow exchangers in each of the reporter peptides for IFN-γ under the reference-stable conditions. The last four helices (C-F) possess the largest number of slowly exchanging amides. Of these helix C exhibited the greatest apparent stability, because there were ~12 consecutive amide sites that exchange slowly (peptides 3", 4, and 5). Indeed, Helix C is the most buried of the helices in the dimer and forms much of the dimer interface area.[44,45] Stretches of helix C were purely slow exchangers (e.g., peptides 4, 5). In contrast, loop regions (peptides 6, 11, and 14) were made up of predominantly fast exchangers, as expected. The results show that helices A and B (peptides 1, 2, and part of 3) were largely solvent exposed. Indeed, these two helices have been reported to be relatively flexible.[45,46] These exchange patterns give a rather detailed picture of local flexibility in different regions of the molecule.

A similar hydrogen exchange analysis was made for IFN-γ aggregates induced by either 1 M GdnHCl or 0.3 M KSCN. The protein was first incubated in H_2O (15 mg/ml

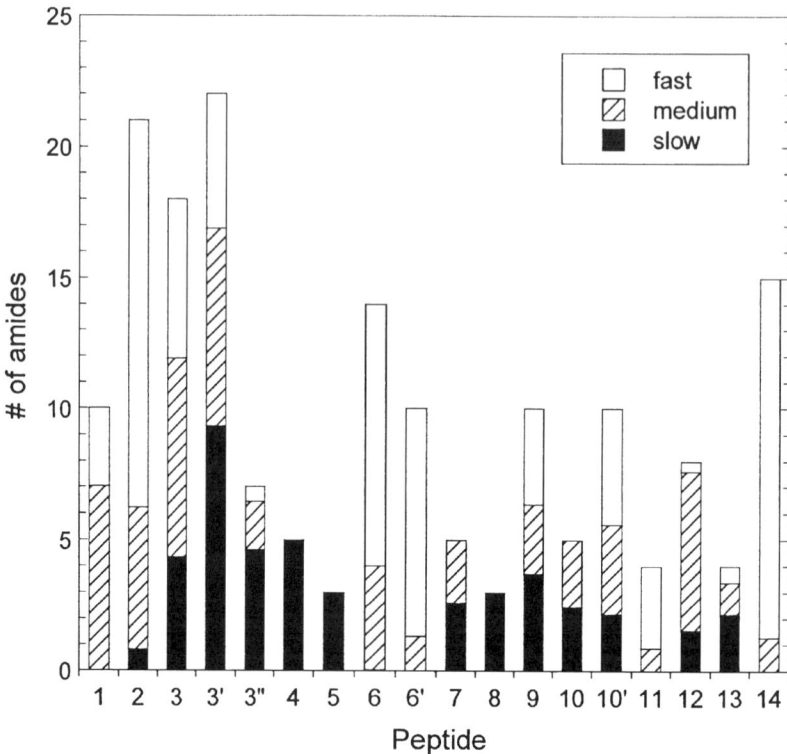

FIGURE 6. Number of fast-, medium-, and slow-exchanging amide sites in each peptide peptide from IFN-undergoing hydrogen exchange under the native conditions of Figure 5. Based on the standard errors of the fit parameters, the uncertainty in the size of each pool was approximately ± 0.5 amides. Reproduced from Tobler and Fernandez,[13] with permission.

protein, 5 mM succinate, pH 5) with one of the two denaturing salts. After 24 hours of incubation the slurry was centrifuged and the supernatant was removed and replaced with an identical buffer (less the protein) made with D_2O and the same chaotrope concentration. Aggregates were then incubated for a series of labeling times, redissolved quickly with 6 M GdnDCl, and digested and analyzed. The use of GdnDCl at this point was to ensure all the protein was redissolved. Control experiments showed that, while there was a loss of some of the deuterium labels, a substantial amount remained, allowing the relative protection patterns of the two aggregates to be compared.

Figure 7 shows the distribution of fast, medium, and slowly exchanging amides for IFN-γ aggregated in the two chaotropic salts. The two aggregates have been reported to have similar secondary structure contents, although somewhat different morphologies—GdnHCl yields a gel, while KSCN aggregates form a white precipitate.[43] We observed these same morphological differences, and Figure 7 shows that their hydrogen exchange properties are quite similar. Compared to the native state, the aggregates had a much smaller fraction of medium-exchanging amides. The distribution of slowly exchanging amides was rather similar, with helix C being the most highly protected helix in the

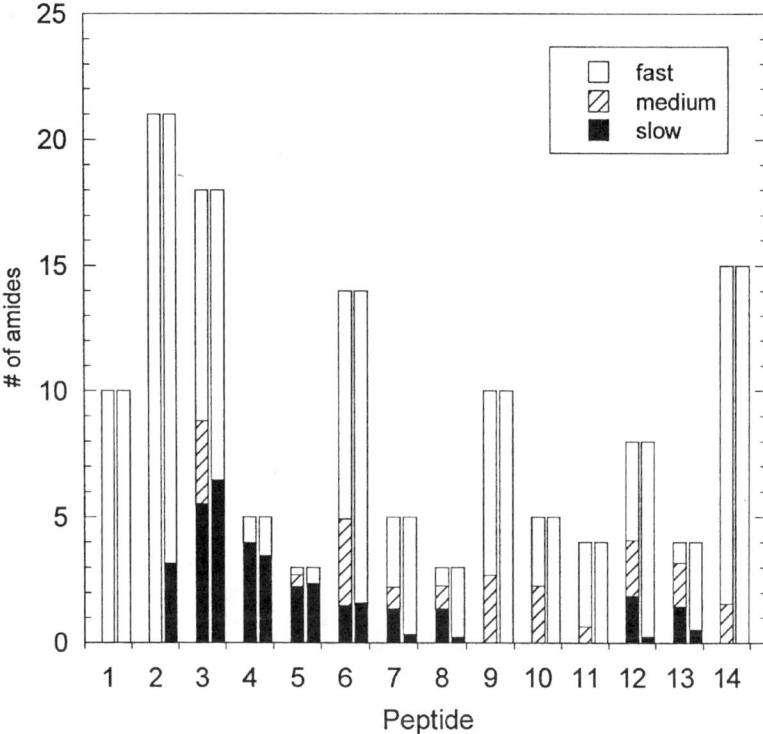

FIGURE 7. Number of fast-, medium-, and slow-exchanging amide sites in each reporter peptide from IFN-γ under aggregating conditions. Aggregation was induced in the presence of either 0.3 M KSCN (left-hand bar in each pair) or 1 M GdnHCl (right-hand bar) in 5 mM succinate, measured pH of 5. Aggregation was allowed to proceed for 24 hr prior to hydrogen-deuterium exchange labeling in the analogous D_2O buffer. Reproduced from Tobler and Fernandez,[13] with permission.

protein. Helices D, E, and F were more solvent exposed in the aggregates than in the native state. Thus, these measurements are in agreement with the prior secondary structure measurements but provide a more detailed picture of the parts of the tertiary structure that are affected. It is also interesting to note that the regions most protected from solvent exposure in the aggregates are those that are most protected in the native state. This suggests that hydrogen exchange patterns measured in the native state may presage regions of proteins likely to be more or less stable under denaturing or aggregating conditions.

To gain insight into the IFN-γ aggregation pathway in the presence of the two different salts, HX experiments were performed at ten-fold lower protein concentrations (2 mg/ml vs. 20 mg/ml), where little or no aggregation occurred (SEC data not shown).[13] Figure 8 shows the resulting HX patterns for the soluble protein in the presence of the two salts. In the presence of 0.3 M KSCN, dilute protein shows an exchange pattern remarkably similar to that of native protein (see Figure 6). In contrast to aggregates, there is a marked difference in the HX patterns for IFN-γ in the presence of the two salts: 1 M GdnDCl severely disrupted the structure of IFN-γ as measured by HX. Figure 8

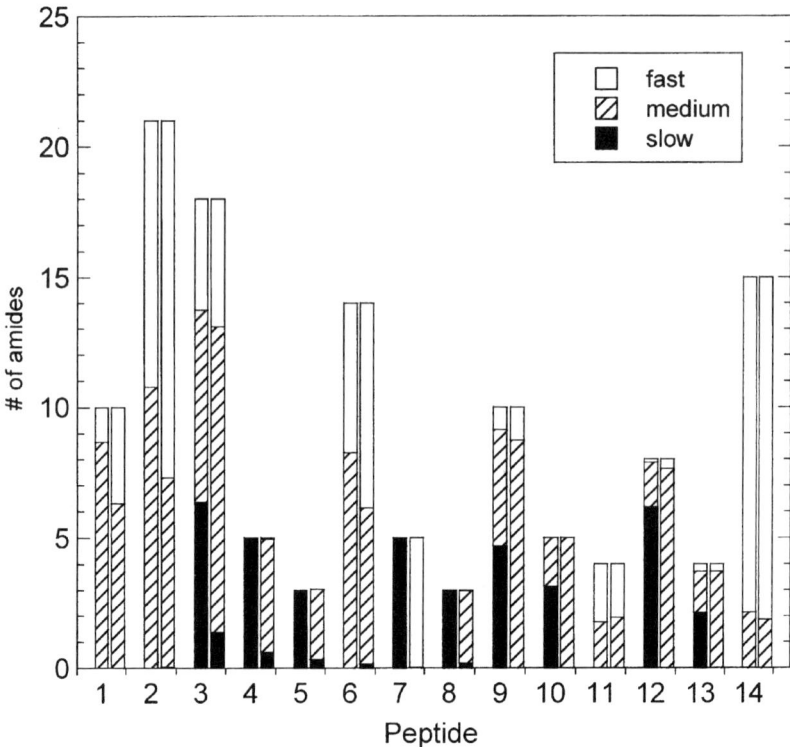

FIGURE 8. Number of fast-, medium-, and slow-exchanging amide sites in reporter peptide from IFN-γ in presence of chaotropes (0.3 M KSCN left-hand bars, 1 M GdnHCl right-hand bars) but at low protein concentration (2 mg/ml) where little or no aggregation occurred. Reproduced from Tobler and Fernandez,[13] with permission.

also suggests that 0.3 M KSCN may have *stabilized* IFN-γ at low protein concentrations, leading to a larger number of slowly exchanging amides in peptides 3, 7, and 12 and smaller or insignificant changes in the other peptides. A measurable slowing of exchange consistent with this also was observed in HX-MS measurements of the whole protein (data not shown). The stabilization may be a nonspecific salt effect, as NaCl gave a comparable degree of stabilization at the same ionic strength (data not shown). Thus, while previous measurements of IFN-γ secondary structure[43] and our HX measurements of tertiary structure show the aggregates are similar, their aggregation pathways appear to be different: GdnHCl (1 M) destabilizes IFN-γ in the solution phase, while KSCN (0.3 M) precipitates the protein in the native form and subsequently destabilizes IFN-γ in the precipitated state. Pulsed-labeling HX measurements of IFN-γ aggregates *after* the aggregates had formed were consistent with these different pathways.[13]

We also have investigated the effects of a multidose preservative, benzyl alcohol, on the destabilization and aggregation of IFN-γ. In benzyl alcohol, IFN-γ can retain its secondary structure, but changes its overall tertiary structure (as measured by CD) and forms soluble aggregates. We performed intact protein HX measurements (protocol in Figure 1A) that showed resolved peaks in the mass spectrum representing unlabeled and deuterated protein, respectively (data not shown). The lower mass peak in such a spectrum represents protein that did not completely unfold at any time during the labeling experiment. Such behavior indicates effective EX1 kinetic behavior for aggregating IFN-γ. Thus, according to Eq. (6), these measurements can reveal changes in unfolding rates for the protein.

Sample hydrogen exchange curves shown in Figure 9 are consistent with first-order kinetics. They show that increases in benzyl alcohol cause a dramatic increase in unfolding rates for the protein. It also was observed that combination of 0.1 M NaCl or KCl combined with benzyl alcohol led to unfolding rates greater than $\sim$ 2/hr (data not shown). Decreasing the surface area/volume ratio by a factor of $\sim$ 4 had no significant effect, indicating that the destabilizing effect of benzyl was not mediated by the container surface. Dynamic light-scattering and size exclusion chromatography showed that a small

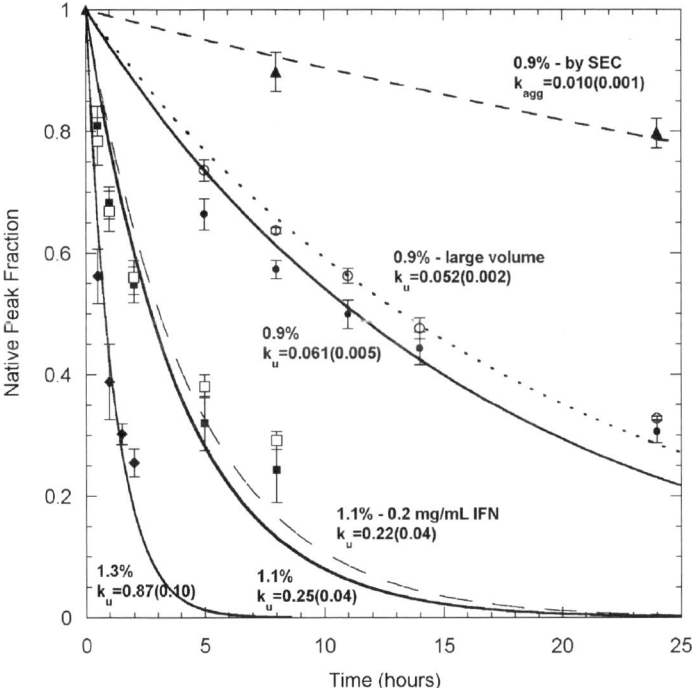

FIGURE 9. Unfolding kinetics for IFN-γ in 5 mM succinate (measured pH 5.0, 20°C) and the indicated concentrations of benzyl alcohol. The native peak fraction was the fraction of the low mass peak area in the bimodal mass spectrum. Error bars were calculated from the standard deviation of replicate experiments ($n = 3$) except for the experiment at 0.2 mg/ml, 1.1% benzyl alcohol ($n = 2$). In that case, the error bars were based on the minimum and maximum values. Apparent first-order unfolding rates from hydrogen exchange, k_u, pseudo-first-order rates of aggregation as measured by SEC, k_{agg}, were determined by nonlinear least squares regression with a first-order rate expression. The dotted curve ("large volume") refers to experiments at 0.9% benzyl alcohol in 10-fold larger container, resulting in approximately a fourfold reduction in surface-area-to-volume ratio. Reproduced from Tobler et al.,[14] with permission from Wiley Periodicals, Inc.

fraction of the protein formed large aggregates during the first few days. Measurements at longer incubation times (up to eight days) showed that a significant fraction of protein was trapped in a structure less protected from hydrogen exchange, but not completely unfolded. This fraction of protein may be responsible for the irreversible loss of activity observed in earlier studies.[42]

4. CONCLUSION

The studies of model and pharmaceutical proteins described above demonstrate the ability of HX-MS to provide information about the structure, thermodynamics, and kinetics of protein unfolding and stability during aggregation with small quantities of sample. While the structural resolution is not as high as that provided by NMR, the advantages in sensitivity, flexibility, and throughput allow a larger parameter space to be evaluated more quickly and for proteins inaccessible to NMR analysis (e.g., large and aggregation prone proteins). The results also show that these two proteins maintain two-state behavior during aggregation, though the "unfolded" states observed may not be fully solvent accessible (e.g., molten globulelike). Further, hydrogen exchange was able to distinguish changes in pathways active in the case of GdnHCl- and KSCN-induced aggregation of IFN-γ. Further, the peptide level HX-MS analysis gave a rather detailed picture of tertiary structure changes in the aggregates, which suggested that native state flexibility and hydrogen exchange patterns could presage regions of unstable proteins susceptible to unfolding during aggregation. Such detailed information could be valuable for engineering proteins themselves to reduce aggregation. In the future, high-throughput HX-MS methods.[26] and resolution at or near the single-residue level achieved with MS/MS[32] or improved proteolytic methods[27] could make this a critical tool for protein engineers. Finally, as mass spectrometry equipment improves and becomes more widely available, such experiments could become more accessible to formulation and process development scientists encountering protein stability issues.

REFERENCES

1. K. A. Dill and H. S. Chan, From Levinthal to pathways to funnels, *Nature Struct. Biol.* **4**(1), 10–9 (1997).
2. H. S. Chan and K. A. Dill, Protein folding in the landscape perspective: chevron plots and non-Arrhenius kinetics, *Proteins* **30**(1), 2–33 (1998).
3. S. B. Ozkan, I. Bahar, and K. A. Dill, Transition states and the meaning of phi-values in protein folding kinetics, *Nature Struct. Biol.* **8**(9), 765–769 (2001).
4. S. B. Ozkan, K. A. Dill, and I. Bahar, Computing the transition state populations in simple protein models, *Biopolymers* **68**(1), 35–46 (2003).
5. A. Hvidt, and S. O. Nielsen, Hydrogen exchange in proteins, *Adv. Protein Chem.* **21**, 287–386 (1966).
6. G. Wagner, and K. Wuthrich, Amide protein exchange and surface conformation of the basic pancreatic trypsin inhibitor in solution. Studies with two-dimensional nuclear magnetic resonance, *J. Mol. Biol.* **160**(2), 343–61 (1982).

7. H. Roder and K. Wüthrich, Protein folding kinetics by combined use of rapid mixing techniques and NMR observation of individual amide protons, *Proteins* **1**(1):34–42 (1986).
8. K. Roder, G. A. Elöve, and S. W. Englander, Structural characterization of folding intermediates in cytochrome *c* by H-exchange labelling and proton NMR, *Nature* **335**, 700–704 (1988).
9. S. W. Englander, Protein folding intermediates and pathways studied by hydrogen exchange, *Annu. Rev. Biophys. Biomol. Struct.* **29**, 213–38 (2000).
10. A. Miranker, C. V. Robinson, S. E. Radford, R. T. Aplin, and C. M. Dobson, Detection of transient protein folding populations by mass spectrometry, *Science* **262**(5135), 896–900 (1993).
11. A. N. Hoofnagle, K. A. Resing, and N. G. Ahn, Protein analysis by hydrogen exchange mass spectrometry, *Annu. Rev. Biophys. Biomol. Struct.* (2003).
12. S. A. Tobler, N. E. Sherman, and E. J. Fernandez, Tracking lysozyme unfolding during salt-induced precipitation with hydrogen exchange and mass spectrometry, *Biotech. Bioeng.* **71**(3), 194–207 (2001).
13. S. A. Tobler and E. J. Fernandez, Structural features of interferon-gamma aggregation revealed by hydrogen exchange, *Protein Sci.* **11**(6), 1340–1352 (2002).
14. S. A. Tobler, B. H. Holmes, M. E. Cromwell, and E. J. Fernandez, Benzyl alcohol-induced aggregation of interferon-γ: A study by hydrogen-deuterium isotope exchange, *J. Pharm. Sci.* **93**, 1605–1617 (2004).
15. I. Kheterpal, S. Zhou, K. D. Cook, and R. Wetzel, A beta-amyloid fibrils possess a core structure highly resistant to hydrogen exchange, *Proc. Natl. Acad. Sci. USA* **97**(25), 13597–13601 (2000).
16. J. H. Ippel, A. Olofsson, J. Schleucher, E. Lundgren, and S. S. Wijmenga, Probing solvent accessibility of amyloid fibrils by solution NMR spectroscopy, *Proc. Natl. Acad. Sci. USA* **99**(13), 8648–8653 (2002).
17. S. S. Wang, S. A. Tobler, T. A. Good, and E. J. Fernandez, Hydrogen exchange-mass spectrometry analysis of beta-amyloid peptide structure, *Biochemistry* **42**(31), 9507–9514 (2003).
18. M. Kraus, M. Bienert, and E. Krause, Hydrogen exchange studies on Alzheimer's amyloid-beta peptides by mass spectrometry using matrix-assisted laser desorption/ionization and electrospray ionization, *Rapid Commun. Mass Spectrom.* **17**(3), 222–228 (2003).
19. Y. S. Kim, J. S. Wall, J. Meyer, C. Murphy, T. W. Randolph, M. C. Manning, A. Solomon, and J. F. Carpenter, Thermodynamic modulation of light chain amyloid fibril formation, *J. Biol. Chem.* **275**(3), 1570–1574 (2000).
20. S. E. Radford, M. Buck, K. D. Topping, C. M. Dobson, and P. A. Evans, Hydrogen exchange in native and denatured states of hen egg-white lysozyme, *Proteins* **14**(2), 237–248 (1992).
21. L. Pershina and A. Hvidt, A study by the hydrogen-exchange method of the complex formed between the basic pancreatic trypsin inhibitor and trypsin, *Eur. J. Biochem.* **48**(2), 339–344 (1974).
22. Y. Paterson, S. W. Englander, and H. Roder, An antibody binding site on cytochrome *c* defined by hydrogen exchange and two-dimensional NMR, *Science* **249**, 755–759 (1990).
23. K. Linderstrøm-Lang, Deterium exchange between peptides and water, *Chem. Soc. (London) Spec. Publ.* **2**, 1–20 (1955).
24. R. S. Molday, S. W. Englander, and R. G. Kallen, Primary structure effects on peptide group hydrogen exchange, *Biochemistry* **11**(2), 150–158 (1972).
25. Y. Bai, J. S. Milne, L. Mayne, and S. W. Englander, Primary structure effects on peptide group hydrogen exchange. *Proteins: Struct. Funct. Genet.* 1993;17:75–86.
26. S. Ghaemmaghami, M. C. Fitzgerald, and T. G. Oas, A quantitative, high-throughput screen for protein stability, *Proc. Natl. Acad. Sci. USA* **97**(15), 8296–8301 (2000).
27. Y. Hamuro, S. J. Coales, M. R. Southern, J. F. Nemeth-Cawley, D. D. Stranz, and P. R. Griffin, Rapid analysis of protein structure and dynamics by hydrogen/deuterium exchange mass spectrometry. *J. Biomol. Tech.* **14**(3), 171–182 (2003).
28. K. Wüthrich and G. Wagner, Nuclear magnetic resonance of labile protons in basic pancreatic trypsin inhibitor. *J. Mol. Biol.* **130**, 1–18 (1979).
29. S. T. Chang and E. J. Fernandez, Probing residue-level unfolding during lysozyme precipitation, *Biotech. Bioeng.* **59**(2), 144–155 (1998).

30. J. L. McNay and E. J. Fernandez, How does a protein unfold on a reversed-phase liquid chromatography surface? *J. Chromatogr. A* **849**(1), 135–148 (1999).
31. Z. Zhang and D. L. Smith, Determination of amide hydrogen exchange by mass spectrometry: A new tool for protein structure elucidation, *Protein Sci.* **2**(4), 522–531 (1993).
32. Y. Z. Deng, H. Pan, and D. L. Smith, Selective isotope labeling demonstrates that hydrogen exchange at individual peptide amide linkages can be determined by collision-induced dissociation mass spectrometry, *J. Am. Chem. Soc.* **121**(9), 1966–1967 (1999).
33. L. Wang and D. L. Smith, Down sizing improves sensitivity 100-fold for hydrogen exchange-mass spectrometry, *Anal. Biochem.* **314**(1), 46–53 (2003).
34. J. G. Mandell, A. M. Falick, and E. A. Komives, Measurement of amide hydrogen exchange by MALDI-TOF mass spectrometry, *Anal. Chem.* **70**(19), 3987–3995 (1998).
35. Y. Z. Deng, Z. Q. Zhang, and D. L. Smith, Comparison of continuous and pulsed labeling amide hydrogen exchange/mass spectrometry for studies of protein dynamics. *J. Am. Soc. Mass Spec.* **10**(8), 675–684 (1999).
36. S. W. Englander, T. R. Sosnick, J. J. Englander, and L. Mayne, Mechanisms and uses of hydrogen exchange, *Curr. Opin. Struct. Biol.* **6**(1), 18–23 (1996).
37. J. Clarke and L. S. Itzhaki, Hydrogen exchange and protein folding, *Curr. Opin. Struct. Biol.* **8**(1), 112–118 (1998).
38. J. G. Mandell, A. M. Falick, and E. A. Komives, Identification of protein-protein interfaces by decreased amide proton solvent accessibility. *Proc. Natl. Acad. Sci. USA* **95**(25), 14705–14710 (1998).
39. H. Ehring, Hydrogen exchange/electrospray ionization mass spectrometry studies of structural features of proteins and protein/protein interactions, *Anal. Biochem.* **267**(2), 252–259 (1999).
40. Y. R. Hsu and T. Arakawa, Structural studies on acid unfolding and refolding of recombinant human interferon gamma, *Biochemistry* **24**(27), 7959–7963 (1985).
41. M. G. Mulkerrin and R. Wetzel, pH dependence of the reversible and irreversible thermal denaturation of gamma interferons, *Biochemistry* **28**(16), 6556–6561 (1989).
42. X. M. Lam, T. W. Patapoff, and T. H. Nguyen, The effect of benzyl alcohol on recombinant human interferon-gamma, *Pharm. Res.* **14**(6), 725–729 (1997).
43. B. S. Kendrick, J. L. Cleland, X. Lam, T. Nguyen, T. W. Randolph, M. C. Manning, and J. F. Carpenter, Aggregation of recombinant human interferon gamma: Kinetics and structural transitions, *J. Pharm. Sci.* **87**(9), 1069–1076 (1998).
44. S. E. Ealick, W. J. Cook, S. Vijay-Kumar, M. Carson, T. L. Nagabhushan, P. P. Trotta, and C. E. Bugg, Three-dimensional structure of recombinant human interferon-gamma, *Science* **252**(5006), 698–702 (1991).
45. S. Grzesiek, H. Dobeli, R. Gentz, G. Garotta, A. M. Labhardt, and A. Bax, 1H, 13C, and 15N NMR backbone assignments and secondary structure of human interferon-gamma, *Biochemistry* **31**(35), 8180–8190 (1992).
46. G. Waschutza, V. Li, T. Schafer, D. Schomburg, C. Villmann, H. Zakaria, and B. Otto, Engineered disulfide bonds in recombinant human interferon-gamma: The impact of the N-terminal helix A and the AB-loop on protein stability, *Protein Eng.* **9**(10), 905–912 (1996).

Application of Spectroscopic and Calorimetric Techniques in Protein Formulation Development

Angela Wilcox[1,2] and Rajesh Krishnamurthy[1]

> The rational development of stable protein formulations requires a detailed understanding of the factors influencing the different routes of protein degradation. Aggregation, one of the major routes of protein degradation, is dependent on, among other factors, the conformational stability of the molecule. Assessing the conformational structure and the factors that affect it, usually via spectroscopic and calorimetric methods, is increasingly used in formulation development to better understand the influence of formulation variables on protein folding and/or aggregation. In this review, we discuss four commonly used techniques—differential scanning calorimetry, isothermal calorimetry, circular dichroism spectroscopy, and infrared spectroscopy—to assess the conformational stability of proteins. While circular dichroism and calorimetry have been used to assess protein stability over a longer period of time, infrared spectroscopy is rapidly gaining acceptance owing to its ability to provide information about the protein in the lyophilized form. The purpose of this review is to provide an overview of the current best practices for processing and analyzing the spectroscopic data along with examples detailing the use of these methods in developing stable protein formulations.

1. INTRODUCTION

While proteins are susceptible to a wide variety of denaturation/degradation reactions, aggregation has been a subject of renewed interest lately due to concerns regarding

---

[1] Human Genome Sciences, 14200 Shady Grove Road, Rockville, MD 20850
[2] To whom correspondence should be addressed: Angela Wilcox, Human Genome Sciences, 14200 Shady Grove Road, Rockville, MD 20850; email: Angela_Wilcox@hgsi.com

its potential for immunogenicity.[1,2] Aggregation leads to the formation of particulate matter[3] and to a loss in efficacy.[4] Aggregation also has been identified as playing an important role in diseases such as bovine spongiform encephalopathy, Alzheimer's disease, Parkinson's disease, and other protein deposition disorders.[5]

A rational approach to minimizing aggregation of proteins requires identification of the underlying factors causing aggregation, use of highly sensitive methods to detect aggregates, and development of effective countermeasures.[6] Protein aggregation can occur quite readily during purification, sterilization, shipping, and storage processes.[7] Physical stresses like freezing, thawing, and shear also play a significant role in promoting aggregation. Aggregation can occur even in solution conditions that greatly favor the native state of the protein and in the absence of stresses.[8] To prevent aggregation-related problems in a clinical setting, one can minimize the presence of aggregates at each step in the process leading up to administration of the therapeutic. Of these steps, aggregation during storage has received the most attention owing to its high potential to adversely influence the safety of the patient.

The most straightforward approach to minimize aggregation during storage, it might seem, would be to monitor the rate of aggregation in different formulations at the intended storage temperature and identify the formulation that reduces aggregation to the greatest extent. However, since a typical protein therapeutic is expected to be stable for longer than 18 months at the intended storage temperature, this approach is of limited use during development. A common strategy has been to employ accelerated conditions of storage such as higher temperature or increased humidity to force the protein to aggregate at an increased rate, and thus select the formulation that minimizes aggregation to the greatest extent. Real time or accelerated stability monitoring using experimental design approaches can be labor- and time-intensive due to the large number of conditions that need to be investigated. Further, these stability studies typically do not provide any insight into the mechanism(s) of aggregation.

Spectroscopic and calorimetric methods are utilized to help overcome these limitations. Spectroscopy and calorimetry have played an important role in understanding the mechanism and thermodynamics of the folding and unfolding of proteins. The experimental approaches used to understand protein folding now are being adapted to provide information about the effect of solution properties on a protein's stability. A protein's aggregation profile, its rate of aggregation, and the onset of aggregation are dependent on the intrinsic stability of the protein as well as the properties of the environment, such as pH, ionic strength, or presence of excipients. Since the intrinsic stability of proteins is an important factor in determining the tendency to aggregate, spectroscopic and calorimetric techniques are useful tools for developing stable protein formulations. These spectroscopic and calorimetric studies provide the framework for rational development of formulation strategies by helping identify the mechanism(s) of aggregation. The primary advantages of the spectroscopic methods are their ability to provide information about the mechanism(s) of aggregation and the opportunity to screen different formulations rapidly for their effect on aggregation.

The objective of this review is to provide an overview of the use of spectroscopy and calorimetry in developing stable protein formulations. This review focuses on the application of four techniques—differential scanning calorimetry (DSC), isothermal calorimetry (ITC), circular dichroism (CD), and infrared spectroscopy (IR)—that aid in understanding the factors influencing aggregation and the use of this knowledge in developing formulations that minimize aggregation of protein therapeutics.

2. AGGREGATION PATHWAYS

When protein aggregates are observed, it is important to determine whether the aggregates are reversible or irreversible, since it influences the strategy needed to minimize aggregation. Reversible aggregates are those aggregates that dissociate to yield the native molecule (within a reasonably short time period) once the stress causing aggregation is removed. In contrast, irreversible aggregates do not dissociate to yield the native molecule upon removal of the stress. In addition, aggregation can be classified as being mediated by noncovalent or covalent interactions. Noncovalent aggregation occurs when one or more native or partially unfolded protein molecules associate, usually via hydrophobic interactions, to form dimers, trimers, and eventually multimers. Proteins also can form covalent aggregates through disulfide-mediated or other covalent interactions. While dimers and smaller multimers often remain in solution, insoluble aggregates are formed when the size of the aggregate becomes too large.[3] The approaches to minimize aggregation described in this chapter are applicable to irreversible aggregates that are formed as a result of noncovalent interactions.

The well-known Lumry-Eyring (LE) model has been used to analyze the aggregation pathways of many proteins.[8] A representation of this model is shown below in Eq. (1), where N represents the native state, A is the aggregation-competent state, and A_m and A_{m+1} are aggregates.[9] The LE model for aggregation involves a reversible first step that leads to the formation of a partially unfolded (or aggregation-competent) state, followed by an irreversible second step that leads to aggregation. The partially unfolded state can retain significant secondary and tertiary structure similar to the native state but without the same compact packing that is characteristic of the native folded state.[8,10]

$$N \xleftrightarrow{k_1} A \\ A_m + A \xrightarrow{k_m} A_{m+1},\qquad(1)$$

According to this model, the kinetics of aggregation are controlled by the equilibrium between the native N and aggregation-competent A states and/or the rate of the irreversible second step. The approaches to minimizing aggregation have thus focused on (1) raising the intrinsic stability of the native state to reduce the formation of the aggregation-competent state, (2) destabilizing the aggregation-competent state, thus shifting the equilibrium to the native state, and (3) reducing the rate of the irreversible reaction that leads to the formation of aggregates.

The nature and properties of the intermediate, aggregation-competent state have been the subject of considerable interest since it represents the first step in the process that leads eventually to aggregation. It is believed that this state is characterized by an increase in surface hydrophobicity[11–14] that favors the intermolecular interactions that result in aggregation.[15] Spectroscopy and calorimetry are used mainly to determine the equilibrium between the native and intermediate states. Other analytical methods such as size exclusion chromatography (SE-HPLC), sodium dodecyl sulfate polyacrylamide gel electrophoresis (SDS-PAGE), and highly sensitive techniques such as ultracentrifugation and field-flow fractionation detect aggregates but do not provide much information on the structural transitions associated with formation of the intermediate, aggregation-competent state.

3. SPECTROSCOPIC AND CALORIMETRIC TECHNIQUES

When analyzing data from spectroscopic and calorimetric methods it generally is assumed that there are only two states, native and denatured. However, most proteins typically consist of several domains, which have similar stabilities. These domains can unfold independently or in a cooperative manner over the experimental conditions employed, which may render the two-state assumption invalid. Numerous studies involving barnase, lysozyme, staphylococcal nuclease, and other model proteins[16] have illustrated the influence of the cooperativity of the domains on the unfolding of the protein. While in most cases decisions regarding the choice of formulation conditions may not be overly influenced by the validity of the assumption, care must be taken to test this assumption.[17] Another consideration when employing these techniques is that concentration usually plays a key role in aggregation. In most cases the aggregation rate and extent of aggregation increases with increasing concentration. The notable exception to this is surface aggregation mediated by interfacial denaturation.[18] The relevance of the concentration used in spectroscopic and calorimetric studies to the concentration of interest during production or storage should be considered before applying the information from these techniques to aggregation prevention strategies.

3.1. Differential Scanning Calorimetry

Differential scanning calorimetry (DSC) measures the specific heat capacity of proteins and their associated complexes under varying conditions. There is a small but measurable change in the heat capacity of the protein upon unfolding.[19] This information is used to derive the thermodynamic parameters associated with the different states of the protein, which is subsequently correlated with structural information.[16]

The ability to measure the thermodynamic parameters directly is a tremendous insight that only calorimetry provides. A common assumption with most other methods that also provide information about the thermodynamics of protein unfolding/refolding is the two-state assumption. The two-state assumption can be tested quite readily via calorimetry. This is performed usually by comparing the effective enthalpy of a two-state process (also referred to as the van't Hoff enthalpy) with the calorimetrically measured enthalpy.[20] The van't Hoff enthalpy is calculated as described in Eq. (2), where T_t is the midpoint of the transition from the folded to the unfolded state, ΔC_p is the excess

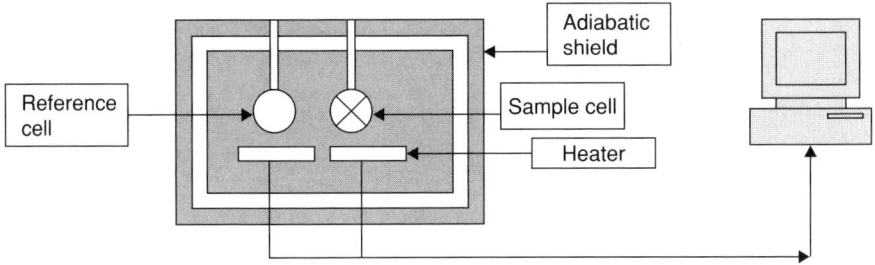

FIGURE 1. Schematic of a DSC instrument.

heat capacity, K^{eff} is the effective equilibrium constant between the native and unfolded states, R is the universal gas constant, and Q is the total heat absorbed/released.

$$\Delta H^{eff} = RT^2 \left(\frac{d \ln K^{eff}}{d \ln T} \right) = 4RT^2 \frac{\Delta C_p(T_t)}{Q}. \tag{2}$$

It is not essential that a two-state assumption be invoked in order to analyze calorimetric data for formulation selection. Recent advances have enabled analysis of calorimetric data to account for any temperature-induced transitions and the possibility that the transition involves an arbitrary number of states.[21] The ability to accurately measure the small changes to the unfolding of the protein (resulting from the cooperativity or noncooperativity of the domains to the overall unfolding) is an advantage that DSC enjoys over other methods used to calculate the energies associated with unfolding.

The basic format of the instrument involves two cells constructed with a material with excellent thermal conductivity properties (Figure 1). The protein-containing solution is loaded into the sample cell. The reference cell contains the solvent in which the protein is dissolved. Initially, both cells are loaded with solvent and the baseline for the experiment is obtained by conducting a scan over the desired temperature range. This is then followed by a scan where the solvent in the sample cell is replaced by the protein-containing solution. A third scan is run if there is a need to assess the reversibility of the transition. This is usually expressed as a percentage of the original signal (second scan) that is recovered in the third scan. The power needed to maintain a zero temperature difference between the two cells is measured and utilized to provide an estimate of the excess heat capacity of the protein at each temperature over the range studied. The typical volume for DSC measurements is 300 µl; however, models requiring even lower volumes and capable of handling samples in a multi-well format recently have been developed. The typical protein concentration used in these studies is 1 to 2 mg/ml.

The most important step in obtaining useful DSC data is accurate filling of the sample and reference cells. Care must be taken to ensure that air bubbles are not introduced. Degassing of the samples prior to filling the cells is highly recommended. Rinsing both reference and sample cells with water followed by buffer also is recommended between protein scans. The procedures for collecting and analyzing data from the DSC are well described by Privalov and Potekhin.[20]

The DSC is used in formulation development to measure the energy of protein unfolding in different solvents with the objective of selecting those that increase the resistance to unfolding.[22-25] The most commonly used parameter in these studies is T_m, the temperature where the excess heat capacity is at a maximum (the midpoint of the folded to unfolded state transition); T_m along with the temperature of the onset and the end of the transition are commonly used as indicators of the stability of the protein in a formulation. Higher values of these parameters signify greater resistance to unfolding, and hence, greater intrinsic stability of the protein. Another commonly used parameter in DSC studies is thermal reversibility, which provides a measure of the susceptibility of the protein to aggregate or precipitate when unfolded. The greater the percent reversibility, the less susceptible the protein is to aggregation upon unfolding in a given formulation. Other thermodynamic parameters such as the van't Hoff enthalpy and the calorimetric enthalpy also have been used in formulation selection as indicators of the amount of energy needed to unfold a protein. In these instances, the larger the enthalpy value, the greater the stability of the native state in a formulation compared to a formulation with a lower enthalpy value.

3.2. Isothermal Titration Calorimetry

Isothermal titration calorimetry (ITC) measures the binding interaction between the protein and a variety of ligands including antibodies,[26] metals,[27] and surfactants.[28] ITC is a method to monitor the energy of binding reactions. Chemical tags or immobilization are not required in order to monitor the reaction. Consequently, it is a direct measurement of the heat change associated with the binding. The parameters that are usually measured include the binding constant (K_{eq}), stoichiometry of the reaction (n), the binding enthalpy (ΔH), and the binding entropy (ΔS). The experiments usually involve the titration of one of the components of the binding reaction (usually the ligand) into a cell containing the second component (usually the protein) at constant temperature.

An ITC instrument contains two identical cells made of a highly efficient thermal conducting material surrounded by an adiabatic jacket[29] (Figure 2). One of the cells

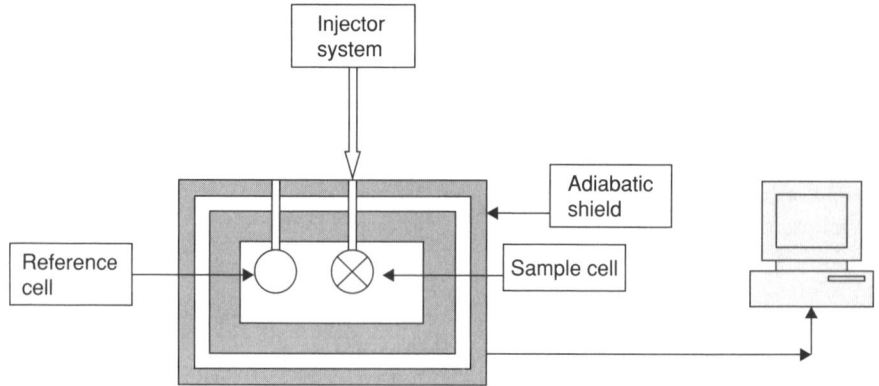

FIGURE 2. Schematic of an ITC instrument.

(sample cell) contains the protein in an appropriate buffer while the other cell (reference cell) contains the buffer. The direct observable parameter measured is the power (time-dependent) needed to maintain equal temperature in both cells. A predefined amount of the ligand is added to the cell. The interaction between the two components results in the release of energy, which is detected by the instrument. The heat absorbed or evolved during a calorimetric titration is proportional to the fraction of bound ligand. Initially, all of the ligand added is bound to the protein, so the energy release (or absorption) is high. As the ligand level in the sample cell increases (i.e., as titration continues), the energy released is lowered.

Accurate concentration estimation, sample delivery, and choice of buffer (enthalpy of ionization must be low) are the key considerations before performing an experiment. The ligand and the protein should be in the same buffer and proper mixing must be ensured. Approximately 1.5 to 2.0 ml is needed to fill a cell with a volume of 1.3 to 1.5 ml. Typically, the molar ratio of ligand to protein following the last injection is approximately 2. About 5 to 10 μl of ligand is added per injection and about 20 such injections will be needed per titration. When the binding constant between the protein and the ligand is high, the concentrations of the protein and the ligand in a titration experiment should be low. This, however, lowers the magnitude of the heat of mixing.

While there are not many published accounts of the use of ITC in formulation development, ITC can be used to determine the minimum level of excipient (which binds to the protein) that is needed to stabilize a protein,[28] the magnitude of the interactions between different components of a protein formulation,[30] and in characterization studies detailing the protein's behavior in the presence of different ligands.

3.3. Circular Dichroism

In normal absorbance spectroscopy, the absorbance of the sample is measured to estimate protein concentration, the local environment near the chromophores, and presence of particulate matter. In circular dichroism (CD), the experimentally measured parameter is the difference in absorbance, ΔA (due to structural asymmetry of the higher-order structures of a protein), of left and right circularly polarized light by the sample of interest (Figure 3). The difference in absorbance in the near-UV CD region (between

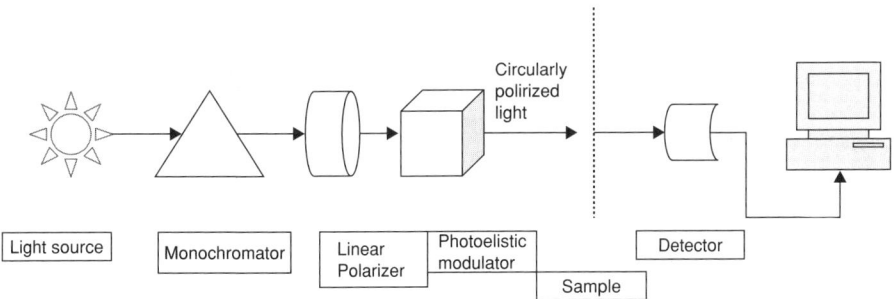

FIGURE 3. Schematic of CD instrument.

250–340 nm) reflects the tertiary and quaternary structures of the protein. The difference in absorbance in the far-UV CD region (190–240 nm) reflects the nature of the secondary structures of the protein. Since both aggregation and denaturation distort the native structure, these structural changes can be monitored using CD. The two main advantages of this technique are that it is highly sensitive to changes in conformation and that the effect of a wide variety of solvents on the protein's stability can be studied.

The Beer-Lambert relationship between absorbance and protein concentration is valid in CD measurements except that the difference in absorbance (ΔA) and extinction ($\Delta \varepsilon$) are used in the equation. The difference in absorbance of left and right circularly polarized light by the sample is measured by the spectropolarimeter (Figure 3) and is expressed as ellipticity Θ (units: deg or mdeg), which equals $32.98 \times \Delta A$. Ellipticity usually is converted to mean residue ellipticity (MRE) to account for differences in protein concentrations across the different samples. The samples are analyzed using quartz cuvettes of varying pathlength (depending on the difference in absorbance). In general, the total absorbance of the buffer and the protein is kept less than 1.0. As a result, certain salts (sodium chloride) are avoided where possible since they contribute to the absorbance. Another consideration is that oxygen absorbs strongly below about 200 nm and is capable of damaging the source lamp. Therefore, extensive purging with nitrogen usually is performed. Cuvettes with pathlengths of 0.1 to 1.0 mm are typically used for far-UV CD measurements, while 10-mm pathlength cuvettes are preferred in the near-UV region. Cells with short pathlength are used for measuring the ellipticity of proteins in the far-UV region because of the relatively high absorption of the aqueous solvent. At these small pathlengths (0.1 mm), the concentration of the protein used in the experiment is typically between 0.2 to 0.5 mg/ml. Cells with longer pathlengths are used for measuring the ellipticity of proteins in the near-UV region because of the very low signal in this region. A reference scan consisting of the buffer alone is obtained to account for possible absorption due to the buffer components followed by the scan involving the protein-containing solution.

The objective of CD experiments is to obtain an estimate of the thermodynamics of the protein unfolding/refolding process by analyzing the equilibrium between the native and unfolded states. Protein unfolding is usually initiated by increasing temperature, varying solvent conditions such as pH, or through the use of denaturants such as urea or guanidine hydrochloride. The free energy of the protein in the presence of denaturant can be obtained when unfolding is performed using a denaturant. This is usually calculated by first determining the equilibrium constant, K, using the measured ellipticity as shown in Eq. (3), where y_N and y_D are the ellipticity of the protein in its fully native and fully denatured states, respectively, and y is the measured ellipticity at a given temperature or denaturant concentration. The parameter K is used to determine the standard Gibbs free energy at a given temperature or denaturant concentration using the relation $\Delta G° = -RT \ln(K)$. The Gibbs free energy of the protein in the absence of the denaturant is obtained by extrapolating the free energy value to zero denaturant concentration. In addition, a parameter referred to as the *m* value, which represents the relationship between the Gibbs free energy and the denaturant concentration, is also obtained. The *m* value is correlated to the change in accessible surface area on unfolding. An increase in accessible surface area usually is associated with increased intermolecular interactions that cause aggregation.[15] The presence of intermediate, unfolded states will reduce the m-value.[17] Together, the m value and $\Delta G(H_2O)$ provide information regarding the conformational stability of the protein. Alternatively, one can use the onset, midpoint, and end of transition (as with temperature) to determine the best conditions for minimizing protein aggregation.

$$K = \frac{y_N - y}{y - y_D}. \qquad (3)$$

At one time, data from CD were linearly transformed to arrive at information regarding the thermodynamics of unfolding. However, nonlinear models can be quite easily used to derive the same information more accurately.[31] For urea-unfolding experiments, free energy estimations are obtained using Eq. (4), where y_F and m_F are the intercept and slope, respectively, of the pretransition baseline (representing folded or native protein), y_U and m_U are the intercept and slope, respectively, of the posttransition baseline (representing unfolded protein), and c_{urea} is the urea concentration.

$$Y = y_F + m_F c_{urea} + (y_U + m_U c_{urea}) \left(\frac{\exp\left(-\Delta G_{H_2O} - c_{urea}RT\right)}{1 + \exp\left(-\Delta G_{H_2O} - c_{urea}RT\right)} \right). \qquad (4)$$

CD has been used in formulation studies to determine the resistance to unfolding of the protein in different solvent conditions with the intent to select those conditions that increase the intrinsic stability of the protein. The resistance is measured by inducing protein unfolding as a result of temperature or by using a chemical denaturant such as guanidinium hydrochloride. Unfolding by increasing temperature provides information on the transition from the "folded" (secondary structure intact) to the "unfolded" (loss of secondary structure) state. As with DSC, the onset, midpoint, and end of the transition are used to provide information on those conditions that stabilize the protein. In contrast to DSC, it is possible to perform titration experiments using CD to evaluate the effect of increasing concentrations of a cosolvent/excipient on the structure of the protein.[23] It also has been used to measure the self-association of proteins.[32]

3.4. Infrared Spectroscopy

Infrared (IR) spectroscopy monitors the changes to the vibrational frequencies of the CO, CN, and the NH bonds in a protein, which play an important role in the formation of secondary structural interactions. The structural information is usually obtained by monitoring protein absorbance in the amide I region (1600–1700 cm^{-1}). This technique is capable of studying the secondary structure of a protein in various states including aqueous, frozen, freeze-dried, as an insoluble precipitate, and adsorbed to surfaces.[33,34] Consequently, it is the only practical technique that is capable of providing structural information regarding the protein in the lyophilized state. This information is utilized in developing freeze-dried formulations that minimize aggregation and/or other degradations.

Figure 4 illustrates a typical IR instrument setup. The infrared spectrum of a protein is obtained using a variety of accessories (transmission and reflectance). The choice of

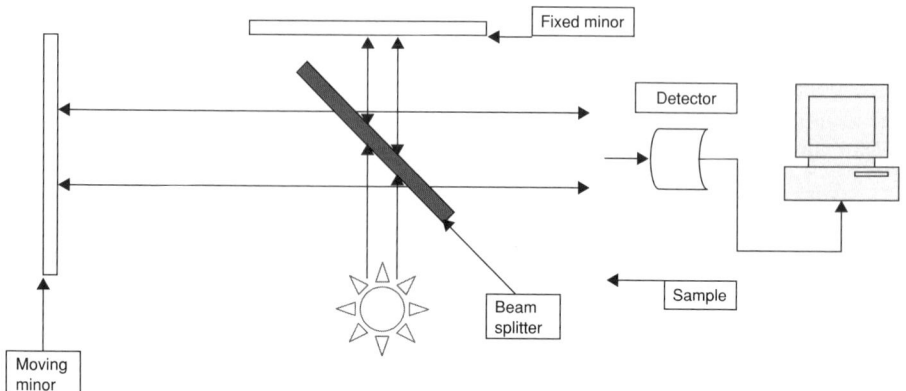

FIGURE 4. Schematic of IR instrument.

the accessory depends on the physical state of the sample and the protein concentration. Water (in the liquid and vapor states) absorbs in the same region as the amide I region.[35] The subtraction of the contributions from water is an important step prior to analyzing data. Alternatively, one might use deuterated water to reduce absorbance interference in the amide I region. This technique is often tedious especially for freeze-dried samples, since the removal of water is variable and can cause denaturation, aggregation, and recovery loss.[35] The spectral profile following the subtraction of water resembles a large broad peak owing to the broad bandwidths that each of the peaks representing the different secondary structural elements possess.[36] Techniques such as second derivative analysis and Fourier self-deconvolution are used to help resolve the features of the spectrum.[36,37]

While a quantitative estimate of the secondary structure of the protein can be obtained, sufficient information for formulation development can be obtained by a visual comparison of the spectrum of the protein in different formulations.[33] Typical studies using IR spectroscopy, especially for freeze-dried formulations, have relied on comparisons between the spectrum of the freeze-dried protein and a spectrum of the protein in the "native" state (usually the liquid sample spectrum before freeze-drying). Additionally, correlation coefficients,[38] area of overlap,[37] and, where a single band helps track changes, the increase/decrease in intensity at a fixed wavelength also are used. For instance, an increase/decrease in intermolecular β-sheet (which usually accompanies aggregation) is typically detected by an increase/decrease in absorbance at 1620 cm^{-1}.

A more recent approach involves measuring the rate of exchange of hydrogen with deuterium to estimate the change in the tertiary structure of the protein.[39] The method relies on the observation that the ND vibrational frequency is different from that of the NH bond. Thus, monitoring the increase in the ND band provides information about the accessibility of the residues. Experiments similar to those employed using DSC and CD (e.g., unfolding curves in the presence of urea or thermal treatment) can be performed to provide information regarding the stability of the protein. This approach has been used to infer the formation of aggregation-competent states.[40,41]

TABLE 1. Comparison of calorimetric and spectroscopic techniques.

	DSC	ITC	CD	IR
Cost	Medium	Medium	Medium	Medium
Ease of operation	Moderate	Easy	Easy	Moderate
Sample volume	0.3 ml	1.5–2.0 ml	0.2–1.0 ml	0.1 ml
Typical protein concentration	1–2 mg/ml	1 mg/ml	0.2 mg/ml	5 mg/ml
Physical state	Solution	Solution	Solution	Solid, solution
Parameter measured	Excess heat capacity	Heat released/absorbed upon binding	Ellipticity	Absorbance or transmittance
Parameters used in formulation development	T_m, reversibility, enthalpy and Gibbs energy of unfolding	Binding enthalpy and stoichiometry	T_m, Gibbs energy of unfolding	Area of overlap, visual comparison

A comparison of the features of the four techniques discussed above is provided in Table 1.

4. FORMULATION STRATEGIES

There are several factors to consider for selection of a suitable formulation that inhibits aggregation. The impact of these factors may be different depending on the physical state of the protein formulation (i.e., liquid, frozen, or freeze-dried). Storage temperature is a very important parameter. Proteins typically are more stable against aggregation at lower temperatures. Each protein is unique, however, as some proteins are particularly sensitive to freezing and thawing stresses or are thermally labile.

The approaches to minimizing protein aggregation can be divided into four broad categories:

1. Raising the intrinsic stability of the protein, i.e., reducing the formation of the intermediate state, which leads to aggregation.
2. Utilizing excipients that are preferentially excluded from the protein's surface in the native state. The presence of these excipients "destabilizes" the intermediate state, thus favoring the reverse reaction in step 1 of the LE model (Eq. 1), and helps maintain the protein in the native state.
3. Freezing or freeze-drying the protein, thereby reducing the rate of the irreversible reaction (step 2 in the LE model, Eq. 1).
4. Providing protection against "external" factors that may induce aggregation through formation of the aggregation-competent state. These include the use of surfactants to protect against interfacial denaturation or reducing shaking-induced aggregation by minimizing the headspace in containers or changing materials of construction.

Several examples of the use of spectroscopy and calorimetry in minimizing aggregation via these four formulation strategies are illustrated in the literature. We describe a few of these examples in detail below and also provide a table listing references to other examples of the use of these techniques for minimizing aggregation.

4.1. Intrinsic Stability

Solution pH plays an important role in formulation development, since proteins are typically only stable against aggregation within a narrow pH range.[8] Maintaining a certain ionic strength may be important if solubility is of concern. These "elementary" formulation variables seek to maximize the conformational stability and thereby retard the transformation of the protein from the native state to the aggregation-competent state. Additives such as caprylate have stabilized albumin by binding to the protein, thereby increasing its intrinsic stability and minimizing aggregation.[42]

Remmele and co-workers provide an excellent example of the use of DSC in screening for solution conditions that minimize protein aggregation by increasing intrinsic stability.[43] They evaluated pH values from 3 to 9 for recombinant human Flt3 ligand and used both T_m and thermal reversibility as parameters for assessing long-term storage stability.

They found that T_m increased as pH increased from 3 to 6; at pH 6 and above, T_m remained relatively constant near 80°C (Figure 5). Therefore, it was inferred that maximum protein stabilization was likely to be attained in the pH range from 6 to 9. To narrow the pH range further, the authors evaluated thermal reversibility of the protein in the pH range from 6 to 9. Thermal reversibility decreased from 96.6% to 15.2%, as pH increased from 6 to 9 (Figure 6), suggesting that a pH closer to 6 was more likely to afford protection against aggregation than a pH closer to 9. To verify the predictions made by DSC, the authors evaluated the stability of Flt3 ligand through an accelerated study conducted at 50°C. These studies indicated that the protein buffered at pH close to 6 was least prone to aggregation at 50°C over a one-month period (Figure 7).

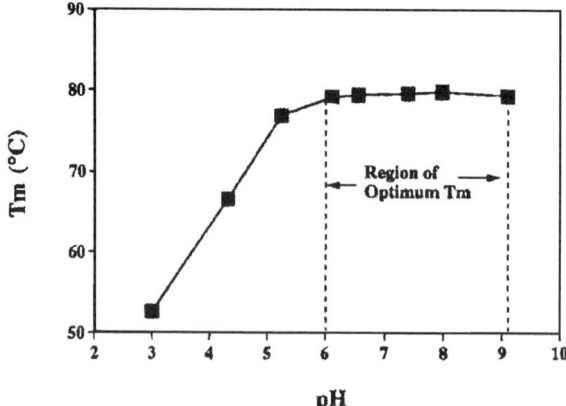

FIGURE 5. T_m as a function of pH for rhFlt3L, derived from DSC experiments. Reproduced with permission from Remmele et al.[43] Copyright (1999) American Chemical Society.

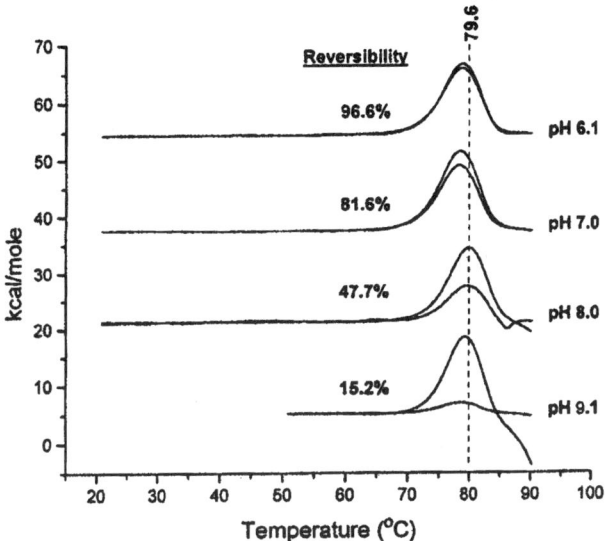

FIGURE 6. Concentration-normalized DSC scans showing the thermal reversibility behavior of rhFlt3L across the pH range of 6.1 to 9.1. Reproduced with permission from Remmele et al.[43] Copyright (1999) American Chemical Society.

This study serves as an example to illustrate the use of DSC in rapidly identifying conditions that promote the intrinsic stability of proteins. By maintaining intrinsic stability, formation of the intermediate, aggregation-competent state was reduced, preventing the formation of protein aggregates.

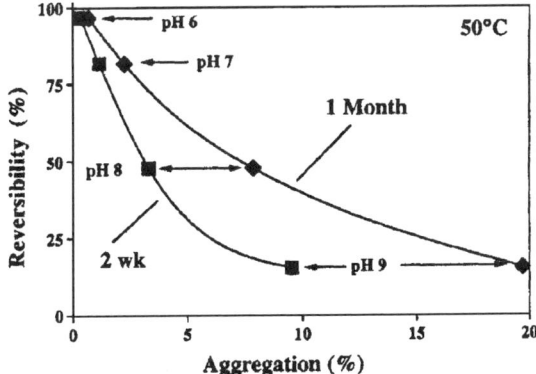

FIGURE 7. Correlation between thermal reversibility and protein stability as measured by the amount of rhFlt3L aggregation. Samples were incubated at 50°C at the indicated pH and for the indicated time. Reproduced with permission from Remmele et al.[43] Copyright (1999) American Chemical Society.

4.2. Preferential Exclusion

Additives that exhibit negative (or excluded) binding to the native state can stabilize proteins, as is seen with sugars, amino acids, and certain salts.[44,45] The negative binding of the additive to the native state results in an increase in the native state's chemical potential. The same ligand is also excluded from the denatured state. Given the larger surface area of most unfolded states, the increase in chemical potential will be even greater for this state. The result is a net increase in the free energy difference between the native and unfolded state, exhibiting itself as a net stabilization relative to denaturation.

Kim et al. describe the use of spectroscopy in studying the effect of sucrose in increasing the stability of proteins.[46] The model proteins ribonuclease A (RNase A), ribonuclease S (RNase S), and horse heart cytochrome c (Cyt C) were examined using far- and near-UV CD spectroscopy and hydrogen-deuterium exchange IR spectroscopy. Guanidine-HCl-induced unfolding curves (Figure 8) and thermal-induced unfolding

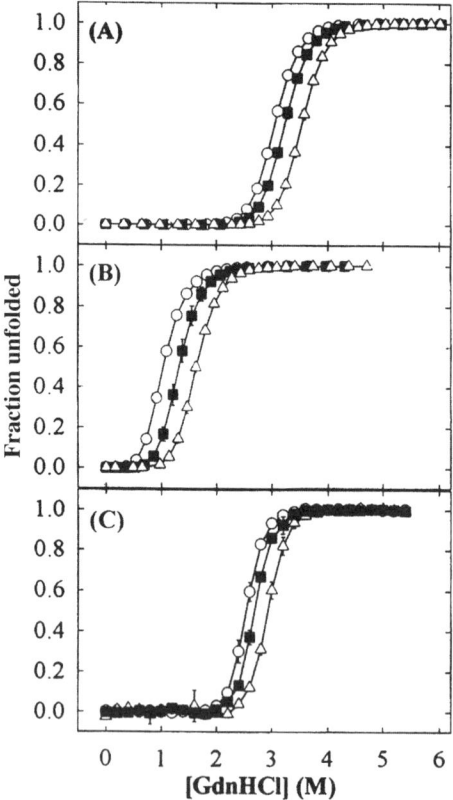

FIGURE 8. Guanidinium chloride unfolding curves for (A) RNase A, (B) RNase S, and (C) Cyt C at 0 M (open circle), 0.5 M (closed square), and 1 M (open triangle) sucrose. Data extracted from measurements taken using CD spectroscopy. Reproduced with permission from Kim et al.[46] Copyright (2003) Cold Spring Harbor Laboratory Press.

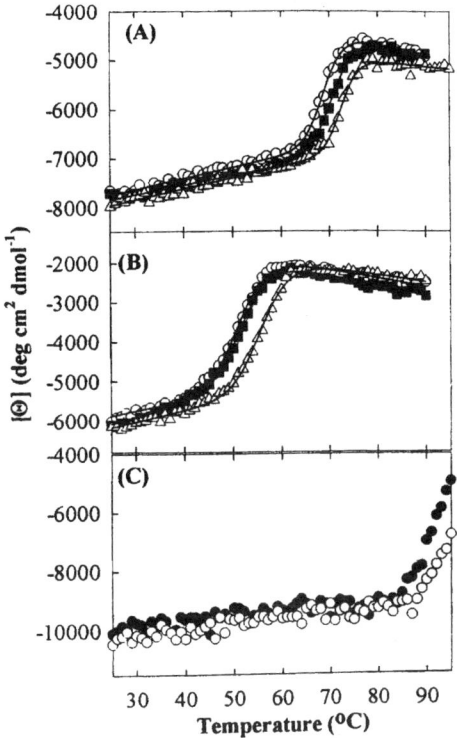

FIGURE 9. Thermally induced unfolding curves obtained by CD spectroscopy for (A) RNase A, (B) RNase S, and (C) Cyt C at 0 M (open circle), 0.5 M (closed square), and 1 M (open triangle) sucrose. Reproduced with permission from Kim et al.[46] Copyright (2003) Cold Spring Harbor Laboratory Press.

curves (Figure 9) were obtained for RNase A, RNase S, and Cyt C at sucrose concentrations of 0, 0.5, and 1.0 M using CD spectroscopy. The thermodynamic stability of the proteins increased with increasing sucrose concentration as evidenced by larger values of C_m (the GdnHCl concentration at the midpoint of the unfolding transition), T_m (the midpoint of the thermal unfolding transition region), and ΔG_u (the free energy of unfolding in water). In addition, hydrogen-exchange studies with IR spectroscopy indicated that addition of 1 M sucrose to the protein-containing solution reduced average HD exchange rates (Figure 10). This suggests that changes to the tertiary structure of the protein were decreased by the addition of sucrose. It appears to demonstrate that sucrose shifts the conformational equilibrium toward the native state of the protein, thus presumably reducing the tendency to aggregate.

This study demonstrates the use of CD spectroscopy and H-D exchange IR spectroscopy in minimizing aggregation through the preferential exclusion strategy. Sucrose, which is excluded from the protein's surface, destabilizes the intermediate (partially unfolded) state of the protein, leading to overall stabilization of the molecule. By maintaining the protein in its native state, aggregation upon storage can presumably be minimized.

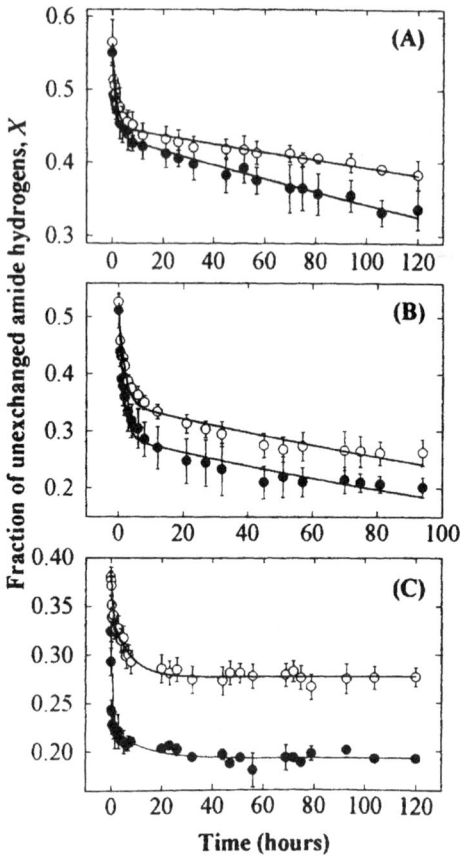

FIGURE 10. Unexchanged amide hydrogens of (A) RNase A, (B) RNase S, and (C) Cyt C in the presence (open circles) or absence (closed circles) of 1 M sucrose. Data obtained from hydrogen-exchange IR spectroscopy. Reproduced with permission from Kim et al.[46] Copyright (2003) Cold Spring Harbor Laboratory Press.

4.3. Freezing and Freeze-Drying

Freezing and freeze-drying strategies rely on the limited mobility of the protein in the frozen or freeze-dried state, which reduces the rate of the reactions involved in aggregation. The success of this approach can be judged by the fact that an overwhelming majority of the protein formulations that have been approved by the Food and Drug Administration (FDA) are freeze-dried. However, these processes expose the protein to freezing- and/or drying-induced stresses, which may induce aggregation.[47] Consequently, care must be taken to use appropriate excipients that help minimize aggregation. Infrared spectroscopy has been used extensively to assess the effectiveness of freeze-drying formulations at minimizing aggregation.

Cleland et al. used IR spectroscopy to determine the effectiveness of different formulations in minimizing structural perturbations during freeze-drying.[48] It is believed that the greater the structural changes following freeze-drying, the greater the likelihood of aggregation upon storage. The authors observed that the amount of aggregated protein decreased with increasing cryoprotectant (sucrose or trehalose) up to 100 mM following freeze-drying and upon storage at 40°C (Figure 11). Further, the extent of aggregation upon storage was least in those formulations that exhibited the least difference (by IR) between the freeze-dried and liquid samples (Figure 12). This and other studies have led to the conclusion that minimization of the difference in secondary structure between the freeze-dried and liquid states is a necessary condition for minimizing aggregation during storage.[33]

This example demonstrates the use of IR spectroscopy in identifying suitable excipient levels that reduce structural perturbations upon freeze-drying, which may lead to aggregation. This, in turn, would enable the successful implementation of the freeze-drying strategy to minimize protein aggregation upon storage.

4.4. External Factors

Proteins are surface-active molecules, i.e., they accumulate at interfaces (air-liquid, liquid-container, etc.). The tendency to adsorb is dependent on the nature of protein, especially those side chains that are on the surface.[49] Adsorbed protein may help promote aggregation in solution as a result of the structural perturbation due to adsorption. As the protein will be exposed to a number of surfaces or interfaces during handling and processing, these pathways for aggregation are critical to consider. Stabilization strategies have relied on preventing the surface of the adsorbent from extensive contact with the protein by adding surfactant,[50] coating the walls of the container with a material to provide a "noncompeting" surface,[51,52] or using different materials of construction.[53] In all of these situations, spectroscopy provides a means to determine the nature and extent of protein adsorption and/or aggregation.

Tzannis and co-workers[34] provide an example of the use of spectroscopy in understanding the cause of biological activity losses of interleukin-2 (IL-2) when exposed to a commonly used delivery device surface. They studied the bioactivity and physical stability of IL-2 upon exposure to a catheter tubing system for delivery via continuous infusion and found a significant reduction in the biological activity of delivered IL-2. SE-HPLC analysis of the delivered protein indicated that the activity losses were not a result of surface-induced aggregation. The IR and CD spectroscopies were performed on adsorbed and delivered IL-2, respectively, and showed extensive structural changes, which are the likely cause of the extensive activity losses. Figure 13 shows the far-UV CD spectra for delivered IL-2 over 24 hours of infusion and exposure to the catheter tubing. The spectra indicate a significant loss of α-helix content as a function of exposure to the catheter tubing. Tzannis et al. determined (via devolution of the CD data ~75%) decrease in α-helix content and an in increase in β-sheet over the first 2 hour of delivery. However, after increased exposure to the catheter tubing, the secondary structure of IL-2 appeared to regain partial α-helix content.

Preservatives often are added to protein formulations to retard microbial growth when multiple doses of the therapeutic need to be administered from a single container, and they represent another example of an external factor that may induce protein aggregation. In general, low levels of preservatives are capable of promoting protein aggregation.

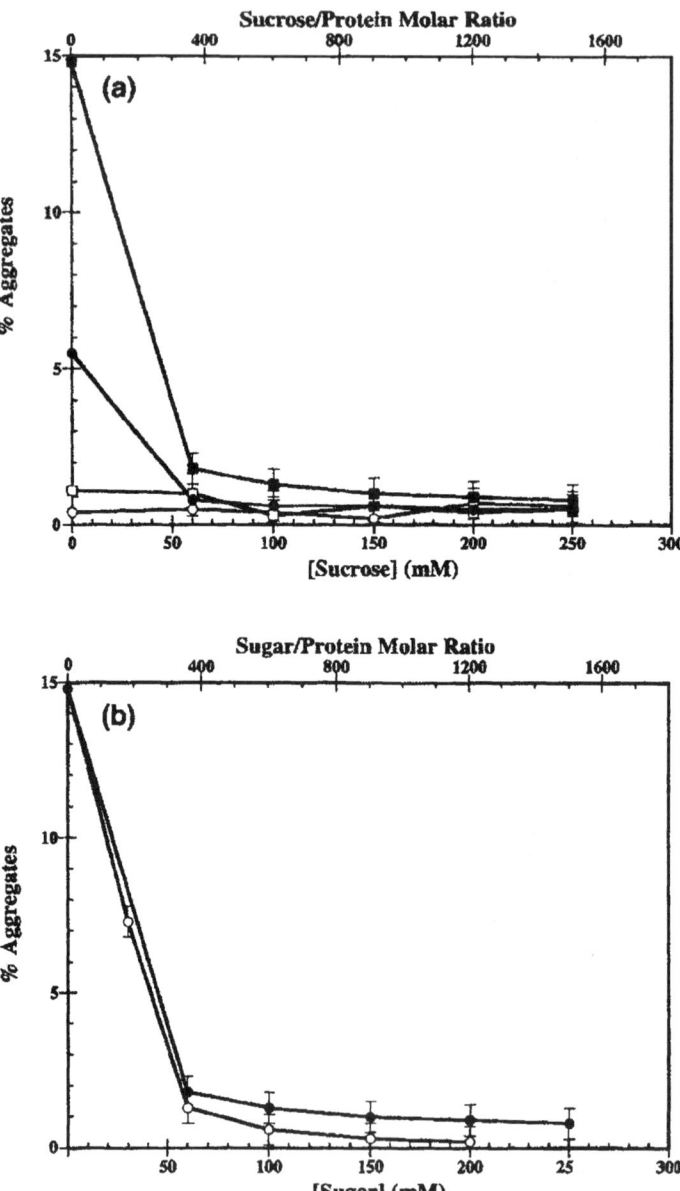

FIGURE 11. Effect of sucrose concentration on aggregation of rhu-MAb HER2. Aggregation was measured by size-exclusion chromatography. (A) Samples were analyzed prior to lyophilization (open circles), immediately after lyophilization (open squares), or after storage of lyophilized protein for 1 month (closed circles) or 3 months (closed squares) at 40°C. (B) Sucrose (closed circles) and trehalose (open circles) provided comparable protection against aggregation upon storage at 40°C for 3 months. Reproduced with permission of Wiley-Liss, Inc., a subsidiary of John Wiley & Sons, Inc. from Cleland et al.[48]

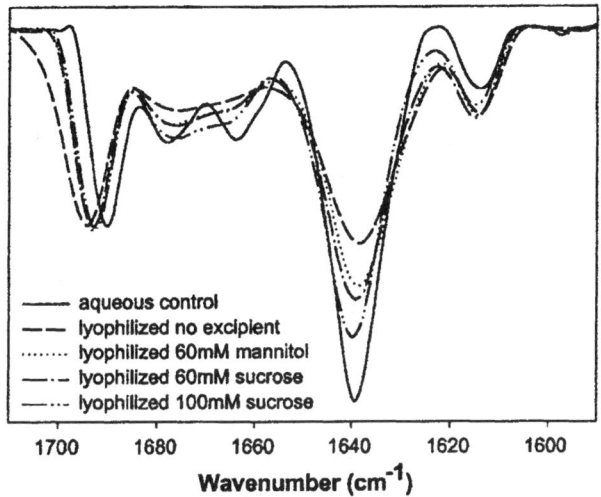

FIGURE 12. Second-derivative IR spectra. The effect of various excipients on structure of rhuMAb HER2, as visualized by second-derivative IR spectra. Reproduced with permission of Wiley-Liss, Inc., a subsidiary of John Wiley & Sons, Inc. from Cleland et al.[48]

The nature and minimum level of preservative needed to retard microbial growth without inducing protein aggregation can be determined using spectroscopy.

Lam and colleagues demonstrated the use of far- and near-UV CD spectroscopy in understanding the effect of benzyl alcohol on the tendency of recombinant human interferon-γ (rhIFN-γ) to aggregate.[54] They compared CD spectra of this protein

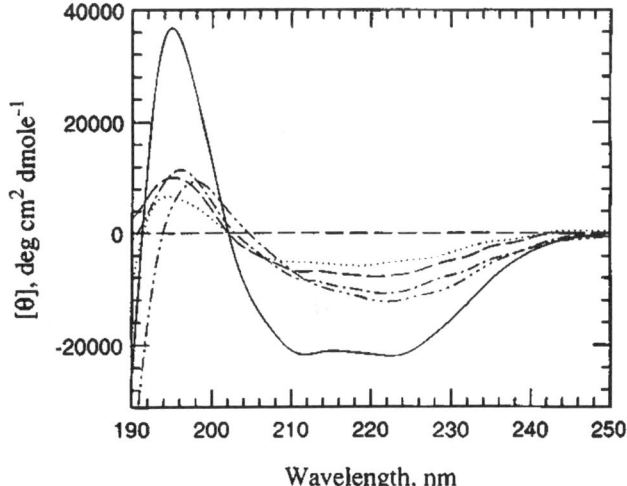

FIGURE 13. Far-UV CD spectra of delivered IL-2 as a function of exposure time to a catheter tubing delivery system. Reprinted with permission from Tzannis et al.[34] Copyright (1996) National Academy of Sciences, USA.

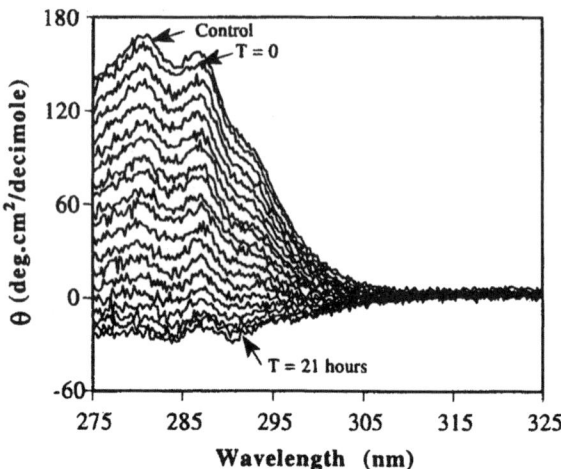

FIGURE 14. Change in near-UV CD spectra of rhIFN-g over time, in the presence of benzyl alcohol. Reproduced from Lam et al.[54]

(at 1 mg/ml) in the far- and near-UV regions in a formulation consisting of 5 mM succinate at pH 5 with 0.9% benzyl alcohol over a 21-hour period at 20°C. They found that the addition of benzyl alcohol induced changes in the tertiary structure of rhIFN-γ githout affecting the secondary structure. Figure 14 shows the near-UV CD spectra for rhIFN-γ over the 21-hour incubation period. Dynamic light scattering was used to track large molecular weight species and it was determined that the tertiary structure changes observed by CD spectroscopy correlated with the formation of large molecular weight species (Figure 15).

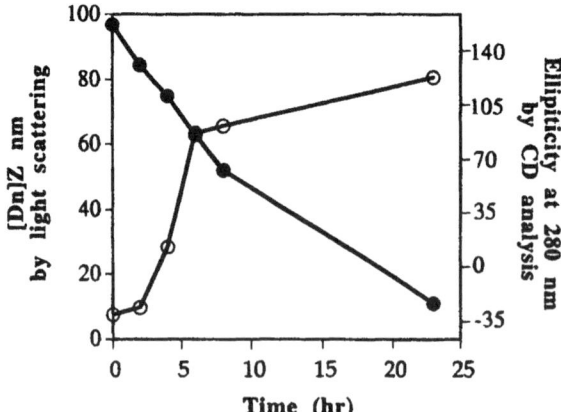

FIGURE 15. Correlation between aggregation, as determined by dynamic light scattering, and change in tertiary structure, as measured by CD analysis, for rhIFN-γ in benzyl alcohol. Reproduced from Lam et al.[54]

TABLE 2. Published reports demonstrating use of calorimetry and/or spectroscopy to develop formulation strategies that minimize aggregation.

Formulation approach	Measurement technique(s)	Reference	Forumulation approach	Measurement technique(s)	Reference
Intrinsic stability	CD	40	Freeze-drying	IR	58
Intrinsic stability	DSC	22	Freeze-drying	IR	39
Intrinsic stability	DSC, CD	55	Freeze-drying	IR	59
Intrinsic stability	DSC, CD, IR	7	Freeze-drying	IR	60
Intrinsic stability	DSC	24	Freeze-drying	IR	61
Intrinsic stability	IR	25	Freeze-drying	IR	62
Preferential exclusion	IR	39	Freeze-drying	IR	63
Preferential exclusion	CD	56	Freeze-drying	IR	64
Preferential exclusion	CD	57	External factors	DSC	22
Preferential exclusion	CD	40	External factors	CD	54
Preferential exclusion	CD, IR	41	External factors	DSC, CD	65
			External factors	IR	66
			External factors	IR	34

These studies illustrate the use of spectroscopy in understanding the solution conditions that favor the formation of the aggregation-competent, intermediate state in scheme 1 of the LE model. Preventing the formation of the intermediate state is clearly desirable in order to minimize aggregation, as has been discussed earlier. Appropriate stabilizing strategies to prevent the formation of the aggregation-competent state (such as addition of surfactants, changing materials of construction, and adding other stabilizers) can be evaluated using spectroscopy as the tool to indicate when adequate protection against denaturation is attained.

Additional examples that describe the use of calorimetry and spectroscopy in developing formulations that minimize aggregation are included in Table 2.

5. SUMMARY

Understanding the factors that influence the route of aggregation is a prerequisite for rational formulation development. In this review, we have discussed the basics of spectroscopic and calorimetric techniques and provided examples from literature describing the use of these techniques in protein formulation development. Spectroscopic and calorimetric techniques are utilized in rapid screening of formulations to assist in the selection of conditions that minimize aggregation. In addition, they provide insight into the subtle structural changes that serve as an early indicator of aggregation.

Detection and inferences about the intermediate state(s) leading to aggregation are not well supported by the analytical techniques used to measure the presence of aggregates, such as SE-HPLC and SDS-PAGE. As a result, these analytical techniques are ill equipped to provide information about the causes of aggregation. Consequently, the spectroscopic and calorimetric techniques fulfill an important need for information about the changes that take place prior to detection of aggregates. This information is critical for rational formulation development. It is only natural to expect that these methods and approaches will continue to improve and enable a better understanding of and a more rapid selection of appropriate formulation conditions.

REFERENCES

1. S. Hermeling, D. J. A. Crommelin, H. Schellekens, and W. Jiskoot, Structure-immunogenicity relationships of therapeutic proteins, *Pharm. Res.* **21**, 897–903 (2004).
2. H. Schellekens, Immunogenicity of therapeutic proteins: Clinical implications and future prospects, *Clin. Ther.* **24**, 1729–1740 (2002).
3. C. E. Glatz, Modeling of aggregation-precipitation phenomena, in: *Stability of Protein Pharmaceuticals, Part A: Chemical and Physical Pathways of Protein Degradation*, ed. T. E. Ahern and M. C. Manning (New York: Plenum Press, 1992), 135–166.
4. W. Wang and D. N. Kelner, Correlation of rFVIII inactivation with aggregation in solution, *Pharm. Res.* **20**, 693–700 (2003).
5. R. W. Carrell and D. W. Lomas, Conformational disease, *Lancet* **350**, 134–138 (1997).
6. R. Krishnamurthy and M. C. Manning, The stability factor: Importance in formulation development, *Curr. Pharm. Biotechnol.* **3**, 361–371 (2002).
7. B. S. Kendrick, J. L. Cleland, X. Lam, T. Nguyen, T. W. Randolph, M. C. Manning, and J. F. Carpenter, Aggregation of recombinant human interferon gamma: Kinetics and structural transitions, *J. Pharm. Sci.* **87**, 1069–1076 (1998).
8. E. Y. Chi, S. Krishnan, T. W. Randolph, and J. F. Carpenter, Physical stability of proteins in aqueous solution: Mechanism and driving forces in nonnative protein aggregation, *Pharm. Res.* **20**(9): 1325–1336 (2003).
9. R. Lumry and H. Eyring, Conformational changes of proteins, *J. Phys. Chem.* **58**, 110–120 (1954).
10. D. Xie and E. Freire, Molecular basis of cooperativity in protein folding. V. Thermodynamic and structural conditions for the stabilization of compact denatured states, *Proteins: Struct. Funct. Genet.* **19**, 291–301 (1994).
11. A. L. Fink, L. J. Calciano, Y. Goto, T. Kurotsu, and D. R. Palleros, Classification of acid denaturation of proteins: Intermediates and unfolded states, *Biochem.* **33**, 12504–12511 (1994).
12. O. B. Ptitsyn, V. E. Bychkova, and V. N. Uversky, Kinetic and equilibrium folding intermediates, *Phil. Trans. Roy. Soc. Lond. B Biol. Sci.* **348**, 35–41 (1995).
13. L. A. Kueltzo and Middaugh, C.R., Structural characterization of bovine granulocyte colony stimulating factor: Effect of temperature and pH, *J. Pharm. Sci.* **92**, 1793–1804 (2003).
14. J. E. Matsuura, A. E. Morris, R. R. Ketchem, E. H. Braswell, R. Klinke, W. R. Gombotz, and R. L. Remmele, Jr., Biophysical characterization of a soluble CD40 ligand (CD154) coiled-coil trimer: Evidence of a reversible acid-denatured molten globule, *Arch. Biochem. Biophy.* **392**(2), 208–218 (2001).
15. A. L. Fink, Protein aggregation: Folding aggregates, inclusion bodies, and amyloid, *Folding Des.* **3**, R9–R23 (1998).
16. G. P. Privalov and P. L. Privalov, Problems and prospects in microcalorimetry of biological macromolecules, *J. Mol. Biol.* **323**, 31–62 (2000).
17. C. N. Pace and K. L. Shaw, Linear extrapolation method of analyzing solvent denaturation curves, *Proteins: Struct. Funct. Genet.* **S4**, 1–7 (2000).
18. M. J. Treuheit, A. A. Kosky, and D. N. Brems, Inverse relationship of protein concentration and aggregation, *Pharm. Res.* **19**, 511–516 (2002).
19. P. L. Privalov, Stability of proteins: Small globular proteins, *Adv. Prot. Chem.* **33**, 167–241 (1979).

20. P. L. Privalov and S. A. Potekhin, Scanning microcalorimetry in studying temperature-induced changes in proteins, *Meth. Enzymol.* **131**, 4–51 (1986).
21. E. Freire, Thermal denaturation methods in the study of protein folding, *Meth. Enzymol.* **259**, 144–168 (1995).
22. R. L. Remmele, Jr., N. S. Nightlinger, S. Srinivasan, and W. R. Gombotz, Interleukin-1 receptor (IL-1R) liquid formulation development using differential scanning calorimetry, *Pharm. Res.* **15**(2), 200–208 (1998).
23. M. Roberge, R. N. A. H. Lewis, F. Shareck, R. Morosoli, D. Kluepfel, C. Dupont, and R. N. McElhaney, Differential scanning calorimetric, circular dichroism, and Fourier transform infrared spectroscopic characterization of the thermal unfolding of Xylanase A from *Streptomyces lividans*, *Proteins: Struct. Funct. Genet.* **50**, 341–354 (2003).
24. M. Cueto, M. J. Dorta, O. Munguia, and M. Llabres, New approach to stability assessment of protein solution formulations by differential scanning calorimetery, *Int. J. Pharm.* **252**, 159–166 (2003).
25. M. Cauchy, S. D'Aoust, B. Dawson, H. Rode, and M. A. Hefford, Thermal stability: A means to assure tertiary structure in therapeutic proteins, *Biological* **30**, 175–185 (2002).
26. K. Tsumoto, K. Ogasahara, Y. Ueda, K. Watanabi, K. Yutani, and I. Kumagai, Role of salt-bridge formation in antigen-antibody interaction. Entropic contribution to the complex between hen egg white lysozyme and its monoclonal antibody HyHEL10, *J. Biol. Chem.* **271**, 32612–32616 (1996).
27. S. L. Clugston, R. Yajima, and J. F. Honek, Investigation of metal binding and activation of *Escherichia coli* glycoxalase I: Kinetic, thermodynamic, and mutagenesis studies, *Biochem. J.* **377**, 309–316 (2004).
28. D. K. Chou, R. Krishnamurthy, T. W. Randolph, J. C. Carpenter, and M. C. Manning, Effects of Tween 20 on the stability of Albutropin during agitation, AAPS National Biotechnology meeting, Boston, 2004.
29. M. M. Pierce, C. S. Raman, and B. T. Nall, Isothermal titration calorimetry of protein-protein interactions, *Methods* **19**, 213–221 (1999).
30. S. W. Dodd, H. A. Havel, P. M. Kovach, C. Lakshminarayan, M. P. Redmon, C. M. Sargeant, G. R. Sullivan, and J. M. Beals, Reversible adsorption of soluble hexameric insulin onto the surface of insulin crystals cocrystallized with protamine: An electrostatic interaction, *Pharm. Res.* **12**, 60–68 (1995).
31. M. M. Santoro and D. W. Bolen, Unfolding free energy changes determined by the linear extrapolation method. 1. Unfolding of phenylmethanesulfonyl α-chymotrypsin using different denaturants, *Biochemistry* **27**, 8063–8068 (1988).
32. S. J. Shire, L. A. Holladay, and E. Rinderknecht, Self-association of human and porcine Relaxin as assessed by analytical ultracentrifugation and circular dichroism, *Biochemistry* **30**, 7703–7711 (1991).
33. J. F. Carpenter, S. J. Prestrelski, and A. Dong, Application of infrared spectroscopy to development of stable lyophilized protein formulations, *Eur. J. Pharm. Biopharm.* **45**, 231–238 (1998).
34. S. T. Tzannis, W. J. M. Hrushesky, P. A. Wood, and T. M. Przybycien, Irreversible inactivation of interleukin-2 in a pump-based delivery environment, *Proc. Nat. Acad. Sci. USA* **93**, 5460–5465 (1996).
35. G. Zuber, S. J. Prestrelski, and K. Benedek, Application of Fourier transform infrared spectroscopy to studies of aqueous protein solutions, *Anal. Biochem.* **207**, 150–156 (1992).
36. A. Dong, S. J. Prestrelski, S. D. Allison, and J. F. Carpenter, Infrared spectroscopic studies of lyophilization- and temperature-induced protein aggregation, *J. Pharm. Sci.* **84**(4), 415–424 (1995).
37. B. S. Kendrick, A. Dong, S. D. Allison, M. C. Manning, and J. C. Carpenter, Quantitation of the area of overlap between second-derivative amide I infrared spectra to determine structural similarity of a protein in different states, *J. Pharm. Sci.* **85**(2), 155–158 (1996).
38. S. J. Prestrelski, T. Arakawa, and J. F. Carpenter, Separation of freezing-and drying-induced denaturation of lyophilized proteins using stress-specific stabilization II. Structural studies using infrared spectroscopy, *Arch. Biochem. Biophys.* **303**(2), 465–473 (1993).

39. B. S. Kendrick, B. S. Chang, T. Arakawa, B. Peterson, T. W. Randolph, M. C. Manning, and J. C. Carpenter, Preferential exclusion of sucrose from recombinant interleukin-1 receptor antagonist: Role in restricted conformational mobility and compaction of native state, *Proc. Natl. Acad. Sci. USA* **94**, 11917–11922 (1997).
40. E. Y. Chi, S. Krishnan, B. S. Kendrick, B.S. Chang, J. F. Carpenter, and T. W. Randolph, Roles of conformational stability and colloidal stability in the aggregation of recombinant human granulocyte colony-stimulating factor, *Prot. Sci.* **12**, 903–913 (2003).
41. S. Krishnan, E. Y. Chi, J. N. Webb, B. S. Chang, D. Shan, M. Goldenberg, M. C. Manning, T. W. Randolph, and J. F. Carpenter, Aggregation of granulocyte colony-stimulating factor under physiological conditions: Characterization and thermodynamic inhibition, *Biochemistry* **41**, 6422–6431 (2002).
42. T. Arakawa and Y. Kita, Stabilizing effects of caprylate and acetyltryptophonate on heat-induced aggregation of bovine serum albumin, *Biochim. Biophys. Acta*, **1479**, 32–36 (2000).
43. R. L. Remmele, Jr., S. D. Bhat, D. H. Phan, and W. R. Gombotz, Minimization of recombinant human Flt3 ligand aggregation at the T_m plateau: A matter of thermal reversibility, *Biochemistry* **38**, 5241–5247 (1999).
44. J. C. Lee and S. N. Timasheff, The stabilization of proteins by sucrose, *J. Biol. Chem.* **256**, 7193–7201 (1981).
45. S. N. Timasheff, Control of protein stability and reactions by weakly interacting cosolvents: The simplicity of the complicated, *Adv. Prot. Chem.* **51**, 355–432 (1998).
46. Y. Kim, L. S. Jones, A. Dong, B. S. Kendrick, B. S. Chang, M. C. Manning, T. W. Randolph, and J. F. Carpenter, Effects of sucrose on conformational equilibria and fluctuations within the native-state ensemble of proteins, *Prot. Sci.* **12**, 1252–1261 (2003).
47. J. F. Carpenter, M. J. Pikal, B. S. Chang, and T. W. Randolph, Rational design of stable lyophilized protein formulations: some practical advice, *Pharm. Res.* **14**, 969–975 (1997).
48. J. L. Cleland, X. Lam, B. Kendrick, J. Yang, T. Yang, D. Overcashier, D. Brooks, C. Hsu, and J. F. Carpenter, A specific molar ratio of stabilizer to protein is required for storage stability of a lyophilized monoclonal antibody, *J. Pharm. Sci.* **90**(3), 310–321 (2000).
49. V. Hlady, J. Buijs, and H. P. Jennissen, Methods for studying protein adsorption, *Meth. Enzymol.* **309**, 402–429 (1999).
50. M. A. Carignano and I. Szleifer, Prevention of protein adsorption by flexible and rigid chain molecules, *Colloids Surf. B: Biointerfaces* **18**, 169–182 (2000).
51. F. Zhang, E. T. Kang, K. G. Neoh, P. Wang, and K. L. Tan, Surface modification of stainless steel by grafting of poly(ethylene glycol) for reduction in protein adsorption, *Biomaterials* **22**, 1541–1548 (2001).
52. M. A. Ruegsegger and R. E. Marchant, Reduced protein adsorption and platelet adhesion by controlled variation of oligomaltose surfactant polymer coatings, *J. Biomed. Mater. Res.* **56**, 159–167 (2001).
53. T. Hasegawa, Y. Iwasaki, and K. Ishihara, Preparation and performance of protein-adsorption-resistant asymmetric porous membrane composed of polysulfone/phospholipid polymer blend, *Biomaterials* **22**, 243–251 (2001).
54. X. M. Lam, T. W. Patapoff, and T. H. Nguyen, The effect of benzyl alcohol on recombinant human interferon-γ, *Pharm. Res.* **14**, 725–729 (1997).
55. B. Chen, T. Arakawa, E. Hsu, L. O. Narhi, T. J. Tressel, and S. L. Chien, Strategies to suppress aggregation of recombinant keratinocyte growth factor during liquid formulation development, *J. Pharm. Sci.* **83**(12), 1657–1661 (1994).
56. Y. Kita and T. Arakawa, Salts and glycine increase reversibility and decrease aggregation during thermal unfolding of ribonuclease-A, *BioSci. Biotechnol. Biochem.* **66**, 880–882 (2002).
57. T. Ueda, M. Nagata, and T. Imoto, Aggregation and chemical reaction in hen lysozyme caused by heating at pH 6 are depressed by osmolytes, sucrose and trehalose, *J. Biochem. (Tokyo)* **130**, 491–496 (2001).
58. B. S. Chang, R. M. Beauvais, A. Dong, and J. F. Carpenter, Physical factors affecting the storage stability of freeze-dried interleukin-1 receptor antagonist: Glass transition and protein conformation, *Arch. Biochem. Biophys.* **331**(2), 249–258 (1996).

59. S. J. Prestrelski, K. A. Pikal, and T. Arakawa, Optimization of lyophilization conditions for recombinant human interleukin-2 by dried state conformational analysis using Fourier-transform infrared spectroscopy, *Pharm. Res.* **12**(9), 1250–1259 (1995).
60. Y. Liao, M. B. Brown, A. Quader, and G. P. Martin, Protective mechanism of stabilizing excipients against dehydration in the freeze-drying of proteins, *Pharm. Res.* **19**(12), 1852–1861 (2002).
61. L. Kreilgaard, S. Frokjaer, J. M. Flink, T. W. Randolph, and J. F. Carpenter, Effect of additives on the stability of *Humicola langinosa* lipase during freeze-drying and storage in the dried solid, *J. Pharm. Sci.* **88**(3), 281–290 (1999).
62. S. D. Allison, M. C. Manning, T. W. Randolph, K. Middleton, A. Davis, and J. F. Carpenter, Optimization of storage stability of lyophilized actin using combinations of disaccharides and dextran, *J. Pharm. Sci.* **89**(2), 199–214 (2000).
63. R. L. Remmele, Jr., C. Stushnoff, and J. F. Carpenter, Real-time *in situ* monitoring of lysozyme during lyophilization using infrared spectroscopy: Dehydration stress in the presence of sucrose, *Pharm. Res.* **14**(11), 1548–1555 (1997).
64. J. D. Andya, C. C. Hsu, and S. J. Shire, Mechanisms of aggregate formation and carbohydrate excipient stabilization of lyophilized humanized monoclonal antibody formulations, *AAPS Pharm. Sci.* **5**(2), 1–11 (2003).
65. J. Fransson, D. Hallen, and E. Florin-Robertsson, Solvent effects on the solubility and physical stability of human insulin-like growth factor I, *Pharm. Res.* **14**, 606–612 (1997).
66. S. D. Webb, J. L. Cleland, J. F. Carpenter, and T. W. Randolph, Effect of annealing lyophilized and spray-lyophilized formulations of recombinant human interferon-gamma, *J. Pharm. Sci.* **92**, 715–729 (2003).

Small-Angle Neutron Scattering as a Probe for Protein Aggregation at Many Length Scales

Susan Krueger,[1,4] Derek Ho,[1,2] and Amos Tsai[3]

Small-angle neutron scattering (SANS) is uniquely suited to the study of biological systems in solution in the size range from 10 Å to 1000 Å. In the case of protein aggregates, SANS is sensitive to structures at all of these length scales, from the total aggregate size down to the monomer size. Thus, it is possible to observe scattering over a wide range of length scales and to follow the change in morphology of a protein system as it assembles into aggregates. In addition, neutrons are especially sensitive to the light elements, such as H, C, N, and O, which are of importance in biological systems. Finally, neutrons are highly penetrating yet nondestructive, so proteins can be studied in solution under conditions that closely mimic their native environment. The radius of gyration and molecular weight of proteins and protein aggregates in solution can be extracted directly from SANS data. With more detailed data analysis involving comparison to model structures, information about overall particle shape and particle size distribution, as well as interparticle interactions, can be obtained. Examples of SANS experiments that have contributed to a better understanding of the mechanism of different types of protein aggregation in solution are presented.

1. INTRODUCTION

Since neutrons are sensitive to the positions of the light elements such as H, C, N, and O, which are of central importance to all biological systems, small-angle neutron

---

[1] NIST Center for Neutron Research, NIST, Gaithersburg, MD 20899
[2] Department of Materials Science, University of Maryland, College Park, MD 20742
[3] Human Genome Sciences, Inc., 14200 Shady Grove Road, Rockville, MD 20850
[4] To whom correspondence should be addressed: Susan Krueger, NIST Center for Neutron Research, Gaithersburg, MD 20899

scattering (SANS) can provide unique information on the structure and function of biological macromolecules. Recent advances in biochemistry, crystallography, and structural NMR have made it possible to prepare greater quantities of deuterium-labeled proteins and to determine an ever-increasing number of high-resolution structures. Thus, SANS also has come into wider use as a complementary tool for comparing the structures in crystal and solution phases and for elucidating the unresolved regions in a crystal structure. Since the measurements are performed in solution, SANS gives unique structural information under conditions that more closely mimic the molecule's natural environment, and thus can provide critical insights in a number of bioengineering areas.

Particularly powerful is the contrast variation technique, in which the isotopic substitution of D for H is routinely used to change the scattering from a macromolecule without affecting its biochemistry. In protein/nucleic acid or protein/lipid complexes, the two components have naturally different neutron scattering length densities. Thus, by measuring the complex in solvents with different H_2O/D_2O ratios, the scattering from one component can be separated from that of the other, providing unique structural information about each component individually, as it is interacting with others in the complex. Similar information can be obtained from multisubunit proteins if one or more subunits can be selectively deuterated, by substituting D for H during synthesis. The technique is extremely effective for the study of structural changes upon binding of nucleotides, lipids, peptides, or cofactors, since the conformations of the macromolecule bound in the complex and free in solution can be compared. This could lead to a better understanding of the function of the molecule in the complex.

It is widely recognized that proteins possess many conformations and structures through their intra- and intermolecule interactions. Comprehension of the interactions occurring in protein solutions is fundamental to understanding protein functions in nature and in all practical processes involving proteins, and the ability to decipher the relationship between protein structure, function, and stability continues to be an important research field in the biological sciences. It happens that SANS also is very sensitive to aggregation effects, making it useful for the studies of aggregating or self-assembling biological systems. In a typical SANS experiment, aggregation is undesirable since the goal is to obtain precise information about the size and shape of a molecule. However, in some cases, the mechanism of aggregation is important for the understanding of the function of the system. It is in these cases that SANS can be used to study the aggregation process itself, and thus, to shed some light on the mechanism of aggregation or self-assembly. It is precisely this sensitivity to aggregation that makes SANS suitable to the study of aggregating protein systems. If the aggregation process is slow enough, intermediate states of aggregation can be studied as well as the initial and final states.

In order to observe small-angle scattering, there must be scattering contrast between the proteins and the surrounding solution. The scattering is proportional to the scattering contrast, $\Delta\rho$, *squared* where

$$\Delta\rho = \rho_p - \rho_s, \tag{1}$$

and ρ_p and ρ_s are the scattering length densities (SLD) of the protein and the solvent, respectively. The SLD is defined as

$$\rho = \frac{\sum_{i=1}^{n} b_i}{V}, \qquad (2)$$

where V is the volume containing n atoms, and b_i is the (bound coherent) scattering length of the ith atom in the volume V. The SLDs and volumes are known for each of the 23 amino acids that can make up a protein. Thus, for proteins, the total SLD is often calculated in terms of the SLDs of the individual amino acids, rather than on an atom-by-atom basis.[1]

Generally, static light scattering and small- angle X-ray scattering (SAXS) provide the same information about a sample as does SANS, i.e., a measurement of the macroscopic scattering cross-section $d\Sigma/d\Omega$. The contrast in light scattering arises from the difference in the light's refractive index for each phase in the sample. The contrast in light scattering is typically much stronger than in SANS, requiring very dilute concentration of particles to avoid multiple light scattering. In addition, the wavelength of light puts a lower limit on the size of the particle that can be measured. Thus, light scattering can be used to estimate the total size or aggregation state of a protein in dilute solution, but it cannot necessarily resolve the monomer structure. The contrast in X-ray scattering arises in the variation in electron density between the phases. The contrast is again stronger for X-rays than neutrons, but thinner samples often mitigate any multiple scattering. X-rays are strongly absorbed by most samples, requiring thin-walled glass capillaries to contain the sample. Also, intense X-rays beams can cause irreversible sample damage, altering the structure and chemistry of the studied solution. This is especially the case for organic compounds such as protein.

Table 1 shows the SLDs for a typical protein in H_2O and in D_2O solvent, as well as the SLDs for H_2O and D_2O. The neutron SLDs also take into account the exchange of labile H atoms with D atoms when the protein is in a solution containing deuterium. Since H and D have the same X-ray slds, H-D exchange has no effect on the protein SLD in that case.

Using the information from Table 1, it is possible to plot the SLD for a typical protein versus. the percentage of D_2O in the solvent. The contrast between the protein and solvent now can be shown visually in Figure 1.

It is clear from the figure that the protein contrast is zero when the solvent contains 40% D_2O (and 60% H_2O). Thus, scattering is not observed from proteins under these

TABLE 1. Neutron SLDs for proteins in H_2O and D_2O

	Mass density (g/cc)	Neutron SLD (cm^{-2}) in H_2O	Neutron SLD (cm^{-2}) in D_2O	X-ray SLD (cm^{-2})
Typical protein	~1.4	1.8×10^{10}	3.0×10^{10}	12.3×10^{10}
H_2O	1.0	-0.56×10^{10}	—	9.4×10^{10}
D_2O	1.1	—	6.4×10^{10}	9.4×10^{10}

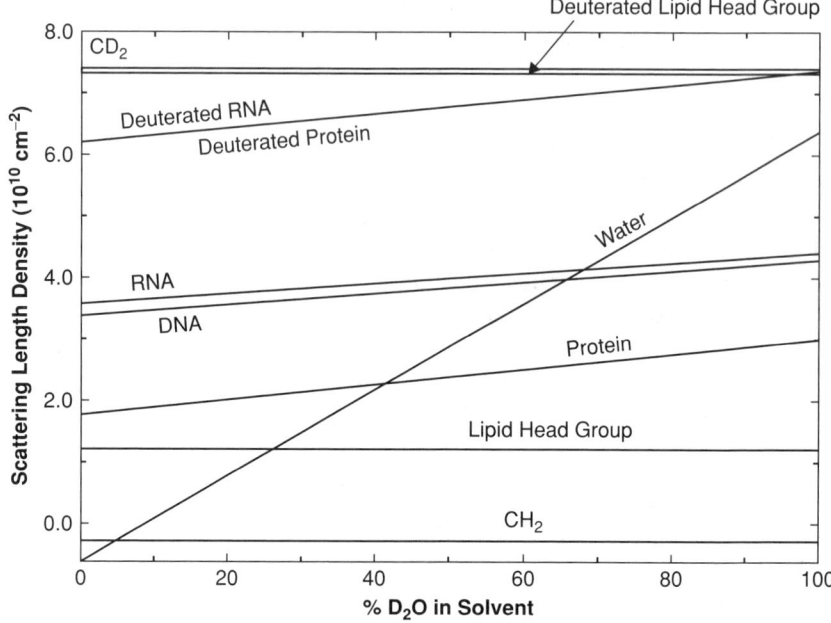

FIGURE 1. Plot of SLD vs. % D_2O for molecules of biological interest.

conditions. However, it is also clear from Fig. 1 that other molecules of biological interest, such as DNA, RNA, deuterated protein, lipid head groups, and hydrocarbon chains in fact, do, have significant contrast in 40% D_2O solvent. Thus, scattering would be observed from those components under the same conditions where it is not observed from proteins. This forms the basis of the contrast variation technique, whereby a complex consisting of two components is measured in solvents with differing D_2O/H_2O ratios in order to suppress the scattering from one component while highlighting that of the other. While this technique will not be discussed in relation to protein aggregation, it is an important application of SANS from biological systems.

This chapter will examine the study of protein aggregation by SANS. It will cover the basic SANS theory and instrumentation that are used to conduct such studies, as well as basic data analysis techniques. Specific examples from the literature will be presented, along with data analysis and structure modeling specific to each case. The examples cover the formation of amyloid fibers by amyloid-β protein and lysozyme, the aggregation of lysozyme in the presence and absence of glycerol and the self-assembly of tubulin monomers into polymers.

2. EXPERIMENTAL METHODS

2.1. Instrument

SANS instrument designs vary depending on the type of neutron source, which is either pulsed, from an accelerator, or steady-state, from a reactor. The description here

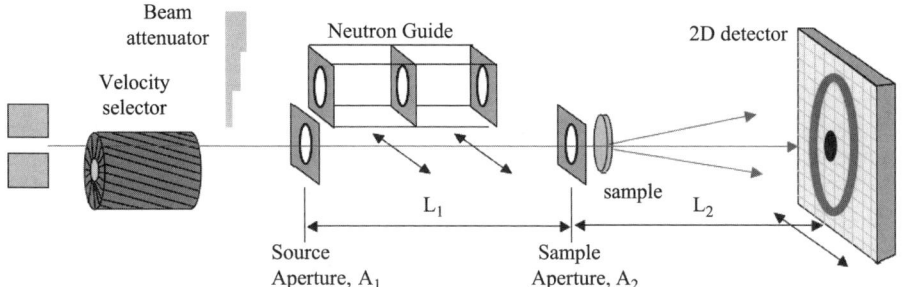

FIGURE 2. Schematic of a 30-m SANS instrument such as the one at the NIST Center for Neutron Research.

is for a SANS instrument on a steady-state source. Neutrons from a reactor source are usually filtered to remove fast neutrons and γ radiation coming from the reactor core. These filtered neutrons are mechanically monochromated by passing through a velocity selector, typically made up of several rotating slotted disks of neutron-absorbing material. The speed of rotation of the velocity selector determines the mean wavelength of the neutrons. Often, the wavelength spread of the beam can be controlled by tilting the velocity selector rotor axis with respect to the neutron beam.

The monochromated beam is collimated by circular apertures inside a presample flight path, which is normally under vacuum. Following the collimation apertures is a sample area where samples to be measured are mounted. Normally, this is a versatile area that supports not only small samples in multisample holders but also large apparatus such as furnaces and displexes. After the sample is the postsample flight path, which is usually under vacuum. At the end of the path is the two-dimensional, position-sensitive neutron detector. Normally, the distance from the sample to the detector can be varied to suit the type of sample being measured. Figure 2 shows a schematic of the 30-m SANS instrument at the NIST Center for Neutron Research,[2] which is an example of a SANS instrument at a reactor source.

2.2. Sample Preparation

Proteins are typically measured in a buffered solution at an appropriate pH. The buffer can be H_2O or D_2O based, or it may contain a mixture of H_2O and D_2O, depending on the goals of the experiment. Protein concentrations are typically between 1 and 5 mg/ml and the sample volume is between 300 and 600 μl, depending on the experimental conditions. Data collection times typically range from 0.5 to 6 hours per sample, depending on the protein size, concentration, and instrument configuration. A typical biology SANS experiment requires 2–4 days of SANS beam time, which is long compared to typical experiments on nonbiological materials.

Raw data obtained from protein samples in solution must be corrected for the scattering from the solvent and sample holder, stray neutrons not scattered from the sample, and electronic noise. The data also are corrected for nonuniformities in the neutron detector and then put on an absolute scale by normalizing to the absolute beam flux at the sample position. Then, the data are ready to be analyzed in order to model the structure of the protein aggregates.

2.3. Data Analysis

2.3.1. Dilute solution The scattering from dilute solutions is expressed in terms of randomly oriented, non-interacting particles. The particles scatter independently and the total scattered intensity is the sum of the scattered intensity from each particle. A number of references describe small-angle scattering theory from dilute solutions in detail.[1,3–5] The measured intensity, corrected for background and put on an absolute scale, can be expressed as

$$I(q) = \frac{d\Sigma(Q)}{d\Omega} = (\Delta\rho)^2 \, V_p^2 \, N_p P(q), \qquad (3)$$

where $\Delta\rho$ is the contrast between the particles and the solvent; V_p is the mean particle volume, N_p is the number of particles per unit volume, and $q = (4\pi/\lambda)\sin(\theta/2)$, where λ is the neutron wavelength and θ is the scattering angle. The $P(q)$ is the scattering form factor for the particles,

$$P(q) = \left| \frac{1}{V_p} \int_{V_p} e^{i\vec{q}\cdot\vec{r}} d\vec{r} \right|^2, \qquad (4)$$

the square of the Fourier transform of the particle shape.

The scattering from dilute protein solutions, at small enough q-values, can be approximated by a simple exponential decay given by

$$I(q) \approx I(0) \exp\left(-\frac{1}{3} q^2 R_g^2\right), \qquad (5)$$

where R_g is the radius of gyration of the protein and $I(0)$ is the scattered intensity at $q=0$. Thus, a plot of $\ln[I(q)]$ versus q^2 will be linear and the slope will be negative and proportional to R_g. This type of plot is called a Guinier plot and it works for any shaped object. It is valid in a q-range of approximately

$$q R_g \leq 1. \qquad (6)$$

If the data are on an absolute scale, $I(0)$ is given by

$$I(0) = (\Delta\rho)^2 \, V_p^2 \, N_p \qquad (7)$$

from Eq. (3), since $P(0) = 1.0$ by definition. Since V_p is proportional to the molecular weight (M_w) of the protein and N_p is proportional to M_w^{-1}, it can be seen from Eq. (7) that

$$I(0) \propto M_w. \qquad (8)$$

Figure 3 shows plots of simulated scattering data from an ellipsoidal-shaped protein monomer and dimer. Note that the scattered intensity at $q=0$ from the dimer is twice that of the monomer.

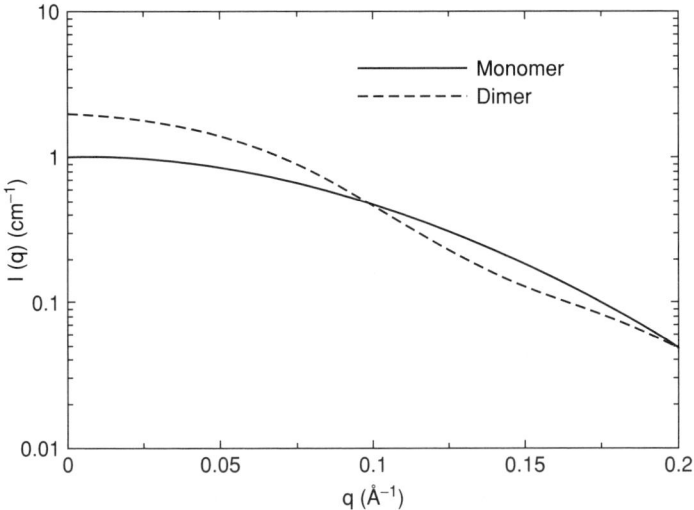

FIGURE 3. Simulated plots of scattering data from a protein monomer and dimer. The scattered intensity at $q = 0$ is proportional to the molecular weight of the molecule.

2.3.2. Large particles If protein aggregates form structures that resemble long rods or thin disks, these are special cases in that the scattering comes mainly from just one or two of the dimensions in the particle. For long, rodlike aggregates in which the length of the rod, L, is much greater than the cross-sectional radius, r_c, the rod makes a contribution to the scattering only when it lies nearly perpendicular to the direction of the vector, **q**. Thus, the axial and cross-sectional factors can be regarded as nearly independent. The scattered intensity can then be written as

$$I(q) = L \frac{\pi}{q} \cdot I_c(q), \tag{9a}$$

where $I_c(q)$ is the contribution from the cross-sectional dimension and $L\pi/q$ is the contribution from the axial dimension. For, $I_c(q)$, the **q** vector lies in the plane of the cross-section so that the problem is now two-dimensional. Similar to the three-dimensional case, $I_c(q)$, can be approximated, at small q values, by

$$I_c(q) \approx I_c(0) \exp\left(-\frac{1}{2} q^2 R_c^2\right), \tag{9b}$$

where R_c is the radius of gyration of the cross section. Thus, a plot of $\ln[I(q) \bullet q]$ versus. q^2 will be linear and the slope will be negative and proportional to R_c. Similarly, for disklike particles, where the thickness, t, is much smaller than the other two dimensions of area, A, the thickness contribution to the scattering can be separated from the planar contribution. Thus, the scattered intensity can be written as

$$I(q) = A \frac{2\pi}{q^2} I_t(q), \tag{10a}$$

where $I_t(q)$ is the contribution from thickness dimension and $2\pi A/q^2$ is the contribution from the planar dimension of area, A. Then, $I_t(q)$, the scattered intensity of the thickness, can be approximated, at small q values, by

$$I_t(q) \approx I_t(0) \exp\left(-q^2 R_t^2\right), \tag{10b}$$

where R_t is the radius of gyration of the thickness. In this case, a plot of $\ln[I(q) \bullet q^2]$ vs. q^2 will be linear and the slope will be negative and proportional to R_t.

The Guinier approximation for globular, rodlike, or disklike particles is useful for predicting the size of protein aggregates. However, its use is limited to the small q region of the data. To analyze the data at q values outside of the Guinier region, the data must be fit to model shapes. For simple shapes, an equation can usually be written for $P(q)$. For instance, $P(q)$ for an ellipsoid with semiaxes, a and b, can be written as

$$P(q) = \int_0^{\pi/2} \Phi^2 \left(qa\sqrt{\cos^2\theta + v^2 \sin^2\theta}\right) \cos\theta d\theta, \tag{11}$$

where θ is the angle between q and the semimajor axis b of the ellipsoid, $v = b/a$ and

$$\Phi^2(x) = 9\left(\frac{\sin x - x \cos x}{x^3}\right)^2. \tag{12}$$

If $a > b$, the object is an oblate ellipsoid (disklike), while the object is a prolate ellipsoid (rodlike) if $a < b$. If the two radii are equal, then the ellipsoid is a sphere. Thus, if $a = b = r$, $v = 1$ and the equation reduces to the form factor for spheres of radius r. Once $P(q)$ is obtained, $I(q)$ is found from Eq. (7).

For more complicated particle shapes, models can be built in real space from a combination of simple shapes and then randomly filled with points using a Monte Carlo method.[6,7] The distance distribution function, $P(r)$, is then calculated by summing all possible pairs of points in the particle to create a histogram-like function, $P(r)$ versus r. By definition, $P(r) = 0$ at $r = 0$, and at $r = D_{max}$, the maximum distance in the particle. Once $P(r)$ has been calculated, the scattered intensity, $I(q)$, can be obtained by a Fourier transform of $P(r)$, since

$$I(q) = 4\pi V_p \int_0^{D_{max}} P(r) \frac{\sin(qr)}{qr} dr. \tag{13}$$

This method also can be used for simple shapes such as spheres, cylinders, and ellipsoids and gives essentially the same results for $I(q)$ as the "form factor method" described above.

2.3.3. Polydispersity Often, aggregated protein systems are polydisperse in size, shape, or both. Such systems can be studied by SANS. However, the interpretation of the data is more difficult, as both the particle shape and the size distribution must be included in the model.

2.3.4. Concentrated solutions As noted above, scattering from dilute solutions is expressed in terms of randomly oriented, noninteracting particles. However, if there are interactions between the particles, then the solution is no longer dilute. In this case, the scattering is given not only by the form factor, $P(q)$, but also by the interparticle structure factor, $S(q)$. In this case, the scattered intensity is written as

$$I(q) = (\Delta\rho)^2 V_p^2 N_p P(q) S(q). \tag{14}$$

The form factor describes the shape of the particle, while the structure factor describes the interaction between particles. Depending on the nature of these interactions, $S(q)$ can describe an attractive or repulsive interaction potential. The $S(q)$ may describe only the effects of the close packing of molecules or it may describe additional electrostatic interactions between molecules. Since proteins do have surface charges under various solvent conditions, such electrostatic interactions often must be taken into account in $S(q)$. There are two main types of potentials that are used to model $S(q)$ from biological systems in solution. The first is the Percus-Yevick hard sphere model[8] and the second is the mean spherical approximation (MSA) model.[9] The hard sphere model is based only on the packing interactions between molecules. On the other hand, the MSA model takes particle surface charge and ionic strength of the solution into account as well. This potential can be modeled as either attractive or repulsive. In a polyelectrolyte solution where the charge on the molecules are not screened (i.e., when there is little or no salt in the solution), evidence of interparticle interactions can be seen even at concentrations that would normally be considered to be in the dilute regime. There are other models for $S(q)$ for specific cases, as will be shown in the example on lysozyme aggregation below.

3. EXAMPLES FROM THE LITERATURE

3.1. Amyloid β-Protein Fibrils

Yong *et al.* used SANS to determine the structure of these Aβ micelles.[10] Increasing evidence supports the hypothesis that amyloid β-protein (Aβ) assembly is a key pathogenic feature of Alzheimer's disease. Thus, understanding the assembly process offers opportunities for the development of strategies for treating this devastating disease. In prior studies, Aβ was found to form micellelike aggregates under acidic conditions. These structures exhibited an average observed hydrodynamic radius of 70 Å. They were found to be in rapid equilibrium with Aβ monomers or low molecular weight oligomers and were centers of fibril nucleation.

The SANS data in this study were fit using a spherocylindrical model (cylinder of length L with hemispherical end caps of radius R) and MATLAB* (Mathworks, Natick, MA) software was used to numerically calculate values of spherocylinders' structure factor within the data-fitting procedure. They reveal that the micellar assemblies comprise 30–50 Aβ monomers and have elongated geometries. The best fit of the data to a uniform spherocylinder yields a radius of $R \sim 24$ Å and cylinder length of $L \sim 110$ Å. These structure parameters remain constant over more than a decade in concentration range. The concentration independence of the length of the cylindrical aggregate indicates the presence of an internal nonrepetitive structure that spans the entire length of the Aβ assembly.

Small variations of the parameters R and L are postulated to be due to the presence of aggregates or fibrils and not to the actual changes in micelle dimensions. In this case, the normalized background-corrected neutron scattering intensity $I(q)$ is the same and only the magnitudes I(0) vary in proportion to the micelle concentration. Indeed, on a logarithmic scale, data curves representing scattering at different concentrations are parallel to each other, meaning that the shapes of the curves are the same.[10] Thus, these Aβ aggregates are not the same as elongated, wormlike micelles that increase in length with concentration.[11,12]

The fact that the Aβ micellelike intermediates in these experiments are elongated objects that do not change size with Aβ concentration has important consequences. There is no increase in the size of the Aβ micelles over greater than a decade of concentration variation, yet at the same time there is compelling evidence indicating that Aβ micelles are in equilibrium with monomers and/or oligomers. The conclusion is that these micellar structures differ significantly from classic cylindrical micelles in that they are not built from repetitive units. One potential model of the structure of the spherocylindrical Aβ micelle is an elongated cylindrical object in which Aβ monomers are stretched along the cylinder axis.

3.2. Lysozyme Amyloid Fibrils

Proteins that are not related to disease also have been shown to form amyloid fibrils.[13] This fibril formation usually occurs under partially denaturing conditions such as low pH,[14] increased temperature,[15,16] or in the presence of alcohols.[17,18] This example exploits the formation of amyloid fibrils by hen egg white lysozyme (HEWL) to try to gain some insights into the mechanism of amyloid fibril formation.[19] Since HEWL forms amyloid fibrils in highly concentrated ethanol solutions, solutions of various concentrations of HEWL in various concentrations of ethanol were prepared and the structures of HEWL in these solutions were investigated by small-angle X-ray scattering and SANS. It was shown that the structural states of HEWL were distinguished as the monomer state, the state of the dimer formation, the state of the protofilament formation, the protofilament state, and the state toward the formation of amyloid fibrils. A phase diagram of these structural states was obtained as a function of protein, water, and ethanol concentrations. Changes in the structural state proceed as the ethanol concentration increases as well as the protein concentration increases.

It was found that under the monomer state the structural changes of HEWL were not gross changes in shape but local conformational changes, and the dimers formed by the association at the end of the long axis of HEWL had an elongated shape. Circular dichroism measurements showed that the large changes in the secondary structures HEWL occurred during dimer formation. The protofilaments were formed by stacking the dimers with their long axis (nearly) perpendicular to and rotated around the protofilament axis to form a helical structure. These protofilaments were characterized by their radius of gyration of the cross section of 24 Å and the mass per unit length of 1600 (±230) Da/Å. It was shown that the changes of the structural states toward the amyloid fibril formation occurred via lateral association of the protofilaments. A pathway of the amyloid fibril formation of HEWL was proposed from these results.

The proposed pathway for amyloid fibril formation is composed of three stages: the formation of the dimers, the formation of the protofilaments, and the formation of the amyloid fibrils. The large structural conversion of the proteins during dimer formation has been pointed out to be a key event in the formation of the amyloid fibrils.[20] This key event lies in the formation of the dimers that are the nuclei for the formation of the protofilaments. The formation of the protofilament is likely to occur via a nucleation-dependent polymerization. The recently proposed "nucleated conformational conversion" (NCC) model[21] for the amyloid fibril formation of the yeast prion protein is consistent with the HEWL data. The formation of the amyloid fibrils of HEWL occurs through the lateral association of the protofilaments. This is consistent with the observations reported for the amyloid fibrils as the bundles of several protofilaments.[17,22] Thus, the proposed pathway of the amyloid formation has features in common with the mechanisms proposed for many different proteins.

3.3. Other Lysozyme Aggregates

Recently, Velev et al.[23] have shown the feasibility of characterizing lysozyme interactions in solutions using SANS. Although studies on protein dynamics affected by a glassy solvent have been intensively performed, relatively few attempts have been made to explore the glassy solvent effects on the structure and the conformation of protein molecules. Thus, SANS experiments now have been performed by two of us (DH and AT) to study the glassy solvent effects on the intermolecule structure and the molecular conformation of proteins via performing SANS measurements on solutions of lysozyme in D_2O with and without glycerol. It was confirmed that the molecules were in an ellipsoidal shape and found that the molecules were packed in bundles. The shape and the size of individual lysozyme molecules remained unchanged under the experimental conditions, while both the long- and short-range structures varied as a function of lysozyme concentration and with the presence of glycerol. More lysozyme molecules were packed closer within a bundle and the correlation length between bundles was larger with glycerol through a decrease in the intermolecule d-spacing and an increase in the averaged number of molecules per bundle as well as an increase in the long-range correlation length between bundles observed, respectively, when glycerol was present. The SANS data show that the samples studied in this work were temperature independent and indicate that even the most dilute sample at 1 mg/ml was not a measurement of the single-particle form factor of noninteracting individual lysozyme molecules, and thus, the intermolecule interaction may be attractive.

Lysozyme from chicken egg white (Sigma*, St. Louis, MO) was used without further purification. The neutron-scattering cross section of hydrogen is ∼10 times greater than that of deuterium. In order to maximize the scattering from the protein relative to the other components in the samples studied in this work, trying to minimize extraneous sources of hydrogen, including the exchangeable protons on the protein itself, would be important and necessary. One gram of lysozyme was dissolved in 10 gm of D_2O and let stand at 4°C overnight in order to D-exchange the exchangeable protons. Afterward, the D-exchanged protein was freeze-dried into a powder. To prepare for the SANS measurements, the freeze-dried powder was added to either pure D_2O or D_2O + deuterated glycerol (at 25% by weight) at the required concentrations of 1, 5, and 50 mg/ml. The solutions were then equilibrated at 4 and 30°C overnight before scattering data were collected. No buffer was used and lysozyme was able to maintain the pD value at 5.5 (meter reading).

SANS experiments over the q range from 0.004 Å^{-1} to 0.331 Å^{-1} were carried out using the 30-m SANS instruments at the NIST Center for Neutron Research.[2] The incident neutron wavelength was $\lambda = 8$ Å with a wavelength resolution of $\Delta\lambda/\lambda = 0.15$. The scattered intensity was corrected for background and parasitic scattering, placed on an absolute level using a calibrated secondary standard, and circularly averaged to yield the scattered intensity I(q) as a function of q. The incoherent background from the pure solvent (D_2O) was measured, corrected by the volume fraction displaced by the dissolved lysozyme, and subtracted from the reduced SANS data. Measurements on the solutions of lysozyme with and without glycerol in D_2O at different lysozyme concentrations and temperatures were conducted.

The lysozyme molecules can be modeled as individual ellipsoids[23] with semi-axes, a and b, respectively, which correspond to the lysozyme dimensions determined crystallographically.[24] If an interaction, most likely attractive, between the molecules is present, the summation of such an interaction over the particle surfaces may give rise to a formation of bundles of lysozyme particles. The contribution of packed particles in a bundle arising from the intermolecule interference to the total coherent scattered intensity is considered as a short-range (local) structure factor $S(q)$. In this case, the total scattered intensity then would be given by Eq. (14). Assuming the next neighbor distance in a bundle of molecules obeys a Gaussian distribution, the $S(q)$ was first calculated by Kratky and Porod[25] to be

$$S(q) = 1 + \frac{2}{N}\sum_{k=1}^{N}(N-k)\cos(kDq)\exp\left[-kq^2\sigma_D^2/2\right], \quad (15)$$

where N corresponds to the total number of particles packed in a bundle and D and σ_D represent the next neighbor center-to-center distance and its Gaussian standard deviation (GSD), respectively.

If a long-range correlation length (ξ) between bundles exists in the system, a Lorentzian square term corresponding to the Debye-Bueche model[26,27] for multiphase systems can be used to describe the low q, long-range behavior ($\sim q^{-4}$), contribution to the total coherent scattered intensity given by Eq. (14).[28–30] Thus, the total coherent scattered intensity becomes

$$I_{total}(q) = \frac{A}{\left(1+q^2\xi^2\right)^2} + I(q), \quad (16)$$

where A is a relative weighting coefficient.

Prior to performing any model fitting, several qualitative observations can be made from the data. Figure 4A demonstrates the SANS data from lysozyme in D_2O solutions at different concentrations at 30°C. The Bragg's peak observed suggests that the lysozyme molecules were packed to form a periodic structure in a way (perhaps bundles) and the shift in the peak position, corresponding to the intermolecule d-spacing within a bundle, toward lower q with decreasing concentration indicates that the d-spacing increases with decreasing concentration. Even the most dilute concentration sample (1 mg/ml) possesses the Bragg's peak, which can be enhanced by normalizing the scattered intensity by the corresponding concentration as shown in Figure 4B, suggesting that the molecules remained in their local packing structure, and consequently the SANS data even at 1 mg/ml are not a measurement of the single-particle form factor of noninteracting individual lysozyme molecules and the interaction between those molecules may be attractive. SANS profiles of 50 mg/ml lysozyme in D_2O and of 50 mg/ml and 5 mg/ml lysozyme in D_2O with glycerol at different temperatures are illustrated in Figures 5A and B, respectively. Essentially, the samples are not sensitive to temperature in the range performed. Nevertheless, the peak position observed from lysozyme in D_2O shifts toward higher q when glycerol is added into the system at the same temperature, indicating that the d-spacing between packed lysozyme molecules is smaller when glycerol is present, as given in Figure 6, for instance. The higher peak intensity in lysozyme with glycerol also suggests that the local structure might be better defined and/or there could be more molecules packed in a bundle. The upturn seen in the relatively low q regime (< 0.015 Å^{-1}) in all the SANS data obtained implies that the bundles may possess a long-range correlation length between them in all the samples studied in this work. Consequently, Eq. (16) should be used to fit the data.

Fitting the SANS data to Eq. (16) yields more quantitative and insightful information regarding the long- and short-range structures in the system. Fitted parameters for lysozyme as a function of concentration are shown in Table 2. The fitted results showed that the parameters obtained from the single-particle form factor, $P(q)$, of lysozyme molecules are essentially constant as a function of temperature and concentration with and without glycerol, indicating that the molecules were not changed in terms of the (ellipsoidal) shape and the size (R_g) under the experimental conditions. The $S(q)$ defined by Eq. (15) changes through an increase in the intermolecule d-spacing (D) and its GSD of the packed molecules within a bundle with decreasing concentration from 50 mg/ml to 5 mg/ml. However, $S(q)$ remains the same in the 5 mg/ml and 1 mg/ml lysozyme in D_2O solutions, suggesting that the corresponding local structure does not change at a sufficiently dilute concentration (≤ 5 mg/ml in this case). The averaged number of molecules per bundle (N) in $S(q)$ decreases while the long-range correlation length (ξ) between bundles increases with decreasing concentration.

The fitted parameters from lysozyme in D_2O with glycerol are shown in Table 3. Comparing the data at the same lysozyme concentration and temperature without and with glycerol, D decreases while N and ξ increase when glycerol is present, indicating that more lysozyme molecules were packed closer within a bundle and the correlation length between bundles was larger with glycerol. The fitting results also suggest that the samples studied in this work are temperature independent in the range

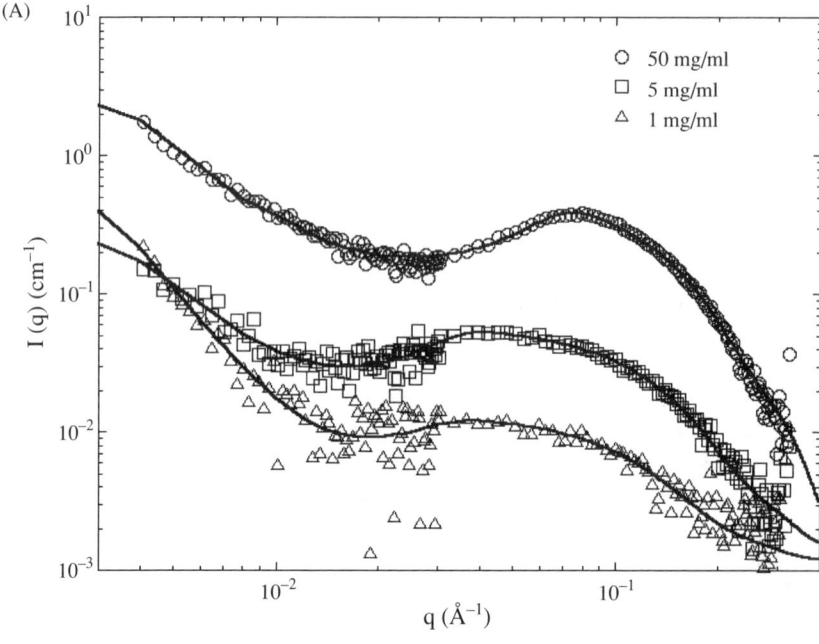

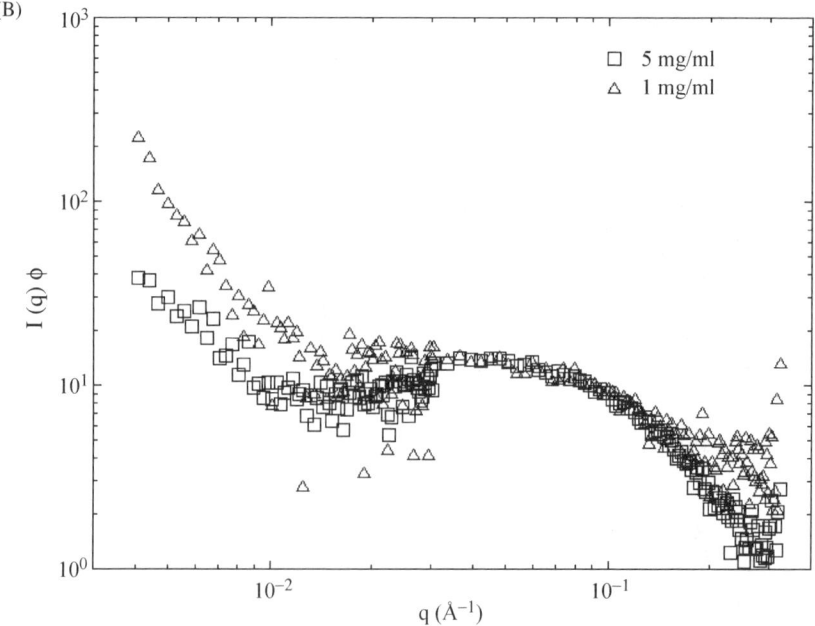

FIGURE 4. (A) SANS profiles of lysozyme in D_2O at different concentrations at 30°C. The solid curves represent the corresponding fitting profiles. (B) Data were normalized by the corresponding volume fraction of lysozyme.

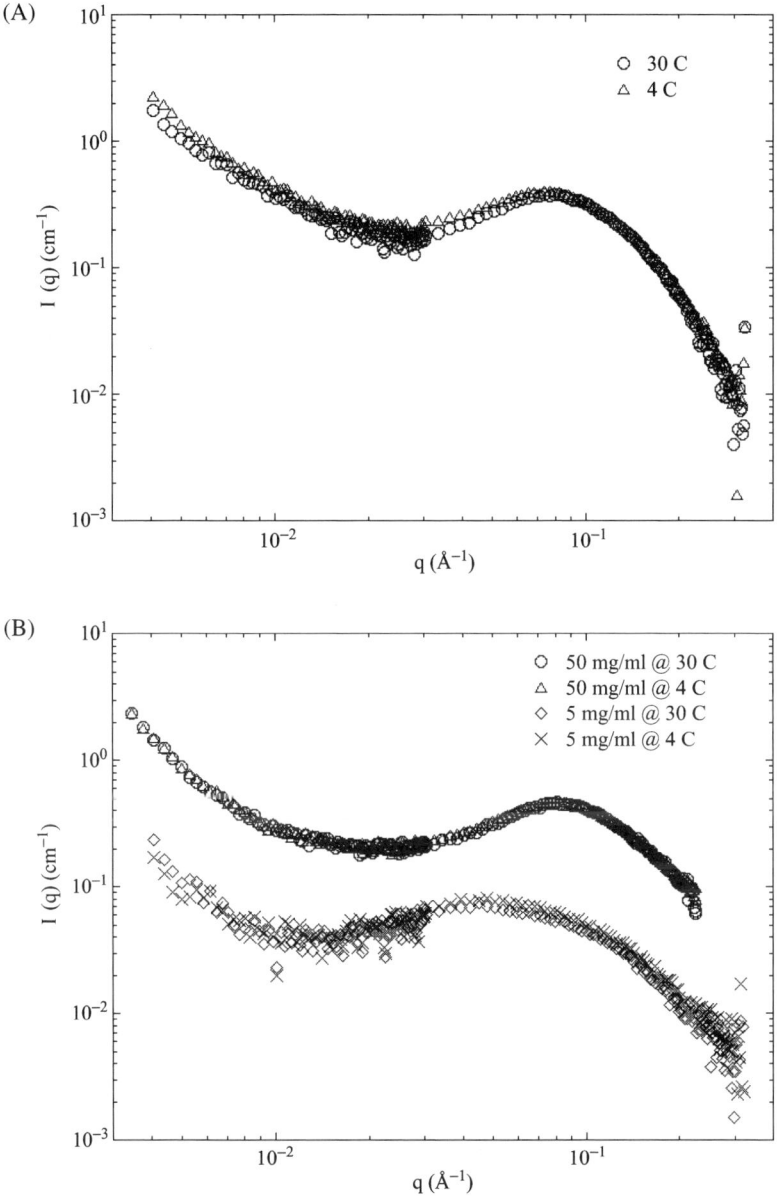

FIGURE 5. (A) SANS profiles of 50 mg/ml lysozyme in D2O at different temperatures. (B) SANS profiles of 50 mg/ml and 5 mg/ml lysozyme + glycerol in D_2O at different temperatures.

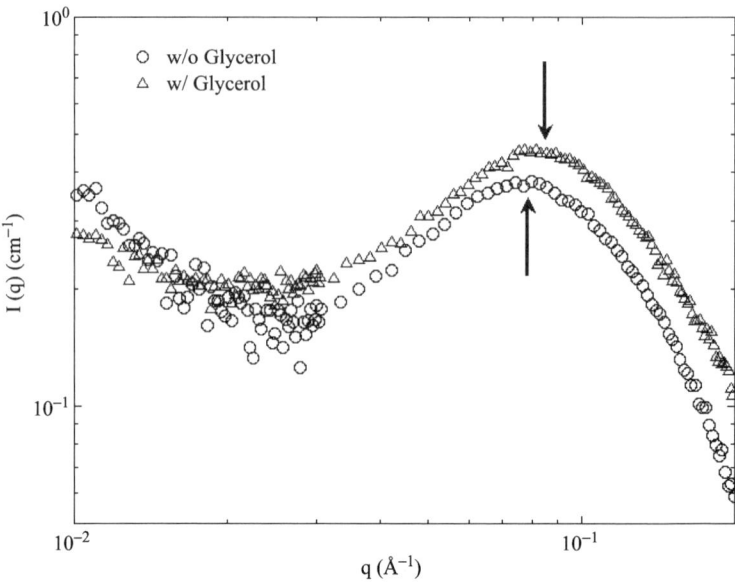

FIGURE 6. SANS profiles of 50 mg/ml lysozyme in D_2O with and without glycerol present in the system at 30°C. Arrows point to the corresponding peak positions.

TABLE 2. Lysozyme in D_2O

Concentration	1 mg/ml	1 mg/ml	5 mg/ml	50 mg/ml	50 mg/ml
T (°C)	30	4	30	30	4
R_g (Å)	17.4 ± 4.7	17.6 ± 3.9	17.6 ± 0.9	15.5 ± 0.4	16.8 ± 0.7
D (Å)	106.7 ± 12.7	109.2 ± 6.1	112.6 ± 0.9	62.0 ± 0.4	60.4 ± 0.4
GSD of D	0.71 ± 0.12	0.69 ± 0.09	0.60 ± 0.02	0.51 ± 0.01	0.49 ± 0.01
ξ (Å)	600.4 ± 45.0	598.6 ± 32.3	204.0 ± 41.8	157.0 ± 6.1	184.5 ± 6.1
No. per bundle	11 ± 5	12 ± 4	22 ± 4	46 ± 2	46 ± 1

Temperature T and the radius of gyration R_g are not fitting parameters. GSD of D (d-spacing) represents the Gaussian standard deviation of the d-spacing between lysozyme molecules packed in a bundle. No. per bundle denotes the averaged total number of lysozyme molecules per bundle.

TABLE 3. Lysozyme + Glycerol in D_2O

Concentration	5 mg/ml	5 mg/ml	50 mg/ml	50 mg/ml
T (°C)	30	4	30	4
R_g (Å)	17.0 ± 1.4	17.6 ± 1.4	15.2 ± 1.0	14.6 ± 0.4
D (Å)	110.6 ± 0.5	107.3 ± 0.3	58.3 ± 0.5	59.0 ± 0.3
GSD of D	0.60 ± 0.02	0.57 ± 0.02	0.51 ± 0.03	0.49 ± 0.01
ξ (Å)	706.8 ± 12.7	1203.1 ± 36.6	540.1 ± 39.4	704.9 ± 32.3
No. per bundle	24 ± 4	24 ± 3	56 ± 3	57 ± 1

Temperature T and the radius of gyration R_g are not fitting parameters. GSD of D (d-spacing) represents the Gaussian standard deviation of the d-spacing between lysozyme molecules packed in a bundle. No. per bundle denotes the averaged total number of lysozyme molecules per bundle.

examined, and the intermolecule interaction is attractive owing to the fact that even the most dilute concentration sample (1 mg/ml) could not be fitted by Eq. (3) alone.

3.4. Tubulin Polymers

SANS has been used to examine taxol-stabilized microtubules (MT) and other tubulin samples in both H_2O and D_2O buffers.[31] Measurements were made at pH/pD values between 6.0 and 7.8, and observed scattered intensities, $I(q)$, have been interpreted in terms of multicomponent models of microtubules and related tubulin polymers. A semiquantitative curve-fitting procedure has been used to estimate the relative amounts of the supramolecular components of the samples. At both pH and pD 7.0 and above, the tubulin polymers are seen to be predominantly microtubules. Although in H_2O buffer the polymer distribution is little changed as the pH varies, when pD is lowered the samples appear to contain an appreciable amount of disklike structures and the average microtubule protofilament number increases from ca. 12.5 at pD $\sim$ 7.0 to ca. 14 at pD $\sim$ 6.0. Such structural change indicates that analysis of microtubule solutions based on H_2O/D_2O contrast variation must be performed with caution, especially at lower pH/pD.

To obtain scattering functions for comparison with data, SANS intensities were simulated using Monte Carlo methods.[6,7] For each component of the scattering assembly, a structure model was constructed from volume elements of suitable size and shape. The molecular volume associated with this structure was randomly filled with points and $P(r)$ was approximated by calculating the distances between all possible pairs of points in the total volume. The scattered intensity, $I(q)$, then was obtained using Eq. (13). The MT were modeled by hollow cylinders, each constructed of a three-start helix formed from tubulin monomers such that the latter provide a continuous wall for the MT.[32–34] The mean cross-sectional radii, which varied in accordance with the number of protofilaments in the MT, were obtained from X-ray diffraction results for 13-protofilament MT[35] and cryoelectron microscopy results for MT consisting of varying numbers of protofilaments.[36] The number of protofilaments, n_f, was allowed to vary between 11 and 14. Mean helical radii values of $R = 102, 110, 118,$ and 127 Å were used for $n_f = 11, 12, 13,$ and 14, respectively. Upon taking the value of the vertical distance (rise) between $\alpha\beta$, dimers along the protofilament to be $p \sim 80$ Å and using the simple formula[34]

$$P = \frac{Sp}{2}, \tag{17}$$

where $S = 3$ is the assumed helix start number, the value of the pitch P for the start helices is estimated to be ~ 120 Å. This value of the helical pitch agrees well with the value of 123.3 Å found by Andreu et al.[32,33] The minor start-helix radius, determining the thickness of the microtubule wall, was assumed to be 35 Å in accordance with the dimensions of the tubulin monomers forming the wall. The model MT were chosen to be of length 1200 Å, which is sufficiently long that length effects become unimportant in the q range probed in the SANS experiments.

Several different structures comprising the scattering assemblies were used when fitting SANS data. The basic units of a MT are paired αβ-tubulin heterodimers and were represented as paired spheres of 40 Å diameter. The end-to-end linear association of dimers along the axis of the MT forms a protofilament. The number of protofilaments determines the radius R. A variable distribution of protofilament numbers was allowed. The model allowed for the possibility of free dimers as well as small aggregates ("oligomers" consisting of two tubulin dimers in a linear array) in the solution. Also included are "disks" consisting of four contiguous substructures, each of which is a three-start helix sliced in half.

The SANS data showed there clearly is little change in the H_2O tubulin samples as a function of pH. The modeled average protofilament number is in good agreement with that previously reported based on SAXS data at pH 6.7.[33] In contrast in D_2O protofilament number is not constant but increases as pD is decreased from 7.0. The difference between D_2O and H_2O samples is reinforced by the results given in Figure 7, which indicate for the D_2O samples a significant increase in the mass fraction of the sample present as MT sheets when pD was decreased. In H_2O, however, the amount of sheet structures remains low at all pH values. Additionally, the present study shows that substitution of D_2O for H_2O alters tubulin polymer structure, even in the presence of taxol, and that these differences are affected by variation of pH/pD. At pH or pD values near 7.0, the polymers observed are almost all MT, but the D_2O and H_2O samples differ

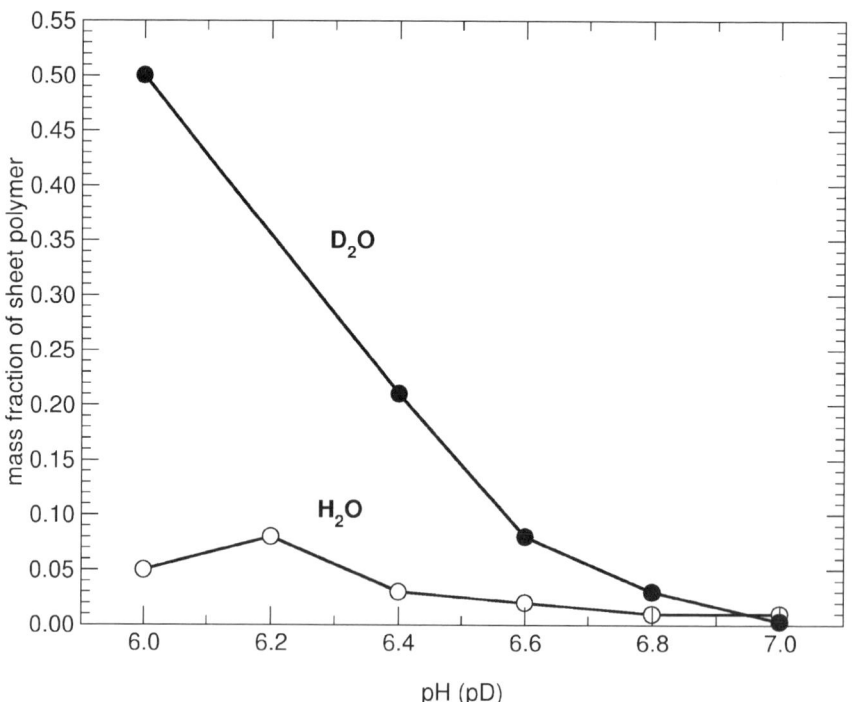

FIGURE 7. The effect of changing the pH or pD on the mass fraction of sheet polymers.

in that D_2O polymers consistently have a slightly larger diameter, indicating a larger number of protofilaments. Taxol is known to lower the average protofilament number from 13 to close to 12.[33] In D_2O, however, the average protofilament number is closer to 13. As pD is lowered, the protofilament number increases further, to about 14 at pD 6.

Several other reports also have demonstrated that the *in vitro* properties of protein molecules can be altered by the presence of D_2O. For example, the conformational stability of β-lactoglobulin is increased by D_2O,[37] and assembled poliovirus capsid particle is protected against heat- or high pH-induced dissociation by D_2O and $MgCl_2$.[38] Polymerization of actin,[39] flagellin,[40] recA,[41] tobacco mosaic virus protein,[42] and tubulin[43-45] have been shown to be promoted and stabilized by D_2O. Final polymer form also can be altered, as in the case of fibrin where polymerization is similar in D_2O and H_2O, but the D_2O gel has a higher degree of lateral association.[46]

Tubulin polymerization is an entropy-driven process. Water molecules that are ordered by hydrophobic surfaces of the tubulin dimer have increased conformational freedom when those surfaces are buried upon polymerization. Hence, it is plausible that the reported promotion of polymerization in D_2O is due primarily to enhanced hydrophobic interactions as suggested by Itoh and Sato.[44] However, it is not clear how this would result in an increase in protofilament number or a pD-dependent protofilament number. Since the protofilament number and sheet fraction increased as pD was lowered from 7 to 6, whereas such change was not seen in H_2O, it seems likely that these changes are due either to enhancement of hydrogen bonding between protofilaments or to D_2O effects on pH-dependent H^+ dissociation from some residue(s) involved in lateral interaction of protofilaments.[31]

Finally, although the possibility of varying D_2O/H_2O to contrast match the components of a biological assembly makes neutron scattering a powerful technique, these results suggest that substitution of D_2O for H_2O in structural studies of proteins should be done with some care. If the structure of the scattering medium does not remain invariant when D_2O is substituted for H_2O, more complicated experimental and analysis methods would be required, as described here, particularly for samples at lower pH/pD where the relative amounts of microtubules and disklike structures seem to vary with D_2O/H_2O.

4. SUMMARY

SANS is a powerful tool for studying the aggregation of proteins in solution. Such studies can provide insight into the interactions between proteins in solution that in turn can aid in the understanding of protein function. This chapter illustrated several examples of the study of protein aggregation in solution. In each case, structure models were derived from the SANS data, along with information obtained from other analysis techniques. Complicated structures can be built from simple shapes in real space, as in the case of the tubulin polymer, to obtain model SANS intensities. Alternatively, scattered neutron intensities can be fit directly by model functions in reciprocal space, as in the case of the amyloid-β and lysozyme aggregates. These model functions can include both intra- and interparticle interactions, allowing for a wide range of possibilities for modeling the protein aggregates. Using information obtained from other analytical and biochemical techniques in addition to the SANS results allows for insightful models into the mechanism of protein aggregation.

NOTES

Certain commercial materials and equipment are identified in this chapter in order to specify adequately the experiment procedure. In no case does such identification imply recommendation by the National Institute of Standards and Technology nor does it imply that the material or equipment identified is necessarily the best available for this purpose.

REFERENCES

1. B. Jacrot, The study of biological structures by neutron scattering from solution, *Rep. Prog. Phys.* **39**, 911–953 (1976).
2. C.J. Glinka, J.G. Barker, B. Hammouda, S. Krueger, J.J. Moyer, and W.J. Orts, The 30 m small-angle neutron scattering instruments at the National Institute of Standards and Technology, *J. Appl. Cryst.* **31**, 430–445 (1998).
3. L.A. Feigin and D.I. Svergun, *Structure Analysis by Small-Angle X-Ray and Neutron Scattering*, (New York: Plenum Press, 1987).
4. A. Guinier and G. Fournet, *Small Angle Scattering of X-Rays* (New York: Wiley, 1955).
5. O. Glatter and O. Kratky, *Small Angle X-ray Scattering*, (New York: Academic Press, 1982).
6. S. Hansen, Calculation of small-angle scattering profiles using Monte Carlo simulation, *J. Appl. Cryst.* **23**, 344–346 (1990).
7. J. Zhou, A. Deyhim, S. Krueger, and S.K. Gregurick, LORES: low resolution shape program for the calculation of small-angle scattering profiles for biological macromolecules in solution, *Comp. Phys. Comm.*, **170**, 186–204.
8. M.S. Wertheim, Analytic solution of the Percus-Yevick equation, *J. Math. Phys.* **5**, 643–651 (1964).
9. J.-P. Hansen and J.B. Hayter, A rescaled MSA structure factor for dilute charged colloidal dispersions, *Mol. Phys.* **46**, 651–656 (1982).
10. W. Yong, A. Lomakin, M.D. Kirkitadze, D.B. Teplow, S.-H. Chen, and G.B. Benekek, Structure determination of micelle-like intermediates in amyloid β-protein fibril assembly by using small angle neutron scattering, *Proc. Natl. Acad. Sci. USA* **99**, 150–154 (2002).
11. H.G. Thomas, A. Lomakin, D. Blankschtein, and G.B. Benedek, Growth of mixed nonionic micelles, *Langmuir* **13**, 209–218 (1997).
12. D. Blankschtein, G.M. Thurston, and G.B. Benedek, Phenomenological theory of equilibrium thermodynamic properties and phase-separation of micellar solutions, *J. Phys. Chem.* **85**, 7268–7288 (1986).
13. C.M. Dobson, Protein misfolding, evolution and disease, *Trends Biochem. Sci.* **9**, 329–332 (1999).
14. J.I. Guijarro, M. Sunde, J.A. Jones, I.D. Campbell, and C.M. Dobson, Amyloid fibril formation by an SH3 domain, *Proc. Natl. Acad. Sci. USA* **95**, 4224–4228 (1998).
15. M.R.H. Krebs, D.K. Wilkins, E.W. Chung, M.C. Pitkeathly, A.K. Chamberlain, J. Zurdo, C.V. Robinson, and C.M Dobson, Formation and seeding of amyloid fibrils from wild-type hen lysozume and a peptide fragment from the β domain, *J. Mol. Biol.* **300**, 541–549 (2000).
16. M. Fändrich, M.A. Fletcher, and C.M. Dobson, Amyloid fibrils from muscle myoglobin, *Nature* **410**, 165–166 (2001).
17. F. Chiti, P. Webster, N. Taddei, K. Clark, M. Stefani, G. Ramponi, and C.M. Dobson, Designing conditions for *in vitro* formation of amyloid proto-filaments and fibrils, *Proc. Natl. Acad. Sci. USA* **96**, 3590–3594 (1999).

18. S. Goda, K. Takano, Y. Yamagata, R. Nagata, H. Akutsu, S. Maki, K. Namba, and K. Yutani, Amyloid protofilaments formation of hen egg white lysozyme in highly concentrated ethanol solution, *Protein Sci.* **9**, 369–375 (2000).
19. Y. Yonezawa, T. S. Tanaka, T. Kubota, K. Wakabayashi, K. Yutani, and S. Fujiwara, An insight into the pathway of the amyloid fibril formation of hen egg white lysozyme obtained from a small-angle x-ray and neutron scattering study, *J. Mol. Biol.* **323**, 237–251 (2002).
20. D.R. Booth, M. Sunde, V. Bellotti, C.V. Robinson, W.L. Hutchinson, P.E. Fraser, P.N. Hawkins, C.M. Dobson, S.E. Radford, C.C.F. Blake, and M.B. Pepys, Instability, unfolding and aggregation of human lysozyme variants underlying amyloid fibrillogenesis, *Nature* **385**, 787–793 (1997).
21. T.R. Serio, A.G. Cashikar, A.S. Kowal, J. Sawicki, J.J. Moslehi, L. Serpell, M.F. Arnsdorf, and S.L. Lindquist, Nucleated conformation conversion and the replication of conformational information by a prion determinant, *Science* **289**, 1317–1321 (2000).
22. L.C. Serpell, M. Sunde, M.D. Benson, G.A. Gennent, M.B. Pepys, and P.E. Fraser. The protofilament substructure of amyloid fibrils, *J. Mol. Biol.* **300**, 1033–1039 (2000).
23. O.D. Velev, E.W. Kaler, and A.M. Lenhoff, Protein interactions in solution characterized by light and neutron scattering: comparison of lysozyme and chymotrypsinogen, *Biophys. J.* **75**, 2682–2697 (1998).
24. T.E. Creighton, *Proteins: Structure and Molecular Properties* (New York: Freeman, 1993).
25. O. Kratky and G. Porod, Diffuse small-angle scattering of X-rays in colloid systems, *J. Colloid Science* **4**, 35–70 (1949).
26. J.S. Higgins and H.C. Benoit, *Polymers and Neutron Scattering* (New York Oxford, 1994).
27. R.-J. Roe, *Methods of X-Ray and Neutron Scattering in Polymer Science* (New York: Oxford, 2000).
28. F. Horkay, A.M. Hecht, and E. Geissler, Fine structure of polymer networks as revealed by solvent swelling, *Macromolecules* **31**, 8851–8856 (1998).
29. C.L. Zhou, E. Hobbie, B.J. Bauer, and C.C Han, *J. Polym. Sci., Polym. Phys. Ed.* **36**, 2745–2750 (1998).
30. B. Hammouda, D.L. Ho, and S. Kline, SANS from poly(ethylene oxide)/water systems, *Macromolecules* **35**, 8578–8585 (2002).
31. D. Sackett, V. Chernomordik, S. Krueger, and R. Nossal, Use of small-angle neutron scattering to study tubulin polymers, *Biomacromolecules* **4**, 461–467 (2003).
32. J.M. Andreu, J. Bordas, J.F. Diaz, J. Garcia de Ancos, R. Gil, F.J. Medrano, E. Nogales, E. Pantos, and E. Towns-Andrews, Low resolution structure of microtubules in solution. Synchrotron x-ray scattering and electron microscopy of taxol-induced microtubules assembled from purified tubulin in comparison with glycerol and MAP-induced microtubules, *J. Mol. Biol.* **226**, 169–184 (1992).
33. J.M. Andreu, J.F. Diaz, R. Gil, J.M. de Pereda, M. Garcia de Lacoba, V. Peyrot, C. Briand, E. Towns-Andrews, and J. Bordas, Solution structure of Taxotere-induced microtubules to 3-nm resolution. The change in protofilament number is linked to the binding of the taxol side chain, *J. Biol. Chem.* **269**, 31785–31792 (1994).
34. F. Metoz, I Arnal and R.H. Wade, Tomography without tilt: three-dimensional imaging of microtubule/motor complexes, *J. Struct. Biol.* **118**, 159–168 (1997).
35. E. Mandelkow, J. Thomas, and C. Cohen, Microtubule structure at low resolution by x-ray diffraction, *Proc. Natl. Acad. Sci. USA* **74**, 3370–3374 (1977).
36. R.H. Wade and D. Chrétien, Cryoelectron microscopy of microtubules, *J. Struct. Biol.* **110**, 1–27 (1993).
37. M. Verheul, S.P. Roefs, and K.G. de Kruif, Aggregation of beta-lactoglobulin and influence of D_2O, *FEBS Lett.* **421**, 273–276 (1998).
38. C.H. Chen, R. Wu, L.G. Roth, S. Guillot, and R. Crainic, Elucidating mechanisms of thermostabilization of poliovirus by D_2O and $MgCl_2$, *Arch. Biochem. Biophys.* **342**, 108–116 (1997).
39. H. Omori, M. Kuroda, H. Naora, H. Takeda, Y. Nio, H. Otani, and K. Tamura, Deuterium oxide (heavy water) accelerates actin assembly in vitro and changes microfilament distribution in cultured cells, *Eur. J. Cell Biol.* **74**, 273–280 (1997).
40. Y. Uratani, Polymerization of *Salmonella* flagellin in water and deuterium oxide media, *J. Biochem.* **75**, 1143–1151 (1974).

41. R.W.H. Ruigrok and E. DiCapua, On the polymerization state of RecA in the absence of DNA, *Biochimie* **73**, 191–197 (1991).
42. M.T. Khalil and M.A. Lauffer, Polymerization-depolymerization of tobacco mosaic virus protein X. Effect of D_2O, *Biochemistry* **6**, 2474–2480 (1967).
43. L.L. Houston, J. Odell, Y.C. Lee, and R.H. Himes, Solvent isotope effects on microtubule polymerization and depolymerization, *J. Mol. Biol.* **87**, 141–146 (1974).
44. T.J. Itoh and H. Sato, The effects of deuterium-oxide (^2H_2O) on the polymerization of tubulin *in vitro*, *Biochim. Biophys. Acta* **800**, 21–27 (1984).
45. D. Panda, G. Chakrabarti, J. Hudson, K. Pigg, H.P. Miller, L. Wilson, and R.H. Himes, Suppression of microtubule dynamic instability and treadmilling by deuterium oxide, *Biochemistry* **39**, 5075–5081 (2000).
46. U. Larsson, Polymerization and gelation of fibronogen in D_2O, *Eur. J. Biochem.* **174**, 139–144 (1988).

Laser Light Scattering as an Indispensable Tool for Probing Protein Aggregation

Regina M. Murphy[1,3] and Christine C. Lee[2]

Protein misfolding and aggregation have emerged as significant problems in two quite different arenas. For the medical community, the spontaneous conversion of soluble proteins or protein fragments into fibrillar polymers with regular cross-β-sheet structure is linked to diseases as diverse as Huntington's, Alzheimer's, type II diabetes, and the prion diseases. For manufacturers of protein products, aggregation of proteins during processing, formulation, or storage results in a reduction in yield and off-specification products. Multiple biophysical tools are required for the elucidation of the mechanisms and pathways by which proteins misfold and aggregate. In this chapter, we turn our attention to one of these tools: laser light scattering. This noninvasive technique provides a means for measuring key size and shape properties of macromolecules in solution. Laser light scattering probes length scales on the order of 1–1000 nm, making it uniquely suited for detecting and characterizing soluble protein aggregates. Laser light scattering comes in two flavors: static and dynamic. From static light-scattering measurements, one can extract useful data on molecular weight and radius of gyration of proteins and protein aggregates in solution. The hydrodynamic diameter is obtained from dynamic light scattering. Given appropriate conditions and with more detailed data analysis, information about particle shape and characteristic dimensions, size distribution and polydispersity of populations of particles, and interparticle interactions can be obtained. Laser light-scattering studies have contributed to many investigations of protein misfolding and aggregation; a few of these studies are discussed.

[1,2] Department of Chemical and Biological Engineering, University of Wisconsin, Madison, WI 53706
[3] To whom correspondence should be addressed: Regina M. Murphy at regina@engr.wisc.edu

1. INTRODUCTION

Protein misfolding and aggregation has emerged as a significant problem in two quite different lines of research. For the medical community, the spontaneous conversion of soluble proteins or protein fragments into fibrillar polymers with regular cross-β-sheet structure is linked to diseases as diverse as Huntington's, Alzheimer's, type II diabetes, and the prion diseases (bovine spongiform encephalitis, scrapie, and variant Creutzfeldt-Jakob, for example). For manufacturers of protein products, aggregation of proteins during processing, formulation, or storage results in a reduction in yield and off-specification products. The Food and Drug Administration is increasingly concerned about adverse immunological responses to even small quantities of aggregated material in protein drugs.

We are interested in the general problem of defining the pathway and kinetics by which monomeric protein and peptides become aggregated. Such knowledge may lead to development of new therapies for treating protein-misfolding diseases and new processing technologies for preventing aggregation during manufacture.

Multiple biophysical tools are required for the study of protein aggregation. Here we turn our attention to one of these tools: laser light scattering. Because larger particles scatter light strongly, detection of the presence of small quantities of aggregated material is much easier and more reliable than with other mass-based methods such as size exclusion chromatography, gel electrophoresis, or filtration. Laser light scattering probes length scales on the order of 1–1000 nm, making it uniquely suited for detecting and characterizing the size and shape of soluble protein aggregates. The technique is noninvasive, data can be collected in real time, no fluorescent or radioactive labels are required, and the sample can be recovered unaltered for further study.

2. EXPERIMENTAL APPARATUS

Both static and dynamic light scattering techniques use the same basic setup: a laser (typically a helium-neon or argon ion laser) is focused on a cylindrical cell containing an optically clear sample. A small portion of the light is scattered; the scattered light intensity is detected at an angle (typically between 15° and 150°) from the incident beam with a sensitive photomultiplier tube or photodiode detector (Figure 1). Some commercial systems include flow cells and most often are used in concert with chromatographic separations; other systems operate in batch mode.

Samples must be optically clear for laser light-scattering experiments. "Dust" or other contaminants scatter light strongly, interfering with data analysis. Removal of large "dust" particles is a crucial step in sample preparation and is particularly difficult with aqueous solutions. We use high-quality filtered and deionized water (Milli-Q) for all buffer preparations and we double-filter buffers immediately prior to use. Besides providing pH control, buffers should contain some salts (10 mM or more); otherwise, at very low ionic strength, electrostatic interactions between particles can lead to cooperative particle motion and anomalous data. Protein samples are prepared in double-filtered buffers and then filtered again directly into scrupulously cleaned light-scattering

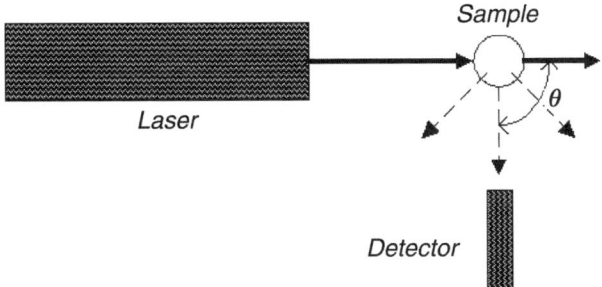

FIGURE 1. Schematic of typical laser light-scattering setup. A laser is focused on a sample in a special light-scattering cuvette or in a flow cell. Most of the light passes through the sample. Scattered light (dashed lines) is detected at an angle from the incident beam. In dynamic light scattering, detection is at a fixed angle θ and fluctuations in the intensity of scattered light are measured. These fluctuations are due to diffusive motion of particles in solution. In static light scattering, the average scattered intensity at multiple angles is detected. The total scattered intensity extrapolated to zero angle is related to the number and molecular mass of the particles in solution; the angular dependence is indicative of particle size and shape.

cuvettes. For particularly demanding applications, some investigators use low-protein-binding Nucleopore membranes and/or use a recirculating filtration system. The sample should be recovered after completion of the light-scattering experiments for accurate protein concentration determination. Typically, sample volumes of 50 µl to 2 ml are required, depending on the instrument setup. Larger sample sizes are generally preferable because scatter from the cuvette-solvent interface is less of a concern; however, with precious biological samples these volumes may not be readily available. Accessible sample concentration depends on the laser power and the detector sensitivity as well as the particle size; protein concentrations in the 0.1–10 mg/ml range are typical, with higher concentrations required for smaller proteins.

Once the cuvette is placed in the sample holder, the laser beam through the sample should be checked by visual examination: the beam should be clean and sharp, with no fuzziness at the edges or "sparkles" in the beam. It is a good idea to prepare a control sample of buffer alone; scattering intensity from the buffer should be only a small fraction of that from the sample and the counts per second should be stable over time; fluctuations are a good indicator that the buffer is "dirty." The sample must be thermally equilibrated prior to beginning data collection.

Two different kinds of light-scattering experiments can be undertaken: dynamic and static. Some commercial instruments are capable of only dynamic or only static measurements, while others accommodate both.

3. ANALYSIS OF LIGHT-SCATTERING DATA

3.1. Static Light Scattering

When a laser beam hits a particle, the electromagnetic radiation perturbs the position and motion of electrons in the particle. The electron re-emits radiation of the same frequency in all directions; this is called "elastic scattering" and is the basic physical phenomenon behind static light scattering (SLS).[1] In SLS, the average intensity of scattered light from the sample $I_s(q)$ is measured as a function of the scattering angle θ (Figure 1). The background (solvent) intensity $I_b(q)$ is subtracted from the sample intensity; this is then normalized by the scattered intensity of a reference solvent such as toluene $I_{ref}(q)$ to adjust for instrument performance. Ideally, $I_{ref}(q)$ shows little or no angular dependence. These intensity measurements are used to calculate the Rayleigh ratio $R_s(q)$

$$R_s(q) = \frac{I_s(q) - I_b(q)}{I_{ref}(q)} \left(\frac{n_s}{n_{ref}}\right)^2 R_{ref}, \quad (1)$$

where n_s is the refractive index of the solvent, n_{ref} is the refractive index of the reference solvent, and R_{ref} is the experimentally determined Rayleigh ratio of the reference solvent. M, the molar mass (g/g mole) of the particles in solution, $P(q)$, the particle-scattering factor, and B_2, the thermodynamic second virial coefficient (g mole/g^2-cm^3), can be determined from $R_s(q)$ per Eq. (2):

$$\frac{Kc}{R_s(q)} = \frac{1}{P(q)M} + 2B_2 c + \ldots, \quad (2)$$

where $K = 4\pi^2 n_s^2 (dn/dc)^2 / N_A \lambda_0^4$, dn/dc is the refractive index increment ($dn/dc = 0.185 \pm 0.005$ cm^3/g for most proteins), λ_0 is the wavelength of the incident beam *in vacuo* (cm), and c is the total solute mass concentration (g/cm^3). For particles very small relative to λ_0, $P(q) \to 1$. For particles of characteristic dimension roughly 1/10 the wavelength of the incident laser beam, we can approximate $P(q)$ as

$$\frac{1}{P(q)} \cong 1 + \frac{q^2 R_g^2}{3}, \quad (3)$$

where R_g is the radius of gyration of the particle and q is the scattering vector:

$$q = \frac{4\pi n_s}{\lambda_0} \sin\left(\frac{\theta}{2}\right). \quad (4)$$

In a standard SLS experiment, samples are prepared at several concentrations and intensity data are collected at multiple angles. The data are plotted as $Kc/R_s(q)$ versus $q^2 + k_s c$, with k_s as an arbitrary stretching coefficient. The rationale for this can be seen by combining Eqs. (2) and (3); this analysis is known as the classical Zimm plot (Figure 2). With this format, extrapolation of the data to $q^2 = 0$ and $c = 0$ yields $1/M$, the slope of the data versus q^2 extrapolated to $c = 0$ gives $R_g^2/3M$, and the slope of the data versus c extrapolated to $q^2 = 0$ gives $2B_2$.

For samples that are aggregating, Zimm plot analysis is problematic: since aggregation is typically a function of concentration, M changes with c. For sufficiently dilute solutions in good solvents, the second virial coefficient term in Eq. (2) can be neglected

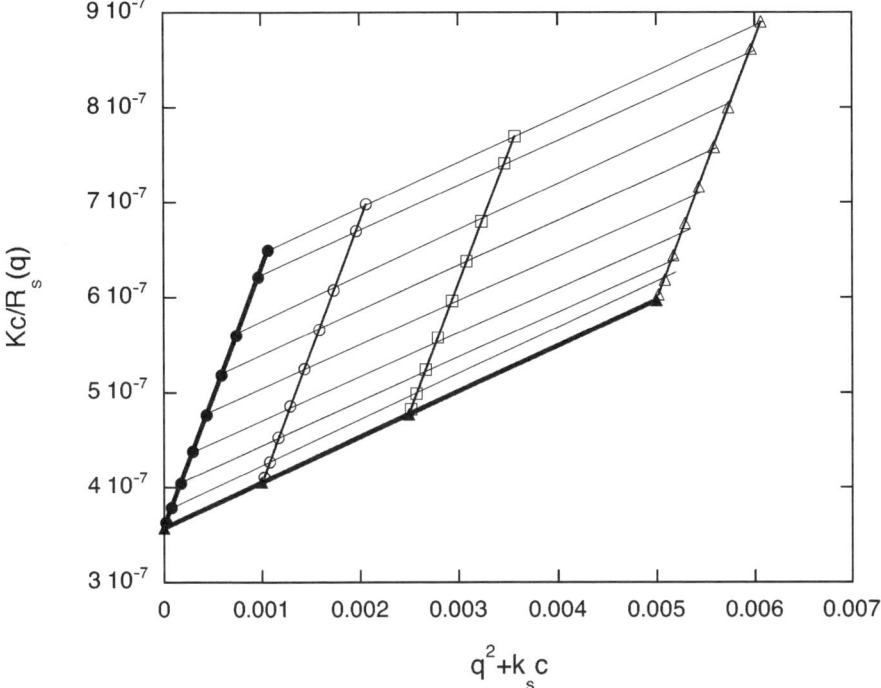

FIGURE 2. Simulated Zimm plot for particle with $M = 2.8 \times 10^6$ g/mol, $R_g = 48$ nm, $B_2 = 1.2 \times 10^{-4}$ mol-cm^3/g^2. "Data" points are plotted at $c = 0.001$ (△), 0.0005 (□), 0.0002 (○) g/cm^3. At each angle, data are extrapolated to $c = 0$ (●). A straight line is fit to the extrapolated points to get R_g from the slope. The data at each concentration are extrapolated to zero angle ($q = 0$, ▲). The slope at zero angle is 2^*B_2, the intercept at zero angle and zero concentration is M.

and an estimate of the apparent molar mass M_{app} can be obtained from extrapolation to $q = 0$ of multiangle data taken at a single concentration.

For larger particles with characteristic dimension approaching λ_0, Eq. (3) is insufficient and higher-order terms in $P(q)$ must be included. In fact, this is advantageous because shape information can be extracted. Figure 3 shows curves of $P(q)$ versus q^2 for different particle morphologies. For large particles ($qR_g > 1$), distinctions between globular versus elongated shapes can be discerned. At typical values of q, $qR_g \sim 1$ corresponds to spheres of radius $R \sim 60$ nm or rods of length $L \sim 140$ nm.

For a polydisperse solution, Eq. (2) needs to be adjusted:

$$\frac{Kc}{R_s(q)} = \frac{1}{P_z(q) M_w} + 2B_2 c + \ldots, \tag{5}$$

where $P_z(q)$ is the z-averaged particle-scattering factor:

$$P_z(q) = \frac{\sum w_i M_i P_i(q)}{\sum w_i M_i} \tag{6}$$

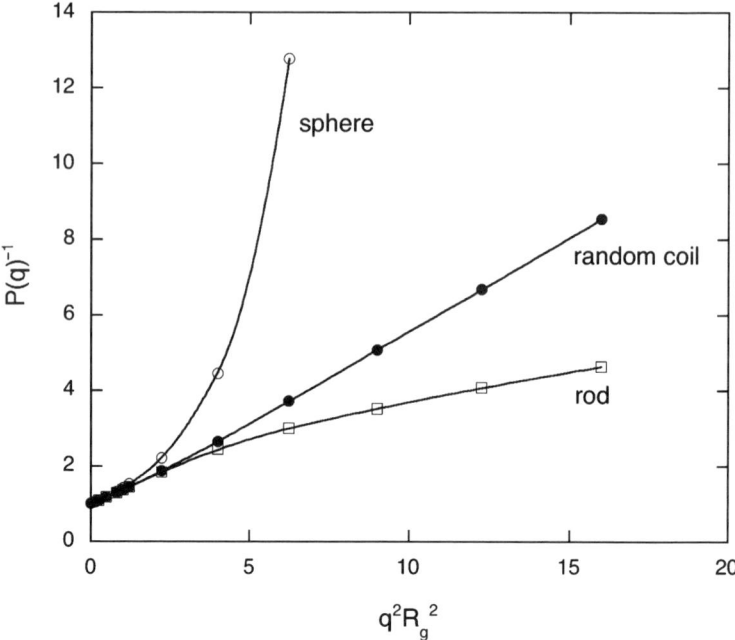

FIGURE 3. Particle-scattering factor as a function of particle shape and size. Curves were calculated from the equations given in Table 1. As $q^2 R_g^2 \to 0$, $P(q)^{-1} \to 1$. The limiting slope = $1/3$ at low $q^2 R_g^2$.

and M_w is the weight-averaged molar mass:

$$M_w = \sum w_i M_i. \tag{7}$$

When data from polydisperse samples are plotted as shown in Figure 2, curved rather than straight lines may result. These curvature effects of polydispersity complicate analysis of scattering data to ascertain particle shape.[2]

3.2. Dynamic Light Scattering

The diffusive motion of particles in solution gives rise to fluctuations in the intensity of the scattered light on the microsecond timescale. In dynamic light scattering (DLS), these fluctuations in scattered light at a single angle (most commonly 90°) are detected and analyzed with an autocorrelator to generate a normalized first-order autocorrelation function $g^{(1)}(\tau)$. (DLS also is called photon correlation spectroscopy, PCS, and quasielastic light scattering, QELS. These are just three different terms for the same technique.) For a dilute solution of monodisperse particles that are small relative to the wavelength of light,

$$g^{(1)}(\tau) = e^{(-\Gamma \tau)} = e^{(-Dq^2\tau)}, \tag{8}$$

where D is the translational diffusion coefficient of the particle, τ is the autocorrelation decay time, and q is the scattering vector as defined in Eq. (4); D is related to the hydrodynamic diameter of the particle d_{sph} by the Stokes-Einstein equation:

$$d_{sph} = \frac{kT}{3\pi\eta D}, \tag{9}$$

where T is temperature, k is the Boltzmann constant, and η is solvent viscosity. The τ is adjusted to allow good sampling of the autocorrelation function and ensure decay to baseline. Typical values for $\Delta\tau$ (decay time per channel) are 0.1 to 40 μsec, and modern autocorrelators have 128 or 256 (or more) channels. Thus, DLS is best suited for measuring diffusion coefficients of $\sim 5 \times 10^{-6}$ to 5×10^{-9} cm²/sec, or particle hydrodynamic diameters of ~ 1 to ~ 1000 nm.

For polydisperse mixtures, $g^{(1)}(\tau)$ is a sum of contributions from various particle sizes. Most commonly, the method of cumulants is used to extract a z-average diffusion coefficient $<D>_z$ and a measure of polydispersity μ_2 by fitting the autocorrelation data to

$$\ln g^{(1)}(\tau) = -\langle D \rangle_z q^2 \tau + \frac{1}{2}\mu_2 \tau^2, \tag{10}$$

where

$$\langle D \rangle_z = \frac{\sum_{\text{all } i} c_i M_i D_i}{\sum_{\text{all } i} c_i M_i} \tag{11}$$

and c_i, M_i, and D_i are the mass concentration, molecular mass, and translational diffusion coefficient, respectively, of particles of size i. The method of cumulants works well for a relatively narrow and unimodal distribution of small particle sizes. If the distribution is broad or multimodal or if the characteristic particle size approaches λ_0, then a more sophisticated data analysis is required. One of the most widely used methods is the constrained regularization package CONTIN. CONTIN finds the "best" weighted distribution $G(\Gamma)$ by fitting the autocorrelation data to

$$g^{(1)}(\tau) = \int_0^\infty G(\Gamma) e^{(-Dq^2\tau)} d\Gamma, \tag{12}$$

with a statistical bias toward smooth distributions. For populations of discrete size,

$$G(\Gamma_i) = \frac{c_i M_i P_i(q)}{\sum_{\text{all } i} c_i M_i P_i(q)}, \tag{13}$$

where $P_i(q)$ is the particle-scattering factor for particle i. Figure 4 shows typical autocorrelation curves for narrow and broad distributions.

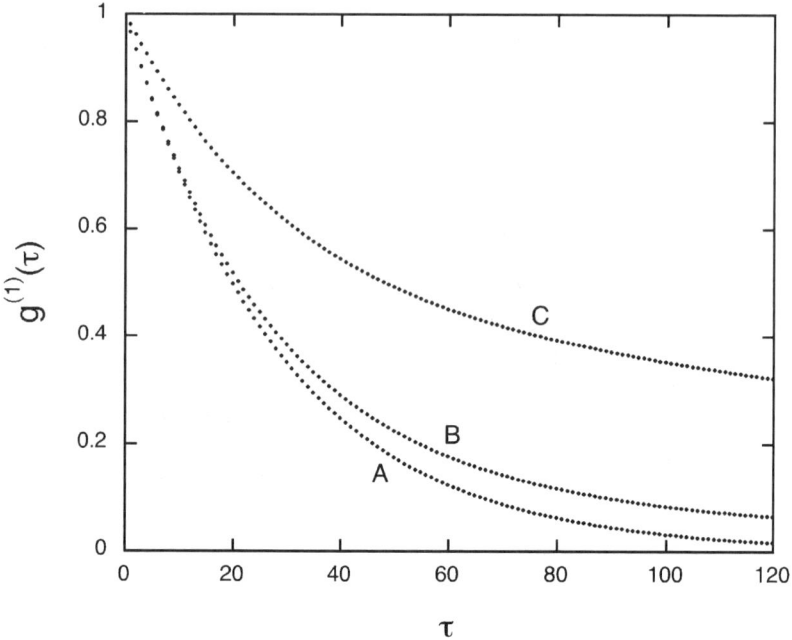

FIGURE 4. Simulated autocorrelation functions. All "data" were simulated at a decay time per channel of 2 μsec and $q = 1.87 \times 10^5$ cm^{-1}. (A) Monodisperse particles with $D = 5 \times 10^{-7}$ cm^2/sec (B) Monomodal distribution of particles with $<D>_z = 5 \times 10^{-7}$ cm^2/sec and moderate polydispersity. (C) Bimodal distribution of particles, with 99 wt % particles of $D = 5 \times 10^{-7}$ cm^2/sec, $M = 150,000$ Da, $P(q) = 1$ plus 1 wt % particles of $D = 5 \times 10^{-8}$ cm^2/sec, $M = 1.5 \times 10^7$ Da, and $P(q) = 0.9$. Notice how sensitive the data are to even small fractions of large particles. For case (C), the decay time should be increased to allow decay to baseline. Some autocorrelators provide an option of multiple decay times, improving the analysis of multimodal distributions.

By collecting DLS data at multiple angles, additional information about the particle size and shape distribution can be obtained. For large aspherical and/or flexible molecules, other diffusional modes (flexing, rotation) may be detected by DLS. For these cases, the reader should consult more specialized literature.

For nondilute solutions, macromolecular interactions affect particle motion. In these cases, the measured D_{app} is related to the dilute-solution D as

$$D_{app} = D\left(1 + k_D c + O\left(c^2\right) + \ldots\right), \qquad (14)$$

where c is the concentration of macromolecules and k_D represents the magnitude and sign of the interaction. When k_D is positive, the interaction between macromolecules is repulsive.

3.3. Combining DLS and SLS Data

Using both SLS and DLS to characterize samples provides the best opportunity to characterize the particle size, shape, and polydispersity. In Table 1 equations for $P(q)$, d_{sph}, and R_g for several common particle shapes are summarized. The ratio $2R_g/d_{sph}$

TABLE 1. Particle-scattering factor, hydrodynamic diameter, and radius of gyration for some simple particle shapes. Equations for $P(q)$ are good for spheres of $R < 100$ nm and for rods of $L < 1000$ nm. Other expressions for more complex particle shapes are given elsewhere.[2-4]

Shape	Characteristic dimensions	$P(q)$	d_{sph}	R_g	$2\,R_g/d_{sph}$
Solid sphere	Radius R $x = qR$	$9\left[\dfrac{\sin x - x\cos x}{x^3}\right]^2$	$2R$	$\sqrt{\dfrac{3}{5}}R$	0.775
Hollow sphere	Radius R $x = qR$	$\left[\dfrac{\sin x}{x}\right]^2$	$2R$	R	1
Thin disk	Radius R $x = qR$	$\dfrac{2}{x^2}\left[1 - \dfrac{J_1(2x)}{x}\right]$ J_1 = Bessel function of order 1		$\dfrac{R}{2}$	
Thin rigid rod	Length L diameter d, $\dfrac{L}{d} \gg 1$ $x = \dfrac{qL}{2}$	$\dfrac{\int_0^{2x}\dfrac{\sin w}{w}dw}{x} - \left[\dfrac{\sin x}{x}\right]^2$	$\dfrac{L}{\ln(L/d) + 0.316}$	$\dfrac{L}{2\sqrt{3}}$	2.84... for $\dfrac{L}{d} = 100$
Random coil	Root-mean-square end-to-end distance $\sqrt{\overline{r^2}}$ $x = \dfrac{q\sqrt{\overline{r^2}}}{\sqrt{6}}$	$\dfrac{2}{x^4}\left[x^2 - 1 + e^{-x^2}\right]$	$\dfrac{1}{8}\sqrt{\dfrac{3\pi\overline{r^2}}{2}}$	$\sqrt{\dfrac{\overline{r^2}}{6}}$	1.5045...

often can be used to distinguish among different particle shapes even when particle size is too small to rely on differences in $P(q)$.

4. LITERATURE EXAMPLES

Misfolding is tightly linked with aggregation through partially unfolded intermediates, whether starting with random-coil peptides or with structured natively folded proteins. Partial unfolding exposes hydrophobic contact sites, which are normally buried in the correctly folded protein, that can participate in intermolecular assembly. Widely used assays for measuring misfolding and/or aggregation include circular dichroism, fluorescence, turbidity, membrane filtration, and size exclusion (gel filtration) chromatography. Each assay has its purpose. Circular dichroism detects changes in secondary structure. Changes in intrinsic fluorescence detect unfolding or environmental changes indicating partial solvent exposure of tryptophans. Extrinsic fluorescence probes (e.g., ANS) are used to detect the presence of solvent-exposed hydrophobic patches, the formation of which is likely to drive further aggregation. With size exclusion chromatography and membrane filtration, the quantity of aggregated material can be measured. Turbidity assays are performed by measuring sample absorbance at wavelengths of 350–450 nm; what is actually measured is the attenuation of the incident beam intensity due to scattering of light by large particles in the sample. The increase in turbidity is linearly proportional to the mass of aggregated material under certain conditions.[5]

Laser light scattering is most useful for gaining information about particle size or morphology prior to precipitation or gelation. Unlike size exclusion chromatography or ultracentrifugation, data in real time can be collected and analyzed, making kinetic studies feasible. Another advantage is that no dilution occurs during data collection. Since light scattering is most sensitive to larger particles, specialized techniques such as fluorescence energy transfer and pulsed-field gradient NMR are more useful for observing small oligomers in solution in the presence of larger aggregates. Although spectrofluorimeters can be set up to detect scattered light intensity as a means to observe aggregation, much more detailed and quantitative information is available only through laser light scattering, as described in the previous section.

Here we discuss a few published cases in which laser light scattering was used effectively to explore specific issues in protein misfolding and aggregation. The primary purpose of this section is to illustrate the various kinds of problems that can be tackled using DLS and SLS.

4.1. Protein Aggregation and Neurodegenerative Disease

4.1.1. β-Amyloid peptide and Alzheimer's disease β-Amyloid peptide (Aβ) is proteolytically cleaved from a membrane-bound precursor protein. Aβ spontaneously assembles into fibrillar aggregates with a cross-β-sheet structure; deposition of Aβ amyloid fibrils as extracellular plaques is one of the definitive diagnostic tests of Alzheimer's disease. There is strong but not conclusive genetic, epidemiological, animal model and *in vitro* tests to indicate that Aβ aggregation is causally linked to the onset and/or progression of this disease. Our group has studied Aβ aggregation kinetics *in vitro* using light scattering and other biophysical tools. In one study, we examined the growth of Aβ aggregates at three concentrations. Aggregation was initiated by diluting urea-denatured Aβ into phosphate-buffered saline.[6] Data from static and dynamic light scattering experiments are shown in Figure 5. Not surprisingly, the scattered intensity was weakest at the

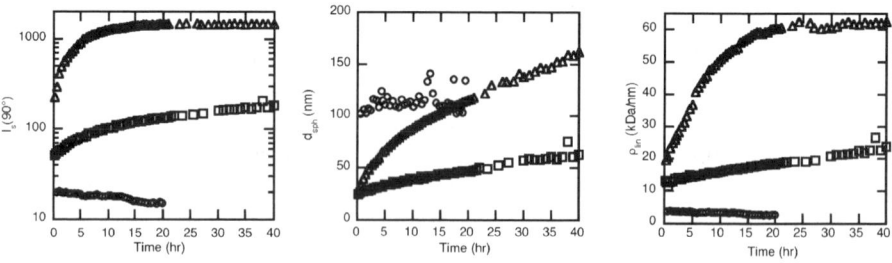

FIGURE 5. Representative light-scattering data showing Aβ aggregation kinetics. The scattered intensity $I_s(90°)$ and the hydrodynamic diameter d_{sph} were measured over time at three concentrations [70 (circles), 140 (squares), and 280 (triangles) μM]. From the combined data, the linear density was calculated.

lowest concentration tested. Only at the higher two concentrations was there an increase in scattered intensity as a function of time, indicative of an increase in average aggregate molar mass with time. Surprisingly, the hydrodynamic size of the aggregates d_{sph} was greatest initially at the lowest concentration; in contrast, the rate of growth of d_{sph} was fastest at the highest concentration. A more detailed analysis of static light scattering data confirmed that aggregates were elongated rigid rods, as would be expected by the fibrillar nature of Aβ aggregates observed by electron and atomic force microscopy. By combining DLS and SLS results we were able to calculate a linear density (molar mass per unit length); linear density was a strong function of concentration (Figure 5). Taken together, these results show that at low concentration Aβ assembles rapidly into a few very long, very thin rodlike aggregates that do not change much with time. At higher concentration, Aβ assembles initially into many, shorter rodlike aggregates that grow both in length and in linear density with time. The initial concentration-dependent behavior is akin to what is observed in crystallization processes and is indicative of an aggregation process wherein initiation (nucleation) is of higher reaction order than is propagation (elongation).

Based on these experimental data, we proposed a detailed model for aggregation that simulates not only the mass of aggregated material but also the size (average molecular weight and length) of the fibrillar aggregates. A simplified schematic is shown in Figure 6.

In related work, we designed a strategy for synthesizing compounds that should bind to and disrupt normal Aβ aggregation, thereby inhibiting its toxicity. One of these compounds, a peptide (KLVFFK$_6$), was studied in detail. We observed a dramatic shift in aggregate morphology in mixtures of KLVFFK$_6$ with Aβ compared to Aβ alone; this was readily detected by plotting SLS data in the so-called Kratky format (Figure 7). By close analysis of the data, we were able to ascertain that KLVFFK$_6$ acted on Aβ by increasing the rate of lateral association of thin filaments into thicker fibrils.[7] Put together with the observation that KLVFFK$_6$ was effective at inhibiting Aβ toxicity in cell culture,[8] these results support the hypothesis that aggregation intermediates, not fully formed fibrils, are the primary Aβ species responsible for toxicity.

4.1.2. Polyglutamine-containing proteins and Huntington's disease Huntington's disease is linked to a CAG/poly-L-glutamine (polyGln) repeat expansion in the gene encoding the huntingtin protein. The polyGln stretch ranges from 6 to 39 residues in unaffected individuals and from 36 to 180 residues in affected individuals. In an effort to elucidate the molecular mechanisms underlying the relationship between polyGln expansion and protein aggregation, Georgalis *et al.*[9] studied glutathione S-transferase-huntingtin fusion proteins that contained either 20 or 51 polyGln extensions (GST-HD20 and GST-HD51). Dynamic light scattering, Western blotting, and electron microscopy were used to examine aggregation and filament formation of GST-HD after site-specific cleavage by trypsin. From light-scattering data, these researchers determined that the apparent diffusion coefficient, D_{app}, for GST-HD20 and GST-HD51 decreased as a function of time. Initially, D_{app} was approximately the same for both constructs, at 1×10^{-7} cm^2/sec for GST-HD20 and 9×10^{-8} cm^2/sec for GST-HD51. However, after 4 hours, D_{app} was approximately five times lower for GST-HD51 than for GST-HD20. Additionally, the normalized second cumulant, an indicator of solution polydispersity, was approximately 100 times greater for GST-HD51 than for GST-HD20. These results indicated that aggregation was much

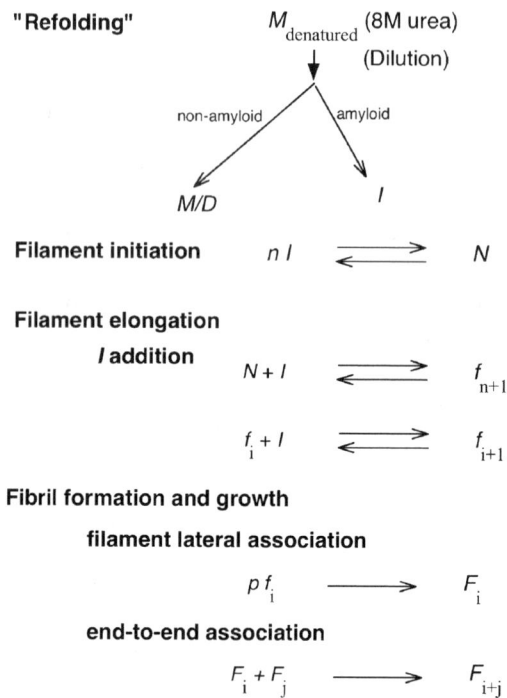

FIGURE 6. Simplified schematic of Aβ aggregation kinetics model. Refolding is initiated by dilution from the urea-denatured state into phosphate-buffered saline; the peptide follows either of two paths to form a non-amyloidogenic and stable monomer/dimer (M/D) or an amyloidogenic dimeric intermediate I. Cooperative association of I into the oligomer N initiates formation of thin filaments f, which then elongate by further addition of I. Thicker fibrils F form by lateral association of filaments and grow lengthwise by end-to-end association.

more pronounced in GST-HD51 compared to GST-HD20. The angular dependence of the diffusion coefficient was determined by plotting $D_{app}(q)$ versus q^2. The observed deviations from linearity for both GST-HD20 and GST-HD51 suggested the formation of fibrillar structures, which was confirmed by electron microscopy. The light-scattering data gave support to a mechanism in which fibrillogenesis was caused by self-aggregation of polyGln sequences that acted as "polar zippers," forming pleated sheets of β-strands that are held together by hydrogen bonding between side-chain amides and the polypeptide backbone.

4.1.3. Prion proteins and "mad cow" disease The conversion of monomeric prion proteins that are α-helical to multimers that are mainly β-sheets is believed to be an important step in the pathogenesis of prion diseases. Sokolowski et al.[10] studied the conformational transition and aggregation of recombinant Syrian hamster prion protein that contained amino acids 90 to 232 (SHaPrP$^{90-232}$). They used dynamic and static light scattering, infrared spectroscopy (FTIR), circular dichroism (CD) spectroscopy, and electron microscopy to examine the behavior of SHaPrP$^{90-232}$ at pH 4.2 and 7 and with varying concentrations of the denaturant guanidium chloride (GdnHCl). Size

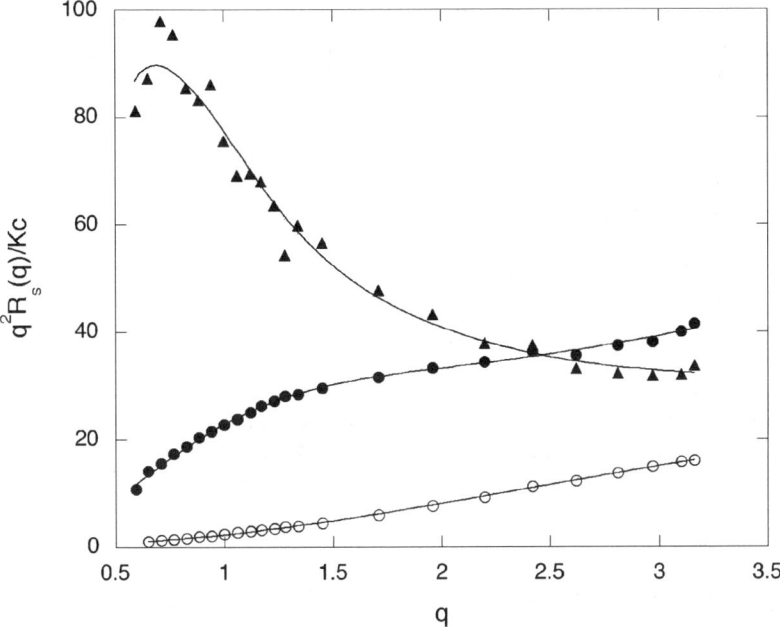

FIGURE 7. Kratky plots showing change in particle morphology. Light scattering from solutions of Aβ alone (open circles) or with KLVFFK$_6$ at 1:2.5 molar ratio (closed circles) or 1:10 molar ratio (triangles) are plotted as $q^2 R_s(q)/Kc$ vs q. The shapes of the curve show the change from relatively short linear rods (Aβ alone), to longer semiflexible chains, to large branched or entangled structures at high KLVFFK$_6$:Aβ ratio.

distributions from dynamic light scattering showed that only one peak of monomers existed at pH 7 and in the absence of GdnHCl (Figure 8). An additional peak existed at pH 4.2, due to a small amount of aggregates. Molecular weights of 16 ± 2 kDa at pH 7 and 18 ± 2 kDa at pH 4.2 were obtained by extrapolating light-scattering data to zero protein concentration. Hydrodynamic radii were 2.36 ± 0.04 nm at pH 7 and 2.66 ± 0.05 nm at pH 4.2. Therefore, the protein was slightly less compact at pH 4.2 than at pH 7.

Aggregation of SHaPrP$^{90-232}$ after addition of 1M GdnHCl at pH 4.2 was examined by light scattering. Analysis of the data indicated that eight monomers formed a stable misfolded oligomer. FTIR spectra showed that the oligomers were stabilized by intermolecular β-sheets with strong hydrogen bonds. The hydrodynamic radii of the oligomers were plotted against their relative molar masses and were seen to have a dimensionality of ∼3, indicating that they were compact spheres. The researchers proposed that the oligomers were composed of eight monomers arranged in a ring. Observed increases in mass and radii with time were due to lengthening of the structures into protofibrils. By combining light-scattering data with FTIR and CD results, the group concluded that the overall reaction order was between 2 and 3 and that oligomer formation occurred in two bimolecular steps with a weakly populated intermediate. A model was proposed in which native protein was transferred to a destabilizing environment (1M GdnHCl, pH = 4.2) that caused it to undergo structural change to state $N_{\alpha'}$, which was less helical

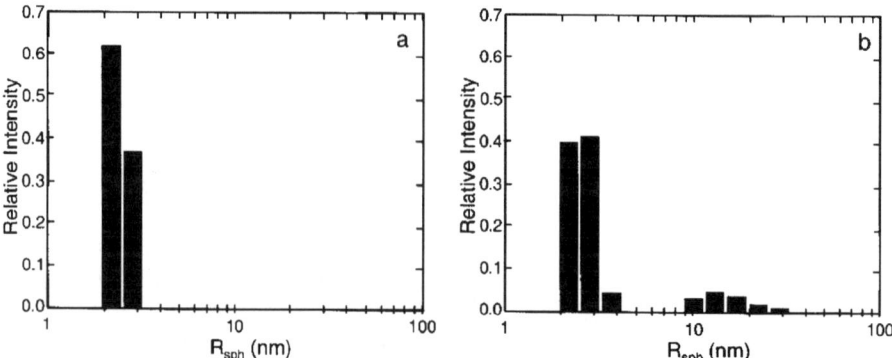

FIGURE 8. Size distributions of SHaPrP$^{90-232}$ at pH 4 and 7. (a) Size distribution at pH 7 and in the absence of GdnHCl. (b) Size distribution at pH 4.2 and in the absence of GdnHCl shows that an extra peak exists due to aggregates. The graph was created from data reported in Sokoloski et al.[10]

and more likely to aggregate (Figure 9). The destabilized proteins associated to form weakly populated intermediates (I_1, I_2) in a series of steps that are not well understood. Once formed, octamers were stable against dissociation and self-associated further into protofibrils.

4.2. Heat-Induced Aggregation

Pots et al.[11] studied the thermal aggregation of the plant protein patatin, in an attempt to elucidate its aggregation mechanism. The rationale for their research was that a better understanding of the kinetics of coagulation could lead to development of protein products that are better suited for application in food systems. Aggregation behavior was studied using DLS and small-angle neutron scattering (SANS). Natively folded patatin was found to have a d_{sph} of 8–10 nm. Patatin was incubated for 6 to 60 min at temperatures ranging between 40 to 65°C and DLS measurements were collected. An increase in the average size with time was readily observed. The aggregation rate of patatin increased at higher temperatures; this phenomenon is commonly considered to be an indicator of hydrophobically driven aggregation. The increased rate coincided with a loss of α-helical structure determined by far-UV circular dichroism. By performing additional experiments on the effect of ionic strength on the rate of aggregation, Pots et al. determined that patatin followed the Smoluchowski-Fuchs mechanism, in which coagulation is limited by both diffusion and reaction. Aggregation was found to follow a second-order dependence on protein concentration, in agreement with that mechanism.

$$N_\alpha \longrightarrow N_{\alpha'} \rightleftharpoons I_1 \rightleftharpoons I_2 \longrightarrow A_\beta$$

FIGURE 9. Proposed model of aggegation of ShaPrP$^{90-232}$ into octamers under moderately denaturing conditions (1 M GdnHCl, pH 4.2). Native structure N_α is destabilized to form $N_{\alpha'}$. In a series of steps, $N_{\alpha'}$ monomers interact to form I_1, which interact with other intermediates to form I_2. I_2 oligomers aggregate to form A_β, which is a ringlike octamer. Redrawn from a figure and data in Sokoloski et al.[10]

Bauer et al.[12] studied heat-induced aggregation of β-lactoglobulin A and B, using a system in which protein samples passed through a size exclusion column before the peaks were analyzed by static light scattering. They detected a mixture of monomers, dimers and trimers at all heating times. After one to several hours, depending on the β-lactoglobulin variant, the monomer and oligomer peaks decreased and large aggregates appeared. Aggregates of β-lactoglobulin A appeared suddenly after several hours; these aggregates had an average R_g of 29 ± 1 nm and M_{app} that increased from 2×10^6 to 3.5×10^6 Da. The β-lactoglobulin B aggregates appeared earlier, grew more gradually, and were generally smaller (R_g of 21 ± 2 nm, M_{app} increasing from 0.8 to 1.3×10^6 Da). These researchers established that the formation of heat-modified small oligomers was an important prerequisite for further aggregation. They determined that aggregation proceeded through the formation of metastable heat-modified dimers and higher oligomers, which then aggregated further.

4.3. Refolding and Aggregation

Panda et al.[13] examined the enzyme rhodanese in order to investigate how the stability and relative size of aggregates formed during protein refolding affect the efficiency of renaturation. Dynamic light scattering was used to obtain size distributions of aggregates at various times after initiating refolding of denatured proteins. Within the first minute of refolding, particles with an average d_{sph} of ~225 nm were detected, in comparison to native rhodanese which has a diameter of ~5.6 nm. Within 44 min, d_{sph} increased to ~330 nm. Additionally, the heterogeneity of the protein sample increased, with the half-height width of the distributions increasing from ~94 nm after 1 min to ~200 nm after 44 min. By performing additional experiments in which refolding rates were shown to be concentration dependent and high hydrostatic pressures were shown to dissociate oligomers, Panda et al. suggested a mechanism for rhodanese folding in which the intermediate, dissociated species could be rescued from an aggregation pathway and returned to a productive folding pathway (Figure 10). After refolding was initiated, rhodanese was postulated to undergo fast compaction to form intermediate I_1, which then rapidly repartitioned to form either I_2 or I_3. The I_3 could directly refold to the active, native

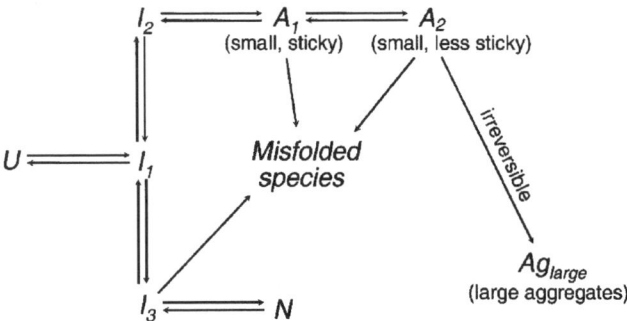

FIGURE 10. Proposed pathway of aggregation of rhodanese during refolding from the denatured state.[13] Intermediates are postulated to be vulnerable to aggregation while traversing the folding pathway.

enzyme, while I_2 had to undergo a slow conformational change to I_1 and then I_3 before forming the active protein. During this process, I_2 was susceptible to aggregation.

Herbst et al.[14] analyzed the kinetics of reactivation from different unfolding intermediates in order to determine the role of folding intermediates as aggregation precursors and to study the rate-limiting step in refolding. The protein studied was firefly luciferase, whose sensitivity to aggregation has made it a popular model for examining the role of chaperones in preventing aggregation. Static and dynamic light-scattering measurements showed that native luciferase had a d_{sph} of 6.5 nm and M of 64 kDa, which agreed with the value calculated from the amino acid composition of the monomer. Luciferase denatured in 1M GdmCl had an initial d_{sph} of 8.6 nm and an initial M of 61 kDa. These values, obtained for protein immediately after being added to denaturing solution, were consistent with an expanded monomer. After 8 hours of incubation in denaturant, the scattering intensity increased 2.5-fold and d_{sph} increased 1.5-fold, suggesting the formation of trimers. At 18–20 hours d_{sph} was 1.6-fold greater than the initial value and no further growth was observed. By combining these data with gel filtration studies, Herbst et al. developed a model for luciferase folding that included two equilibrium intermediates (I_1 and I_2) and a kinetic intermediate, I_{fast}, an enzymatically inactive species that was able to reactivate rapidly to the native state when denaturant was diluted (Figure 11). I_1, I_2, and unfolded protein (U) were in rapid equilibrium with each other. Reactivation for each of these species upon addition of renaturing buffer occurred unusually slowly and did not follow first-order kinetics, suggesting that the same long-lived conformation was adopted for each of the species prior to undergoing the slow conversion to I_{fast}. The long-lived conformation functioned as a kinetic trap. The slow conversion of I_{fast} to I_1 was not due to global unfolding but rather to local denaturing that released a small part of the amino-terminal domain of luciferase and allowed it to form nonnative interactions that needed to be broken for luciferase to fold into its native conformation. In vivo, luciferase was able to avoid this kinetic trap through a fast folding throughway that is thought to involve chaperones.

Intermediates I_1 and I_2 also served as aggregation precursors. Rapid aggregation stopped when the concentration of the intermediate dropped below a critical value. The

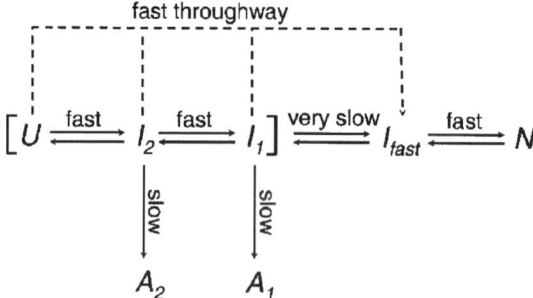

FIGURE 11. Proposed pathway of aggregation of luciferase during refolding from the denatured state.[14] Unfolded protein U and intermediates I_1 and I_2 equilibrate rapidly before slowly converting to I_{fast}, which is able to reactivate to the native protein N. I_1 and I_2 accumulated and aggregated during the slow conversion to I^{fast}. In vivo, the fast throughway could be accessed through the use of chaperones.

observed concentration dependence of aggregation was characteristic for a reaction order greater than 2 and was consistent with a nucleated polymerization mechanism, in which the reversible formation of small oligomers nucleated irreversible aggregation.

4.4. Protein Aggregation and Chaperones

Abgar et al.[15] used dynamic light scattering, along with equilibrium sedimentation and gel filtration, to analyze the chaperone function of α-crystallin, a bovine lens protein that is part of the heat-shock protein superfamily. The interaction between α-crystallin and reduced-and-destabilized lysozyme was studied at 25°C and 37°C. DLS was used to examine the change in the apparent d_{sph} of α-crystallin/lysozyme complexes as a function of time. At 25°C and a 5:1 ratio of α-crystallin to lysozyme, a small population of large particles with diameters of ∼60 nm and greater existed after 12 hours of incubation. When the ratio of α-crystallin to lysozyme was decreased to a 4:1 ratio, the rate of formation of large particles increased. After 4 hours of incubation, the population of particles was composed of one group with a diameter of ∼21 nm and a second group with a diameter of ∼55 nm. After 12 hours, both the size and number of the larger particles increased, resulting in an overall average diameter of ∼78 nm.

These experimental results suggested that α-crystallin bound to the destabilized lysozyme to prevent nonspecific aggregation. The unstable α-crystallin/lysozyme complexes that were formed resulted in reorganization and interparticle exchange of the α-crystallin peptides and destabilized lysozyme peptides such that large soluble particles were formed. Under conditions unfavorable for the chaperone activity of α-crystallin, new populations of very large particles were created from association of several α-crystallin/lysozyme complexes. The associations may be due to interactions between the destabilized lysozyme. Since the complexes at 37°C were smaller than those at 25°C, the researchers concluded that α-crystallin was a better chaperone at the higher temperature.

4.5. Protein Aggregation in Ethanol Solutions

Tanaka et al.[16] used DLS to study denaturation and aggregation of hen egg lysozyme in a water-ethanol solution. Drastic changes in the solution structure occurred when the ethanol concentration was increased to over 70% (v/v). The autocorrelation data were fitted to a stretched exponential function. The researchers found that behavior at lower ethanol concentrations [0, 45, and 63% (v/v)] differed greatly from that at higher concentrations [72, 81, and 90% (v/v)]. At 0% (v/v) ethanol, only a fast-relaxation mode was observed. At 45 and 63% (v/v) ethanol, the autocorrelation function had two shoulders, corresponding to fast and slow-relaxation modes. The researchers proposed that the slow mode was due to clusters or concentration fluctuations caused by the repulsive interactions of lysozyme monomers, akin to similar behavior observed in solutions of polyelectrolytes with no added salt. At even higher ethanol concentrations, diffusive motion slowed considerably. Tanaka et al. proposed that this large decrease was due to entanglement of lysozyme aggregates. At 90% ethanol, diffusive motion slowed over time, and after a month the solution formed a transparent gel, which X-ray diffraction indicated was due to the formation of intermolecular β-sheets.

5. SUMMARY

Static and dynamic light scattering provide the best way to examine particle size, shape, and growth rates in aggregating protein systems prior to precipitation. The techniques are noninvasive, do not require chemical modification of the protein under study, and data can be taken in real time. If sample preparation and data interpretation are undertaken with care, it is possible to gain important insights into aggregation mechanisms. Why is it important to characterize the particle size and kinetics of protein aggregation? For protein misfolding diseases such as Alzheimer's, there is continuing uncertainty about the size of the toxic species and whether the toxicity is found in the final aggregate product or an intermediate in the aggregation pathway. Quantitative kinetic studies provide a means for interpreting biological responses and for validating or refuting hypotheses regarding the mechanism of toxicity. Further, such studies guide the design of therapeutic compounds to prevent formation of toxic species. In manufacturing of proteins, light scattering is exquisitely sensitive to very small quantities of aggregates, quantities that may escape detection by methods such as size exclusion chromatography. Additionally, light scattering is a very useful tool for evaluating mechanisms of aggregation that can aid in developing strategies for controlling aggregation during processing, formulation, and storage. We anticipate continued growth in the application of this technique as one of several important biophysical tools in the study of misfolding and aggregation of proteins.

REFERENCES

1. S. Timasheff and R. Townend, Light scattering: in *Physical Principles and Techniques of Protein Chemistry*, Part B, ed. by S. J. Leach (New York: Academic Press, 1970).
2. W. Burchard, Static and dynamic light scattering from branched polymers and biopolymers, *Adv. Polymer Sci.* **48**, 1–124 (1983).
3. E. P. Geiduschek and A. Holtzer, Application of light scattering to biological systems: deoxyribonucleic acid and the muscle proteins, *Adv. Biol. Med. Phys.* **6**, 431–551 (1958).
4. M. M. Tirado and J. G. de la Torre, Translational friction coefficients of rigid, symmetric top macromolecules. Application to circular cylinders, *J. Chem. Phys.* **71**, 2581–2587 (1979).
5. J. M. Andreu and S. N. Timasheff, The measurement of cooperative protein self-assembly by turbidity and other techniques, *Meth. Enzymol.* **130**, 47–59 (1986).
6. M. M. Pallitto and R. M. Murphy, A mathematical model of the kinetics of β-amyloid fibril growth from the denatured state, *Biophys. J.* **81**, 1805–1822 (2001).
7. J. R. Kim and R. M. Murphy, Mechanism of accelerated assembly of β-amyloid filaments into fibrils by KLVFFK$_6$, *Biophys. J.* **86**, 3194–3203 (2004).
8. T. L. Lowe, A. Strzelec, L. L. Kiessling and R. M. Murphy, Structure-function relationships for inhibitors of β-amyloid toxicity containing the recognition sequence KLVFF, *Biochemistry* **40**, 7882–7889 (2001).

9. Y., Georgalis, E. B. Starikov, B. Hollenbach, R. Lurz, E. Scherzinger, W. Saenger, H. Lehrach, and E. E. Wanker, Huntingtin aggregation monitored by dynamic light scattering, *Proc. Natl. Acad. Sci. USA* **95**: 6118–6121 (1998).
10. F. Sokolowski, A. J. Modler, R. Masuch, D. Zirwer, M. Baier, G. Lutsch, D. A. Moss, K. Gast, and D. Naumann, Formation of critical oligomers is a key event during conformational transition of recombinant Syrian hamster prion protein, *J. Biol. Chem.* **278**, 40481–40492 (2003).
11. A. M. Pots, E. ten Grotenhuis, H. Grappen, A. G. J. Voragen, and K. G. de Kruif, Thermal aggregation of patatin studied *in situ*, *J. Ag. Food Chem.* **47**, 4600–4605 (1999).
12. R. Bauer, R. Carrotta, C. Rischel, and L. Ogendal, Characterization and isolation of intermediates in β-lactoglobulin heat aggregation at high pH, *Biophys. J.* **79**, 1030–1038 (2000).
13. M. Panda, B. M. Gorovits, and P.M. Horowitz., Productive and nonproductive intermediates in the folding of denatured rhodanese, *J. Biol. Chem.* **275**, 63–70 (2000).
14. R. Herbst, K. Gast, and R. Seckler, R, Folding of firefly (*Photinus pyralis*) luciferase: aggregation and reactivation of unfolding intermediates, *Biochemistry* **37**, 6586–6597 (1998).
15. S. Abgar, J. Vanhoudt, T. Aerts, and J. Clauwaert, Study of the chaperoning mechanism of bovine lens α-crystallin, a member of the α-small heat shock superfamily, *Biophys. J.* **80**, 1986–1995 (2000).
16. S. Tanaka, Y. Oda, M. Ataka, K. Onuma, S. Fujiwara, and Y. Yonezawa, Denaturation and aggregation of hen egg lysozyme in aqueous ethanol solution studied by dynamic light scattering, *Biopolymers* **59**, 370–379 (2001).

X-ray Diffraction for Characterizing Structure in Protein Aggregates

Hideyo Inouye,[1] Deepak Sharma,[1] and Daniel A. Kirschner[1,2]

Protein molecules often aggregate in a hierarchical manner to form large assemblies that are characteristic of amyloid diseases: from linear arrays of molecules that constitute protofilaments to lateral aggregation of molecules and protofilaments that form sheets and fibrils. Here we review the forces involved in such aggregation and the X-ray patterns that are obtained from the aggregates. Examples of the different types of aggregates that we describe include assemblies having no lattice structure (hydrated prion peptide PrP90-145), two-dimensional lateral associations of helical fibrils (betabellin), and one-dimensional stacking of platelike structures (PrP106-122). Experimental procedures we describe include sample preparation techniques that allow us to manipulate protein aggregation (including vapor hydration, solubilization followed by drying, and fibril alignment using a magnetic field) and the correction factors needed to analyze properly the observed intensity data from fiber and powder diffraction.

1. INTRODUCTION

Protein aggregation refers to the assembly of protein molecules into macromolecular forms via protein-protein interactions. The macromolecular assemblies have varied geometries, including a three-dimensional lattice (as in single crystals), sheetlike structures in a two-dimensional lattice, and fibrillar structures having an axial one-dimensional or helical-array in the fibril direction. The protein aggregates often show a structural hierarchy[1,2] from a limited crystal size (i.e., crystallites) to the assembly of crystallites into a sheet or fibrillar structure. Protofilaments that are formed from linearly arrayed crystallites can associate laterally to form larger fibrils. A nucleation-dependent mechanism[3-7] has been used to account for the observed kinetics of aggregation and crystallization. The conditions involving protein-protein interactions leading

---

[1] Biology Department, Boston College, Chestnut Hill, MA 02467-3811
[2] To whom correspondence should be addressed: Daniel A. Kirschner, Biology Department, Boston College, Chestnut Hill, MA 02467-3811, email: kirschnd@bc.edu

to aggregation/coagulation and crystallization are described by the Derjaguin-Landau-Verwey-Overbeek (DLVO) theory of colloid stability[8–12] and by the second virial coefficient.[13]

Analysis of X-ray diffraction patterns can provide information about the size, shape, and distribution of the protein aggregates and those of the constituent protein molecules. The medium- and low-angle region of the pattern informs about protein aggregation, e.g., the ~40 Å- and ~100 Å-sizes, respectively, of the amyloid protofilament and fibril[1] and the interparticle interference between tetramers of myelin P0-glycoprotein in a membrane mimetic.[12] The wide-angle region in X-ray patterns informs about the conformation of the constituent proteins, e.g., α-helical structure is defined by pitch P (5.41 Å) and rise per unit h (1.50 Å), the polypeptide backbone helix in antiparallel β-pleated sheets is described by $P = 6.95$ Å and $h = 3.47$ Å (with the distance between pleated sheets ~5–10 Å[14,15]), and the DNA helix by $P = 34$ Å and $h = 3.4$ Å.[16]

In this review, we provide technical and practical descriptions involving the analysis of protein aggregates using the method of X-ray diffraction. Structural studies such as these are relevant to pathology and biopolymer-based fabrication, i.e., amyloidosis,[1,7,17] inclusion-body formation,[18] hemoglobin fibril formation,[19] cataract formation,[20] assembly of bionanotube materials,[21–23] DNA-templated dye aggregation,[24] and carbon nanotube assembly.[25]

2. FUNDAMENTALS

2.1. Diffraction Method

X-ray diffraction patterns of protein aggregates show a mixture of sharp reflections, broad reflections, and diffuse bands. Furthermore, these intensities are not concentrated at finite positions but are distributed along angular and radial directions. It is useful, therefore, to describe the total intensity as a function of reciprocal coordinates. In kinematic theory[26–30] the total intensity is given as a function of the reciprocal coordinate vector $\vec{R}$,

$$I(\vec{R}) = N[<F^2(\vec{R})> - <F(\vec{R})>^2] + \frac{1}{v}<F(\vec{R})>^2 \left[Z(\vec{R})^* \left| \sum (\vec{R}) \right|^2 \right]. \qquad (1)$$

The symbols <> and * are the averaging and convolution operations. The first term corresponds to the diffuse scattering and the second term corresponds to the Bragg reflections. $Z(\vec{R})$ is the interference function; $|\sum(\vec{R})|$ arises from the Fourier transform of the step function expression for the extent of the lattice; $F(\vec{R})$ is the Fourier transform of a unit structure; N is the number of lattice points; and v is the unit volume.

For single crystals the first term on the right-hand side of Eq. (1) is zero and the interference function in the second term defines the discrete positions of the Bragg reflections at $R = ha^* + kb^* + lc^*$, where h, k, l are integers and a^*, b^*, c^* are reciprocal vectors forming the unit cell.

In fiber or powder diffraction, the intensity $I(R)$ is further cylindrically or spherically averaged.[31] Kinematic theory has been applied to analyzing the disorder in membrane stacking[29,32,33] and in fibrils.[1,34–37] The distribution function $z(r)$ (or distance statistic term, which is the radial distribution in spherical or cylindrical coordinates) and its Fourier transform $Z(R)$ (interference function or structure factor*) describe how the objects are aggregated. If the particles are dispersed in solution, the interference function is unity, in which case the scattered intensity arises from the structure amplitude of a single particle $\langle F^2(R) \rangle$. If the particles are arrayed, the interference function from the lattice results in successive sharp Bragg reflections that sample the structure amplitude. When the lattice is disordered or liquidlike, paracrystalline theory predicts sharp interference peaks at lower scattering angle and increasingly broader peaks at higher angle. The following indicates how X-ray diffraction can be used to probe the hierarchical nature of the aggregates.

2.1.1. Spherical particle aggregation Solution scattering is best described in the spherical coordinates (r, ϕ, φ). For a spherically symmetric particle where the diffuse term is zero, the averaged intensity function, written in terms of the radial component of the reciprocal spherical coordinate R, is

$$I(R) = NF^2(R)Z(R), \qquad (2)$$

where N is the number of particles, $F(R)$ is the particle structure factor, and $Z(R)$ is the interference term. By using the radial distribution function $g(r)$, $Z(R)$ can be written as

$$Z(R) = 1 + (1/v) \int 4\pi r^2 (g(r) - 1) \sin(2\pi r R)/(2\pi r R) dr. \qquad (3)$$

The radial distribution is related to the surface charge of a particle, which influences protein-protein interactions in the liquid model.[12,38,39] The interference term at origin $Z(0)$ also is related to the second virial coefficient, which is an indicator of the crystallizability of proteins.[13] According to the Debye equation the structure amplitude is given by the constituents of the particle as

$$F^2(R) = \sum_m \sum_n f_m(R) f_n(R) \sin(2\pi r_{mn} R)/(2\pi r_{mn} R), \qquad (4)$$

where $f_m(R)$ is a structure factor of the mth constituent protoparticle and $r_{mn} = |r_m - r_n|$.[12] For example, this type of formulation indicates from low-angle X-ray scattering that tetramers of myelin P0 glycoprotein constitute single particles in solution.[12] Moreover, the interference function indicates that the particle aggregation depends on protein concentration, pH, and ionic strength.

2.1.2. Cylindrical particle aggregation For a fibrillar object the cylindrical axis is often along the fibril direction. The cylindrically averaged intensity is written as

$$I(R, Z) = N(<F^2> - <F>^2) + <F>^2 \sum_j \sum_k J_0(2\pi r_{jk} R)$$
$$<F^2> = \sum_m \sum_n f_m f_n J_0(2\pi r_{mn} R) \exp(i 2\pi z_{mn} Z) \qquad (5)$$
$$<F>^2 = \left| \sum_m f_m J_0(2\pi r_m R) \exp(i 2\pi z_m Z) \right|^2,$$

where F is the Fourier transform of the electron density distribution of the unit object, $<>$ refers to statistical averaging, and N is the number of lattice points.[28] If a unit object is cylindrically symmetric, the first term is zero and $<F^2> = <F>^2$. This formulation has been used for analyzing lateral aggregation for many different types of fibrils including actin, collagen, feather keratin, flagella,[40,41] and more recently amyloid fibrils.[1,42,43]

2.1.3. Stacking of plates Consider the case of infinitely extended plates, each one of width a, and stacking in direction r with period d. The intensity is restricted to the reciprocal coordinate R parallel to the r direction. This one-dimensional intensity distribution expressed in terms of the structure factor $F(R)$ and the interference function $Z(R)$ is

$$I(R) = NF^2(R)Z(R), \qquad (6)$$

where $F(R) = a \operatorname{sinc}(\pi a R)$. Assuming that the stacking is characterized by paracrystalline disorder of the second kind (i.e., where there is a statistical distribution of the period d), the interference function becomes

$$Z(R) = \frac{1 - |H(R)|^2}{1 + |H(R)|^2 - 2|H(R)|\cos(2\pi dR)}, \qquad (7)$$

where $H(R) = \exp(-2\pi^2 R^2 \Delta^2)$ is the Fourier transform of the Gaussian distribution function $h(r)$ for the nearest lattice point and Δ is the deviation in $h(r)$.[28,29,44] If the disorder parameter is large, only a single interference peak may be observed,[45] followed by broad intensity maxima from the structure amplitude. This type of formulation has been used for analyzing the structure of the nerve myelin membrane,[29] the slablike structure of Alzheimer's β-amyloid analogues,[1] and prion-related peptides.[45]

2.2. Forces Involved in Aggregation

2.2.1. DLVO theory applied to platelike structures In order to become aggregated the particles (plates) must become sufficiently close. The interplate distance can be evaluated using DLVO theory of colloid stability that includes van der Waals attractive and electrostatic repulsive long-range forces. We previously have used this approach to account for membrane-membrane interactions in nerve myelin as a function of pH and ionic strength[10,11] and for the oligomerization of myelin P0-glycoprotein in solution.[12] We do not consider here the short-range interaction between specific amino acids that is due to ion-pairing and hydrogen bonding.

At equilibrium van der Waals attraction is balanced against repulsion that has both electrostatic (F_r) and short-range hydration (F_h) components. The electrostatic force is given by

$$F_r = 2nkT\{\cosh[e\phi(d_w/2)/kT] - 1\}, \tag{8}$$

where n is the univalent electrolyte concentration, e is the elementary charge, k is the Boltzmann constant, T is the absolute temperature, d_w is the separation between plates or membranes, and $\phi(d_w/2)$ is the potential at the midpoint between surfaces.[8] For a pair of plates the potential at the midpoint between them is approximately equal to twice that at the same distance from the surface of an isolated plate.[8] The potential is given as a function of distance from the surface of the isolated plate having surface charge σ as

$$\tanh[Y(x)/4] = \tanh[Y(x_s)/4]\exp(-\kappa x)$$
$$\sigma = (\varepsilon kT/2\pi e)\kappa \sinh[Y(x_s)/2], \tag{9}$$

where $Y(x) = e\phi(x)/kT$ and the Debye parameter $\kappa = [8\pi e^2 n/(\varepsilon kT)]^{1/2}$. The repulsion due to hydration is expressed in the exponential form

$$F_h = K\exp(-d_w/L), \tag{10}$$

where $L = 1.93$ Å, $K = 1 \times 10^{11}$ for the force (units of dyne/cm^2), and d_w is the separation between plates.[46] The van der Waals attraction F_a is

$$F_a = (H/6\pi)[1/d_w^3 - 2/(d_w + d_{ex})^3 + 1/(d_w + 2d_{ex})^3], \tag{11}$$

where H is the Hamaker coefficient (5×10^{-14} erg) and d_{ex} is the plate thickness (or exclusion length).[8]

Applying this analysis to the case of myelin membranes (Figure 1), we see that the lowest energy minimum corresponds to the isoelectric pH (pI) where the electrostatic repulsion force between membranes is zero and the hydration repulsion is balanced against van der Waals attraction. At pH > pI or pH < pI, electrostatic repulsion drives the energy minimum, which becomes shallower and broader, to larger membrane separations (reflecting less stability in membrane packing). If the repulsive hydration force is abolished, for example, by dehydrating reagents such as acetone, alcohol, or salts of a lyotropic series, much more stable aggregates will be formed by the balance between electrostatic repulsion and van der Waals attraction. Consequently, the separation between the dehydrated membranes would be much smaller (Figure 1).

2.2.2. Potential function and lattice disorder The breadth or curvature of the potential energy minimum (Figure 1) indicates the stability of that particular membrane separation, e.g., a broader minimum indicates less stability. Moreover, the observed broadening of

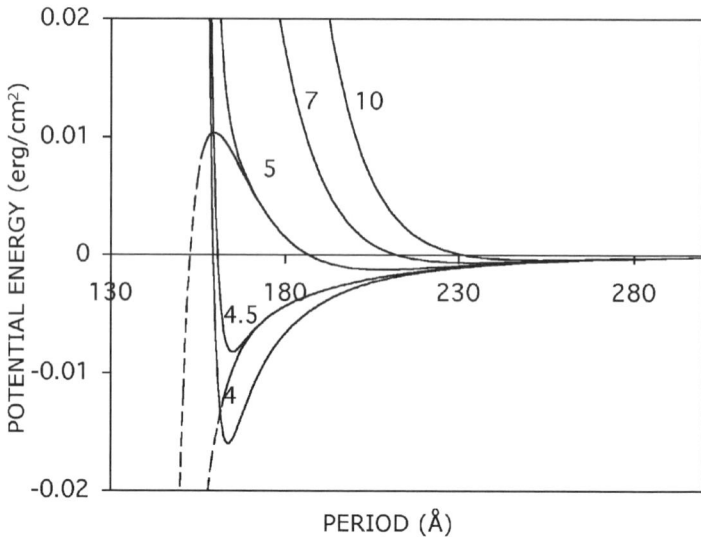

FIGURE 1. Total potential energy curves at different pH (4, 4.5, 5, 7, and 10) as a function of myelin period at constant ionic strength (0.06) for a model of the interaction between extracellular surfaces of peripheral nerve myelin.[12] The intermembrane distance that gives the energy minimum is expressed as the myelin period observed by X-ray diffraction. Note that the energy minima get broader and shallower as the period increases. (*Dashed line*) Potential energy curves at pH 4.5 and 5 without including hydration repulsion. The height of the energy maximum becomes much larger at higher pH due to stronger electrostatic repulsion, while at lower pH the maximum is abolished due to stronger van der Waals attraction.

the Bragg reflections is likely caused by fluctuation of the width of the water layer between membranes. If a harmonic oscillation is used to describe the potential function $V(d)$ at the equilibrium period d_e, then

$$V(d) = V_e + A(d - d_e)^2/2, \tag{12}$$

where V_e is the energy at d_e and A is a measure of the curvature of the potential energy minimum at position d_e. The probability of period d follows the Boltzmann distribution

$$\begin{aligned}P(d) &= \exp[-V(d)S_m/kT] \\ &= P(d_e)\exp[-AS_m(d-d_e)^2/(2kT)],\end{aligned} \tag{13}$$

where S_m is the area of the interacting membrane surfaces. The integral width $2d_i$ of the Gaussian $P(d)$ is given by

$$2d_i = (2\pi kT/AS_m)^{1/2}, \tag{14}$$

The integral width $2d_i$ of the Gaussian curve can be regarded as lattice disorder Δd in paracrystalline theory. The interference function $Z(R)$ of such a disordered lattice is

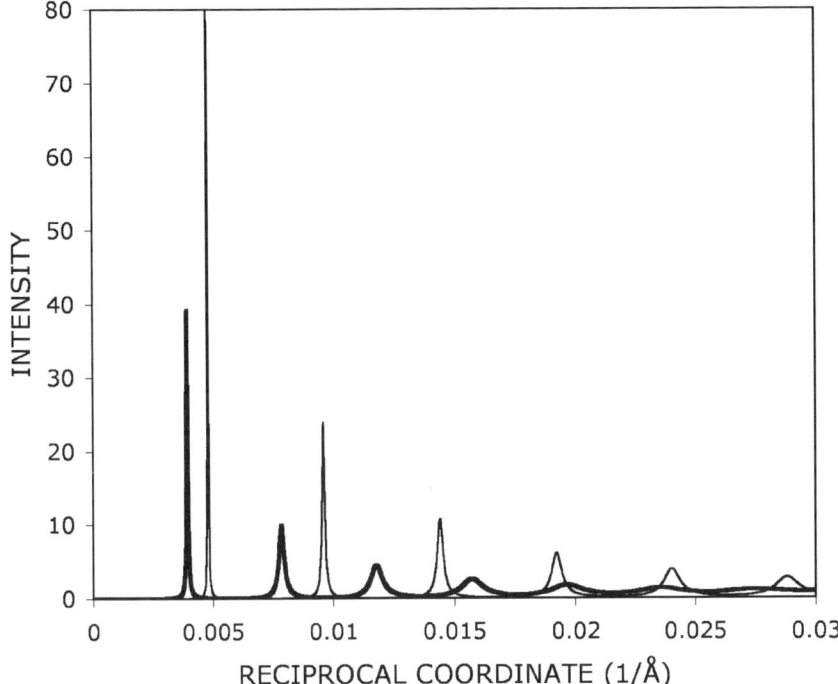

FIGURE 2. Interference function for paracrystalline lattice model of myelin at pH 5 (*thinner line*) and at pH 10 (*thicker line*) at constant ionic strength 0.06. The area of the interacting surfaces (S_m) was chosen as 2.16×10^{-9} cm^2; other conditions are the same as for Figure 1.

given by

$$Z(R) = \frac{1 - [H(R)]^2}{1 + [H(R)]^2 - 2[H(R)]\cos(2\pi d R)}$$

$$H(R) = \exp(-2\pi^2 R^2 \Delta^2), \qquad (15)$$

$H(R)$ is the Fourier transform of the Gaussian distribution function $h(r)$ for the neighboring lattice point and Δ is the deviation of $h(r)$.[26,28,29,35] For the myelin example described above, the potential function gives d_e and Δd as 208 Å and 6.8 Å, respectively, at pH 5 and as 254 Å and 12.9 Å, respectively, at pH 10. The calculated interference function (Figure 2) shows that a lattice having a greater deviation at pH 10 has broader reflections than at pH 5; both cases show greater broadening for wider-angle reflections.

2.2.3. *Coagulation condition* The coagulation condition for protein particles is accounted for by DLVO theory.[12] Assuming a particle is spherical with diameter σ, the electrostatic and attractive potential V_r and V_a can be expressed[8] as a function of the distance between their centers r by

$$V_r = \pi \varepsilon_0 \varepsilon \sigma^2 \psi_0^2 \exp[-\kappa(r - \sigma)]/r, \qquad (16)$$

$$V_a = H\{2/(s^2 - 4) + 2/s^2 + \ln[(s^2 - 4)/s^2]\}/6, \qquad (17)$$

where H is the Hamaker coefficient (as above) and $s = r/(\sigma/2)$. The total potential $V_t = V_r - V_a$. The critical ionic condition for the coagulation satisfies $V_t = 0$ and $dV_t/dr = 0$. From the first condition,

$$\pi \varepsilon_0 \varepsilon \sigma^2 \psi_0^2 \exp[-\kappa(r - \sigma)]/r = H\sigma/[24(r - \sigma)]. \qquad (18)$$

From the second condition,

$$\kappa + 1/r = 1/(r - \sigma). \qquad (19)$$

From these two equations, the ionic condition for an abrupt coagulation can be derived. Figure 3 shows application of this method for myelin protein P0 in a low concentration of sodium dodecyl sulfate (P0-SDS particles), where there is a surface charge of +10 for an 80 Å-diameter sphere.

2.2.4. *Force involved in aggregation: Second virial coefficient* Proteins that can be successfully crystallized exhibit in unsaturated solution a narrow distribution of the second

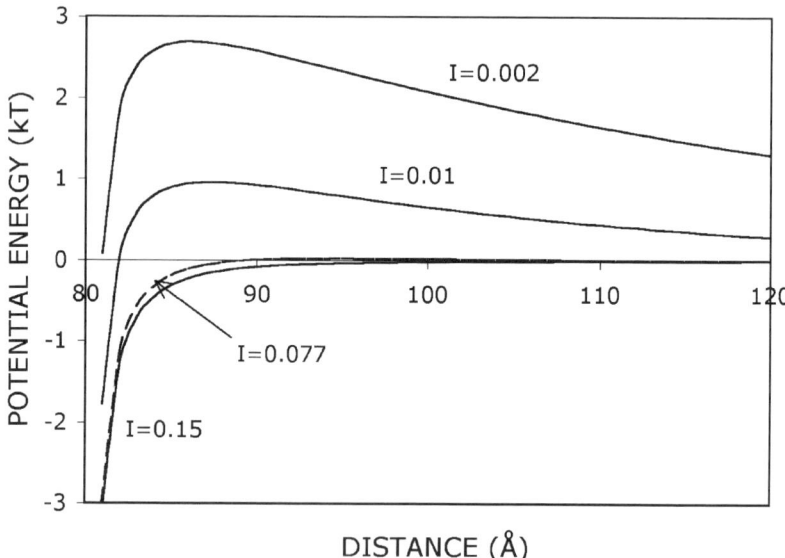

FIGURE 3. Total pair potential (including repulsion and van der Waals attraction) as a function of distance (in Å) between centers for a pair of spheres, at different ionic strengths ($I = 0.002$–0.15). The critical ionic strength for coagulation is 0.077 (*dashed line*). The parameters are similar to those used for the P0-SDS solutions.[12]

virial coefficient as measured from light scattering.[13] The slightly negative value of the coefficient (-1×10^{-4} to -8×10^{-4} mole cm^3/g^2) indicates that the protein molecules attract each other weakly. For supersaturated solutions, the coefficient is similar to that in unsaturated solutions.[47,48] The second virial coefficient is related to the interference function at the origin, and when measured from X-ray diffraction the coefficient is similar to those measured by osmotic pressure and quasielastic light scattering, for example, as shown for γ-crystallin solutions.[49] The osmotic pressure Π can be expanded as a function of particle concentration c (g/cm^3) with virial coefficient A_n, where n refers to the number of terms, according to

$$\Pi/(cRT) = 1/M + A_2 c + A_3 c^2 + \ldots \quad (20)$$

M is the particle mass in Da, T is the absolute temperature, and R is the gas constant (8.31 J-K^{-1}mole^{-1}). The interference function at the origin $S(0)$ is related to the osmotic pressure by

$$S(0) = (RT/M)(d\Pi/dc)^{-1}. \quad (21)$$

Assuming

$$\Pi/(cRT) = (1/M) + A_2 c, \quad (22)$$

then

$$1/S(0) = 1 + 2A_2 Mc, \quad (23)$$

where A_2 is the second virial coefficient in mole-cm^3/g^2, M is molecular weight in Da, and c is macromolecular concentration (g/cm^3). The values c can be obtained from the density of the particle (d) and volume fraction of particles φ according to $c = d\phi$. In this way, the second virial coefficient A_2 can be estimated from the X-ray diffraction.

3. EXPERIMENTAL

3.1. Samples

Protein aggregates or fibrils for X-ray diffraction analysis either can be isolated and purified from *ex vivo* tissue or prepared from synthetic (or recombinant) peptides corresponding in sequence to naturally occurring or *de novo*-designed polypeptides. The synthetic peptides are analyzed under three different conditions: lyophilized, vapor-hydrated, and solubilized/dried. Examining structure in lyophilized peptides informs us of the presence of residual secondary structure and salts that could influence the way in which the peptide subsequently assembles.[50] Lyophilized peptide is gently packed into a thin-walled glass capillary (0.7 mm diameter; Charles A. Supper Co., South Natick, MA) to form a 1-mm-thick disk. To maintain a dry environment inside the tube a small piece of desiccant (anhydrous $CaSO_4$) is placed in the mouth of the capillary before sealing. For vapor-hydration, peptide is gently packed in the capillary as above and then a small volume of water is placed in the mouth of the tube, which is then sealed. The sample is left at room temperature for 7–10 days to allow equilibration against the 100% relative humidity in the tube. Examining structure in vapor-hydrated peptide indicates whether it has a greater propensity to form "classic" fibrillar (amyloidlike) species or ribbon platelike structures, or to remain unaltered by water.[50] For preparing solubilized/dried peptide assemblies, lyophilized peptide is first dissolved in ultra-pure 18 MΩ water (U.S. Filter, Lowell, MA), or appropriate buffer at 5–10 mg/ml, or in organic solvents (e.g., 50% acetonitrile, fluorinated ethanols, etc.). After being vortexed, the solution is briefly centrifuged at 16,000 g to sediment any precipitate. About 2–4 μl of the supernatant is then slowly drawn into a 0.5 to 0.7-mm-diameter siliconized, thin-walled glass capillary tube (Charles A. Supper Co., South Natick, MA). All glass capillaries are thoroughly cleaned using Chromerge® (Sigma, St. Louis MO), rinsed, and then siliconized. Rendering the inner surface of the capillary hydrophobic with dimethyldichlorosilane (e.g., Sigmacote® ; Sigma) helps to prevent the peptide from being adsorbed to the glass surface and to ensure a flat meniscus as the solution becomes concentrated and the sample forms a uniform pellet. The capillary tubes containing peptide solution are sealed at their narrow end and placed in a 2-Tesla (T) permanent magnet[51] (Charles A. Supper Co.), which can promote fibril alignment by the diamagnetic anisotropy of peptide bonds and aromatic groups. The wide end of the capillary is sealed with wax through which a pinhole is punched using a hot needle. The peptide solution is then allowed gradually to dry over the course of 1–2 weeks under ambient temperature and humidity. Slower drying is achieved by leaving the tubes in the cold. The drying is monitored by periodic observation under an optical microscope equipped with a polarizer and analyzer (Figure 4). When the peptide solution has dried to a small, uniform disk, the capillary tube is removed from the magnetic field and transferred to the sample holder for analysis by X-ray diffraction.

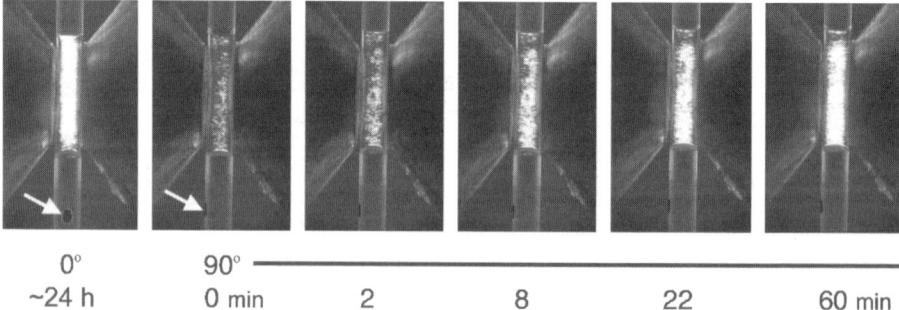

FIGURE 4. Magnetic orientation in a 2-Tesla field (horizontal) of a solution of an amyloidogenic PHF tau hexapeptide,[71] viewed between crossed polarizers. The pole pieces are on the right and left in each panel. The arrow indicates a reference mark on the 0.7 mm diameter capillary. The gradual development of birefringence is observed from time 0 to 60 min after the sample was equilibrated for 24 hr and became oriented, and then was rotated about its axis from 0° to 90°.

We have found that samples that develop birefringence subsequently exhibit oriented X-ray patterns that are indicative of fibril orientation. Even before preparing peptides in solution for magnetic orientation, one can ascertain the likelihood of orientation by using a simple "drop test" in which the birefringence of a 0.5–1 μl drop of peptide solution is monitored during drying.[43] An increase in its birefringence is indicative of the propensity of the peptide assemblies to align in an external 2-T magnetic field, thus yielding highly oriented samples. X-ray patterns recorded from doubly oriented samples in different directions relative to the X-ray beam show that the fibrils align parallel to the magnetic field: for fibrils having the cross-β conformation, this means that the hydrogen bonds are parallel to the field and the polypeptide chains perpendicular to the field.[52] Even for peptides that do not become magnetically oriented, the fibrillar assemblies often are partially oriented owing to surface tension effects that cause the fibers to become positioned parallel to the solution-air interface. We also have demonstrated that a comparison of X-ray results from lyophilized and solubilized/dried samples can be particularly useful for assessing the effect of potential inhibitors of amyloid assembly.[53] Alternative methods to magnetic orientation include using the stretch-frame procedure[54] or a shearing force that occurs when a gelled peptide solution is pushed from the larger to the narrower portion of the capillary.[72] Diffraction patterns from dehydrated peptide assemblies formed in the presence of buffer salts will show wide-angle spacings characteristic of salts; however, exposure of the pellet to water vapor will help to diminish the intensity of such reflections and reduce their potential masking of reflections from the protein assemblies.

3.2. X-ray Data Collection and Densitometry

X-ray patterns for the aggregates can be recorded in a conventional laboratory setting using nickel-filtered and double-mirror focused CuKα radiation from a rotating anode X-ray generator (for example, Elliott GX-20 with 200-μm focal spot). A helium tunnel placed in the X-ray path reduces air scatter. We traditionally have recorded patterns on flat direct-exposure X-ray film (Eastman Kodak Co., Rochester, NY) using exposures of 1–3 days; however, film increasingly is being replaced by electronic detectors that have a considerably greater sensitivity and dynamic range and preclude the necessity of densitometry.[55–57] Useful specimen-to-film distances are 70–90 mm and can be calibrated to ±0.05 mm using the known spacing of calcite (3.035 Å)[58] or silver behenate. The Bragg spacings of reflections are measured either visually off the films using a 6 × optical comparator or from digitized tracings. Digitization of the patterns can be carried out using any of a variety of film scanners and at 25- to 100-μm pixel (corresponding to ~1000–250 dpi). The scanned image can be viewed using the public domain NIH Image program (developed by the U.S. National Institutes of Health and available on the Internet at http://rsb.info.nih.gov/nih-image/). Accurate quantitation of the intensity requires using the known optical density of a Calibrated Step Tablet (Eastman Kodak Company, Rochester, NY 14650) to determine the characterization curve (approximated as a polynomial curve) that relates the optical density of the film (0–2 OD) to the 8-bit or 16-bit scanner readout.

A significant part of the uncertainty in the calculated Bragg spacing arises from uncertainty in the measurement of the distance between the positions of the incident beam and the reflections on the flat film. Unless specified, the uncertainty in spacing does not refer to sample-to-sample variation. If the integral width of the direct beam is w in film space and the distance between the peak position and the direct beam position is l, the two positions at l and $(l + w)$ may be indistinguishable. The deviation in Bragg spacing Δd[31,59] is given by

$$\Delta d = w(\lambda/4s)(\cos\theta \cos^2 2\theta)/\sin^2\theta, \quad (24)$$

where the peak distances l and $(l + w)$ are measured from the origin [corresponding to the Bragg spacings d and $(d + \Delta d)$, respectively], s is specimen-to-film distance, λ is wavelength, and 2θ is scattering angle. For example, if s is 90 mm, w is 300 μm, and λ is 1.542 Å, then Δd is estimated to be 0.04 Å for the 4.7 Å reflection and 0.4 Å for the 10 Å reflection in the characteristic β-sheet diffraction pattern. In our own experiments, the accuracy of the observed Bragg spacing is not better than three significant digits.

3.3. Data Reduction

Before calculating the structure amplitudes, the raw intensity data are corrected by Lorentz-type geometric and polarization factors.[60] The Lorentz factors L for powder and fiber diffraction patterns have been published,[61] but still must be modified according to the beam geometry (line or point), film (flat or curved), integration of the intensity (arc integration or pixel-by-pixel correction), and fiber tilt and disorientation.[1,30] The following is an example of a correction factor in the case of flat film (parameters are shown in Table 1 and Figure 5).

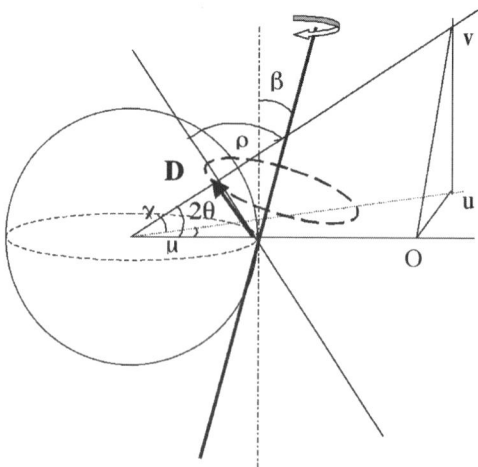

FIGURE 5. Figure 5.Schematic drawing of the reciprocal vector, Ewald sphere, and flat film. The parameters are indicated in Table 1.

TABLE 1. Glossary

Symbol	Meaning
D	Position vector in reciprocal space
R, Φ, Z	Cylindrical coordinates in reciprocal space
X, Y, Z	Cartesian coordinates in reciprocal space
μ, χ	Longitude and latitude of intersection of scattered X-ray with Ewald sphere
λ	X-ray wavelength, which is inverse of the radius of Ewald sphere
β	Inclination of fiber axis from the normal to the incident X-ray beam (positive value directs away from the Ewald sphere)
u, v	Cartesian coordinates on flat film
s	Specimen-to-film distance
2θ	Scattering-angle
ρ	Angle between fiber axis and position vector D
$2\Delta\rho$	Disorientation angle (integral width of Gaussian distribution)
w	Angle between the reflection and the equator on flat film
ΔR	Integral width of intensity along R direction for the reflection
F	Structure factor
f	Atomic scattering factor
I	Intensity
P	Polarization factor
B	Debye factor

The relation between the flat film Cartesian coordinate (u, v) and reciprocal cylindrical vector (R, Z) is

$$\begin{aligned} u &= s \tan \mu = s(\tan 2\theta)(1 - G^2)^{1/2} \\ v &= s \sin \chi / \cos \mu \cos \chi = s(\tan 2\theta)G \\ \sin \chi &= (\sin 2\theta)G \\ \cos \mu \cos \chi &= \cos 2\theta, \end{aligned} \quad (25)$$

where

$$\begin{aligned} G(\rho) &= \cos(\rho)/(\cos\theta \cos\beta) + \tan\theta \tan\beta \\ R &= D \sin\rho, \quad Z = D \cos\rho. \end{aligned} \quad (26)$$

The parameter $G(\rho)$ plays a central role in determining the condition of X-ray diffraction. If $|G| < 1$, the reflection of the reciprocal coordinate (R, Z) intersects the Ewald sphere and will give scattering on the film.[30] The Lorentz L and polarization factors P are defined according to $I = F^2 PL$, where I is the integral intensity along the scanning direction and over an arc and F is the structure factor.[61] The polarization factor for the unpolarized incident beam striking the sample is $P = (1 + \cos^2 2\theta)/2$, and further correction is needed when using a polarized synchrotron beam or the beam reflected by a monochromator crystal.[62] The Lorentz factor for fiber diffraction is

$$L = \frac{\lambda[a \sin G(\rho - \Delta\rho) - a \sin G(\rho + \Delta\rho)]}{4\pi \sin\theta \sin\rho \sin\Delta\rho}. \quad (27)$$

If the intensity data are measured within a narrow strip scan of the film, the measured peak intensity I_p is related to the integral intensity I by

$$I_p = I/(2D\Delta\rho) = F^2 PL/(2D\Delta\rho). \tag{28}$$

The fibril tilt and disorientation used in the Lorentz factor can be determined.[1,16] The case for meridional reflections and a pixel-by-pixel correction also have been documented.[1]

For powder diffraction the Lorentz factor[61] is given as

$$L = \lambda/(4\sin\theta). \tag{29}$$

When the spherically distributed intensity I_s is recorded on flat film with a point-focus incident beam and is measured along the radial direction ($R = 2\sin\theta/\lambda$), the structure factor F is related to I_s by

$$I_s = F^2 LP/(2\pi R) = F^2 P/(4\pi R^2). \tag{30}$$

4. EXAMPLES

4.1. PrP90-145 Solution Scattering: Monolayer β-Sheet

Human prion diseases are manifested as sporadic, familial, or infectious. A "protein-only" scenario in which there is a structural alteration from an α-helical cellular form of the prion (PrP^C) to its abnormal and infectious form (PrP^{Sc}) or protease-resistant form (PrP^{res}) in a β-sheet conformation[63] underlies the molecular mechanism. "Sporadic" refers to the spontaneous structural transformation of PrP^C → PrP^{Sc}. "Familial" denotes that a specific mutation may facilitate the conversion, for example, in Gerstmann-Sträussler-Scheinker (GSS: P102L, P105L, A117V, Y145Stop, F198S, Q217R), familial Creutzfeldt-Jakob disease (CJD: V180I, E200K, M232R), and fatal familial insomnia (FFI: D178N). "Infectious" prion diseases include kuru, iatrogenic CJD, variant CJD (vCJD) for humans, bovine spongiform encephalopathy (BSE), scrapie for sheep, etc. While the 90-145 domain of prion protein PrP is flexibly disordered in solution as shown by NMR, dried assemblies of the PrP90-145 peptide show a cross-β, fibril structure with a sharp 4.7 Å and broad ∼10 Å reflections.[30,59] Hydrated samples, however, show a series of five diffuse intensity maxima at Bragg spacings of 10–50 Å (Figure 6), which may be identified as an intermediate structure between the flexible chain in solution and the stacked β-sheets in the dried assemblies. The size of the object is derived from the correlation function $z(r)$, which is directly calculated from the total intensity:

$$z(r) = 4\pi \int R^2 I(R) \sin(2\pi r R)/(2\pi r R) dR, \tag{31}$$

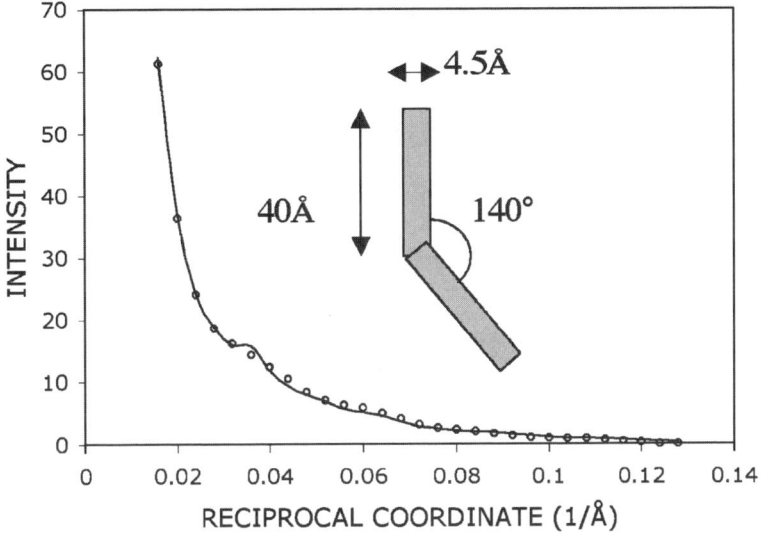

FIGURE 6. Observed (circle) and calculated (solid line) intensities for hydrated PrP90-145 peptide. The model indicates two 4.5 Å × 40 Å-long rectangles with a 140° angle between them.[30]

where r and R are radial components in spherical coordinates in real and reciprocal space, respectively. The $z(r)$ of the observed intensity gives a minimum value at $r \sim 40$ Å, indicating that the greatest extent of the particle is about 40 Å. Based on the Guinier approximation, the small-angle scattering for monodispersed particles is

$$I(R) = I(0) \exp(-4\pi^2 R^2 r_g^2/3), \tag{32}$$

where r_g is the radius of gyration. The plot of $\ln I(R)$ as a function of R^2 gives a linear curve from which r_g is estimated to be 17.7 Å. Since the 4.7 Å reflection is sharp, the observed small-angle scattering arises from the spherical distribution of the two-dimensional intensity function normal to the H-bonding direction. From the absence of the $\sim$10 Å-intersheet reflection, the width of the rectangular scattering object should be $\sim$3 Å as derived from the backbone thickness.[15] From the exclusion length given above, the chain size should be at most 40–50 Å. A model consisting of a pair of rectangular objects having dimensions of a_0 and b_0 and unit density (Figure 7) was considered. The center of one of the rectangles was placed on the origin (Cartesian coordinates) and the other one was tilted by angle $\alpha°$ and placed at (a_t, b_t). The structure factor[30] is then expressed as

$$\begin{aligned} F(\vec{R}) &= F_1(\vec{R}) + F_2(\vec{R}) \exp(i 2\pi \vec{r}_t \cdot \vec{R}) \\ \vec{r}_t &= a_t \vec{e}_1 + b_t \vec{e}_2, \vec{R} = a' \vec{e}_1' + b' \vec{e}_2'. \end{aligned} \tag{33}$$

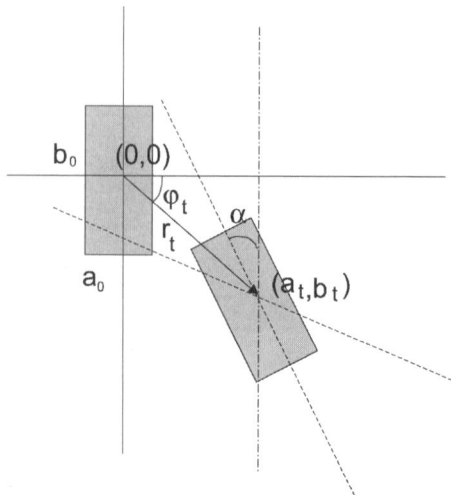

FIGURE 7. Schematic drawing of the two rectangle model. The structure factor is calculated using the given parameters.

F_1 and F_2 are structure factors of the normal and tilted rectangles, respectively, and are given by

$$F_1(a', b') = a_0 b_0 \operatorname{sinc}(\pi a_0 a') \operatorname{sinc}(\pi b_0 b')$$
$$F_2(a', b') = a_0 b_0 \operatorname{sinc}[\pi a_0(a' \cos \alpha - b' \sin \alpha)] \operatorname{sinc}[\pi b_0(a' \sin \alpha + b' \cos \alpha)]. \quad (34)$$

The spherically averaged intensity, obtained numerically, is

$$I_s(R) = \sum \sum I(a', b')/R^2$$
$$R^2 = a'^2 + b'^2. \quad (35)$$

To fit the calculated intensity to the observed one (Figure 6), the parameters were optimized by nonlinear least-squares.[29] Three different initial models consisting of a pair of rectangles were considered: rectangles arranged parallel, sequential (or tandem), and diagonal with respect to one another. The initial parameters in the third model were: $a_0 = 3$ Å, $b_0 = 40$ Å, $\alpha = 40°$, $r_t = 50$ Å, and $\phi_t = 90°$. Optimization of these parameters by nonlinear least-squares gave 4.7% as the best R-factor, with $a_0 = 4.5$ (0.01) Å, $b_0 = 39.5$ (0.03) Å, $\alpha = 36.8$ (0.03)°, $r_t = 34.6$ (0.03) Å, and $\phi_t = 77.3$ (0.06)°. This model indicated two 4.5 Å × 40 Å-long rectangles with a 140° angle between them. By identifying a single-layer β-sheet the model suggests that the α-helical PrP[C] is first folded as a single β-sheet during the conversion to the β-sheet-rich PrP[Sc]. The structure is similar to the single layer β-sheets in β-helical peptides and proteins.[21,64,65]

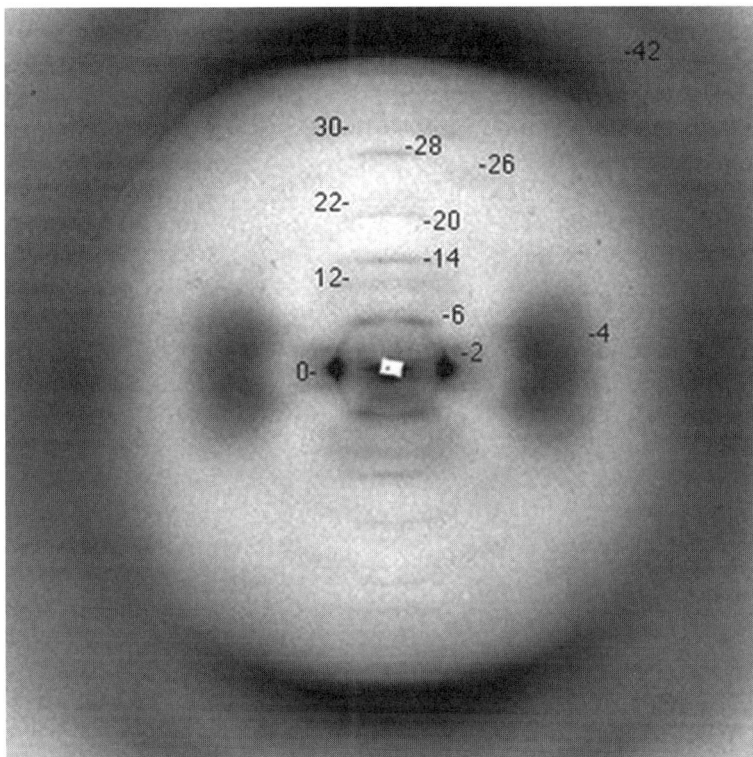

FIGURE 8. X-ray diffraction of B15D (dried from 50% ACN) with the incident beam perpendicular to the fiber axis (vertical).[37] The reciprocal coordinate in the fibril direction for a helical array is defined by $l/c = n/P + m/h$ where c is period, P is pitch, and h is rise per unit. The values here are $c = P = 199$ Å and h is 28 Å. The strong meridional reflection at 4.76 Å therefore is indexed as ($l = 42$, $n = 0$) and the neighboring off-meridional reflections at 4.91 Å and 4.12 Å are indexed as ($l = 40$, $n = -2$) and ($l = 48$, $n = -1$), respectively. The average helix radius 10 Å is estimated from the off-meridional peak positions for layer lines, $l = 2, 4, 6$, and 40, which correspond to Bessel terms $n = 2, 4, -1$, and -2. The image presented here has been contrast-enhanced for clarity.

4.2. Betabellin: Helical Array Constitutes Fibril

Betabellins (beta-sandwich bell-shaped proteins) are a series of proteins designed *de novo* to model protein folding.[66] The 64-residue, disulfide-bridged, dimer ("B15D") forms an amyloidogenic fibril[66] and gives a fiber diffraction pattern (Figure 8) indicative of helical arrays of β-sandwiches.[37] In the wide-angle region there are many meridional layer lines that are interpreted as arising from a discrete double-helical array with selection rule $l = n + 7m$, where l, n, and m are integers. The axial period and helical pitch are 199 Å, the rise per unit is 28 Å, and the average helical radius is 10 Å. This double-helix model is equivalent to a reverse-handed, single helix with half the period and defined by the selection rule $l = -3n + 7m$.

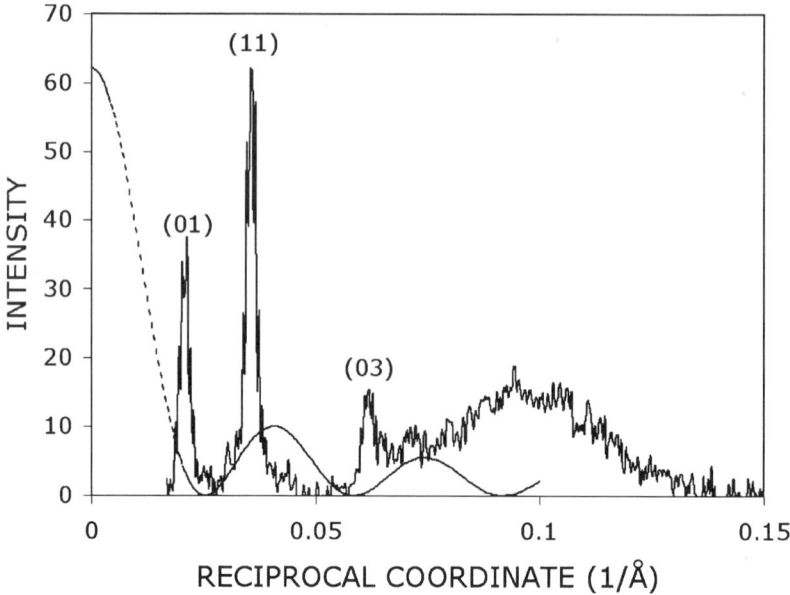

FIGURE 9. Observed equatorial intensities (normalized) as a function of radial coordinate (1/Å) for a betabellin sample. The unit cell constant for the hexagonal lattice is $a = 55.6$ Å. The three peaks at 47.7 Å, 28.1 Å and 16.1 Å, correspond to reflections (01), (11), and (03). *Dotted line*: The intensity of the initial phase model for a 30 Å-diameter tube.

The diffraction pattern of one of the betabellin samples shows sharp, equatorial Bragg peaks that correspond to reflections from a two-dimensional hexagonal lattice (Figure 9). Assuming a tubular cylinder as an initial model leads to an electron density map that resembles the axial projection of the helical model (Figure 10). Low-angle reflections that also can be indexed by a two-dimensional lattice have been observed for prion-related peptides.[30,45] By contrast, assemblies from other amyloidogenic peptides, including Aβ analogues[1,43] and transthyretin,[42,67] show a series of low-angle reflections that are more consistent with either a restricted[‡] two-dimensional or circular lattice.[1] For such assemblies, the number of protofilaments ranges from 3 to 6 (e.g., 4–6 for Aβ11–28[1]; 3–5 for Aβ1–40[43]; and 4 for transthyretin.[42,67]

4.3. PrP106-122: One-Dimensional Stacking of Sheetlike Structure

The sequence that has been proposed to be the core, hydrophobic domain of PrP consists of residues KTNMKHMAGAAAAGAVV, and the corresponding peptide, which is designated PrP106–122, shows different X-ray diffraction depending on hydration (dried vs. vapor hydrated[45]). The reflections that are slightly accentuated on the meridian in the diffraction pattern of the dried sample (Figure 11) suggest that the β-chain direction is parallel to the equator and is the cylindrical axis. On the equator (Figure 12) the first intensity maximum arises from the one-dimensional interference function $Z(R)$ having

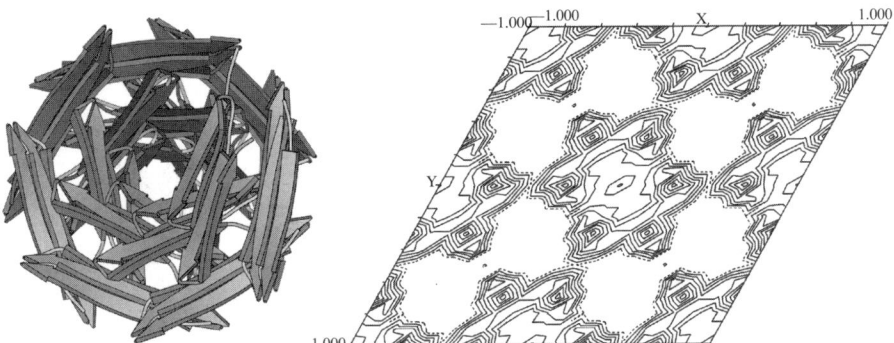

FIGURE 10. (*Left*) Cross-sectional view of the double-helical B15D model (MOLSCRIPT representation[68]). The fibril structure in the helical array is defined by the selection rule (see text and Figure 8 legend for details). (*Right*) Electron density projection along the fibril direction as derived from the observed equatorial reflections and the phases of the tubular model (Figure 9) (XtalView[69]). Note the similarity between the tubelike electron density and the B15D fibril cross section.

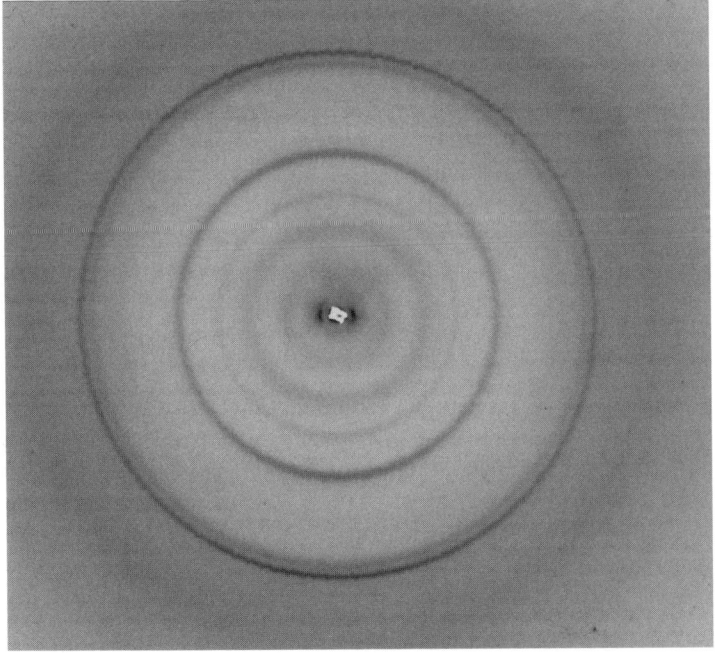

FIGURE 11. X-ray diffraction from assemblies of prion-related peptides. SHa106–122, dried from 50% ACN in water. On the meridian there is a slight accentuation of sharp reflections at 7.17 Å (002). 4.56 Å (201) and 4.73 Å (200), whereas on the equator there is a very strong intensity peak at 73 Å and successive broad and weak intensity maxima at 23 Å and 13 Å (see also Figure 12).

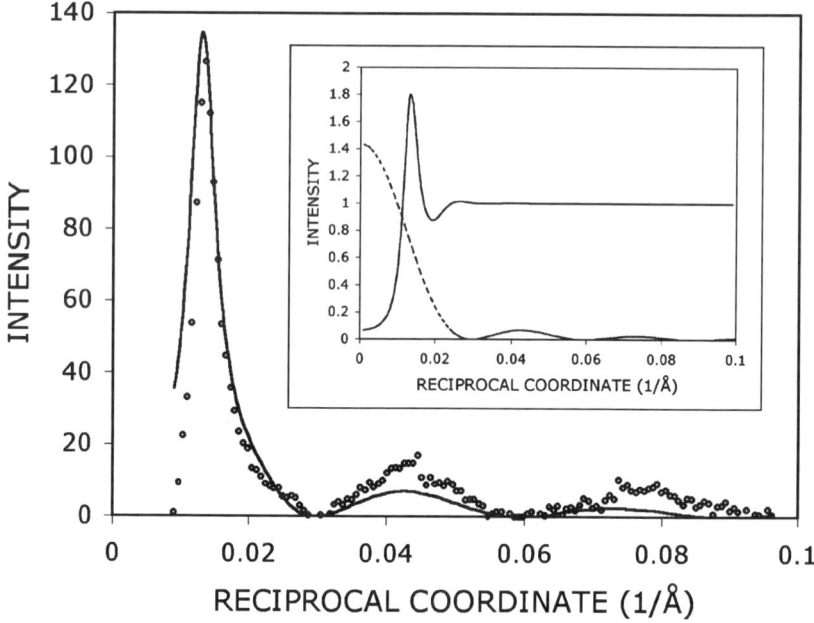

FIGURE 12. Analysis of equatorial diffraction data from SHa106–122 (Figure 11). The observed (*circles*) and calculated (*continuous line*) intensity curves were normalized so that the area under each was 1. The R-factor for the comparison was 41%. The interference function was calculated according to one-dimensional paracrystalline theory,[26] with a lattice period of 72.6 Å and lattice disorder of 18.5 Å (*continuous curve*). The structure factor (*broken curve*) was calculated from the Fourier transform of a solid slab of thickness 33.6 Å and arbitrarily scaled.

period d and the subsequent reflections arise from the Fourier transform $F(R)$ of a slab of thickness a, where R is the reciprocal coordinate in the direction normal to the slab surface. Values for plate thickness, lattice disorder (Δ), which were derived by searching for the minimum residual between the observed and calculated intensities, are 33.6 Å and 18.5 Å, respectively, where the stacking distance is 72.6 Å (Figure 12). The sample-to-sample variation in stacking period of the plates is likely due to different hydration. To account for the swelling behavior between the neighboring plates, the amino acid residues at the edge of the plate should be polar. The period therefore is likely determined by the balance between the electrostatic repulsion and van der Waals attractive forces as defined by the DLVO theory.

A similar platelike structure also is observed for assemblies formed by Aβ amyloid analogues[1] and SAA-related peptides.[50] In X-ray diffraction patterns from Aβ protein analogues, including Aβ18–28, Aβ17–28, and Aβ15–28,[1] the equatorial reflections are indexed one-dimensionally with periods ∼40 Å. The meridional reflections all are indexed by a two-dimensional lattice of β-sheets in the hydrogen-bonding and intersheet directions, suggesting that the plate stacking is along the β-chain direction. Since these fragments of Aβ do not contain His13 and His14 and inclusion of these residues results in a fibrillar structure, we suggested that the addition of His residues weakens rather than strengthens intersheet interactions, and thus promotes the formation of protofilaments and their lateral aggregation into larger assemblies or fibrils.[1]

5. CONCLUSIONS

X-ray diffraction at low- and medium-range resolution ($>\sim 10$ Å Bragg spacing) reveals protein aggregates having various structural morphologies, including sphere, fibril, and sheets. These macromolecular assemblies form a noncrystalline, disordered lattice that can be described using paracrystalline theory or a liquid model. X-ray diffraction at wide angles (spacings $<\sim 10$ Å), on the other hand, provides details on the structural unit from which the aggregate is built, e.g., the β-sheet for amyloid protofilaments. The particular type of protein aggregation is mediated by protein-protein interactions. A balance between electrostatic and/or hydration repulsion forces and the van der Waals attractive force likely accounts for the separation between the constituent objects; these forces can be used to estimate the pH and ionic strength for particle coagulation. Second virial coefficients as measured by X-ray diffraction have been used for determining the media conditions required for protein crystallization,[49] and DLVO theory has been used for predicting the myelin swelling and compaction by using the measured chemical composition of proteins and lipids in the membrane assemblies.[10,11] Recently, there has been considerable interest focused on the fabrication of molecular assemblies in the forms of nanotubes[21–23] and plates[2] and on the design of therapeutic reagents to inhibit amyloid fibril formation.[53,70,73] X-ray diffraction study and analysis of protein-protein interaction as summarized here can provide insights into the mechanism of protein aggregation processes and possible strategies for therapeutic intervention.

ACKNOWLEDGMENTS

We thank our colleagues Jeremy Bond, Sean Deverin, and Leonid Shinchuk for contributing to the fiber diffraction, Drs. Amareth Lim and Catherine Costello for collaborating on betabellin, and Drs. Michael Baldwin, Haydn Ball, Robert Fletterick, Fred Cohen, Stanley Prusiner, and Mario Salmona on the prion study. We thank Dr. Warren Goux for providing us with PHF tau peptides. Research at Boston College was supported by an Alzheimer's Association/T.L.L. Temple Discovery Award (to DAK) and by institutional support from Boston College. DAK also acknowledges a Fulbright Senior Research Scholar Award from the Binational US-Italian Fulbright Scholar Program during the preparation of this manuscript.

NOTES

∗. The terminology mainly used in the field of small-angle scattering.
‡. By "restricted" we mean that the lattice is not infinitely extended.

REFERENCES

1. H. Inouye, P. E. Fraser, and D. A. Kirschner, Structure of β-crystallite assemblies formed by Alzheimer β-amyloid protein analogues: Analysis by x-ray diffraction, *Biophys. J.* **64**(2), 502–519 (1993).
2. A. Aggeli, I. A. Nyrkova, M. Bell, R. Harding, L. Carrick, T. C. B. McLeish, A. N. Semenov, and N. Boden, Hierarchical self-assembly of chiral rod-like molecules as a model for peptide β-sheet tapes, ribbons, fibrils, and fibers, *Proc. Natl. Acad. Sci. USA* **98**(21), 11857–11862 (2001).
3. F. Oosawa and S. Asakura, *Thermodynamics of the Polymerization of Protein* (New York: Academic Press, 1975).
4. M. Ataka and M. Asai, Analysis of the nucleation and crystal growth kinetics of lysozyme by a theory of self-assembly, *Biophys. J.* **58**(3), 807–811 (1990).
5. F. Ferrone, Analysis of protein aggregation kinetics, *Meth. Enzymol.* **309**, 256–274 (1999).
6. H. Inouye and D. A. Kirschner, Aβ fibrillogenesis: Kinetic parameters for fibril formation from Congo red binding, *J. Struct. Biol.* **130**(2–3), 123–129 (2000).
7. R. M. Murphy, Peptide aggregation in neurodegenerative disease, *Annu. Rev. Biomed. Eng.* **4**, 155–174 (2002).
8. E. J. W. Verwey and J. Th. G. Overbeek, *Theory of the Stability of Lyophobic Colloids* (New York: Elsevier, 1948).
9. B. W. Ninham and V. A. Parsegian, Electrostatic potential between surfaces bearing ionizable groups in ionic equilibrium with physiologic saline solution, *J. Theor. Biol.* **31** (3), 405–428 (1971).
10. H. Inouye and D. A. Kirschner, Membrane interactions in nerve myelin. I. Determination of surface charge from effects of pH and ionic strength on period, *Biophys. J.* **53**(2), 235–246 (1988).
11. H. Inouye and D. A. Kirschner, Membrane interactions in nerve myelin. II. Determination of surface charge from biochemical data, *Biophys. J.* **53**(2), 247–260 (1988).
12. H. Inouye, H. Tsuruta, J. Sedzik, K. Uyemura, and D. A. Kirschner, Tetrameric assembly of full-sequence protein zero myelin glycoprotein by synchrotron x-ray scattering, *Biophys. J.* **76**(1), 423–437 (1999).
13. A. George and W. W. Wilson, Predicting protein crystallization from a dilute solution property, *Acta Crystallogr.* **D50**, 361–365 (1994).
14. R. E. Dickerson and I. Geis, *The Structure and Action of Proteins* (Menlo Park, CA: Benjamin/Cummings Publishing Company, 1969).
15. R. D. B. Fraser and T. P. MacRae, *Conformation in Fibrous Proteins* (New York: Academic Press, 1973).
16. R. E. Franklin and R. G. Gosling, Molecular configuration in sodium thymonucleate, *Nature* **171**, 740–741 (1953).
17. D. A. Kirschner, H. Inouye, L. K. Duffy, A. Sinclair, M. Lind, and D. J. Selkoe, Synthetic peptide homologous to β protein from Alzheimer disease forms amyloid-like fibrils in vitro, *Proc. Natl. Acad. Sci. USA* **84**(19), 6953–6957 (1987).
18. M. F. Perutz, T. Johnson, M. Suzuki, and J. T. Finch, Glutamine repeats as polar zippers: Their possible role in inherited neurodegenerative diseases, *Proc. Natl. Acad. Sci. USA* **91**(12), 5355–5358. (1994).
19. L. Makowski and B. Magdoff-Fairchild, Polymorphism of sickle cell hemoglobin aggregates: Structural basis for limited radial growth, *Science.* **234**(4781), 1228–1231 (1986).
20. M. Delaye and A. Tardieu, Short-range order of crystallin proteins accounts for eye lens transparency, *Nature* **302**(5907), 415–417 (1983).
21. M. F. Perutz, J. T. Finch, J. Berriman and A. Lesk, Amyloid fibers are water-filled nanotubes, *Proc. Natl. Acad. Sci. USA* **99**(8), 5591–5595 (2002).
22. T. Scheibel, R. Parthasarathy, G. Sawicki, X-M. Lin, H. Jaeger, and S. L. Lindquist, Conducting nanowires built by controlled self-assembly of amyloid fibers and selective metal deposition, *Proc. Natl. Acad. Soc. USA* **100**(8), 4527–4532 (2003).
23. C. Valery, M. Paternostre, B. Robert, T. Gulik-Krzywicki, T. Narayanan, J. C. Dedieu, G. Keller, M. L. Torres, R. Cherif-Cheikh, P. Calvo, and F. Artzner, Biomimetic organization: Octapeptide

self-assembly into nanotubes of viral capsid-like dimension, *Proc. Natl. Acad. Sci. USA* **100**(18), 10258–10262 (2003).
24. R. A. Garoff, E. A. Litzinger, R. E. Connor, I. Fishman, and B. A. Armitage, Helical aggregation of cyanine dyes on DNA templates: Effect of dye structure on formation of homo- and heteroaggregates, *Langmuir* **18**(16), 6330–6337 (2002).
25. K. Keren, R. S. Berman, E. Buchstab, U. Sivan, and E. Braun, DNA-templated carbon nanotube field-effect transistor, *Science* **302**(5649), 1380–1382 (2003).
26. R. Hosemann and S. N. Bagchi, *Direct Analysis of Diffraction by Matter* (Amsterdam: North-Holland Publishing Co., 1962).
27. A. Guinier, *X-ray Diffraction* (New York: W. H. Freeman and Company, 1963).
28. B. K. Vainshtein, *Diffraction of X-Rays by Chain Molecules* (Amsterdam: Elsevier, 1966).
29. H. Inouye, J. Karthigasan, and D. A. Kirschner, Membrane structure in isolated and intact myelins, *Biophys. J.* **56**(1), 129–137 (1989).
30. H. Inouye and D. A. Kirschner, X-ray diffraction analysis of scrapie prion: Intermediate and folded structures in a peptide containing two putative α-helices, *J. Mol. Biol.* **268** (2), 375–389 (1997).
31. H. Inouye and D. A. Kirschner, Refined fibril structures: The hydrophobic core in Alzheimer's amyloid β-protein and prion as revealed by X-ray diffraction, in *Ciba Foundation Symposium No. 199. The Nature and Origin of Amyloid Fibrils*, ed. G. R. Bock and J. A. Goode (New York: John Wiley & Sons, 1996, 22–35).
32. S. Schwartz, J. E. Cain, E. A. Dratz, and J. K. Blasie, An analysis of lamellar X-ray diffraction from disordered membrane multilayers with application to data from retinal rod outer segments, *Biophys. J.* **15**(12), 1201–1233 (1975).
33. A. E. Blaurock and J. C, Nelander, Disorder in nerve myelin: Analysis of the diffuse x-ray scattering, *J. Mol. Biol.* **103**(2), 421–431 (1976).
34. S. Tanaka and S. Naya, A theory of x-ray scattering by disordered polymer crystals, *J. Phys. Soc. Japan.* **26**(4), 982–993 (1969).
35. H. Inouye X-ray scattering from a discrete helix with cumulative angular and translational disorders, *Acta Crystallogr.* **A50**, 644–646 (1994).
36. W. J. Stroud and R. P. Millane, Diffraction by disordered polycrystalline fibers, *Acta Crystallogr.* **A51**, 771–790 (1995).
37. H. Inouye, J. E. Bond, S. P. Deverin, A. Lim, C. E. Costello, and D. A. Kirschner, Molecular Organization of amyloid protofilament-like assembly of betabellin 15D: Helical array of β-sandwiches, *Biophys. J.* **83**(3), 1716–1727 (2002).
38. J. B. Hayter and J. Penfold, An analytic structure factor for macro-ion solutions, *Mol. Phys.* **42**, 109–118 (1981).
39. J. B. Hayter and J.-P. Hansen, The structure factor of charged colloidal dispersions at any density, *Institut Laue-Langevin Report* No. 82HA14T (1982).
40. R. E. Burge, The structure of bacterial flagella: The packing of the polypeptide chains within a flagellum, *Proc. R. Soc. (Lond.)* **A260**, 558–573 (1961).
41. C. R. Worthington and H. Inouye, X-ray diffraction study of the cornea. *Int. J. Biol. Macromol.* **7**(1), 2–8 (1985).
42. C. Blake and L. Serpell, Synchrotron x-ray studies suggest that the core of the transthyretin amyloid fibril is a continuous β-sheet helix, *Structure* **4**(8), 989–998 (1996).
43. S. B. Malinchik, H. Inouye, K. E. Szumowski, and D. A. Kirschner, Structural analysis of Alzheimer's β(1-40) amyloid: Protofilament assembly of tubular fibrils, *Biophys. J.* **74**(1), 537–545 (1998).
44. H. Inouye and D. A. Kirschner, X-ray fibre diffraction analysis of assemblies formed by prion-related peptides: Polymorphism of the heterodimer interface between PrP^C and PrP^{Sc}, *Fibre Diffraction Review* **11**, 102–112 (2003).
45. H. Inouye, J. Bond, M. A. Baldwin, H. L. Ball, S. B. Prusiner, and D. A. Kirschner, Structural changes in a hydrophobic domain of the prion protein induced by hydration and by Ala → Val and Pro → Leu substitutions, *J. Mol. Biol.* **300**(5), 1283–1296 (2000).

46. D. M. LeNeveu and R. P. Rand, Measurement and modification of forces between lecithin bilayers, *Biophys. J.* **18**(2), 209–230 (1977).
47. M. Muschol and F. Rosenberger, Interactions in undersaturated and supersaturated lysozyme solutions: Static and dynamic light scattering results, *J. Chem. Phys.* **103**(24), 10424–10432 (1995).
48. O. D. Velev, E. W. Kaler, and A. M. Lenhoff, Protein interactions in solution characterized by light and neutron scattering: Comparison of lysozyme and chymotrypsinogen, *Biophys. J.* **75**(6), 2682–2697 (1998).
49. F. Bonneté, M. Malfois, S. Finet, A. Tardieu, S. Lafont, and S. Veesler, Different tools to study interaction potentials in γ-crystallin solutions: Relevance to crystal growth, *Acta Crystallogr.* **D53**, 438–447 (1997).
50. D. A. Kirschner, R. Elliott-Bryant, K. E. Szumowski, W. A. Gonnerman, M. S. Kindy, J. D. Sipe, and E. S. Cathcart, *In vitro* amyloid fibril formation by synthetic peptides corresponding to the amino terminus of apoSAA isoforms from amyloid-susceptible and amyloid-resistant mice, *J. Struct. Biol.* **124**(1), 88–98. (1998).
51. R. Oldenbourgh and W. C. Phillips, Small permanent magnet for fields up to 2.6T, *Rev. Sci. Instrum.* **57**(9), 2362–2365. (1986).
52. P. E. Fraser, L. K. Duffy, M. B. O'Malley, J. Nguyen, H. Inouye, and D. A. Kirschner, Morphology and antibody recognition of synthetic β-amyloid peptides, *J. Neurosci. Res.* **28**(4), 474–485 (1991).
53. G. M. Castillo, D. A. Kirschner, A. G. Yee, and A. D. Snow, Electron microscopy and x-ray diffraction studies further confirm the efficacy of PTI-00703 (Cat's Claw derivative) as a potential inhibitor of Alzheimer's β-amyloid protein fibrillogenesis, in *Alzheimer's Disease: Advances in Etiology, Pathogenesis and Therapeutics*, ed. K. Iqbal, S. S. Sisodia, and B. Winblad (New York: John Wiley & Sons, 2001, 449–460).
54. M. Sunde, L. C. Serpell, M. Bartlam, P. E. Fraser, M. B. Pepys, and C. C. F. Blake, Common core structure of amyloid fibrils by synchrotron x-ray diffraction, *J. Mol. Biol.* **273**(3), 729–739. (1997).
55. Y. Amemiya, Imaging plates for use with synchrotron radiation, *J. Synchrotron Rad.* **2**, 13–21 (1995).
56. W. C. Phillips, M. Stanton, A. Stewart, H. Qian, C. Ingersoll, and R. M. Sweet, Multiple CCD detector for macromolecular x-ray crystallography, *J. Appl. Crystallogr.* **33**(2), 243–251 (2000).
57. S. M. Gruner, M. W. Tate, and E. F. Eikenberry, Charge-coupled device area x-ray detectors, *Rev. Sci. Instr.* **73**(8), 2815–2842 (2002).
58. D. L. Graf, Crystallographic tables for the rhombohedral carbonates, *Am. Mineral.* **46**(11), 1283–1316 (1961).
59. J. T. Nguyen, H. Inouye, M. A. Baldwin, R. J. Fletterick, F. E. Cohen, S. B. Prusiner, and D. A. Kirschner, X-ray diffraction of scrapie prion rods and PrP peptides, *J. Mol. Biol.* **252**(4), 412–422 (1995).
60. R. D. B. Fraser, T. P. MacRae, A. Miller, and R. J. Rowlands, Digital processing of fibre diffraction pattern, *J. Appl. Crystallogr.* **9**, 81–94 (1976).
61. R. J. Cella, B. Lee, and R. E. Hughes, Lorentz and orientation factors in fiber X-ray diffraction analysis, *Acta Crystallogr.* **A26**, 118–124 (1970).
62. R. Kahn, R. Fourme, A. Gadet, J. Janin, C. Dumas, and D. André, Macromolecular crystallography with synchrotron radiation: Photographic data collection and polarization correction, *J. Appl. Crystallogr.* **15**(3), 330–337. (1982).
63. S. B. Prusiner, Molecular biology of prion diseases, *Science* **252**(5012), 1515–1522, (1991).
64. A. Aggeli, M. Bell, N. Boden, J. N. Keen, P. F. Knowles, T. C. B. McLeish, M. Pitkeathly, and S. E. Radford, Responsive gels formed by the spontaneous self-assembly of peptides into polymeric β-sheet tapes, *Nature* **386**(6622), 259–262 (1997).
65. H. Wille, M. D. Michelitsch, V. Guénebaut, S. Supattapone, A. Serban, F. E. Cohen, D. A. Agard, and S. B. Prusiner, Structural studies of the scrapie prion protein by electron crystallography, *Proc. Natl. Acad. Sci USA* **99**(6), 3563–3568 (2002).
66. A. Lim, A. M. Makhov, J. Bond, H. Inouye, L. H. Connors, J. D. Griffith, B. W. Erickson, D. A. Kirschner, and C. E. Costello, Betabellins 15D and 16D, *de novo* designed β-sandwich proteins that have amyloidogenic properties, *J. Struct. Biol.* **130**(2-3), 363–370 (2000).

67. H. Inouye, F. S. Domingues, A. M. Damas, M. J. Saraiva, E. Lundgren, O. Sandgren, and D. A. Kirschner, Analysis of x-ray diffraction patterns from amyloid of biopsied vitreous humor and kidney of transthyretin (TTR) Met30 familial amyloidotic polyneuropathy (FAP) patients: Axially arrayed TTR monomers constitute the protofilament, *Amyloid* **5**(3), 163–174 (1998).
68. P. J. Kraulis, MOLSCRIPT: A program to produce both detailed and schematic plots of protein structures, *J. Appl. Crystallogr.* **24**(5), 946–950 (1991).
69. D. E. McRee, XtalView/Xfit—A versatile program for manipulating atomic coordinates and electron density, *J. Struct. Biol.* **125**(23), 156–165 (1999).
70. H. Inouye, J. T. Nguyen, P. E. Fraser, L. M. Shinchuk, A. B. Packard, and D. A. Kirschner, Histidine residues underline Congo red binding to Aβ analogs, *Amyloid* **7**(3), 179–188 (2000).
71. H. Inouye, D. Sharma, W.J. Goux, and D.A. Kirschner, Structure of core domain of fibril-forming PHF/tau fragments, *Biophys. J.* **90**(5), 1774–1789 (2006).
72. L. M. Shinchuk, D. Sharma, S. E. Blondelle, N. Reixach, H. Inouye, and D. A. Kirschner, poly(L-alanine) expansions form core β-sheets that nucleate amyloid assembly, *Proteins: Structure, Function, and Bioinformatics*, **61**(3), 579–589 (2005).
73. M. Findeis, Approaches to discovery and characterization of inhibitors of amyloid β-peptide polymerization, *Biochim. Biophys. Acta* **1502**(1), 76–84.

Glass Dynamics and the Preservation of Proteins

Christopher L. Soles,[1,3] Amos M. Tsai,[2] and Marcus T. Cicerone[1]

Proteins can be stored, with their biochemical functionality preserved for extended periods under nonphysiological conditions, by enveloping them in certain viscous glass-forming compounds such as trehalose or glycerol. However, the relevant variables that are critical to stabilization are not completely understood, meaning that effective preservation is achieved more often by trial and error than by rational design. While it is widely felt that the viscous preservant must mimic water in terms of thermodynamic interactions with the protein, the precise role of the dynamic interactions between the biomolecule and the preservant still is not clear. Here we present incoherent neutron scattering and biochemical activity measurements giving strong evidence that an effective protein preservation matrix is characterized by suppression of the high-frequency ($\omega \approx 200$ MHz and faster) dynamics. We contend that suppressing these nanosecond and faster motions in the glass impedes the protein dynamics at a similar timescale, thereby conferring stability to the protein. To the best of our knowledge, this is the first direct evidence indicating that these high-frequency dynamics are relevant for protein preservation. This new perspective also sheds light on behavior of proteins in glass that was previously seen as aberrant; namely, the observation of slower protein dynamics in liquid versus glassy environments and the lack of a strict correlation between the glass transition temperature of a glass and its ability to serve as a protein preservation medium.

1. INTRODUCTION

Protein structure plays a critical role in determining enzymatic activity. The details by which a proteinaceous macromolecule with a given primary amino acid sequence assembles into more complicated secondary and tertiary structures plays a major role in establishing the biochemical functionality of that protein. The central theme in this volume is secondary or tertiary structural deviations through misfolding or aggregation events and how these in turn affect biochemical functionality of the protein. While these are primarily static structural deviations, it is important to realize that dynamic pathways are required to transform the protein from its natural to a deviant state. The ability to affect the dynamics of these pathways is therefore nearly tantamount to controlling

---

[1] NIST Polymers Division, 100 Bureau Drive, Gaithersburg, MD 20899-8541
[2] Human Genome Sciences, 14200 Shady Grove Road, Rockville, MD 20850
[3] To whom correspondence should be addressed: Christopher L. Soles, NIST Polymers Division, 100 Bureau Drive, Gaithersburg, MD 20899-8541; email:soles@nist.gov

protein stability. If the dynamic processes responsible for the misfolding or aggregation events can be suppressed, the protein can be stabilized.

In this chapter we describe inelastic neutron scattering (INS) as a powerful tool to quantify dynamics on the time and length scales that are relevant for stabilizing proteins. This will be done by way of example, using sugar or polyalchohol mixtures to preserve biochemical activity in model protein systems. It is generally understood that the highly viscous nature of these sugars/polyalcohols in comparison to physiological environments is critical for imparting stability against denaturation/aggregation. Here we use INS to elucidate how the nanosecond to picosecond relaxations and/or vibrations in the preservation media couple to the dynamics of the protein and impart stability. While these motions are faster than the timescales traditionally associated with biological processes (generally on the order of milliseconds or slower), their suppression is interpreted as a "shutting down" of the precursor motions that initiate the dynamic pathways that ultimately lead to denaturation, aggregation, or misfolding on biologically relevant time scales.

2. INELASTIC NEUTRON SCATTERING

INS is a versatile and powerful tool for quantifying the dynamics in biological and soft materials. The basis of the technique is illustrated schematically in Figure 1. A beam of neutrons with a narrow, well-defined energy distribution is focused onto the sample of interest. Different INS spectrometers vary in their ability to monochromate the incident neutron, but typically (for the sake of illustration) the beam can be focused into a narrow Gaussian energy distribution centered on E_0. The solid lines in Figure 1 indicate such a distribution. When the incident beam of neutrons collides with the atomic nuclei in the sample, the strong nuclear interactions scatter the neutrons in various directions. The energy of the neutrons after scattering changes in comparison to the initial distribution. Some of the neutrons gain or lose energy through the dynamic interactions with the sample, while others are scattered elastically with no energy exchange. Figure 1 shows that there are two general types of INS events. The left panel depicts scattering from a vibrating atomic nucleus, the motion confined within a well-defined potential energy minimum. The well-defined minimum leads to a distinct peak in the scattered energies.

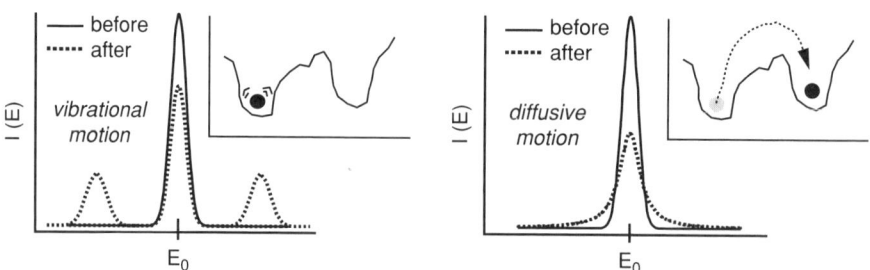

FIGURE 1. A cartoon depicting the two general types of energy exchange between a sample and a neutron for vibrational (left) and relaxational (right) molecular motions.

The shift in scattered peak from the incident energy indicates the energy of the excitation; this type of scattering is referred to as pure inelastic scattering. The second type of scattering is depicted in the right panel of Figure 1 where there is a diffusive motion of the atomic nucleus from one location to another. This results in a diffuse broadening of the incident energy distribution. In Figure 1 the narrow Gaussian distribution broadens into a Lorentzian. Conventionally this diffuse broadening is referred to as quasielastic scattering (QES), although it is technically still a form of INS. By understanding the energy exchange between the sample and the neutrons, it is possible to perceive the atomic or molecular dynamics in the sample.

Inelastic scattering from light and X-rays, as well as neutrons, is useful for measuring sample dynamics. However, there are several attributes that make neutron scattering ideal for quantifying dynamics in biological and/or soft materials. First, the wavelength of most neutrons is between 1 Å and 10 Å, commensurate with many of the interatomic or intermolecular distances in soft, organic materials. The length scale of the motions over which a motion is probed is given by the scattering vector Q

$$Q = \frac{4\pi}{\lambda} \sin \theta, \quad (1)$$

where θ is scattering angle and λ is the neutron wavelength. With most INS instruments it is reasonable to probe scattering over a range of angles that encompass a reciprocal space domain of $0.2 \text{ Å}^{-1} < Q < 2.0 \text{ Å}^{-1}$. In real space that corresponds to length scales approximately 3 Å to 30 Å, well suited for characterizing organic materials.

The second property of neutrons that makes them useful for measuring dynamics is their energy range. Most cold neutrons have energies that are on the order of a few millielectron volts (meV), comparable to the energy of the solid-state excitations or dynamic processes in soft and biological materials. This means that when a neutron gains or loses energy from a dynamic interaction with the sample, the change in the energy of the neutron is usually a significant fraction of the initial neutron energy. This means it is relatively easy to ascertain if a neutron has changed energy upon scattering. This is in contrast to inelastic X-ray scattering where the incident beams are on the order of several kiloelectron volts (keV). To discriminate a millielectron volts (meV) change in the energy of a keV source requires extremely sensitive energy discrimination.

The final reason why neutrons are ideal for measuring dynamics is related to the scattering cross section of the elements common to soft and biological materials. Typically these materials are composed of elements like C, O, N, Ca, P, and H. With X-rays, the scattering is dominated by the heavier elements and the scattering cross section increases with atomic number. However, the strength of the nuclear interactions changes in more of a random manner with atomic number. Fortunately, H has the biggest scattering cross section of any of the elements. This is shown pictorially in Figure 2 where the size of the circle represents the magnitude of the scattering cross section; H-rich moieties will dominate an inelastic scattering experiment. This is quite useful because H dynamics often are difficult to quantify with complimentary optical and X-ray scattering techniques that often are insensitive to the lighter elements. The utility of neutron scattering is further enhanced by the fact that the isotopic switch from hydrogen (H) to deuterium

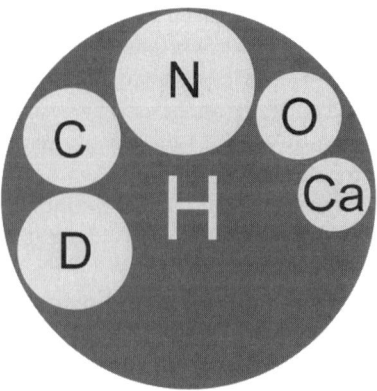

FIGURE 2. Spheres with their areas scaled relative to the total neutron scattering cross section of a few representative isotopes relevant to biological or soft materials. Hydrogen has a significantly larger scattering cross section than any other element.

(D) greatly reduces this massive scattering cross section. From a chemistry point of view, the switch from H to D is usually trivial in terms of physical properties, meaning that powerful isotopic-labeling schemes can be devised. By selectively replacing certain H with D, one can study the dynamics isolated to a certain species or within a given region of a biological molecule. This is somewhat analogous to the isomorphic replacement schemes used in protein crystallography where heavy metal tags are attached to certain regions of a protein to enable structural characterization with X-ray scattering. A similar analogy also could be made with the radioisotope labeling schemes often used in nuclear magnetic resonance (NMR) measurements.

2.1. Instrumentation

INS experiments require access to a neutron-scattering facility. While there are only a handful of these in most major countries, access is often encouraged and most users find the facilities open to the scientific and research communities in general. The measurements described in this chapter were performed at the NIST Center for Neutron Research (NCNR). The NCNR is a federally funded facility that is open to the general public. The user community at the NCNR spans academic, industrial, and government scientists from both the United States and all over the world. Information about utilizing the NCNR facilities can be found on-line at http://www.ncnr.nist.gov. This chapter will discuss INS from two types of spectrometers at the NCNR: a time-of-flight spectrometer (TOF) and a high flux backscattering spectrometer (HFBS). Both of these are instruments are commonly found at most major neutron-scattering facilities across the word, so the discussion is not limited to the NCNR. In this section we briefly review the way in which these two instruments operate.

The TOF spectrometer is conceptually straightforward. It operates on the principle that when scattered neutrons gain energy, they speed up. Likewise, when neutrons lose energy, they slow down. By measuring the time of flight of the scattered neutron across

GLASS DYNAMICS AND PROTEIN PRESERVATION

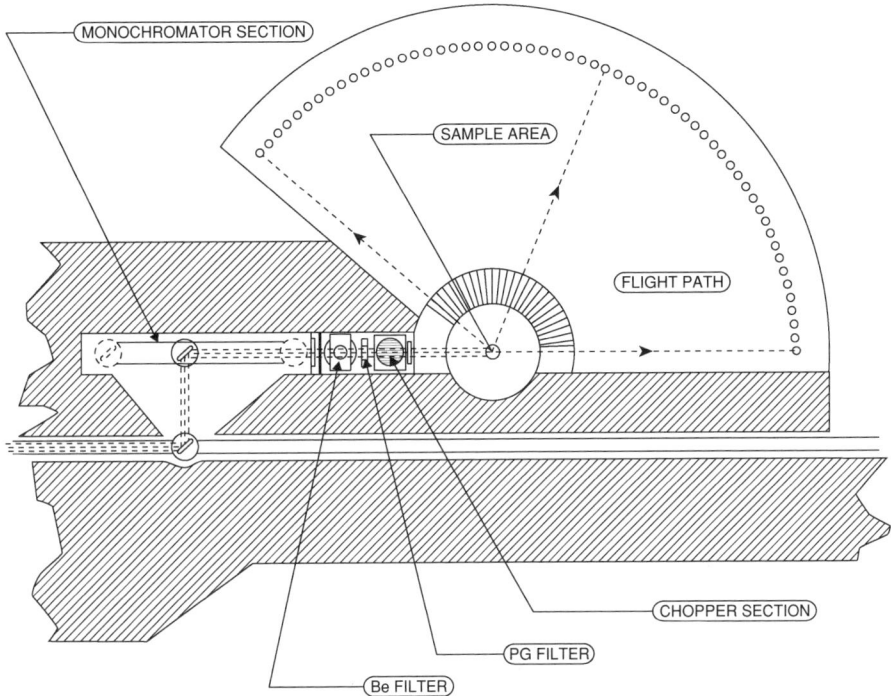

FIGURE 3. A schematic representation of a typical time-of-flight inelastic neutron spectrometer.

a fixed length from the sample to the detector, the energy of the scattered neutron can be determined. A schematic in Figure 3 depicts how a typical TOF spectrometer operates. The neutron beam enters a monochromator that only allows neutrons of a well-defined wavelength, and therefore a well-defined energy distribution, to pass through. This ideally monoenergetic beam of neutrons then goes into a "chopper" that periodically releases pulses of neutrons onto the sample. The distance from the chopper to the sample and then from the sample to the detectors is accurately known for each TOF spectrometer. Since the energy or speed of the incident beam of neutrons is known (defined by the monochromator), it is straightforward to predict how long it should take a given pulse of neutrons to leave the chopper, scatter from the sample elastically, and then reach the detector. If the actual time of arrival at the detector is sooner than predicted, the neutrons have gained energy. If the time of arrival is later, the neutrons have lost energy through the scattering event. As Figure 3 shows, there are actually several detectors equidistant from the sample, spread out in a semicircle. This allows the dynamics to be probed over a large Q range, or range of different length scales.

In this chapter we will discuss measurements made on the Fermi-Chopper TOF spectrometer at the NIST Center for Neutron Research. The Fermi-Chopper instrument is capable of collimating an incident energy distribution to a full width at half maximum (FWHM) of 130 μeV. This energy resolution means that the instrument is sensitive to motions faster than approximately 7 GHz. A better energy resolution would mean that

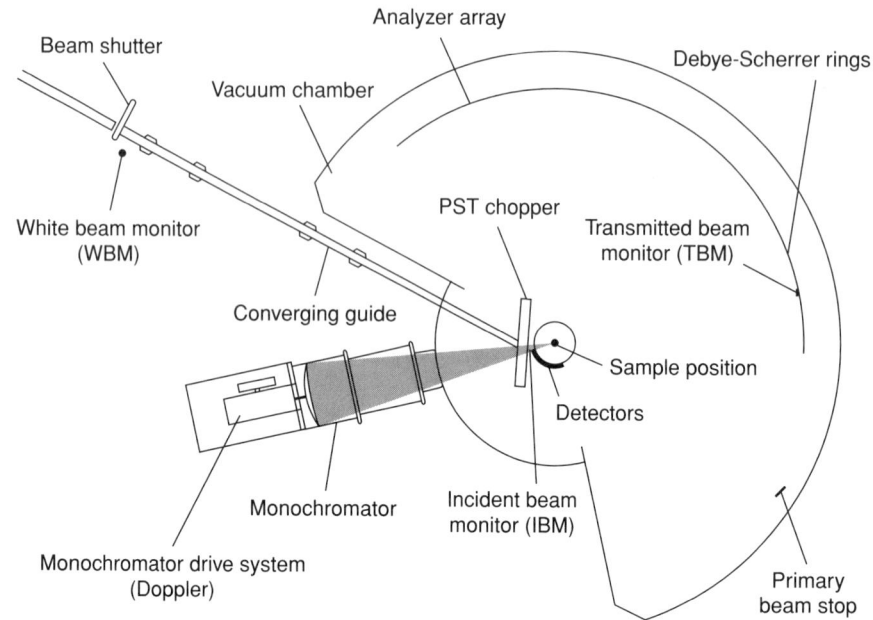

FIGURE 4. A schematic representation of the NIST high flux backscattering spectrometer.

slower motions could be detected. As a scale of reference, these motions are a few orders of magnitude faster than the KHz to MHz processes that can be probed by NMR. It also may be useful to think of these energy resolutions in terms of wave numbers. The Fermi-Chopper TOF spectrometer can see modes higher in energy than approximately 1 cm^{-1} (1 meV ≈ 8 cm^{-1}).

The HFBS spectrometer is a fixed final energy spectrometer (Figure 4). A beam of neutrons travels down the converging guide and bounces backward off of a phase space chopper, toward the monochromater. The phase space chopper allows only those neutrons with a wavelength of 6.271 Å to pass onto the monochromater. The monochromater reflects these 6.271 Å neutrons back toward the phase space chopper and into the sample. When the neutrons hit the sample, they scatter at different angles into the Debye-Scherrer ring of reflectors. The Debye-Scherrer rings also reflect only those neutrons with a wavelength of 6.271 Å back toward a bank of detectors that resides just behind and slightly above the sample. Given that all of the neutron optics in this system are designed for 6.271 Å neutrons, only elastically scattered neutrons reach the detectors when the monochromator is static. The key to detecting dynamics (inelastic neutrons) with the HFBS spectrometer is that the monochromator can oscillate back and forth into the incident neutron beam. This Doppler-shifts the reflected neutrons; some slightly increase in energy and some slightly decrease. If the frequency and stroke of the Doppler drive oscillation is known, it is possible to calculate the broadened energy distribution of the Doppler-shifted, initially monochromatic neutron beam. However, only those Doppler-shifted neutrons that change back to their original incident 6.271 Å wavelength after scattering are able to reflect off of the Debye-Scherrer ring and into the detectors. From this it is possible to determine the energy distribution of the scattered neutrons.

The HFBS spectrometer is capable of detecting much smaller neutron energy exchanges with the sample, and therefore slower dynamics than the TOF spectrometer. The incident energy beam can be collimated to 0.85 μeV FWHM in terms of an elastic energy resolution. This means that motions faster than 205 MHz can be seen by the HFBS spectrometer; slower motions appear as static. This energy resolution is much closer to the frequencies accessible through a NMR measurement. In terms of wave numbers, the HFBS is sensitive to modes of 0.007 cm^{-1} and higher in energy.

2.2. Data Interpretation

To understand the nature of the motion in a biological or soft material, the Q and ω (frequency) dependence of the scattered neutrons must be modeled to extract the characteristic lengths and timescales of the motion. That Q is the space Fourier transform of the relevant length scales while ω is the Fourier transform of the timescales of the motion into the frequency (or energy) domain. The Q and ω dependence of the scattered neutrons contains all the spatial and temporal characteristics information of the dynamics that fall within the energy and Q resolution of the spectrometer. It can be challenging to extract the proper details of the motion given the inverse problem (loss of phase information) in the scattering process. Therefore, the experimental data must be modeled and inherent assumptions or limitations often are implied by the nature of the model. A common starting point for most dynamic models is the one-phonon approximation:

$$S_{inc}(Q, \omega) = \frac{3N\eta}{2M} e^{-2W} Q^2 \frac{n(\omega) + 1}{\omega} g(\omega), \qquad (2)$$

where $S_{inc}(Q,\omega)$ is proportional to the number of neutrons (i.e., intensity) scattered at a wave vector Q with a frequency ω, $g(\omega)$ is the density of states, $n(\omega) + 1$ is the Bose population factor, and e^{-2W} is the Debye-Waller factor. In this expression, $W = (1/6)Q^2\langle u^2 \rangle$, with $\langle u^2 \rangle$ denoting the mean-square atomic displacement. The prefactor $3N\eta/2M$ contains all the information about the total number of scattering nuclei in the sample and their representative scattering cross sections. The process of choosing an appropriate model for the motion is beyond the scope of this chapter. However, there are several textbooks dedicated to this subject.[1–3] In the following we present a few of the most simple models, which are sufficient to illustrate the power of the technique. Before describing these neutron-scattering measurements, it is important to review some basic facts regarding protein preservation.

3. SUGARS AND PROTEIN PRESERVATION

Over a decade ago it was discovered that carbohydrate glass plays a central role in anhydrobiosis.[4,5] Since then preservation of biological agents in nominally dry carbohydrate glasses has become a problem of both technological and scientific interest. Rapidly growing sectors of the pharmaceutical and tissue engineering industries are predicated on delivering functional proteins into the body from a dry state. It is further anticipated that proteinaceous pharmaceuticals will account for half of all the new drugs in the next 10 to 20 years.[6] Likewise, it is understood that tissue scaffolds for regenerative medicine need to contain stabilized signaling proteins or DNA. Protein preservation is becoming a critical technology. However, despite the technological significance of this issue, the mechanisms by which viscous sugars impart dehydration stability are not fully understood. Rational design and the optimization of bioprotective glasses for protein stabilization require a more detailed understanding of the underlying mechanisms of preservation.

There appear to be two intrinsic characteristics of a hydrophilic glass that are important for protein stabilization: an amorphous structure and retarded dynamics relative to physiological conditions. It also is realized that the hydrogen bonding environment of the preservant must act as a "substitute" for water in terms of the protein's local environment.[7–10] Such a substitution necessitates the amorphous character of the glass. This is evidenced by the fact that proteins are not stabilized in the crystalline phase of an otherwise effective preservant.[11,12]

The slow relaxation dynamics in a polyhydroxyl glass relative to physiological timescales are believed to likewise retard potential adverse chemical reactions of the protein, such as deamidation and/or oxidation of peptide side chains. The slow dynamics also are thought to retard motion of the protein itself, limiting unwanted processes such as denaturation and aggregation. This follows from the notion of a viscous coupling between the preservant and the protein. Significant progress has been made in understanding the relationship between protein reaction kinetics and host (solvent) dynamics in the low to intermediate viscosity range.[13–17] However, the nature and importance of the protein–solvent dynamic interactions are still an open question in the high-viscosity regime, like those encountered in a glassy preservation medium.[18–22]

The uncertainties in the relationship between protein and solvent dynamics become even less clear when anhydrous biopreservation is considered. In the context of pharmaceutical lyoprotective glasses (i.e., when protein stability is of central interest), dynamics-related discussions often are couched in terms of the glass transition temperature (T_g) of the lyoprotective host. It has been suggested that higher T_g sugars should make better preservants than others with lower T_g.[23] The underlying assumption here is that dynamics are more severely retarded deeper in the glassy state. Indeed, a positive correlation between stability and T_g has been reported,[24–27] but there also are counterexamples illustrating that a high T_g alone is not sufficient to maximize stability.[28,29] We illustrate here that the frequency and length scale details of the dynamics in the glass are relevant for stability, and that T_g alone cannot be used to predict these relevant dynamics in a meaningful way.

Previously we demonstrated that small amounts of a low-T_g diluent such as glycerol added to a the common bioprotective glass trehalose substantially can increase the stability of both horse radish peroxidase (HRP) and yeast alcohol dehydrogenase (YADH) freeze-dried in these binary sugar mixtures.[30] This increase in protein stability was somewhat striking because the added glycerol also reduces the T_g of the lyoprotective trehalose host. However, glasses in general exhibit rich dynamics, over many decades in frequency, which cannot be described by simple parameters such as T_g, or even viscosity. The data presented here affirms that T_g of the biopreservation glass alone does not predict the degree of stability that a glass can impart to a protein. Rather, the amplitude of the local, high-frequency dynamics of the glass measured by INS correlates with the ability to impart protein stability. These results were described in detail,[31] but a brief recap is provided here for illustration. INS is used to show that small amounts of diluent suppress local relaxation and "stiffen" collective vibrations that occur with timescales of a nanosecond and faster. The stiffening or suppression of these fast, local motions correlates precisely with the effectiveness of the glass at stabilizing labile proteins. This strong correlation suggests that dynamics of the protein and glass remain coupled.

3.1. Dynamics of Pure Trehalose and Trehalose-Glycerol Mixtures

We start by studying the dynamics of the pure trehalose and glycerol mixtures, even before protein is added to the system. Glycerol, d_8-glycerol, and α-α (d) trehalose were obtained from Sigma-Aldrich (St. Louis, MO)* and used without further purification. The neutron-scattering samples were prepared by freeze-drying aqueous solutions of trehalose and glycerol. The individual concentrations of trehalose and glycerol varied, but the combined concentration was maintained at a mass fraction of 0.20. Control samples were made from solutions also containing 0.100 mol/L $CaCl_2$, 300 μg/ml Tween, and 0.050 mol/L histidine buffer, pH 6.0. These additives are consistent with the previous enzyme activity measurements.[30] All solutions were prepared using milliQ deionized water (18 MΩcm) and then freeze-dried overnight. By measuring the final mass, we confirmed that sample is not lost during the freeze-drying process.

The freeze-dried glasses were handled in one of two ways. In an initial set of experiments, using the hydrogenous glycerol, the glasses were briefly exposed to air (a few minutes) while loading the neutron scattering sample cells. Potential moisture uptake for these samples was not quantified, but this brief exposure to ambient air also was consistent with the protocol for the enzyme stability studies published previously.[30] In subsequent experiments using d_8-glycerol, the vials containing the freeze-dried glasses were backfilled with argon and sealed in the freeze-drier. Those samples were then loaded into the hermetically sealed neutron-scattering sample cells under an argon atmosphere without any exposure to ambient air.

The elastic incoherent neutron-scattering measurements were performed at the NIST Center for Neutron Research on the high-flux backscattering (HFBS) spectrometer,[32] located on the NG2 beam line. The HFBS spectrometer operates with an incident neutron wavelength of 6.27 Å and an energy resolution of 0.85 μV full width at half maximum. The momentum transfer (Q) range of the spectrometer is (0.25 to 1.75) Å^{-1}. In these experiments the spectrometer is run in the "fixed window" mode where the Doppler drive remains static; only the elastic-scattering intensity is recorded as a function of Q while the sample is heated at 1 K/min from 40 K to just above the calorimetric T_g. Since the total scattering (elastic plus inelastic) is conserved, a decrease in the elastic intensity infers an increase in the inelastic. For our samples the scattering is dominated by hydrogen, which has an incoherent scattering cross section approximately 20 times greater than carbon, oxygen, nitrogen, sulfur, or most other elements common to biological systems. Given that only the elastic intensities ($\omega = 0$) are considered, the the Q dependence of the incoherent elastic-scattering intensity I_{inc} [proportional to $S(Q,\omega)$ in Eq. (2)] can be approximated in terms of just a Debye-Waller factor (a harmonic oscillator model) where the hydrogen weighted mean-square atomic displacement $\langle u^2 \rangle$ is given by:

$$I_{inc}(Q) \propto \exp\left(-\frac{Q^2 \langle u^2 \rangle}{3}\right). \tag{3}$$

With this simplification, a plot of $\ln(I_{inc})$ versus Q^2 is linear and the slope is proportional to $\langle u^2 \rangle$.

Figure 5 shows the Q dependence of the elastic-scattering intensity from a pure trehalose at a few different temperatures. The I_{inc} is normalized by the scattering intensity at the lowest temperature available, which is 40 K in these experiments. As we are interested primarily in dynamics, and therefore the incoherent scattering, this normalization also helps remove small artifacts of coherent scattering (static structure) that might affect the Q variations of the data. This normalization means that the slope of $\ln(I_{inc})$ versus Q^2 is zero at 40 K, i.e., $\langle u^2 \rangle = 0$ at 40 K. We compensate for this effect by fitting a straight line through the $\langle u^2 \rangle$ versus T data between 40 K and 200 K and vertically offsetting the

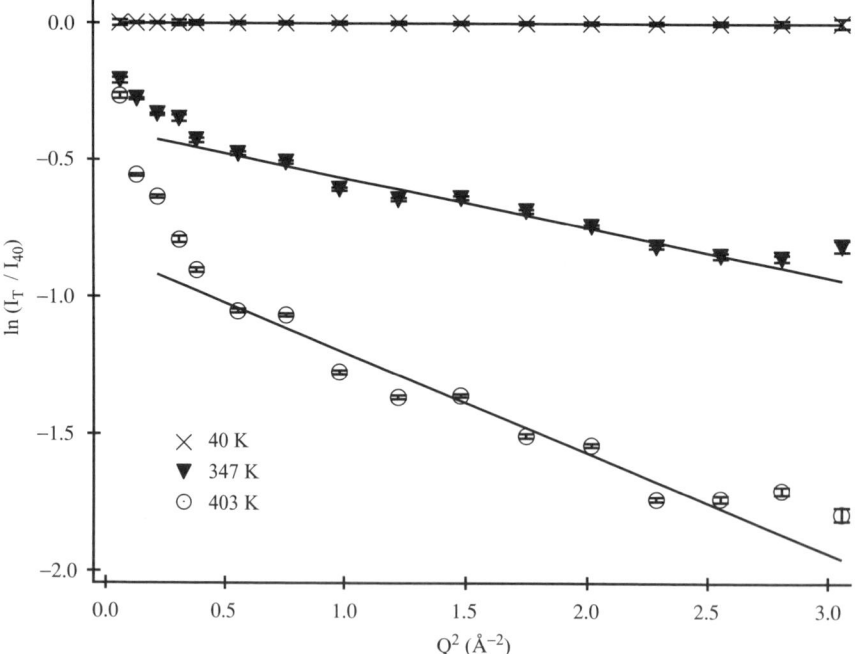

FIGURE 5. The natural log of the incoherent elastic neutron-scattering intensities (normalized by their value at 40 K) are presented as a function of Q^2 for three different temperatures in pure trehalose. In this representation the slope, indicated by the linear fit in the $(0.5 < Q^2 < 3.0)$ Å^{-2} range, is proportional to $\langle u^2 \rangle$ in accordance to Eq. (3).

data so that the $\langle u^2 \rangle = 0$ intercept occurs at $T = 0$ K. At temperatures above 40 K there is a reduction of the elastic scattering due to the thermal motion. As seen in the linear fits of Figure 5, this leads to an increase in the slope, or an increase in $\langle u^2 \rangle$. It is important to realize that the 0.85 μeV resolution of the HFBS spectrometer means that only those motions faster than approximately 205 MHz, or a few nanoseconds in the time domain, give rise to a increase of $\langle u^2 \rangle$; slower motions appear as static.

The Debye-Waller formalism is a harmonic approximation, typically an oversimplification in soft matter at ambient T. This is evident by the fact that a single linear function does not fit the data in Figure 5; the dependence is almost bilinear with break in the slope near $Q^2 = 0.5$ Å^{-2}. Nonlinear dependencies of $\ln(I_{inc})$ on Q^2 have been reported in both polymer systems[33] and biological systems,[34,35] and there have been attempts to calculate $\langle u^2 \rangle$ using more complicated models. These models generally include both a Gaussian (harmonic) and a non-Gaussian component, with the latter being thermally activated and dominant at low Q. In our view one must fit the data over a very broad range of Q in order to reliably separate the local harmonic motions (high Q) from the long-range anharmonic contributions (low Q). The HFBS spectrometer is not well-suited for this separation as it does not access a sufficiently large region of Q space. Therefore we simplify the situation and only use the harmonic approximation for the high Q data, beyond $Q^2 = 0.5$ Å^{-2}, where the motions are smaller and the harmonic approximation is most appropriate. Despite its simplicity, the harmonic oscillator approximation has proven useful in qualitatively analyzing several biological systems.[36] The small set of lower Q data (below $Q^2 = 0.5$ Å^{-2}) that we neglect contains potentially interesting information about longer-range motions in the system, but attempts are not made to interpret these data because of the limited Q range; the significance of the fits would be marginal here.

Figure 6 displays the $\langle u^2 \rangle$ values obtained by applying the Debye-Waller formalism to the elastic incoherent neutron scattering for the five freeze-dried trehalose glasses diluted with d$_8$-glycerol; the mass fractions ($\phi_{glycerol}$) range from 0 to 0.20 in increments of 0.05. These glasses were prepared from H$_2$O, so only the nonexchangeable protons remained deuterated (i.e., each glycerol contains 3 –OH groups, not –OD). We emphasize that none of these glasses contain protein. The previous stability studies[30] using these same glass formulations were performed with glass-to-protein mass ratios greater than 10^6:1. Such small amounts of protein would be undetectable by neutron scattering and including protein at a detectable level would completely change the character of the glass. The glasses studied here also do not contain the salt buffer or surfactant that is used in the stability studies. However, control experiments demonstrate that the presence of the salt and surfactant at levels consistent with the activity measurements do not have a noticeable effect on $\langle u^2 \rangle$.

We see in Figure 6 that $\langle u^2 \rangle$ evolves more or less linearly with T below 200 K for all samples, as is characteristic for a harmonic solid. Within this framework, the equipartition theorem ($kT/2$ per degree of freedom, where k is Boltzmann's constant) is used to calculate an effective spring constant κ; stiffer harmonic vibrations give a larger κ or a smaller temperature dependence of $\langle u^2 \rangle$. The inset to Figure 6 displays κ values derived from the slope of linear fits to the $\langle u^2 \rangle$ data in the range 40 to 200 K. We see that the trehalose with $\phi_{glycerol} = 0.05$ provides the "stiffest" glassy environment in the low-temperature regime. Independent of the harmonic spring constant model

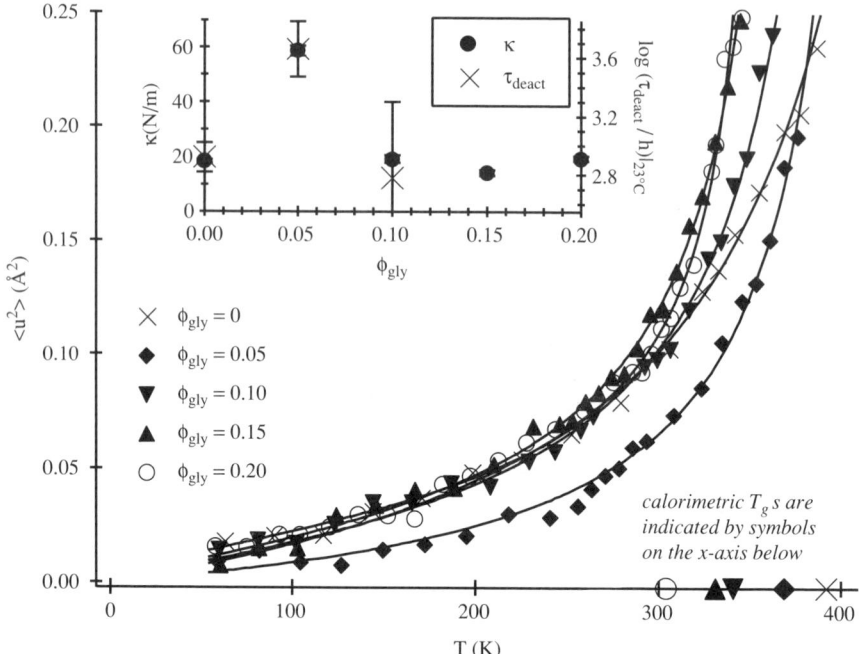

FIGURE 6. Thermal evolution of $\langle u^2 \rangle$ is shown for the trehalose glasses containing the indicated mass fractions of glycerol. Notice that $\langle u^2 \rangle$ is most dramatically suppressed at $\phi_{glycerol} = 0.05$. The low temperature (below 250 K) evolution is parameterized in terms of a harmonic oscillator spring constant κ, shown in the inset (left axis and filled circles). The sample with $\phi_{glycerol} = 0.05$ also shows the most strongly stiffened vibrations. The inset also displays the protein stability lifetime data τ_{deact} (right axis and Xs) for HRP, described later in the text below. The binary glass with the stiffest κ also provides the best protein stability.

this qualitatively reflects the observation that $\langle u^2 \rangle$ is the most strongly suppressed in the sample with $\phi_{glycerol} = 0.05$. This is true not only in the low T linear regime, but also extending out to nearly 350 K. The harmonic oscillator assumption implicit in the determination of κ is of course not appropriate above 200 K. However, as suggested by Zaccaï[36] it is still useful to think of reduced $\langle u^2 \rangle$ at these higher T qualitatively as an increased effective "environmental" spring constant of the soft medium.

Aside from the nonmonotonic variations of κ, the apparent glass transitions of these binary glass mixtures vary as expected with the glycerol content. In Figure 6 the five data markers lying directly on the x-axis indicate the calorimetric T_g monotonically shifting to lower T with increasing glycerol content. Likewise, the strong upturn in $\langle u^2 \rangle$ with T gets pushed to lower temperatures with increasing glycerol content. This upturn is anticipated near T_g.[37,38] Both observations are consistent with the lower T_g of glycerol in comparison to trehalose. Likewise, we note that the amplitude of $\langle u^2 \rangle$ at T_g also generally decreases with T_g. For example, $\langle u^2 \rangle|_{T=T_g} = 0.25$ Å^2 for pure trehalose ($T_g = 392$ K) and decreases toward $\langle u^2 \rangle|_{T=T_g} = 0.11$ Å^2 for trehalose with $\phi_{glycerol} = 0.20$ ($T_g = 304$ K). These monotonic variations with $\phi_{glycerol}$ in relation to T_g are in stark contrast with the environmental spring constant which displays a pronounced maximum at $\phi_{glycerol} = 0.05$.

It is very striking that the nonmonotonic variation in κ precisely reflects the protein stability data, as shown in the inset of Figure 6. The protein deactivation lifetimes are about an order of magnitude longer (notice the logarithmic presentation of the lifetime data) at the same $\phi_{glycerol} = 0.05$. These stability data will be described in greater detail below. However, it becomes immediately obvious that a higher T_g alone is not sufficient for enhanced protein stability. It appears that local dynamics play an important role in stabilization.

Elastic-scattering experiments also were performed on trehalose glasses diluted with hydrogenated glycerol at the same $\phi_{glycerol}$ loadings. These samples were briefly exposed to atmosphere while being transferred to the neutron-scattering cell, consistent with the protein stability data.[30] Like the samples with the deuterated glycerol, the same maximum suppression of $\langle u^2 \rangle$ occurs at $\phi_{glycerol} = 0.05$. The magnitude of the low T harmonic stiffening at $\phi_{glycerol} = 0.05$ is somewhat less (approximately 35 N/m versus 59 N/m in Figure 3). However, this also was consistent with a slight suppression in all the κ values for the hydrogenous glycerol samples relative to their deuterated analogs. It is not clear whether this small difference is an isotopic effect or due the absorption of atmospheric water in the hydrogenated samples during their brief exposure to the ambient atmosphere. There have been reports that small amounts of moisture can soften similar glasses in the frequency window of the HFBS spectrometer.[39]

Incoherent inelastic neutron-scattering measurements also were performed at the NIST Center for Neutron Research, using the Fermi-Chopper TOF spectrometer on the NG6 beam line. Figure 7 displays a typical inelastic neutron scattering spectrum, in this case for pure trehalose at 100 K. The spectra are adjusted for the different macroscopic scattering cross sections, corrected for detector efficiency with a vanadium standard, summed over the accessible Q range of the instrument (essentially the same Q range as the HFBS spectrometer), and Bose-scaled to account for trivial temperature variations in the density of states. The inset of Figure 7 shows trehalose spectra at 100 K and 296 K; the Bose-scaling results in overlap of the data at higher energies as expected. The spectrum in the main portion of Figure 7 appears to contain both types of INS depicted in Figure 1. There is a broad but distinct inelastic scattering peak at approximately (−4 to −5) meV, known as a boson peak. Likewise, Figure 7 also shows evidence of the QES broadening that reflects diffusive or relaxational motions [the instrumental resolution function (IRF) is indicated by the short dashed line]. To interpret the data, we must fit the two components to separate the vibrational and relaxational contributions.

The boson peak energies are extracted from the inelastic spectra in order to compare peak positions between different glasses. This requires fitting the spectra with several components, as illustrated in Figure 7 for the 100 K spectrum. These components include a 130 μeV instrumental resolution function (IRF) is indicated by the narrow Gaussian centered at 0 meV, a broad quasielastic scattering component also centered at 0 meV (QES, a Lorentzian line shape) due to fast relaxations, and a broad, asymmetric boson peak (BP) with a distinct maximum between (−4 and −5) meV. At 100 K the relaxations are not strong meaning that the QES and BP are well separated. In this limit, fitting the boson peak position is model independent. Boson peaks have been fitted with a number

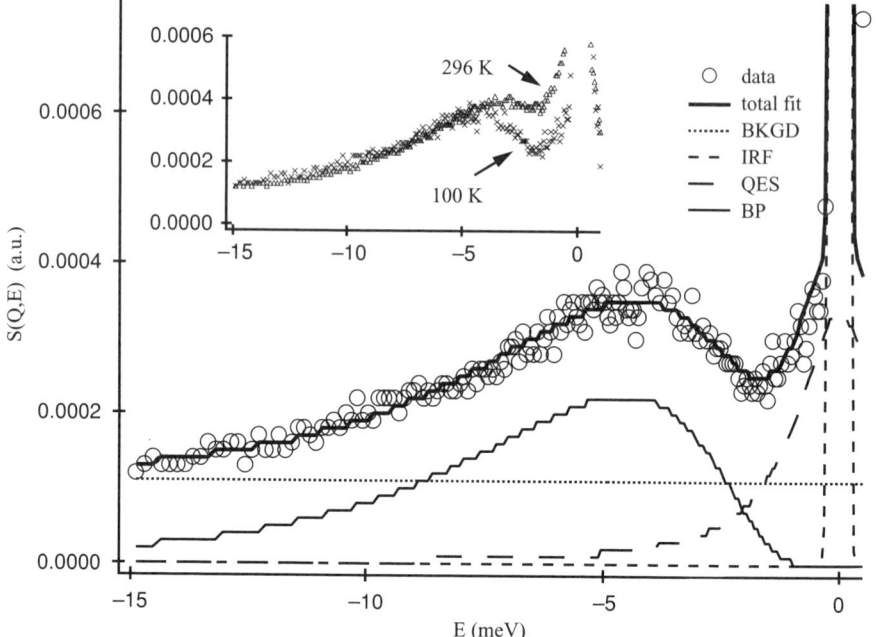

FIGURE 7. INS spectra for the pure trehalose sample. The data are plotted along with various components of the fit, as discussed in the text, for the 100 K spectrum. *Inset*, Comparison of 100 K and 296 K spectra, revealing that the boson peak softens and the QES increases both with increasing temperature, while the high-frequency tails of the Bose-scaled spectra coincide.

of physical models and/or empirical functions. Here we parameterize the line shape with a model-independent lognormal distribution function[40]:

$$P_{LN}(x) = \frac{1}{\sqrt{2\pi\sigma^2}} \frac{1}{|x|} \exp\left[\frac{-(\ln|x| - \mu)^2}{2\sigma^2}\right], \quad (4)$$

where μ and σ are the center and width of the asymmetric distribution. These are related to the peak maximum (mode) through the relationship $x_{peak} = \exp(\mu - \sigma^2)$; this mode is reported as the boson peak position. In addition to the IRF Gaussian, QES Lorentzian, and BP lognormal, a flat background is needed to fit the data. Physically, this background represents the Debye-level density of states that reaches a plateau at $E = 0$ meV. The height of this plateau depends on the velocity of sound, which has not been measured for these materials. In the absence of this knowledge, we assign the intensity of this background to equal the total signal intensity at −15 meV in the high-frequency tail of the boson peak. The inset of Figure 7 shows that the high frequency tails of the Bose-scaled spectra coincide in this region, suggesting that this assignment is reasonable.

The Lorentzian-shaped QES reflects fast relaxations or diffusive-type motions. This QES contribution increases with T as the matrix softens, as expected for a diffusive

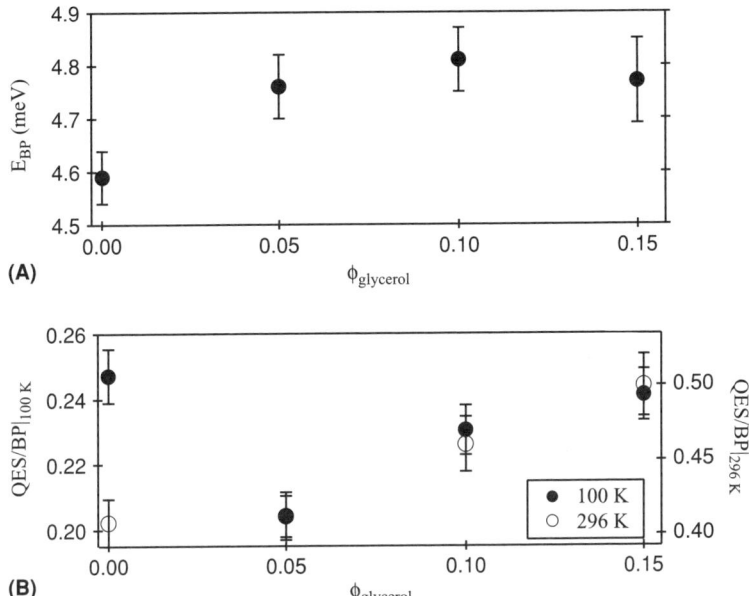

FIGURE 8. Effect of glycerol content. (A) Variation of the boson peak energy with glycerol content. (B) The ratio of the fast quasielastic (QES) relaxations to the similar frequency collective vibrations (BP) as a function of the glycerol content. Generally, vibrations are most effectively stiffened and the relaxations most strongly suppressed at $\phi_{glycerol} = 0.05$.

process. The broad BP excitation near ~5 meV at 100 K also softens to lower energies with increasing T (see Figure 7, inset). The BP is a more or less a ubiquitous feature for glass-forming materials below T_g, stemming from a collective excitations/vibrations, not relaxations like the QES. These collective vibrations are lower in frequency than the optic modes typically associated with infrared and/or Raman spectroscopy, but higher in frequency (and more localized) than the acoustic modes or phonons. While the exact nature of the boson peak is still uncertain, estimates indicate that somewhere between 10 to 100 atoms participate in this high-frequency collective mode.[41,42]

Figure 8A shows the 100 K BP energies (E_{BP}) of trehalose as a function of the glycerol content. The BP position in pure trehalose lies at (4.59 ± 0.05) meV, and like the $\langle u^2 \rangle$ data, stiffens significantly at $\phi_{glycerol} = 0.05$. Further additions of glycerol soften the BP to lower energies, but this is not as pronounced as the effect in κ from the $\langle u^2 \rangle$ data. At $\phi_{glycerol} = 0.10$ the BP position is the same to within error, if not slightly greater, than the $\phi_{glycerol} = 0.05$ sample; the BP softening does not become appreciable until $\phi_{glycerol} = 0.15$. In passing we mention that the $\phi_{glycerol} = 0.20$ BP stiffened dramatically to (5.20 ± 0.08) meV (not shown here). We do not completely understand this notable stiffening, but suspect that it is due to phase separation of the glycerol and trehalose; 5.2 meV is consistent with the BP of pure glycerol.[43] Likewise, we occasionally see evidence for phase separation in samples with $\phi_{glycerol} \geq 0.15$. X-ray diffraction measurements reveal that at these high glycerol loadings the trehalose can crystallize quickly (on the order of

10s of minutes) upon exposure to ambient relative humidity. However, it did not appear that phase separation was an issue in the any of HFBS experiments or in the protein stability studies. Crystallization results in an extremely pronounced decrease of $\langle u^2 \rangle$ and loss of biochemical activity.

The trend of E_{BP} with glycerol content is qualitatively similar at room temperature (where the protein stability measurements were made) to the 100 K data, with just an overall softening of the BP or shift of E_{BP} to lower energies. We do not present the room temperature E_{BP} peak assignments here because the softening and the increasing of the QES intensities convolute the peak fitting as can be seen in the inset of Figure 7. Rather the higher T inelastic-scattering data are presented in terms of the ratio between the QES and BP component, as shown Figure 8B. The impetus for this is the following: The collective BP vibrations can be viewed as cooperative attempts at fast molecular relaxations, like a diffusive motion, whereas the QES reflect successful relaxations. Therefore the QES/BP ratio roughly indicates the "efficiency" with which fast relaxations occur in the glass. We estimate this ratio in a model-independent way by using the minimum in $S(Q, E)$ as a dividing line between the inelastic and quasielastic components. This minimum occurs at -1.70 and -1.85 meV in the 100 K and 296 K data, respectively (see Figure 7). We integrate the scattering on the high-energy side of the minimum, out to -15 meV, to estimate the BP component and likewise the total scattering between the minimum and -0.35 meV to estimate the total QES contribution. Before determining this ratio the flat Debyelike background was removed. However, neglecting this subtraction makes no difference in the general trend between the different data sets.

At 100 K, these fast relaxations are most strongly suppressed at $\phi_{glycerol} = 0.05$ (Figure 8B), exactly the same as the stiffening of κ seen with the $\langle u^2 \rangle$ data. It appears that this trend breaks down at room temperature; the pure trehalose and the $\phi_{glycerol} = 0.05$ glasses appear to be equal in terms of their relative ability to suppress fast relaxations. However, the overall amplitude of room temperature scattering is also much lower in the $\phi_{glycerol} = 0.05$ sample. Thus, the $\phi_{glycerol} = 0.05$ glass exhibits less relaxation at room temperature.

These results illustrate that small amounts of glycerol suppress the amplitude of the fast (205 MHz and faster) dynamics in the glassy trehalose. This is despite the fact that the glycerol acts as a plasticizer and reduces the glass transition temperature. However, such behavior is not entirely unexpected given that small-molecule diluents often plasticize T_g while simultaneously antiplasticizing the sub-T_g dynamics in the kHz frequency range; this is seen with sugars[44,45] and synthetic polymers[46,47] alike. The dynamic signatures of antiplasticization are qualitatively consistent with the suppressed dynamics witnessed here. This effect also appears to be general. We have observed similar effects with several disaccharides and polymeric sugars and a range of small-molecule diluents, including glycerol, dimethylsulfoxide, propylene glycol, ethylene glycol, and oligomeric polyethylene glycol.

3.2. Dynamics of Lysozyme-Glycerol Mixtures

Thus far all of the neutron-scattering measurements have been on the glassy preservation media without protein. By correlating the activity data on protein-containing glasses with the dynamics of the protein-free glass, we were able to conclude that the dynamics of the glass were coupled to the protein. However, it would be reassuring to directly measure this dynamic coupling. In most protein-preservant systems this is difficult because the protein loading levels into the glassy matrix are quite small, typically micromolar concentrations. With neutron scattering it is difficult to measure just the protein dynamics at this level, even if the glass or preservation matrix is completely deuterated (recall Figure 2; H dominates the neutron scattering); the amount of scattering from such a small amount hydrogenous protein is insignificant. With most proteins higher levels of protein loading in the glass are not feasible as they naturally lead to aggregation; the glass can accommodate only low levels of biologically active protein. However, the lysozyme-glycerol system provides a nice exception to this rule. It is now understood that lysozyme can fold into its native structure and remain biologically active in pure glycerol.[48] This allows us to achieve very high loadings of glycerol without aggregation problems.

Lysozyme powder was obtained directly from Sigma. Since the incoherent scattering cross section for deuterium is nearly 20 times less than that of hydrogen, we can selectively study the dynamics of the hydrogenous lysozyme *in situ* in the preservation media by using the deuterated glycerol (pure glycerol is also frequently used as a lyoprotective host). As some of the protons in lysozyme will exchange readily with the deuterated glycerol, the lysozyme was first dissolved in D_2O to preempt these unwanted exchange events that would lead to hydrogenated glycerol. A total of 400 mg of the partially deuterated lysozyme was then used to prepare each of the neutron-scattering samples. For the dry sample, the partially deuterated glycerol and D_2O solution was directly lyophilized into a powder. Equilibrating this freeze-dried lysozyme over a saturated K_2SO_4/D_2O solution at room temperature created the wet sample. Under these conditions, the freeze-dried lysozyme picked up approximately 28–30% D_2O by mass. For the glycerol-lysozyme mixture, equal masses of deuterated glycerol and D-exchanged lysozyme were codissolved with the D_2O and then freeze-dried into a powder. For comparison, wet lysozyme without glycerol also was freeze-dried into a dry, agglomerated powder.

Figure 9A displays how $\langle u^2 \rangle$ evolves with temperature in the glycerol-preserved lysozyme, a moisture-saturated (as received) lysozyme, and a moisture-free, freeze-dried lysozyme powder. It should be emphasized that these $\langle u^2 \rangle$ values directly reflect the dynamic of the lysozyme and not the preservation media. Using an analysis similar to Figure 5, the effective spring constant κ can be extracted from the low-temperature region between 150 K and 250 K. Here we find that κ stiffens significantly with the addition of glycerol ($\kappa = 7.6 \pm 0.3$ N/m for the wet lysozyme), stiffens slightly when the protein is dried into a powder ($\kappa = 8.8 \pm 0.2$ N/m), and stiffens significantly ($\kappa = 12.0 \pm 0.3$ N/m) in the presence of glycerol. In other words, an effective preservant stiffens or suppresses protein dynamics relative to aqueous (i.e., biological) conditions. The same message is obtained from Fermi-Chopper TOF measurements of these lysozyme samples. Figure 9B shows that the presence of the boson peak significantly stiffens the collective boson peak to higher energy vibrations and suppresses the QES scattering due to relaxations. Again, because of deuterium masking, these boson peaks reflect collective excitations within the protein, not the matrix. While we do not have direct activity measurements of the lysozyme under these conditions, the measurements clearly indicate that an effective preservant directly suppresses the dynamics of the protein.

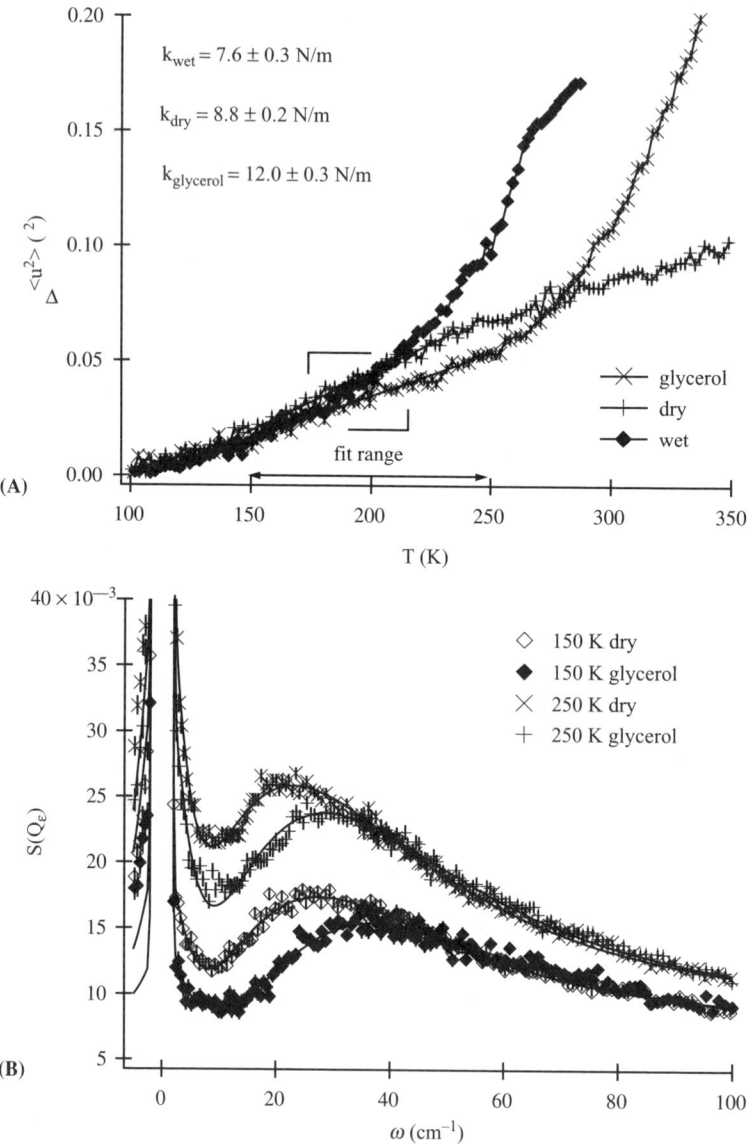

FIGURE 9. Analysis of lysozyme in glycerol. (A) Evolution of $\langle u^2 \rangle$ with temperature for moisture-saturated lysozyme, freeze-dried lysozyme, and lysozyme freeze-dried in deuterated glycerol. The presence of glycerol stiffens the lysozyme relative to the native and dried states and shifts the dynamic transition (in the wet lysozyme) to higher temperature. (B) At low temperature, glycerol also stiffens the collective boson peak vibrations of the protein and suppresses the QES scattering.

One may find it remarkable that the enzyme deactivation lifetimes correlate with the dynamics measured by incoherent neutron scattering; the deactivation measurements are made slowly, over hundreds of hours, while the neutron scattering is sensitive to processes on the order of nano- to picoseconds, a separation of at least 15 decades in time. However, these are not the first correlations between the frequency and amplitude of $\langle u^2 \rangle$ and biological activity in general. Parak et al.[49] showed that proteins undergo a dynamic transition, T_d in the vicinity of 200 K, from a regime of purely harmonic to anharmonic motions. Such a dynamic transition is observed in the data shown in Figure 9B with the wet lysozyme. Brooks et al.[50] suggested that these anharmonic atomic fluctuations act as the lubricants that enable conformational fluctuations on a physiological timescale. This idea was supported by the discovery that T_d coincides with the temperature for the onset of biological activity.[51] Likewise, studies on bacteriorhodopsin reveal a positive correlation between local of biological activity in the membrane[52] and the regions of large-amplitude fluctuations reflected in $\langle u^2 \rangle$.[53]

4. CONCLUSION

Recently, Cordone et al.[54] noticed that trehalose, a well-known protein preservation media, suppressed the $\langle u^2 \rangle$ in myoglobin significantly in comparison to an aqueous environment. The same effect is shown in Figure 9A with the lysozyme-glycerol system. This led them and others[36,55-57] to speculate that an effective lyophilization medium will suppress fast dynamics of the protein reflected in $\langle u^2 \rangle$. However, until now there has been a lack of a direct correlation to supporting biological activity measurements. The data presented here provide this link for the first time. We see a marked positive correlation between improved protein preservation and suppression in $\langle u^2 \rangle$ of the preservation glass itself.

Utilizing this correlation between high-frequency glass dynamics and protein preservation would constitute a paradigm shift in the way in which potential preservation schemes are designed and evaluated. Yet we would be remiss not to reiterate the importance of the thermodynamic interactions. Thermodynamic compatibility in many ways, is the first cut; the preservant must not denature, deactivate, or destroy the protein. This is precisely why often there are good correlations with thermodynamic parameters. For example, Miller and de Pablo[58] showed a correlation between the effectiveness of lyoprotective glasses and heat of solution; connections between hydrogen bonding and effective stabilization are well established.[7] However, thermodynamics cannot be used to predict kinetics, and visa versa. This is why the dynamics must be assessed independently to identify which of the thermodynamically compatible preservation schemes will perform the best. Both the thermodynamics and kinetics must be considered when optimizing protein-stabilizing systems.

In closing we also recognize that using neutron spectroscopy to assess protein stabilization can be difficult in that neutron-scattering facilities are not widely available. However, comparable information can be gleamed form low-frequency Raman scattering, a technique that is far more amenable to the general laboratory setting. Caliskan and coworkers[56] recently demonstrate that low-frequency Raman can be used to characterize the boson peak variations and integrated quasielastic intensities (a parameter related to our $\langle u^2 \rangle$ measurements here) as a function of T for lysozyme preserved in glycerol and trehalose. While this study lacked direct protein activity measurements, they present compelling arguments, based on time-dependent geminate CO recombination rates for myoglobin in both glycerol and trehalose,[21] that protein stability and fast dynamics are correlated. Their supposition is identical to the conclusion here; enhanced protein stability is achieved by stiffened collective vibrations and suppressed fast relaxations on the nano- to picosecond timescale. In fact, their results explain the previously paradoxical observation that over a certain low temperature range, liquid glycerol is more effective at suppressing biological activity than glassy trehalose. The simple notion of viscosity as the primary factor controlling protein dynamics does not explain this observation. Beyond glass transitions and macroscopic viscosities, high-frequency dynamic coupling on timescales of a nanosecond and faster must be considered.

NOTE

* Certain commercial equipment and materials are identified in this chapter in order to specify adequately the experimental procedure. In no case does such identification imply recommendation by the National Institute of Standards and Technology nor does it imply the material or equipment identified is necessarily the best available for this purpose.

REFERENCES

1. M. Bée, *Quasielastic Neutron Scattering* (Philadelphia, PA: Adam Hilger Press, 1988).
2. J. S. Higgins and H. C. Benoît, *Polymers and Neutron Scattering* (Oxford: Carendon Press, 1994).
3. R. J. Roe, *Methods of X-ray and Neutron Scattering in Polymer Science* (Oxford: University Press, Oxford, 2000).
4. J. F. Carpenter, L. M. Crowe, and J. H. Crowe, Stabilization of phosphofructokinase with sugars during freeze-drying: characterization of enhanced protection in the presence of divalent cations, *Biochim. Biophys. Acta* **923**, 109–115 (1987).
5. R. Mouradian, C. Womersley, L. M. Crowe, and J. H. Crowe, Preservation of functional integrity during long term storage of a biological membrane, *Biochim. Biophys. Acta* **778**, 615–617 (1984).
6. C. M. Henery, The next pharmaceutical century, *Chem. Eng. News* **78**, 85–100 (2000).
7. S. D. Allison, B. Chang, T. W. Randolph, and J. F. Carpenter, Hydrogen bonding between sugar and protein is responsible for inhibition of dehydration-induced protein unfolding, *Arch. Biochem. Biophys.* **365**, 289–298 (1999).
8. J. L. Cleland, X. Lam, B. Kendrick, J. Yang, T. H. Yang, D. Overcashier, D. Brooks, C. Hsu, and J. F. Carpenter, A specific molar ratio of stabilizer to protein is required for storage stability of a lyophilized monoclonal antibody, *J. Pharm. Sci.* **90**, 310–321 (2001).
9. H. R. Costantino, K. G. Carrasquillo, R. A. Cordero, M. Mumenthaler, C. C. Hsu, and K. Griebenow, Effect of excipients on the stability and structure of lyophilized recombinant human growth hormone, *J. Pharm. Sci.* **87**, 1412–1420 (1998).
10. K. Tanaka, T. Takeda, and K. Miyajima, Cryoprotective effect of saccharides on denaturation of catalase by freeze-drying, *Chem. Pharm. Bull.* **39**, 1091–1094 (1991).
11. K. Izutsu, S. Yoshioka, and T. Terao, Effect of mannitol crystallinity on the stabilization of enzymes during freeze-drying, *Chem. Pharm. Bull. (Tokyo)* **42**, 5–8 (1994).

12. M. J. Pikal and D. R. Rigsbee, The stability of insulin in crystalline and amorphous solids: observation of greater stability for the amorphous form, *Pharm. Res.* **14**, 1379–1387 (1997).
13. A. Ansari, C. M. Jones, E. R. Henry, J. Hofrichter, and W. A. Eaton, The role of solvent viscosity in the dynamics of protein conformational changes, *Science* **256**, 1796–1798 (1992).
14. D. Beece, L. Eisenstein, H. Frauenfelder, D. Good, M. C. Marden, L. Reinisch, A. H. Reynolds, L. B. Sorensen, and K. T. Yue, Solvent viscosity and protein dynamics, *Biochemistry* **19**, 5147–5157(1980).
15. W. Doster, Viscosity scaling and protein dynamics, *Biophys. Chem.* **17**, 97–103 (1983).
16. B. Gavish, Position-dependent viscosity effects on rate coefficients, *Phys. Rev. Lett.* **44**, 1160–1163 (1980).
17. T. Kleinert, W. Doster, H. Leyser, W. Petry, V. Schwarz, and M. Settles, Solvent composition and viscosity effects on the kinetics of CO binding to horse myoglobin, *Biochemistry* **37**, 717–733 (1998).
18. D. S. Gottfried, E. S. Peterson, A. G. Sheikh, J. Q. Wang, M. Yang, and J. M. Friedman, Evidence for damped hemoglobin dynamics in a room temperature trehalose glass, *J. Phys. Chem.* **100**, 12034–12042 (1996).
19. S. J. Hagen, J. Hofrichter, and W. A. Eaton, Protein reaction kinetics in a room-temperature glass, *Science* **269**, 959–962 (1995).
20. H. Lichtenegger, W. Doster, T. Kleinert, A. Birk, B. Sepiol, and G. Vogl, Heme-solvent coupling: a Mossbauer study of myoglobin in sucrose, *Biophys. J.* **76**, 414–422 (1999).
21. G. M. Sastryand and N. Agmon, Trehalose prevents myoglobin collapse and preserves its internal mobility, *Biochemistry* **36**, 7097–7108 (1997).
22. J. Schlichter, J. Friedrich, L. Herenyi, and J. Fidy, Trehalose effect on low temperature protein dynamics: fluctuation and relaxation phenomena, *Biophys. J.* **80**, 2011–2017 (2001).
23. J. L. Green and C. A. Angell, Phase-relations and vitrification in saccharide-water solutions and the trehalose anomaly, *J. Phys. Chem.* **93**, 2880–2882 (1989).
24. L. N. Bell, M. J. Hageman, and L. M. Muraoka, Thermally induced denaturation of lyophilized bovine somatotropin and lysozyme as impacted by moisture and excipients, *J. Pharm. Sci.* **84**, 707–712 (1995).
25. J. Buitink, I. J. van den Dries, F. A. Hoekstra, M. Alberda, and M. A. Hemminga, High critical temperature above T_g may contribute to the stability of biological systems, *Biophys. J.* **79**, 1119–1128 (2000).
26. S. P. Duddu, G. Zhang, and P. R. Dal Monte, The relationship between protein aggregation and molecular mobility below the glass transition temperature of lyophilized formulations containing a monoclonal antibody, *Pharm. Res.* **14**, 596–600 (1997).
27. W. Wang, Lyophilization and development of solid protein pharmaceuticals, *Int. J. Pharm.* **203**, 1–60 (2000).
28. M. P. Buera, S. Rossi, S. Moreno, and J. Chirife, DSC confirmation that vitrification is not necessary for stabilization of the restriction enzyme EcoRI dried with saccharides, *Biotechnol. Prog.* **15**, 577–579 (1999).
29. P. Davidson and W. Q. Sun, Effect of sucrose/raffinose mass ratios on the stability of co-lyophilized protein during storage above the T_g, *Pharm. Res.* **18**, 474–479 (2001).
30. M. T. Cicerone, A. Tellington, L. Trost, and A. P. Sokolov, Substantially Improved stability of biological agents in dried form, *BioProcess Int.* **1**, 36–47 (2003).
31. M. T. Cicerone and C. L. Soles, Fast dynamics and stabilization of proteins: binary glasses of trehalose and glycerol, *Biophys. J.* **86**, 3836–3845 (2004).
32. P. M. Gehring and D. A. Neumann, Backscattering spectroscopy at the NIST Center for Neutron Research, *Physica B* **241**, 64–70 (1997).
33. B. Frick, and L. J. Fetters, Methyl-group dynamics in glassy polyisoprene—A neutron backscattering investigation, *Macromolecules* **27**, 974–980 (1994).
34. A. Paciaroni, S. Cinelli, and G. Onori, Effect of the environment on the protein dynamical transition: a neutron scattering study, *Biophys J.* **83**, 1157–1164 (2002).
35. M. Settles and W. Doster, Anomalous diffusion of adsorbed water: A neutron scattering study of hydrated myoglobin, *Farad. Disc.* **103**, 269–279 (1996).

36. G. Zaccaï, How soft is a protein? A protein dynamics force constant measured by neutron scattering, *Science* **288**, 1604–1607 (2000).
37. C. A. Angell, K. L. Ngai, G. B. McKenna, P. F. McMillan, and S. W. Martin, Relaxation in glass-forming liquids and amorphous solids, *J. Appl. Phys.* **88**, 3113–3157 (2000).
38. B. Frick and D. Richter, The microscopic basis of the glass-transition in polymers from neutron-scattering studies, *Science* **267**, 1939–1945 (1995).
39. P. B. Conrad and J. J. de Pablo, Computer simulation of the cryoprotectant disaccharide alpha,alpha-trehalose in aqueous solution, *J. Phys. Chem. A* **103**, 4049–4055 (1999).
40. V. K. Malinovsky, V. N. Novikov, and A. P. Sokolov, Log-normal spectrum of low-energy vibrational excitations in glasses, *Phys. Lett. A* **153**, 63–66 (1991).
41. U. Buchenau, Y. M. Galperin, V. L. Gurevich, and H. R. Schober, Anharmonic potentials and vibrational localization in glasses, *Phys. Rev. B* **43**, 5039–5045 (1991).
42. O. Yamamuro, K. Harabe, T. Matsuo, K. Takeda, I. Tsukushi, and T. Kanaya, Boson peaks of glassy mono- and polyalcohols studied by inelastic neutron scattering, *J. Phys. Conden. Matt.* **12**, 5143–5154 (2000).
43. A. P. Sokolov, A. Kisliuk, D. Quitmann, A. Kudlik, and E. Rossler, The dynamics of strong and fragile glass formers—vibrational and relaxation contributions, *J. Non-Cryst. Solids* **172**, 138–153 (1994).
44. D. Lourdin, H. Bizot, and P. Colonna, Correlation between static mechanical properties of starch-glycerol materials and low-temperature relaxation, *Macromol. Symp.* **114**, 179–185 (1997).
45. T. R. Noel, R. Parker, and S. G. Ring, A comparative study of the dielectric relaxation behaviour of glucose, maltose, and their mixtures with water in the liquid and glassy states, *Carb. Res.* **282**, 193–206 (1996).
46. P. Bergquist, Y. Zhu, A. A. Jones, and P. T. Inglefield, Plasticization and antiplasticization in polycarbonates: The role of diluent motion, *Macromolecules* **32**, 7925–7931 (1999).
47. R. Casalini, K. L. Ngai, C. G. Robertson, and C. M. Roland, Alpha- and beta-relaxations in neat and antiplasticized polybutadiene, *J. Poly. Sci. B: Poly. Phys.* **38**, 1841–1847 (2000).
48. R. V. Rariy and A. M. Klibanov, Correct protein folding in glycerol, *Proc. Natl. Acad. Sci. USA* **94**, 13520–13523 (1997).
49. F. Parak, E. W. Knapp, and D. Kucheida, Protein dynamics—Mossbauer spectroscopy on deoxymyoglobin crystals, *J. Mol. Biol.* **161**, 177–194 (1982).
50. C. L. Brooks, M. Karplus, and B. M. Pettitt, Proteins: a theoretical perspective of dynamics, structure, and thermodynamics, *Adv. Chem. Phys.* 1–200 (1988).
51. W. Doster, S. Cusack, and W. Petry, Dynamical transition of myoglobin revealed by inelastic neutron scattering, *Nature* **337**, 754–756 (1989).
52. S. Subramaniam, M. Gerstein, D. Oesterhelt, and R. Henderson, Electron diffraction analysis of structural changes in the photocycle of bacteriorhodopsin, *EMBO J.* **12**, 1–8 (1993).
53. V. Reat, H. Patzelt, M. Ferrand, C. Pfister, D. Oesterhelt, and G. Zaccai, Dynamics of different functional parts of bacteriorhodopsin: H-H-2 labeling and neutron scattering, *Proc. Natl. Acad. Sci. USA* **95**, 4970–4975 (1998).
54. L. Cordone, M. Ferrand, E. Vitrano, and G. Zaccai, Harmonic behavior of trehalose-coated carbonmonoxy-myoglobin at high temperature, *Biophys. J.* **76**, 1043–1047 (1999).
55. C. Branca, S. Magazu, G. Maisano, and F. Migliardo, Vibrational and relaxational contributions in disaccharide/H_2O glass formers, *Phys. Rev. B* **64**, art-224204 (2001).
56. G. Caliskan, A. Kisliuk, A. M. Tsai, C. L. Soles, and A. P. Sokolov, Protein dynamics in viscous solvents, *J. Chem. Phys.* **118**, 4230–4236 (2003).
57. A. M. Tsai, D. A. Neumann, and L. N. Bell, Molecular dynamics of solid-state lysozyme as affected by glycerol and water: A neutron scattering study, *Biophys. J.* **79**, 2728–2732 (2000).
58. D. P. Miller and J. J. de Pablo, Calorimetric solution properties of simple saccharides and their significance for the stabilization of biological structure and function, *J. Phys. Chem. B* **104**, 8876–8883 (2000).

Part IV
Fundamental Studies in Model Systems

Folding and Misfolding as a Function of Polypeptide Chain Elongation

Conformational Trends and Implications for Intracellular Events

Silvia Cavagnero[1,2] and Nese Kurt[1]

Protein folding in the intracellular environment is dictated by several driving interactions whose marvelous interplay is essential for maintaining the correct functioning of healthy living organisms. Physical forces directing polypeptide chains toward specific conformational ensembles start acting during ribosome-assisted biosynthesis, due to slow translation timescales. These noncovalent forces determine and modulate the polypeptide conformational progression toward the native state, as nascent chains elongate cotranslationally. In order to separate intrinsic polypeptide conformational trends from additional effects due to complex cellular components (such as cotranslationally active chaperones, the ribosome, and molecular crowding agents), a number of recent studies have focused on the structural characterization of purified N-terminal polypeptides of increasing length. These peptides bear the amino acid sequence of known soluble single-domain proteins. Key insights have emerged, leading to unveiling intrinsic conformational trends toward unfolding, native-like folding and misfolding. The results depend on primary structure, degree of chain elongation, and specific amino acid physical properties. In addition to revealing poorly understood yet biologically relevant aspects of protein folding and misfolding in the absence of denaturing agents, the above work justifies nature's choice to overcome the dangers of cotranslational (and immediately posttranslational) misfolding by an appropriate support machinery.

---

[1] Department of Chemistry, University of Wisconsin, 1101 University Avenue, Madison WI 53706
[2] To whom correspondence should be addressed: Silvia Cavagnero, Department of Chemistry, University of Wisconsin, 1101 University Avenue, Madison, WI 53706; e-mail: cavagnero@chem.wisc.edu

1. INTRODUCTION

Significant progress in understanding the *in vitro* folding mechanisms of full-length proteins has been made in recent years. *In vivo*, however, proteins are synthesized within the ribosomal machinery starting from their N-terminus, and they vectorially grow toward their C-terminus. The development of structure and folding during chain elongation remains poorly explored. The physical forces driving the conformational search culminating into protein folding are already active on polypeptide chains during biosynthesis. A typical gene for a single domain protein undergoes translation on the minute timescale, while secondary structure formation and global folding are known to occur on the submillisecond[1,2] and second[1,3] timescales, respectively. Given the slow translation rates relative to typical folding times, it is likely that polypeptide chains are able to sample most of their available conformational space cotranslationally. The dimensionality of a nascent polypeptide's free energy landscape increases with chain length. Whether this progressive conformational search leads to any degree of cotranslational folding or misfolding is currently a subject of intense debate.

The environment experienced by nascent polypeptide chains is very complex owing to intracellular factors such as chaperones, molecular crowding, and the ribosome. This complexity makes it difficult to determine and evaluate the contribution of different variables to the overall nascent polypeptide conformation. Assuming that some of these variables can be separated and independently assessed, *in vitro* experiments on simplified model systems are potentially able to address the contribution of individual variables to nascent polypeptide conformation. Purified N-terminal polypeptides of increasing length, up to full-length protein, have been used for this purpose. Investigation of these truncated polypeptides in solution promises to serve as a reference for future studies directly addressing how ribosome-bound nascent polypeptide conformation is affected by the complex parameters present in the cellular environment.

The conformational properties of protein fragments have been investigated thoroughly in the past. However, rather than systematically addressing N-terminal polypeptides of increasing length, most efforts have focused on isolated secondary or supersecondary structure elements derived from *any* portion of the parent protein. These fragments often have been studied in connection with investigations aimed at exploring the role of locally versus globally driven structure formation in protein folding. In contrast, the present review specifically focuses on N-terminal fragments comprising a fraction of the amino acid sequence of known proteins. The primary goal of this work is to summarize and critically evaluate preferred structural features and trends as a function of chain elongation. Purified N-terminal polypeptides bearing part of the amino acid sequence of soluble and structurally well-characterized proteins have been employed as model systems for this purpose. All the polypeptides discussed here belong to single-domain proteins, since these can be viewed as minimal autonomously folding units.

The next section summarizes the available results on N-terminal polypeptides derived from four proteins whose chain elongation has been systematically analyzed. Attention then will be turned to the structural characterization of a few individual N-terminal fragments whose chain elongation (up to full-length protein) has not been methodically explored (Section 3). Examples pointing to the absence of folding, partial folding, and chain misfolding will be analyzed. Section 4 highlights qualitative relations between the known conformation of incomplete polypeptides, their chain length, degree of hydrophobicity, fraction of polar and nonpolar solvent-accessible surface areas, and folding entropy. We will outline semiquantitative correlations between intrinsic chain properties and observed conformational ensembles. Finally, the implications for intracellular events will be briefly discussed, with emphasis on Nature's ingenious strategies to overcome the dangers of cotranslational misfolding and proteolytic degradation.

2. SYSTEMATIC STUDIES ON POLYPEPTIDE CHAIN ELONGATION

In the early 1960s, it was discovered that proteins are synthesized *in vivo* from their N-terminal ends.[4,5] Soon thereafter, Phillips suggested that the N-terminal nascent chain may start folding while the protein is still being synthesized.[6] As a response to this hypothesis, Taniuchi and Anfinsen studied N-terminal fragments of the protein staphylococcal nuclease (SNase) *in vitro*. They reported that no nativelike conformation of the 149-residue SNase can be generated until the polypeptide chain has been extended beyond residue 126.[7] Although the circular dichroic spectrum of SNase(1-126) was not that of a random coil, it was very difficult to investigate any residual structure by the experimental techniques available at that time. Since these pioneering early efforts, systematic conformational analysis on N-terminal fragments of increasing length was performed with a more extensive set of spectroscopic tools for SNase and a few other proteins, as detailed below.

2.1. Staphyloccocal Nuclease

Staphylococcal nuclease (SNase) is one of the primary models employed in the *in vitro* protein-folding field. The initial investigations on its folding mechanisms date back to Anfinsen.[8,9] SNase is a small 149-residue protein that folds efficiently and reversibly. According to its crystal structure, the overall tertiary structure is composed of a twisted five-stranded β-barrel and three α-helices[10] (Figure 1). The β- and α-subdomains of SNase pack against each other along the active site cleft. The β-barrel is the first structural unit to form upon kinetic refolding, while the α-helical subunit is largely disordered in the earliest detectable folding intermediate.[11–13]

Large N-terminal fragments of SNase were suggested as model systems to investigate the poorly populated denatured state under physiological conditions.[14] Deleting residues from the C-terminal end aimed to destabilize the native state, with minimal disturbance to the denatured state of interest. Four large fragments, SNase(1-103), SNase(1-112), SNase(1-128), and SNase(1-136), were expressed as inclusion bodies in *Escherichia coli*.[14] Once purified and refolded, these species were soluble and well behaved, with no detectable aggregation. The evolution of secondary and tertiary structure contents of the fragments as a function of chain elongation is illustrated in Figures 2 and 3, respectively. These species display a significant amount of residual secondary structure, which increases as the chain gets longer. Tertiary structure slowly builds up and dramatically increases upon incorporation of the final 10% of the chain.

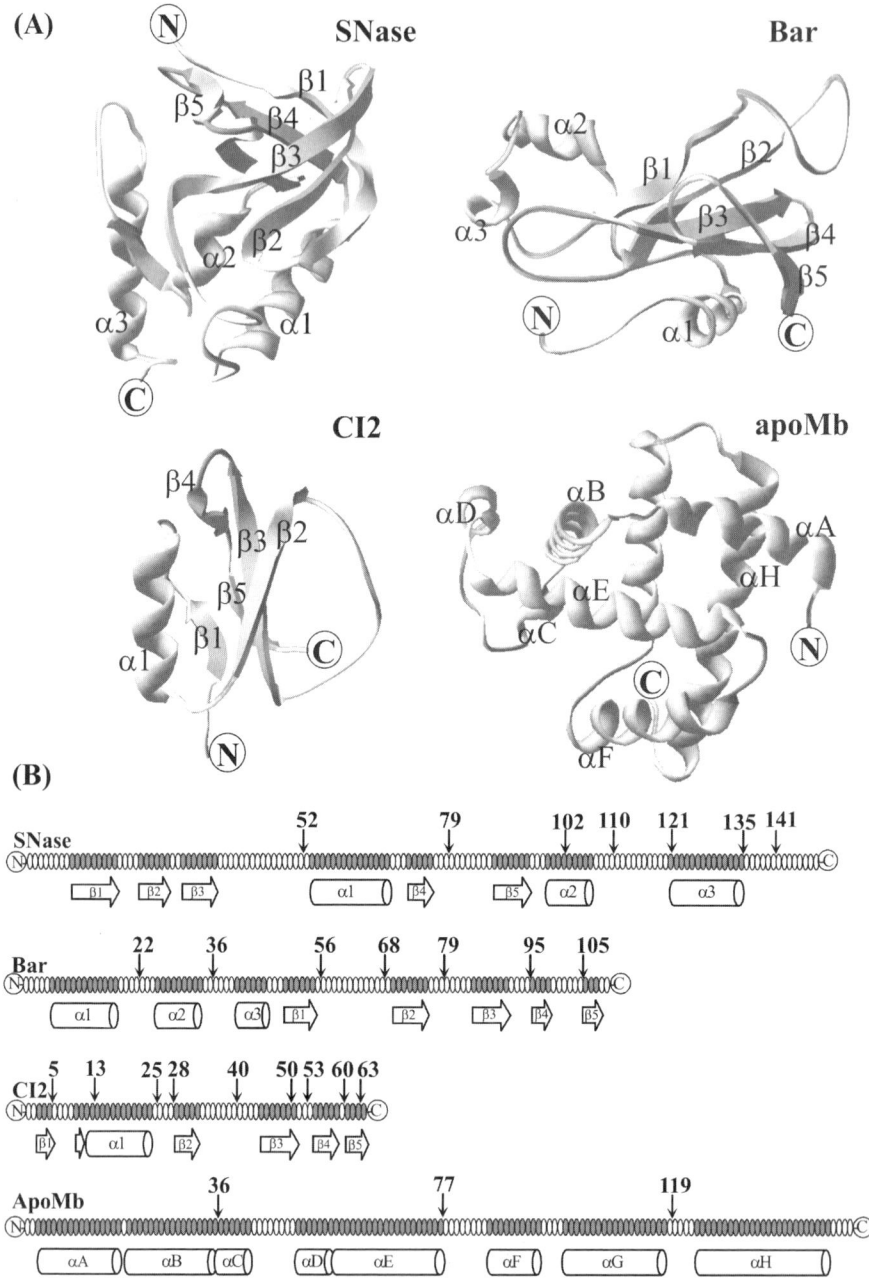

FIGURE 1. Structural features of model proteins. (A) Ribbon diagrams illustrating the three-dimensional structures of staphylococcal nuclease (SNase), barnase (Bar), chymotrypsin inhibitor 2 (CI2), and apomyoglobin (apoMb). The major secondary structure units are labeled. (B) Secondary structure maps for the above four proteins. α-Helices and β-strands are colored in gray along the sequence and labeled by block cylinders and arrows, respectively. The C-termini of the polypeptide fragments discussed in the text are shown by vertical arrows. N and C denote N- and C-termini, respectively.

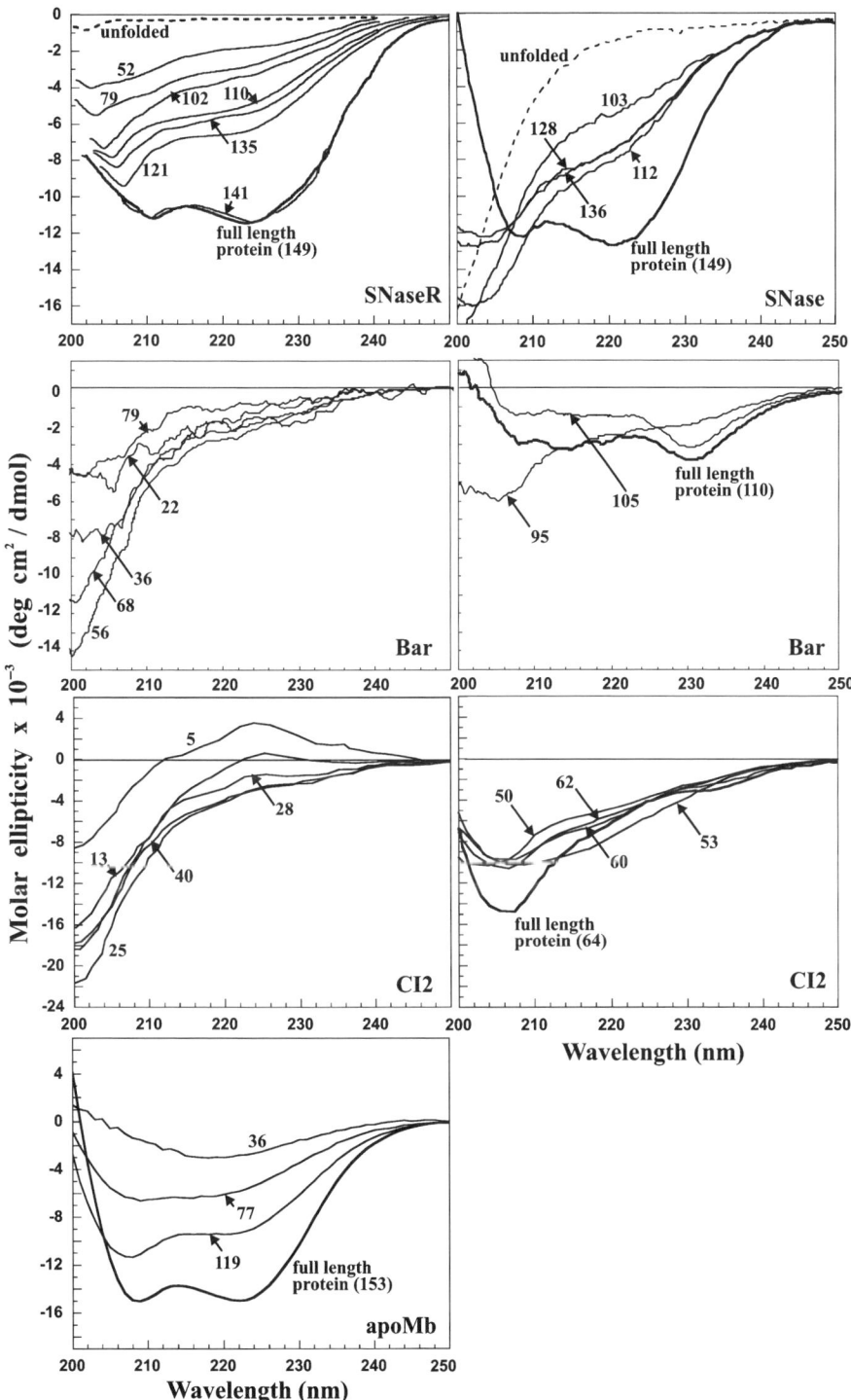

FIGURE 2. Far-UV circular dichroism spectra of SNase, its variant SNaseR, Bar, CI2, apoMb full-length proteins, and their N-terminal fragments. Fragment lengths are indicated on the plots. The "unfolded" state notation in the SNaseR and SNase plots refers to the chemically unfolded short fragments SnaseR52, SNaseR79, and a tryptic digest of SNase, respectively.

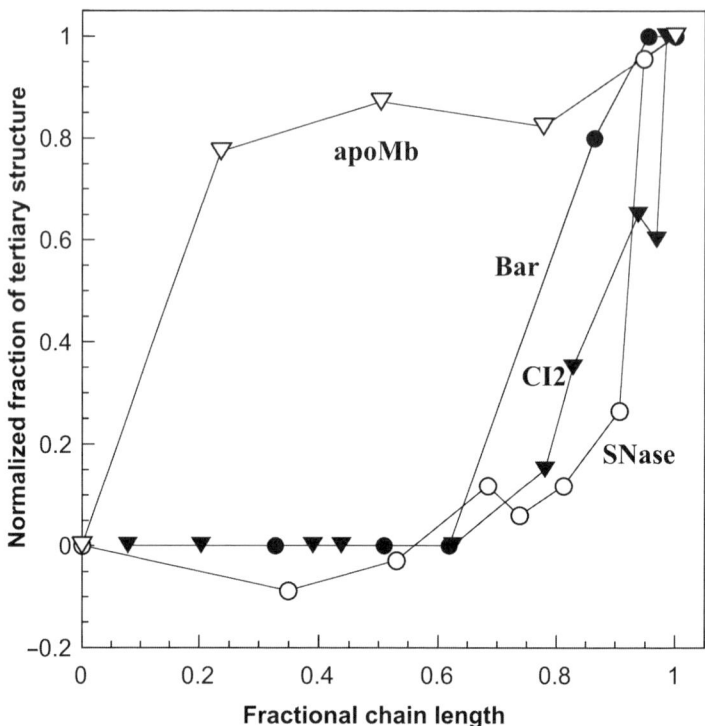

FIGURE 3. Evolution of tertiary structure as a function of chain length in SNaseR, Bar, CI2, and apoMb. Near-UV CD data (SNaseR) and Trp fluorescence wavelength maxima (other proteins) are normalized such that values of 0 and 1 correspond to unfolded and full-length protein, respectively.

SNase(1-128) and SNase(1-136) display significant and full catalytic activity, respectively. They have rather compact, nonrandom, and disordered structures. Binding of pdTp, a potent nuclease inhibitor, brings SNase(1-136) to its folded conformation.[15] This suggests that the missing C-terminal 13 amino acid region contains little or no information necessary for foldability, but is important for the stability of the native conformation.

In order to further investigate the effect of the last few residues on SNase stability, systematic stepwise deletions of individual C-terminal residues were made.[16] Spectroscopic analysis of these fragments revealed that deletions up to eight residues have minimal effect on the protein's structure and function. These amino acids are disordered and undetectable by X-ray crystallography. Deletion of Ser141 and Trp140 causes a significant loss of secondary structure and globular character, resulting in a compact denatured structure similar to SNase(1-136). Therefore these two residues play a key role in promoting nativelike structure formation.

The N-terminal large fragments, SNase(1-110), SNase(1-121), and SNase(1-135), were found to be in a partially folded state by NMR, with significant nativelike structure in the hydrophobic β-barrel region.[17]

SNaseR is an analog of SNase with six additional disordered N-terminal residues, which have no discernible effect on the remainder of the molecule.[18] Similarly to SNase, the secondary structure content of the SNaseR progressively increases as chain lengthens (Figure 2).[19,20] All SNaseR fragments have good solubility and display no observable self-association. The short SNaseR(52) and SNaseR(79) polypeptides contain some residual nativelike secondary structure,[21] which progressively increases to include some of the nativelike β-sheet found in the N-terminal subdomain of the parent protein (Figure 2). SNaseR(121) lacks the residues corresponding to the last C-terminal α-helix portion of the chain (Figure 1). This fragment forms a solvent-exposed fluctuating hydrophobic core[20] due to the lack of residues 121-135, corresponding to a critical α-helix, which buries the hydrophobic core in the full-length protein. Surprisingly, as chain length progresses up to the residues corresponding to the C-terminal helix [SNaseR(1-135)], only additional disordered structure is detected (Figure 2). The unstructured 121-135 residues become helical upon addition of the key C-terminal residues 140 and 141. The newly formed α3 helix packs against the β-subdomain and it stabilizes the overall conformation. This results in a dramatic increase in both the secondary and tertiary structure content at the very end of the polypeptide chain (Figures 2 and 3).

Although compact tertiary structure is detected only in fragments with near-complete chain lengths, local burial of nonpolar surface starts at shorter chain lengths. Hydrophobic interaction chromatography suggests that the nonpolar side chains get increasingly buried in SNaseR(1-121) up to full-length protein.[19]

In summary, all the investigated N-terminal fragments of SNase are water soluble and show no observable nonnative misfolding or aggregation. The shortest fragments are mainly coillike and their secondary structure content increases proportionally with chain length. The large fragments are compact and partially folded, with more nativelike structure in the N-terminal β-barrel region. Upon polypeptide chain elongation from 135 to 141 residues, there is a remarkable concurrent increase in the secondary and tertiary structure content. These amino acids, which are very close to the C-terminus, provide a dramatic increase in the structural content. The last eight C-terminal residues of the full-length protein (amino acids 142-149) are seemingly redundant and contribute minimally to both structure and function.

2.2. Barnase

Barnase (Bar) is a monomeric 110-residue ribonuclease with mixed α/β secondary structure (20% α-helix and 21% β-sheet) and no disulfide bridges (Figure 1). Full-length barnase has an *in vitro* kinetic folding intermediate with substantial population of the C-terminus of the major α-helix and the β-sheet centered around the β-hairpin (residues 92-95).[22] A number of N-terminal fragments of barnase were studied in the Fersht group by a variety of biophysical techniques.[23]

The evolution of secondary and tertiary structures of this protein is shown in Figures 2 and 3, respectively. All fragments up to Bar(1-95) show no significant protection against hydrogen/deuterium exchange, indicating that the hydrogen bonds are short-lived. A very small amount of helical secondary structure can be detected by far-UV CD analysis of the mainly disordered small fragments up to Bar(1-95) (Figure 2). NOE-based NMR studies provide more detailed structural information. These small fragments are mainly disordered, with increasing population of helical structure in residues corresponding to the first N-terminal helix, which is stabilized by nonnative contacts as the chain elongates. Detailed analysis on Bar(1-36) shows that only 20% of native helix1 is present.[24] The C-terminal region is consistently highly disordered in each incomplete chain fragment.

Bar(1-95) displays conformational heterogeneity and tendency toward aggregation, preventing detailed NMR characterization. The ^{1}D ^{1}H-NMR spectrum of Bar(1-95) points to a largely disordered structure, probably devoid of a significant degree of compaction.

Bar(1-105), lacking only 5 C-terminal residues, is the first compact and folded Bar fragment. It has a lower melting temperature than the parent full-length species, suggesting lower thermodynamic stability. Detailed NMR characterization reveals that the three α-helices are fully formed and the binding loop has a nativelike structure, in accordance with the ribonuclease activity of Bar(1-105). Structural regions making contacts with the missing C-terminal β-strand in the full-length protein are quite disordered. Most of the β-sheet scaffold is present in Bar(1-105), and the degree of disorder in the β-strands is proportional with their proximity to the missing C-terminal region of full-length Bar.

In summary, the small fragments of Bar are mainly unstructured and contain very little residual secondary structure. There are several local nonnative hydrophobic contacts. The polypeptide chain, however, is not compact. As chain length increases, the secondary structure content gets moderately larger (Figure 2) while the C-terminal region remains highly disordered in all the fragments. Some conformational heterogeneity has been detected in Bar(1-79) and Bar(1-95). The latter fragment is prone to self-association. There is large increase in tertiary structure content for the longest fragments, which approach the length of the complete chain (Figure 3). A nearly full-length fragment, Bar(1-105), is compact and folded with significant nativelike secondary and tertiary structure content and a disordered C-terminus.

2.3. Chymotrypsin Inhibitor 2

Chymotrypsin inhibitor 2 (CI2) is an 83-residue protein, which comprises a single α-helix, six β-strands, and an extended active site loop[25] (Figure 1). The first 19 residues are unstructured and undetectable by X-ray crystallography[26] and solution NMR.[25] The N-terminally truncated, 64-residue CI2, which will be discussed here, is one of the simplest model proteins amenable to folding and stability studies, since it folds cooperatively and has no disulfide bonds.

The (1-5) to (1-28) CI2 fragments are largely disordered in aqueous solution.[27] No stable secondary structure is present in CI2(1-5) and CI2(1-13) except for some possible weak residual β-turn (Figure 2).[28] CI2(1-25) and CI2(1-28) contain the sequence that forms the α-helix in the intact protein. However they do not display any significant secondary structure. NMR studies reveal a nonnative interaction of the hydrophobic Trp5 with Thr3 in all four fragments and other weak nonnative contacts among nonpolar residues. In full-length CI2, Trp5 interacts only with residues distant in sequence. In the absence of these C-terminal residues, the large hydrophobic side chain of Trp is averted from the solvent by nonnative contacts with nearby residues.

CI2(1-40) displays no nativelike tertiary structure.[29] The gel filtration profile is consistent with the exclusive presence of a relatively small-sized self-associated species.[30] On the other hand, NOE and secondary chemical shift analysis point to the presence of locally driven nonnative buried hydrophobic clusters and a very small amount of α-helical structure.[31] The C-terminus of CI2(1-40) contains the only highly hydrophobic region in this protein (residues 28-40). The above information suggests that CI2(1-40) may form a dimer or some other monodisperse small self-associated species, thus allowing the burial of its hydrophobic residues.

Significant variations in the far-UV CD spectrum are observed upon further chain elongation by 10 residues. CI2(1-50) has no stable tertiary interactions and its secondary structure is partially in place. The gel filtration profile points to conformational heterogeneity. CI2(1-53) provides the minimal chain length displaying nativelike tertiary interactions, with a fully formed minicore (residues 32, 34, 50), and a fluctuating α-helix.[30] Residues 50 and 51 are involved in some critical contacts that link major structural elements. This results into nativelike tertiary structure. The secondary structure content is significantly enhanced by the stabilizing tertiary interactions. Gel filtration and NMR amide proton line widths suggest the presence of conformational heterogeneity.[30]

The NMR spectral features of larger fragments, from CI2(1-53) to CI2(1-63), are undistinguishable from that of full-length CI2. Far-UV CD shows a progressive approach to full-length-like secondary structure (Figure 2). CI2(1-60) has a partially folded structure, with a fully formed α-helix and residues 51-57 defining a nativelike β-hairpin. The α-helix is significantly populated only when the chain is sufficiently elongated to establish the necessary stabilizing tertiary contacts. The structural features of CI2(1-62) are intermediate between those of the partially folded CI2(1-60) and the fully folded CI2(1-63).[32] The CI2(1-63) has a compact nativelike structure but it is less stable than full-length CI2. The C-terminal residues 60 to 64 are critical for the complete packing of the major hydrophobic core and important for the folding/unfolding cooperativity.[30]

In summary, short CI2 fragments are highly unstructured but contain some local and largely nonnative hydrophobic contacts, whose number increases with chain length. Fragments including the 28-40 hydrophobic region [CI2(1-40) to (1-53)] display conformational heterogeneity and possibly form small self-associated species. Larger fragments starting with CI2(1-53) have a significant fraction of nativelike tertiary structure (Figure 3). A concurrent further increase in nativelike secondary and tertiary structure is observed beyond this chain length, as full-length protein (i.e., 64 residues) is approached.

2.4. Apomyoglobin

Apomyoglobin (apoMb) is an all α-helical, 153-residue single-domain protein whose structure and biochemical properties are well characterized (Figure 1). The structures of both the *Sperm whale* and *Horse* variants, as well as those of apomyoglobins from other species, share the well-known globin fold. This comprises eight helices, denoted as A to H, from the N- to the C- terminus, respectively. Full-length apomyoglobin, i.e., the heme-depleted version of the protein, has well-characterized folding pathways. It folds by a molten globular intermediate comprising mainly its A, G and H helices.[33]

Due to its exclusively helical nature, interesting topology and relatively small size, this protein often has been employed as a model system in structure prediction and protein folding investigations.

Cavagnero and co-workers have studied specific aspects of apoMb chain elongation by systematic analysis of N-terminal fragments of increasing length, starting from apoMb(1-36) up to the 153 amino acid full-length chain.[34] The secondary structure of the polypeptides progressively evolves as the chain elongates (Figure 2). The short apoMb(1-36) fragment comprises the N-terminus and all the intervening residues up to the native C helix. Unlike the fragments for all the other proteins reviewed here, this polypeptide has fully nonnative conformation and it has a predominantly β-sheet secondary structure. This behavior may be related to the highly hydrophobic nature of the N-terminal A helix. Previous studies by Dyson and Wright addressed peptides corresponding to individual apomyoglobin helices.[35,36] The N-terminal A helix was found to be extremely insoluble in water and most organic solvents. The CD spectrum of the A helix peptide, apoMb(1-17), shows a predominant β structure in pure methanol. A fragment spanning the first two helices, apoMb(4-34), is only soluble in methanol.

The longer apoMb(1-77) and apoMb(1-119) fragments investigated by Cavagnero and co-workers include residues up to the native E helix and G helix, respectively.[34] These truncated polypeptides have a mixed β-sheet/α-helical character, with an increasing fraction of α-helical content and a decreasing fraction of β-sheet structure as the chain elongates. These structural changes lead to progressive variations in the spectroscopic features as a function of chain elongation. For instance, the far-UV CD signal of the small 36-amino acid fragment is dominated by a singlet centered at 210 nm, while the full-length protein signal has an intense doublet, typical of the predominantly α-helical native structure (Figure 2). The β-sheet to α-helix progression also was assessed and confirmed by FTIR spectroscopy. Spectral deconvolution was possible and the relative contributions of the different secondary structures could be assessed for each of the available chain lengths. A significant amount of α-helical content is acquired upon addition of the amino acids corresponding to the last C-terminal H helix. This finding highlights the importance of the C-terminal amino acids to promote chain helicity and nativelike structural features by cooperative tertiary interactions. In addition, incorporation of the residues to the last C-terminal helix causes a major switch from misfolded and heavily aggregated conformation to monomeric and native structure.

All of the apoMb N-terminal polypeptides, unlike the incomplete chains of the other proteins examined in this review, are heavily self-associated in solution. The polypeptides are generally polydisperse in size and assume a higher-order association state at shorter chain lengths: ApoMb(1-36) and apoMb(1-77) form very large supramolecular assemblies, which are not able to elute through the size exclusion column. The ApoMb(1-119) is mostly present as a mixture of dimers and trimers, and full-length apoMb elutes as a single peak. Unlike the other three proteins discussed above, the Trp fluorescence maximum wavelengths indicate a high amount of tertiary structure for all fragments (Figure 3). This is due to burial of nonpolar chains by the self-association present at all incomplete chain lengths. Electron microscopy and thioflavin T fluorescence assays indicate that apoMb(1-36) and apoMb(1-77) give rise to amyloidlike species with protofibrillar morphology. It is remarkable that potentially amyloidogenic species are generated from polypeptides bearing the amino acid sequence of a nonpathogenic, very abundant and ubiquitous mammalian protein under very mild conditions, i.e., room temperature and neutral pH. A plot of the fraction of solvent-accessible nonpolar surface area of the fully extended apoMb polypeptides as a function of chain length points to a qualitative link between the observed self-association and the nonpolar nature of the polypeptide chains, under conditions where the nonpolar amino acids cannot be buried by intramolecular folding.

It is worth noting that self-association leading to amyloidlike species has been observed either in truncated or modified polypeptide chains (e.g., the apoMb fragments discussed here, [34] the prion and Aβ proteins[37–39] under *physiological* conditions, or alternatively, in full-length proteins under *nonphysiological* conditions,[40,41] including wild-type myoglobin[42])

In summary, the above investigations on apoMb chain elongation highlight the tendency of hydrophobic N-terminal peptides toward self-association or association with other nonpolar molecules, when a large portion of nonpolar amino acids becomes solvent-exposed. This phenomenon may in general occur to elongating polypeptide chains belonging to proteins that, like apoMb, have strong hydrophobic cores. Finally, apoMb achieves nativelike secondary and tertiary structure only upon incorporation of the last 34 C-terminal amino acids.

3. N-TERMINAL FRAGMENTS FROM OTHER PROTEINS: UNIFYING THEMES

In addition to the systematic analysis of elongating N terminal fragments summarized above, there is accumulated knowledge on selected fragments derived from a few other proteins. The purpose of these studies has been to identify secondary structure contributions to protein folding.[43] In some cases, complementary N- and C-terminal polypeptides have been tested individually and then combined in solution to monitor tertiary structure formation by fragment complementation.[44–47] Some of the above investigations underscore fundamental trends pertaining to structure evolution as a function of chain elongation.

C-Terminal residues play an important role in triggering folding. The systematic chain elongation studies in Section 2 show that a large increase in nativelike structure occurs as the last few C-terminal amino acids are added to N-terminal polypeptides (Figures 2 and 3). An analogous investigation on a long ribonuclease A fragment shows that truncation of four C-terminal residues results in the complete loss of nativelike conformation.[48] Interestingly, computational studies[49] and matching experimental immunological tests[50] suggest that the kinetic folding nucleation site of ribonuclease A is localized within the last few C-terminal amino acids. Therefore the final C-terminal residues of a protein play a key role in folding.

N-Terminal β-hairpins tend to display quasi-native folding. In order to exhibit some degree of folding, truncated protein fragments need to preserve some structural stability despite the missing contacts with the C-terminal portion of the chain. Contiguous β-strands have a high propensity for such behavior. For instance, N-terminal fragments from ubiquitin,[51] protein-G (B1 domain),[52] ferredoxin,[53] and carp granulin-1[54] form nativelike β-hairpins. The intrinsic propensity to populate β-hairpin conformational space is strongly encoded within the amino acid sequence. While bearing an overall quasi-native fold, these fragments undergo some nonnative conformational rearrangements to compensate for lost contacts and minimize exposure of nonpolar side chains to water.

Domain swapping is a strategy to bury nonpolar surface. Domain swapping leads to different degrees of self-association and it is an ingenious tactic employed by Nature to create species with minimally perturbed quasi-native multimeric structure.[55] Open-ended swapping is potentially involved in protein misfolding leading to aggregation and amyloid fibril formation.[56,57] The presence of a strained or generally unfavorable loop conformation in a full-length monomeric protein is thought to enhance the likelihood of domain swapping.[58,59] This is mediated by the increased kinetic accessibility of extended loop conformations in the domain-swapped structures. In the case of N-terminal polypeptides, domain swapping allows effective burial of solvent-exposed nonpolar surface generated as a result of chain truncation. An interesting example is given by the (1-51) N-terminal fragment of the 76-residue ubiquitin. This species forms a symmetrical dimer with intermolecular strand-swapping serving as a replacement for the lost interstrand contacts originally present in the full-length protein.[60]

Local burial of hydrophobic surface may occur in disordered chains. As seen in Section 2, truncated fragments may bear some residual structure even if they are largely unfolded. On the other hand, no such residual structure has been detected (by CD and NMR) in some N-terminal fragments of the SH2 domain[61] and *E. coli* thioredoxin.[45,62] Heat capacity measurements highlight the important fact that burial of nonpolar surface may occur even in the absence of any significant degree of ordered secondary or tertiary structure.[45,62,63]

Heme cofactors are buried from aqueous solvent. Heme iron has been used as a spectroscopic reporter for an N-terminal cytochrome c fragment.[64] Metal coordination includes two axially bound histidines, native His18 and a misligated His26 or His33. Despite its substantially disordered nature, this fragment effectively shields heme from solvent by adopting a nonnative iron coordination.

4. DETERMINANTS OF INTRINSIC CONFORMATIONAL TRENDS IN INCOMPLETE CHAINS

The experimental analysis of protein N-terminal fragments of increasing length described in Sections 2 and 3 highlights some peculiar trends. Despite the relatively small available database, a few common features emerge. With the exception of apomyoglobin, a protein with a strong hydrophobic core, polypeptides assume a compact tertiary structure only very late upon polypeptide chain elongation (Figure 3). This is consistent with the presence of either dynamic extended chains or locally fluctuating molten globular conformational ensembles, until nearly all the missing C-terminal residues have been incorporated. Therefore, nativelike topology is achieved late, when the native full-length protein sequence is close to being complete. On the other hand, far-UV CD (Figure

2), NMR, and other spectroscopic techniques show that noncompact fragments in some cases attain local nativelike secondary structure (as seen in SNase) even at short chain lengths. This secondary structure evolves with chain elongation in the case of SNase. In other cases (Bar and CI2), only negligible secondary structure is detected until most of the chain has been synthesized. However, nativelike and nonnative local contacts between hydrophobic residues highlight the trend to bury nonpolar surface while preserving a sufficient degree of chain entropy. In the case of apoMb, nonnative misfolded β-sheet accompanied by chain compaction is observed at short chain length, with gradual progression toward nativelike secondary structure as the chain elongates. For all proteins, a significant increase in the secondary structure content is achieved when the last few amino acids are incorporated into the chain. Very interestingly, the above suggests that the energy landscape of an elongating polypeptide evolves from a relatively flat profile to funnellike shape only late, as sequence evolves.

Just as the amino acid sequence of a full-length protein determines its native structure under given experimental conditions (temperature, solvent, etc.), the amino acid sequence of N-terminal incomplete chains determines their conformational fate. At any given chain length, conformational sampling leads to populating the most stable kinetically accessible conformations. As a result, the polypeptide chain may undergo several different structural fates, which then evolve as a function of chain elongation. Briefly, truncated protein fragments can either remain largely unfolded, become folded (with a conformation resembling that of the full-length native state at the corresponding chain length), or undergo misfolding.

This section attempts to address the origins of the structural behavior of incomplete polypeptide chains. Our focus here is to highlight any correlations between the tendency to bury nonpolar surface area, an important driving force toward folding and misfolding, and the experimentally observed conformational behavior of the N-terminal fragments and full-length chains. Toward this end, we will outline the main energetic contributions to folding and misfolding. The role of these contributions then will be revisited in the context of the folding/misfolding chain-length dependence of the four proteins (SNase, Bar, CI2, apoMb) reviewed in Section 2.

4.1. Driving Forces Leading to Folding and Misfolding

Full-length protein structure formation is driven by the contribution of several noncovalent forces, involving hydrogen bonds, van der Waals, electrostatic, and dipolar interactions.[65] Extensive inspection of full-length globular protein structures reveals that hydrophobic side chains are preferentially buried from the solvent[66–68] and typically form a tightly packed core.[69–71] The surface area that residues bury upon folding is strongly related to their hydrophobicity.[72] The amount of buried surface correlates with protein molecular weight and is mostly composed of hydrophobic groups that become buried between the secondary structure units of native proteins.[71,72] Burial of nonpolar amino acid side chains away from water minimizes unfavorable interactions. This phenomenon, better known as the hydrophobic effect,[66] results from a complex combination of forces of dipolar nature, in addition to van der Waals and hydrogen-bonding interactions involving both polypeptide chain and solvent. While the origins of the hydrophobic effect are still the subject of lively scientific debate, it now is widely recognized that this effect is one of the dominant contributors to the thermodynamic stability and the kinetic conformational sampling leading to native proteins.[65,68]

The hydrophobic effect has well-known characteristic thermodynamic signatures that are relevant to understanding some aspects of the polypeptide structural trends as chain length elongates. For instance, the tendency to bury nonpolar solvent-accessible surface area (SASA) is the main determinant to the surprisingly high-heat-capacity decrease generally observed upon folding.[73–75] This massive reduction in heat capacity associated with protein folding, ΔC_f°, can be partitioned into nonpolar and polar components:

$$\Delta C_f^\circ \approx \Delta C_{np}^\circ + \Delta C_p^\circ = 0.32 \Delta A_{np} - 0.14 \Delta A_p. \quad (1)$$

The heat capacity terms are in e.u. (i.e., cal mol^{-1} K^{-1}), and ΔA_{np} and ΔA_p refer to the amount of nonpolar and polar surface area that gets buried upon folding, in Å^2.[76] The contribution from the burial of nonpolar surface, ΔA_{np}, is always the dominant term.[74] In addition, the standard folding entropy, ΔS_f°, arises from two contributions, i.e., the burial of nonpolar surface upon folding, ΔS_{np}°, and the reduction of configurational chain entropy, ΔS_{ch}°, that accompanies protein folding:

$$\Delta S_f^\circ = \Delta S_{np}^\circ + \Delta S_{ch}^\circ. \quad (2)$$

The ΔS_{np}° term is related to ΔC_{np}°:

$$\Delta S_{np}^\circ = \Delta C_{np}^\circ \ln\left(\frac{T}{T_s}\right), \quad (3)$$

where T_s is a reference temperature at which ΔS_{np} is zero. The T_s is set equal to 386 K, from studies on the transfer of liquid hydrocarbons to aqueous phase.[74] The ΔC_{np}° depends on ΔA_{np} as seen in Eq. (1). It is known that ΔS_{ch}° amounts to ca –5.6 e.u./residue for globular proteins.[74,77,78] Hence, ΔS_f° can be predicted from the nonpolar surface area buried upon folding, computed from structural data.

The standard enthalpy change for folding, ΔH_f°, typically arises from a combination of three terms:

$$\Delta H_f^\circ = \Delta H_{np}^\circ + \Delta H_p^\circ + \Delta H_{other}^\circ. \quad (4)$$

The ΔH_{np}° is due to the burial of nonpolar surface upon folding, ΔH_p° is due to the burial of polar surface, and ΔH_{other}° term arises from contributions of other nature, including hydrogen bonding. The ΔH_{np}° and ΔH_p° are directly related to the nonpolar and polar components of the folding heat capacity:

$$\Delta H_f^\circ = \Delta C_{np}^\circ \left(T - T_{H,np}\right) + \Delta C_p^\circ \left(T - T_{H,p}\right) + \Delta H_{other}^\circ. \quad (5)$$

The $T_{H,np}$ and $T_{H,p}$ are reference temperatures at which the nonpolar and polar contributions to the standard folding enthalpy are zero (or a constant value), respectively.[79]

Liquid hydrocarbon models provide estimates of $T_{H,np}$ while organic amide models are used to derive the parameters for the polar contribution.[80] The ΔC_{np}° and ΔC_p° can be predicted according to Eq. (1). However, it is not possible to reliably estimate ΔH_{other}° *a priori*. This term includes the effect of hydrogen bonds, as well as contributions of other nature. Calculations based on experimental data on a number of proteins show that ΔH_{other}° is large and positive, and it makes a significant contribution to the overall folding enthalpy. Accurate values for this term are essential to estimate ΔH_f° and ΔG_f°. Protein stability results from a delicate balance between ΔH_f° and $T\Delta S_f^\circ$, and errors in ΔH_f° lead to significant errors in calculated ΔG_f values. Therefore, we deliberately omit any specific discussions on folding Gibbs energies and enthalpies here, since no experimental free energy data are available for incomplete chains and no reliable values can be predicted.

4.2. Local and Overall Polypeptide Chain Properties

A polypeptide chain folds intramolecularly if, at any given chain length, there is one (or more) kinetically accessible monomeric low-energy conformation containing a significant amount of structure. The availability of such conformation(s) is well established in the case of full-length proteins but difficult to predict for incomplete chains. The following discussion focuses on the role of chain polarity and the hydrophobic tendency of nonpolar groups to become buried.

Hydrophobicity scores provide a semiquantitative measure of the degree of nonpolar character of individual amino acids. Here we employ the hydropathy scale proposed by Kyte and Doolittle, based on experimental water-vapor transfer free energies of amino acids and the interior-exterior distribution of amino acid side chains in native protein structures.[81] The hydropathy scores for the main proteins reviewed here are displayed in Figure 4. The regions with high positive scores in the plots, showing local hydrophobic clusters, correspond to mostly buried regions in the full-length native structure (enclosed by dashed circles). The central strand of the extended SNase β-sheet is an example of such buried amino acid regions. There are no regions of high local hydrophobicity in the Bar sequence. The CI2 has a single hydrophobic stretch corresponding to β-strand 2 and half of the extended active site loop. On the other hand, apoMb has several regions with high hydropathy scores. These correspond to the highly hydrophobic native helices A, B, E, and G.

While hydropathy provides a good illustration of the distribution of polar and nonpolar amino acids within a given amino acid sequence, a better estimate of the *overall* degree of hydrophobicity for the entire polypeptide at a given chain length is provided by the cumulative hydropathy parameter. Cumulative hydropathies as a function of chain length are displayed in Figure 5A. In general, protein cumulative hydropathy values tend to decrease at longer chain lengths, since there is a statistically larger fraction of hydrophilic residues in proteins. Hence, the contribution of these residues to cumulative hydropathies increases with chain length due to the additive nature of this score. Local peaks are due to the presence of clustered hydrophobic amino acids. The relatively long nonpolar region in CI2 leads to a dramatic increase of its cumulative hydropathy score. On the other hand, the plots for the other proteins are characterized by negative scores

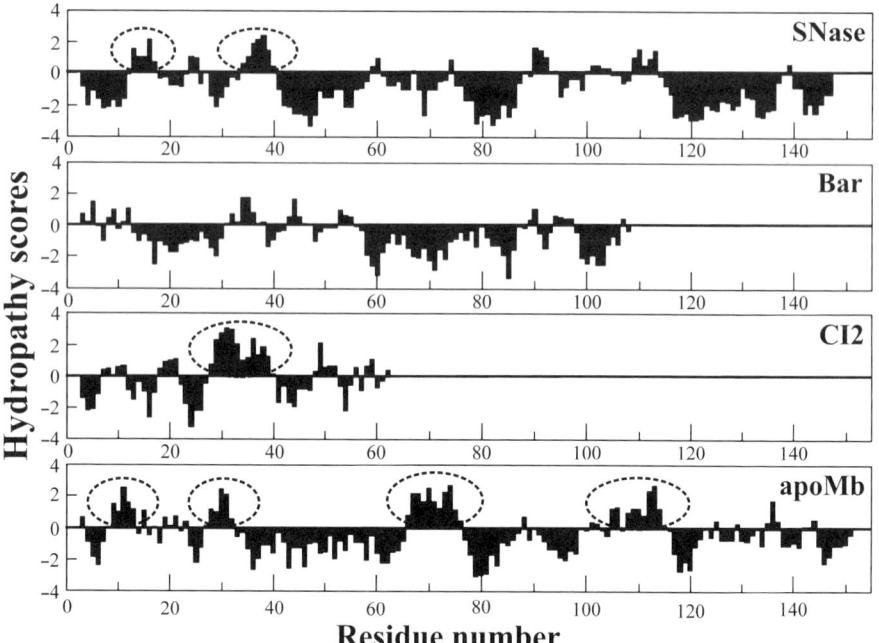

FIGURE 4. Hydropathy scores according to Kyte and Doolittle[75] for SNase, Bar, apoMb, and CI2. The values are averaged over a five-residue sliding window. Positive and negative scores are proportional to the degree of local hydrophobicity and hydrophilicity, respectively. Regions of high hydrophobicity are highlighted by dashed ellipses.

with local fluctuations toward higher cumulative hydropathies in correspondence of the nonpolar regions. Cumulative hydropathy scores depend, as expected, on a protein's total chain length. A more direct parameter, which allows more straightforward comparisons among polypeptide sequences of different length, is the mean residue hydropathy, as shown in Figure 5B. Scores at short chain lengths are determined by a smaller number of residues and the weights of individual residues decrease as chain elongates. This accounts for the large score fluctuations at short chain lengths.

The SNase sequence is quite hydrophilic at all chain lengths, except for a nonpolar region close to residue 40, caused by the hydrophobic nature of the central β-strands. The first few N-terminal residues in Bar are somewhat hydrophobic, but the mean hydropathy score quickly decreases and then levels off as more hydrophilic residues are incorporated into the sequence. The large fraction of CI2 hydrophobic residues, arising from its large cluster of nonpolar amino acids, makes this protein the most hydrophobic among the four full-length sequences examined here. The nonpolar nature of the residues corresponding to the apoMb A and B helices is responsible for the high mean residue hydropathy scores at short chain lengths, while the hydrophobic E and G helices lead to the two local maxima observed at longer chain lengths (close to residues 75 and 115, respectively). It is noteworthy that the apoMb(1-77) and apoMb(1-119) fragments, which exhibit pronounced misfolding, correspond to these two maxima, and thus have some clustered hydrophobic amino acids near their C-termini. The investigated short

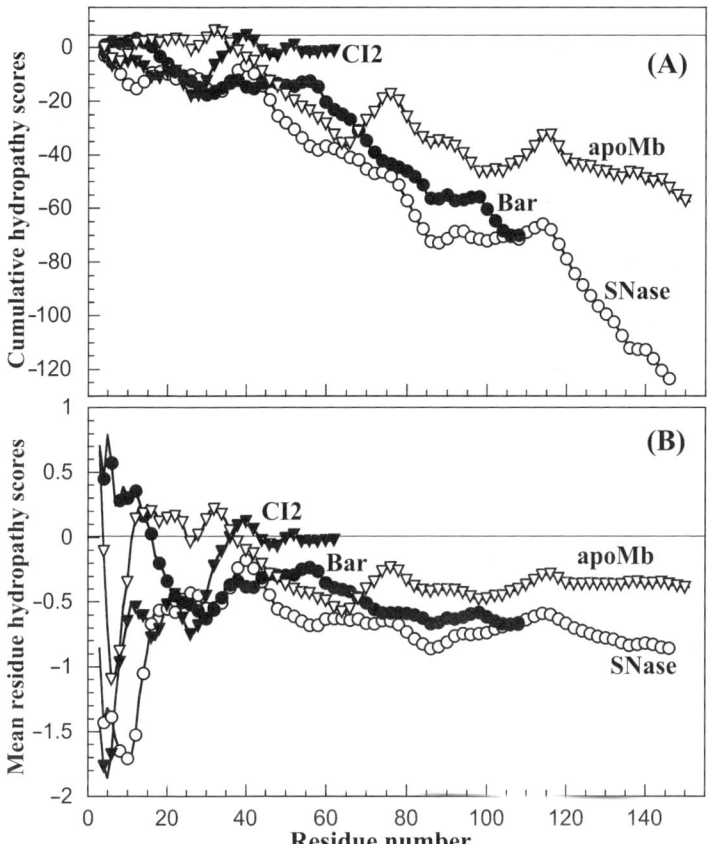

FIGURE 5. (A) Cumulative and (B) mean residue cumulative hydropathy scores (Kyte-Doolittle) for the amino acid sequences of SNase, Bar, apoMb, and CI2.

fragments of SNase (52, 79), CI2 (5, 13, 25, 28) and Bar (22, 36) are quite hydrophilic by all three hydropathy criteria described above and are found to be largely unfolded by biophysical characterization. On the other hand, the short apoMb fragment (36) has a strikingly higher hydrophobicity. This fragment misfolds to form high-order oligomers that aggregate into amyloidlike species.

Hence, the degree of hydrophobicity of amino acid sequences, expressed in terms of hydropathy-related parameters, correlates well with the observed tendency toward misfolding, for short fragments whose chain length extends up to about the first half of the amino acid sequence. The longer fragments display a more complex behavior and require the more detailed analysis presented below.

4.3. Accessible Conformations at Different Chain Lengths

In general, the solvent-exposed surface area of truncated fragments is expected to have more nonpolar character than the full-length protein. Some of the residues normally buried within the core of a globular protein become accessible to the solvent as a result of chain truncation. The presence of an increased amount of nonpolar SASA in a truncated fragment tends to destabilize the native polypeptide chain conformation and promote the burial of nonpolar surface. Thus N-terminal fragments are likely to either assume a predominantly unfolded state or to attain conformations bearing some nonnative character. Significantly unfolded conformations are expected to be favorable only if the polypeptide sequence is very hydrophilic. If, on the other hand, the sequence has a large fraction of nonpolar character, some hydrophobic segments may become buried intramolecularly. If the fraction of nonpolar SASA is particularly high or if the conformational search does not lead to the burial of nonpolar surface at the intramolecular level, there is a high chance for intermolecular association to take place. Depending on the extent of nonpolar content in a polypeptide chain, either soluble low-order oligomers or higher-order associated states, sometimes leading to aggregation and amyloidlike species, are generated.

The predicted fractions of nonpolar SASA for full-length proteins and their incomplete chains are compared to the main experimental results discussed in Section 3.[82] The SASA values were calculated using the Surface Racer program.[83] The effect of chain truncation was considered on two limiting species: (1) the unfolded state represented by a fully extended conformation, and (2) the folded state in the hypothetical event that N-terminal fragments were to retain an entirely nativelike conformation. Regardless of whether these species are actually significantly populated in solution, this comparison serves the purpose of illustrating the trends that the polypeptide chains assume a nativelike conformation, at any given chain length. Furthermore, specific SASA values depend on chain length and are expected to increase as chain elongates. Therefore, in order to facilitate a direct comparison between different chain lengths, it is desirable to consider the *fraction* of nonpolar SASA relative to the total SASA. It also is worth mentioning that, not surprisingly, the fractional nonpolar SASA for unfolded chains have nearly identical features as the Kyte-Doolittle hydropathy plots (Figure 5).

The short N-terminal fragments, which are generally unlikely to fold into nativelike conformations, are discussed in the context of mean residue hydropathy scores in the previous section. The present discussion, on the other hand, focuses on the longer fragments only, whose chain lengths cover approximately the right-hand sides of the SASA plots in Figure 6. In all four sequences, the fraction of nonpolar SASA for the unfolded conformations fluctuates somewhat as chain elongates and it finally settles to a constant value, ranging between 0.61 and 0.64, which is quite similar for all four sequences. The fractional nonpolar SASAs for the native conformation, on the other hand, generally decreases as the missing C-terminal portion gets shorter and more hydrophobic surface gets buried into the protein core.

These plots display a remarkable change in slope at longer chain lengths, corresponding to a more effective burial of nonpolar surface, which achieves its minimum value as the polypeptide reaches its full-length. This change in slope takes place at different chain lengths for all four proteins (shaded areas in Figure 6) and it provides a sensor for the chain length range where burial of nonpolar surface is most efficient as nativelike topology is approached. Within this chain length range, the folded conformation is expected to be preferentially populated. This is possibly due to thermodynamic stabilization and improved kinetic accessibility of the nativelike conformation. The experimental findings discussed in Section 2 are remarkably consistent with this trend, in that the population of nativelike

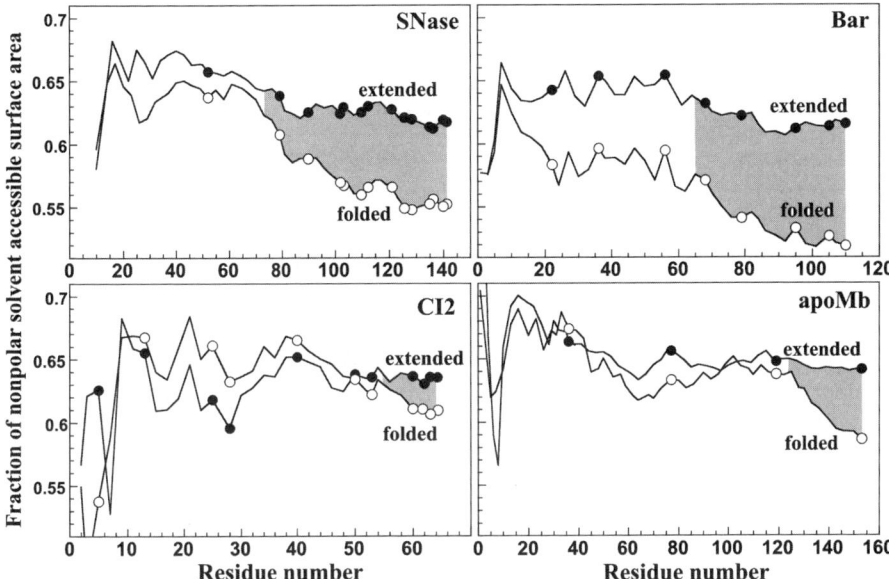

FIGURE 6. Calculated fractions of nonpolar solvent accessible surface area (SASA) for SNase, Bar, apoMb, and CI2 as a function of chain length. Two limiting cases are shown, i.e., fully extended unfolded (•) and nativelike folded (o) conformations. The shaded regions indicate a sharp slope change in the curves for the folded conformations.

secondary and tertiary structures progressively increases as the fragments get longer, with a sudden sharper turn as chain lengths become close to their full length value. The experimental tertiary structure data, mainly based on wavelength shifts as a function of chain elongation, suggest that chain compaction occurs only when the polypeptide length gets extremely close to its full length value, for the proteins where data are available for several fragments missing only the last few residues (i.e., SNase, Bar, and CI2). This is in slight contrast with the trend based on the SASA values of Figure 5, which predicts a more gradual shift toward nativelike conformation, for SNase and Bar. However, predictions and experimental findings can be reconciled by considering that fairly dynamic molten globular structures are observed before nativelike compaction. These dynamic molten globules retain the ability to sample nativelike conformations while preserving a sufficient degree of chain entropy, which favorably contributes to thermodynamic stability. This prenative-like state also may provide an important kinetic advantage in Nature to allow a sufficient degree of conformational sampling, possibly coupled with chaperone extrusion. The cotranslational addition of the very last few amino acids then promotes compaction toward complete nativelike folding. The optimal burial of nonpolar surface at late stages of elongation has no directional preference, suggesting that this was probably not a determining factor for the selection of N to C directionality of polypeptide translation.[84]

A few additional comments are in order regarding specific proteins. The fraction of nonpolar SASAs is quite small for the folded SNase fragments longer than about half the full-length chain (< 0.6). This fraction is significantly larger for comparable fragments of CI2 and apoMb. Not surprisingly, the highly soluble SNase sequence displays a significant degree of cotranslational nativelike folding even at early chain lengths. The fraction of nonpolar SASAs for the folded Bar fragments also is quite small early in sequence. This suggests a general high propensity by the polypeptide chain to bury nonpolar surface. However, this protein is unable to assume nativelike features early on its sequence. Bar fragments minimize their free energy in solution by burying nonpolar surface and at the same time sampling highly dynamic local non-native conformations. Quite unexpectedly, Bar(1-95) shows some aggregation. At this chain length, the last β-strand is missing. The C-terminal extended β-sheet may become significantly destabilized and the hydrophobic core at the active site cleft and/or between the β-sheet and the N-terminal helix may become excessively solvent-exposed. Even if the overall fractional nonpolar SASAs are not very high, there may be a significant local density of exposed hydrophobic side chains. This may well be responsible for the observed intermolecular association. The coarse-grained SASA analysis presented here does not provide a sufficient level of detail to properly predict or rationalize this behavior.

Surprisingly, the fractional nonpolar SASA of nativelike CI2 at long chain lengths is larger than that of the SNase and Bar fragments and their full-length native states. The fraction of nonpolar SASA of CI2 has a maximum close to residue 40 for both the nativelike and fully extended conformations. This is due to a relatively long nonpolar segment in the sequence (circled region in Figure 4). As seen in Section 2, experimental evidence suggests that some CI2 fragments containing this group of highly nonpolar residues give rise to water-soluble low-order multimers. Hence, the highly nonpolar character of the CI2 fragments analyzed here is consistent with the experimentally observed tendency toward self-association. On the other hand, naturally occurring wild-type CI2 has 19 additional disordered residues at its N-terminus (which are not included in the CI2 analysis presented here). These amino acids are very hydrophilic and ensure that the biologically significant CI2 is as hydrophilic as typical water-soluble proteins. Interestingly, under special conditions even full-length CI2 might be forming multimeric species: the biologically relevant form of CI2 (i.e., crystal structure biological unit; PDB code 2CI2) is thought to be a hexamer.[26,85]

The apoMb(1-77) and apoMb(1-119) fragments have nonpolar regions at their C termini corresponding to local maxima in the hydropathy plots. These species display a significantly hydrophobic SASA in both extended and native conformations. The severely hydrophobic characteristics of both primary structures and tertiary conformations leads to the intense intermolecular association observed experimentally. The SASA plots suggest that some of the fragments that have not yet been experimentally explored might display high solubility and possibly nativelike folding.

A final interesting correlation with the experimentally observed behavior is provided by the estimated folding entropy ΔS_f° as a function of chain elongation. The corresponding plot, generated according to Eq. (3), is shown in Figure 7. The ΔS_f° values have been calculated at the representative temperature of 25°C. The folding entropy is dominated by the nonpolar burial contribution ΔS_{np}° and the conformational chain entropy ΔS_{ch}° in globular proteins. The nonpolar contribution favors folding (>0) while the chain entropy contribution opposes it (<0). These two terms are typically quite similar in absolute value and they nearly compensate each other. The fine balance between nonpolar and chain entropy leads to the observed overall folding entropy. Similar effects also are expected to take place in the folding transition state ensemble. Figure 7

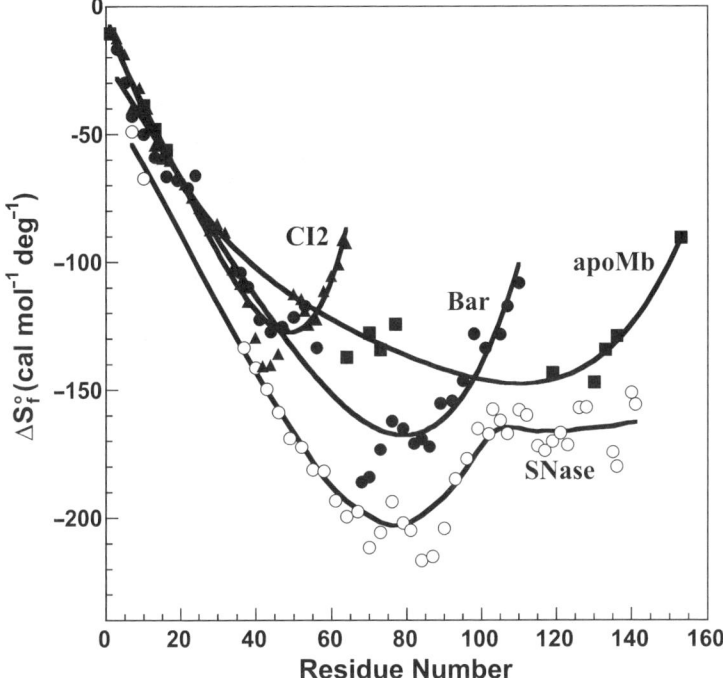

FIGURE 7. Standard entropy change for folding ΔS_f° (cal/mol-°C) as a function of chain length for SNase, Bar, apoMb, and CI2. The solid lines are to guide the eye. Values are calculated at 25°C, assuming a nativelike three-dimensional structure at all chain lengths.

shows that all the four proteins examined here display an inversion in the slope of the ΔS_f° curve versus chain length, toward the end of the polypeptide chain. This slope inversion reflects a variation in the nonpolar entropy term, ΔS_{np}°, suggesting that a favorable contribution to the burial of nonpolar surface gets triggered at long chain lengths. This contribution helps promoting native-like conformations when the amino acid sequence is close to being complete. This behavior is consistent with the fractional SASA trends of Figure 6. The overall unfavorable sign of the folding entropy at all chain lengths simply reflects the specific simulated temperature.

In brief, the above analysis on fractional nonpolar SASAs of elongating chains reveals that folding is expected to become most favorable at long chain lengths, when nativelike topology provides maximally efficient burial of nonpolar surface (Figure 6). The standard folding entropy ΔS_f° has a minimum at about 80% of the chain length and then undergoes a dramatic increase until full chain length is achieved (Figure 7). This effect is due to the contribution of nonpolar folding entropy.

4.4. Role of C-Terminal Residues

As seen, the fractions of nonpolar SASA for nativelike polypeptide conformations of the proteins (Figure 6) exhibit a remarkable sudden change in slope towards the C-terminal region. Similar trends are followed by the standard entropy of folding (Figure 7).

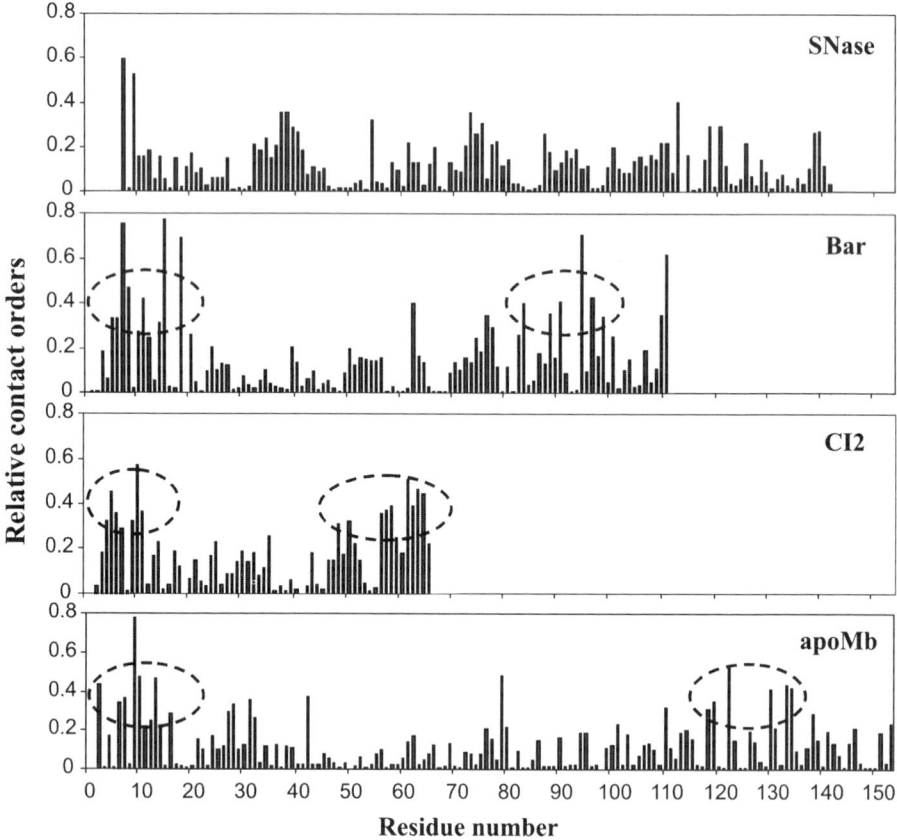

FIGURE 8. Relative contact order profiles for the full-length native states of SNase, Bar, apoMb, and CI2. The dashed ellipses enclose regions (spanning more than four amino acids) with high contact order near the N- and C-termini of the sequences.

A sharp increase in the tertiary structure content also is observed at the very end of the chain for the proteins that do not undergo significant misfolding as a function of chain elongation (Figure 3). These trends suggest that the C-terminal region is very important for effectively burying nonpolar surface and channeling folding toward a nativelike conformation. In CI2 and apoMb, deletion of the very last portion of the chain exposes a significant amount of hydrophobic surface.

In order to further assess the importance of the C-terminal region, we have examined the relative contact order for each residue in its native state conformation (Figure 8). The relative contact order (CO_r) is defined as

$$CO_r = \frac{1}{NL} \sum \Delta S_{ij} \qquad (6)$$

for each residue. The N is the total number of contacts for any given particular residue and L is the total number of amino acids in the protein. The ΔS_{ij} is the sequence separation between contacting residues i and j. The CO_r provides an estimate on the degree of long-range character of the interactions pertaining to each particular residue. In order to eliminate the effect of polypeptide chain length, this parameter is divided by the number of amino acids. The CO_r values were calculated using a known algorithm.[86,87]

The contact order profiles display regions of high CO_r close to the Bar, CI2, and apoMb N- and C-terminal regions. Upon inspection of the native structures (see Figure 1), it easily can be verified that these two regions are spatially close to each other. This leads to the high CO_r values. These regions are in Bar, the N-terminal α-helix and the C-terminal β-sheet forming the main hydrophobic core; in CI2, the N- and C-terminal β-strands that are part of the main hydrophobic core; and in apoMb, the C-terminal H helix that packs against the hydrophobic N-terminal A helix. In case the C-terminal region is missing, as in truncated chains, then these important stabilizing tertiary contacts are lost. Moreover, hydrophobic SASA is exposed. Accordingly, the N-terminal fragments of these proteins cannot achieve a nativelike folded state until the very end of the chain has been incorporated. Relatively stable tertiary interactions are seen first in CI2(1-53); Bar(1-95) displays self-association while only Bar(1-105) can fold into a nativelike conformation; and apoMb(1-119), which lacks the C-terminal H helix, misfolds with intermolecular association. On the other hand, there are no regions of high-contact order at the N- and C-termini of SNase. The two chain ends do not make any contacts in the native state conformation (see Figure 1). In accordance with the above trends, SNase fragments polpulate a significant degree of secondary structure, comprising the N-terminal β-sheet, even at relatively short chain lengths.

In summary, the nonpolar characteristics of the N-terminal fragments generally correlate well with their experimentally observed conformational characteristics. A more detailed analysis requires addressing other complicating factors, including electrostatic interactions, the detailed nature of nonnative contacts that bury hydrophobic side chains, the chance of domain swapping, and others. Nevertheless, the relatively simple analysis based on hydrophobicity and contact order concepts help understanding some broad conformational trends, e.g., why some fragments tend to associate intermolecularly to form heavy aggregates while others do not. The chain-length dependence of a protein's conformation is of great biological relevance toward ultimately understanding the intrinsic conformational tendencies of protein fragments in the cell. Nature's ingenious strategies to deal with the improper solution behavior, frequently displayed by incomplete polypeptide chains, are briefly summarized below.

5. IMPLICATIONS FOR INTRACELLULAR EVENTS

During protein synthesis in the cell, newly made polypeptide chains travel through the ribosomal exit tunnel and become exposed to the complex cellular environment. There is sufficient time for these exposed chains to sample the accessible conformational space, given the relatively slow translation timescales in both prokaryotic and eukaryotic cells. The *in vitro* model system investigations performed so far on isolated N-terminal chains support the idea that, depending on their amino acid sequence, N-terminal fragments have an intrinsic conformational tendency to either fold into a nativelike conformation, stay largely unfolded, or assume a misfolded conformation. However, these model studies isolate polypeptides from their natural context. The environment experienced by the growing polypeptide chains in the cell is more formidably complex and additional factors in the cell affect the folding behavior. The most important of these cell-related parameters are molecular crowding, geometrical constraints, and the action of cotranslationally active chaperones.

The intracellular environment is much more crowded than the buffered aqueous solutions used in *in vitro* experiments, due to the presence of numerous intramolecular components that occupy 20–30% of the cellular volume.[88–90] Molecular crowding, according to excluded volume theory,[88] favors states that exclude the minimum volume from access by the other species present in solution. Thus, compact states are stabilized relative to more unfolded ones. This results in an increased equilibrium constant for folding and association and leads to perturbation of protein folding rates.[91,92] Hence, the risk of aggregation is greatly increased by the effect of molecular crowding in the cell.[93]

During biosynthesis, newly synthesized N-terminal polypeptide chains exposed to the cellular environment are physically bound to the ribosome.[94,95] This imposes geometrical constraints on the chain and it restricts the accessible conformational space. Additionally, tethering to the ribosome and the presence of the ribosomal exit tunnel provides some protection from association with nearby chains. Electrostatic and other kinds of interactions between the negatively charged ribosome surface[94] and the emerging polypeptide chains also may play a role.

Nature has devised ingenious and intricate support systems to help polypeptide chains survive in the cellular jungle and most importantly to ensure the proper functioning and vitality of the cell. The most important of these support systems are the proteolysis and chaperone networks. In addition to growing nascent chains, incomplete proteins arise in the cell due to nonsense mutations, incorporation of puromycin, premature translation termination, or proteolytic cleavage.[96] Degradation of such fragments is of vital importance since they can be highly toxic due to their tendency to form intracellular aggregates. It also is likely that a significant fraction of newly synthesized proteins is degraded soon after synthesis,[97,98] as it fails to achieve native conformations in the hostile cellular milieu, despite the intense protective action of molecular chaperones.

Chaperones are helper molecules that interact with newly synthesized (or stress-denatured) proteins to prevent their misfolding and/or help their correct folding (for reviews, refs. 99 and 100). Ribosome-tethered chaperones that meet newly synthesized polypeptide chains as they emerge from the ribosome have been identified in both prokayotes (trigger factor in *E. coli*) and eukaryotes (nascent-polypeptide-associated complex).[101–103] The ubiquitous Hsp70 and Hsp60 chaperone machines play an essential role in assisting folding of newly synthesized proteins, co-translationally and posttranslationally.[104] Hsp70 chaperones bind to exposed hydrophobic regions in unfolded or partially folded conformations of the nascent chain and act as safeguards against the potential risk for misfolding during and soon after biosynthesis. Molecular chaperones also serve the purpose of keeping proteins in an extended conformation for translocation across organelle membranes, disassembling protein aggregates, preventing aggregation, and helping proteolysis, secretion, and catalysis of proline isomerization.[104,105] Hence, the chaperone network constitutes an intricate support mechanism for elongating polypeptide chains. This amazing machinery ensures the proper folding, delivery to the proper cell compartment, and functioning of polypeptides and proteins inside the cell.

ACKNOWLEDGMENTS

We would like to thank Ruth Saecker for helpful discussions, and Robert L. Baldwin and Alexander Tropsha for providing manuscript preprints. This work was supported by the National Science Foundation (grant 0215368), the National Institutes of Health (grant 5R01 GM068535), the Milwaukee Foundation (Shaw Scientist Award), the Hereditary Disease Foundation and the American Heart Association (postdoctoral fellowship to NK).

REFERENCES

1. J. K. Myers and T. G. Oas, Mechanisms of fast protein folding, *Annu. Rev. Biochem.* **71**, 783–815 (2002).
2. W. A. Eaton, The physics of protein folding, *Phys. World* **12**, 39–44 (1999).
3. S. E. Radford, Protein folding: progress made and promises ahead, *Trends Biochem. Sci.* **25**(12), 611–618 (2000).
4. H. M. Dintzis, Assembly of the peptide chains of hemoglobin, *P. Natl. Acad. Sci. USA* **47**(3), 247–261 (1961).
5. R. E. Canfield and C. B. Anfinsen, Nonuniform labeling of egg white lysozyme, *Biochemistry* **2**(5), 1073–1078 (1963).
6. D. C. Phillips, The hen egg-white lysozyme molecule, *Proc. Natl. Acad. Sci. USA* **57**(3), 483–495 (1967).
7. H. Taniuchi and C. B. Anfinsen, An experimental approach to the study of the folding of staphylococcal nuclease, *J. Biol. Chem.* **244**(14), 3864–3875 (1969).
8. A. N. Schechter, R. F. Chen, and C. B. Anfinsen, Kinetics of folding of staphylococcal nuclease, *Science* **167**(919), 886–887 (1970).
9. H. F. Epstein, A. N. Schechter, R. F. Chen, and C. B. Anfinsen, Folding of staphylococcal nuclease: kinetic studies of two processes in acid denaturation, *J. Mol. Biol.* **60**(3), 499–508 (1971).
10. T. R. Hynes and R. O. Fox, The crystal structure of staphylococcal nuclease refined at 1.7 angstroms resolution, *Proteins* **10**, 92–105 (1991).
11. J. H. Carra, E. A. Anderson, and P. L. Privalov, Three-state thermodynamic analysis of the denaturation of staphylococcal nuclease mutants, *Biochemistry* **33**(10842–10850) (1994).
12. M. D. Jacobs and R. O. Fox, Staphylococcal nuclease folding intermediate characterized by hydrogen exchange and NMR spectroscopy, *Proc. Natl. Acad. Sci. USA* **91**, 449–453 (1994).
13. W. F. Walkenhorst, S. M. Green, and H. Roder, Kinetic evidence for folding and unfolding intermediates in staphylococcal nuclease, *Biochemistry* **36**, 5795–5808 (1997).
14. D. Shortle and A. K. Meeker, Residual structure in large fragments of staphylococcal nuclease: Effects of amino acid substitution, *Biochemistry* **28**, 936–944 (1989).
15. J. M. Flanagan, M. Kataoka, D. Shortle, and D. M. Engelman, Truncated staphylococcal nuclease is compact but disordered, *Proc. Natl. Acad. Sci. USA* **89**, 748–752 (1992).
16. S. Hirano, K. Mihara, Y. Yamazaki, H. Kamikubo, Y. Imamoto, and M. Kataoka, Role of C-terminal region of staphylococcal nuclease for foldability, stability, and activity, *Proteins* **49**, 255–265 (2002).
17. Y. Feng, L. Dongsheng, and J. Wang, Native-like partially folded conformations and folding process revealed in the N-terminal large fragments of staphylococcal nuclease: A study by NMR spectroscopy, *J. Mol. Biol.* **330**, 821–837 (2003).

18. P. A. Evans, R. A. Kautz, R. O. Fox, and C. M. Dobson, A magnetization-transfer nuclear magnetic resonance study of the folding of staphylococcal nuclease, *Biochemistry* **28**, 362–370 (1989).
19. G. Jing, B. Zhou, L. Xie, L. Li-jun, and Z. Liu, Comparative studies of the conformation of the N-terminal fragments of staphylococcal nuclease R in solution, *Biochim. Biophys. Acta* **1250**, 189–196 (1995).
20. B. Zhou, K. Tian, and G. Jing, An *in vitro* peptide folding model suggests the presence of the molten globule state during nascent peptide folding, *Protein Eng.* **13**(1), 35–39 (2000).
21. K. Tian, B. Zhou, F. Geng, and G. Jing, Folding of SNase R begins early during synthesis: the conformational feature of two short N-terminal fragments of staphylococcal nuclease R, *Int. J. Biol. Macromol.* **23**, 199–206 (1998).
22. M. Bycroft, A. Matouschek, J. T. Kellis, L. Serrano, and A. R. Fersht, Detection and characterization of a folding intermediate in barnase by NMR, *Nature* **346**, 488–490 (1990).
23. J. L. Neira and A. R. Fersht, Exploring the folding funnel of a polypeptide chain by biophysical studies on protein fragments, *J. Mol. Biol.* **285**, 1309–1333 (1999).
24. J. Sancho, J. L. Neira, and A. R. Fersht, An N-terminal fragment of barnase has residual helical structure similar to that in a refolding intermediate, *J. Mol. Biol.* **224**, 749–758 (1992).
25. S. Ludvigsen, H. Shen, M. Kjaer, J. C. Madsen, and F. M. Pouslen, Refinement of three-dimensional solution structure of barley serine protease inhibitor 2 and comparison with the structures in crystals, *J. Mol. Biol.* **222**, 621–635 (1991).
26. C. A. McPhalen and M. N. James, Crystal and molecular structure of the serine proteinase inhibitor CI-2 from barley seeds, *Biochemistry* **26**(1), 261–269 (1987).
27. G. de Prat Gay, J. Ruiz-Sanz, J. L. Neira, L. S. Itzhaki, and A. R. Fersht, Folding of a nascent polypeptide chain *in vitro*: Cooperative formation of structure in a protein module, *Proc. Natl. Acad. Sci. USA* **92**, 3683–3686 (1995).
28. L. S. Itzhaki, J. L. Neira, J. Ruiz-Sanz, G. de Prat Gay, and A. R. Fersht, Search for nucleation sites in smaller fragments of chymotrypsin inhibitor 2, *J. Mol. Biol.* **254**, 289–304 (1995).
29. G. de Prat Gay and A. R. Fersht, Generation of a family of protein fragments for structure-folding studies. 1. Folding complementation of two fragments of chymotrypsin inhibitor-2 formed by cleavage at its unique methionine residue, *Biochemistry* **33**, 7957–7963 (1994).
30. G. de Prat Gay, J. Ruiz-Sanz, J. L. Neira, F. J. Corrales, D. E. Otzen, A. G. Ladurner, and A. R. Fersht, Conformational pathway of the polypeptide chain of chymotrypsin inhibitor-2 growing from its N terminus *in vitro*. Parallels with the protein folding pathway, *J. Mol. Biol.* **254**, 968–979 (1995).
31. G. de Prat Gay, J. Ruiz-Sanz, B. Davis, and A. R. Fersht, The structure of the transition state for the association of two fragments of the barley chymotrypsin inhibitor 2 to generate native-like protein: Implications for mechanism of protein folding, *P. Natl. Acad. Sci. USA* **91**, 10943–10946 (1994).
32. G. de Prat Gay, Spectroscopic characterization of the growing polypeptide chain of the barley chymotrypsin inhibitor-2, *Arch. Biochem. Biophys.* **335**(1), 1–7 (1996).
33. P. A. Jennings and P. E. Wright, Formation of a molten globule intermediate early in the kinetic folding pathway of apomyoglobin, *Science* **262**(5135), 892–896 (1993).
34. C. C. Chow, C. Chow, V. Rhagunathan, T. Huppert, E. Kimball, and S. Cavagnero, The chain length dependence of apomyoglobin folding: structural evolution from misfolded sheets to native helices, *Biochemistry* **42**(23), 7090–7099 (2003).
35. M. T. Reymond, G. Merutka, H. J. Dyson, and P. E. Wright, Folding propensities of peptide fragments of myoglobin, *Protein Sci.* **6**, 706–716 (1997).
36. J. P. Waltho, V. A. Feher, G. Merutka, H. J. Dyson, and P. E. Wright, Peptide models of protein-folding initiation sites.1. Secondary structure formation by peptides corresponding to the G-helice and H-helice of myoglobin, *Biochemistry* **32**(25), 6337–6347 (1993).
37. C. Hetz and C. Soto, Protein misfolding and disease: the case of prion disorders, *Cell. Mol. Life Sci.* **60**(1), 133–143 (2003).
38. E. Zerovnik, Amyloid-fibril formation—Proposed mechanisms and relevance to conformational disease, *Eur. J. Biochem.* **269**(14), 3362–3371 (2002).

39. A. J. Thompson and C. J. Barrow, Protein conformational misfolding and amyloid formation: Characteristics of a new class of disorders that include Alzheimer's and prion diseases, *Curr. Med. Chem.* **9**(19), 1751–1762 (2002).
40. C. M. Dobson, Principles of protein folding, misfolding and aggregation, *Sem. Cell Dev. Biol.* **15**(1), 3–16 (2004).
41. C. M. Dobson, Protein misfolding diseases: Getting out of shape, *Nature* **418**(6899), 729–730 (2002).
42. M. Fandrich, M. A. Fletcher, and C. M. Dobson, Amyloid fibrils from muscle myoglobin—Even an ordinary globular protein can assume a rogue guise if conditions are right, *Nature* **410**(6825), 165–166 (2001).
43. H. J. Dyson and P. E. Wright, Insights into protein folding by NMR, *Annu. Rev. Phys. Chem.* **47**, 369–395 (1996).
44. M. L. Tasayco and J. Carey, Ordered self-assembly of polypeptide fragments to form native-like dimeric Trp repressor, *Science* **255**(5044), 594–597 (1992).
45. X.-M. Yang, W.-F. Yu, J.-H. Li, J. Fuchs, J. Rizo, and M. L. Tasayco, NMR evidence for the reassembly of an αvβ domain after cleavage of an α-helix: implications for protein design, *J. Am. Chem. Soc.* **120**, 7985–7986 (1998).
46. G. R. Parr and H. Taniuchi, Ordered complexes of cytochrome c fragments. Kinetics of formation of the reduced (ferrous) forms, *J. Biol. Chem.* **256**(1), 125–132 (1981).
47. M. Juillerat and H. Taniuchi, Conformational dynamics of a biologically-active 3-fragments complex of horse cytochrome c, *Proc. Natl. Acad. Sci. USA* **79**(6), 1825–1829 (1982).
48. H. Taniuchi, Formation of randomly paired disulfide bonds in des-(121–124)-ribonuclease after reduction and reoxidation, *J. Biol. Chem.* **245**(20), 5459–5468 (1970).
49. R. R. Matheson and H. A. Scheraga, A method for predicting nucleation sites for protein folding based on hydrophobic contacts, *Macromolecules* **11**(4), 819–829 (1978).
50. L. G. Chavez and H. A. Scheraga, Immunological determination of the order of folding of portions of the molecule during air oxidation of reduced ribonuclease, *Biochemistry* **16**, 1849 (1977).
51. R. Zerella, P. A. Evans, J. M. C. Ionides, L. C. Packman, B. W. Trotter, J. P. Mackay, and D. H. Williams, Autonomous folding of a peptide corresponding to the N-terminal β-hairpin from ubiquitin, *Protein Sci.* **8**(6), 1320–1331 (1999).
52. M. S. Searle, R. Zerella, D. H. Williams, and L. C. Packman, Native-like β-hairpin structure in an isolated fragment from ferrodoxin: NMR and CD studies of solvent effects on the N-terminal 20 residues, *Protein Eng.* **9**(7), 559–565 (1996).
53. F. J. Blanco, M. A. Jimenez, A. Pineda, M. Rico, J. Santoro, and J. Nieto, NMR solution structure of the isolated N-terminal fragment of protein-G B1 domain. Evidence of trifluoroethanol induced native-like β-hairpin formation, *Biochemistry* **33**, 6004–6014 (1994).
54. R. Hrabal, Z. Chen, S. James, H. P. J. Bennett, and F. Ni, The hairpin stack fold, a novel protein architecture for a new family of protein growth factors, *Nat. Struct. Biol.* **3**, 747–752 (1996).
55. F. Rousseau, J. W. H. Schymkowitz, and L. S. Itzhaki, The unfolding story of three-dimensional domain swapping, *Structure* **11**, 243–251 (2003).
56. Y. Liu, G. Gotte, M. Libonati, and D. Eisenberg, A domain-swapped RNase A dimer with implications for amyloid formation, *Nat. Struct. Biol.* **8**, 211–214 (2001).
57. M. P. Schlunegger, M. J. Bennett, and D. Eisenberg, Oligomer formation by 3D domain swapping: a model for protein assembly and misassembly, *Adv. Protein Chem.* **50**, 61–122 (1997).
58. F. Rousseau, J. W. H. Schymkowitz, H. R. Wilkinson, and L. S. Itzhaki, Three-dimensional domain swapping in p13suc1 occurs in the unfolded state and is controlled by conserved proline residues, *Proc. Natl. Acad. Sci. USA* **98**(10), 5596–5601 (2001).
59. S. M. Green, A. G. Gittis, A. K. Meeker, and E. E. Lattman, One-step evolution of a dimer from a monomeric protein, *Nat. Struct. Biol.* **2**, 746–751 (1995).
60. D. Bolton, P. A. Evans, K. Stott, and R. W. Broadhurst, Structure and properties of a dimeric N-terminal fragment of human ubiquitin, *J. Mol. Biol.* **314**, 773–787 (2001).

61. D. D. Ojennus, M. R. Fleissner, and D. S. Wuttke, Reconstitution of a native-like SH2 domain from disordered peptide fragments examined by multidimensional heteronuclear NMR, *Protein Sci.* **10**, 2162–2175 (2001).
62. R. E. Georgescu, M. M. Garcia-Mira, M. L. Tasayco, and J. M. Sanchez-Ruiz, Heat capacity analysis of oxidized *Escherichia coli* thioredoxin fragments (1-73, 74-108) and their noncovalent complex, *Eur. J. Biochem.* **268**, 1477–1485 (2001).
63. C. Mendoza, F. Figueirido, and M. L. Tasayco, DSC studies of a family of natively disordered fragments from *Escherichia coli* thioredoxin: Surface burial in intrinsic coils, *Biochemistry* **42**, 3349–3358 (2003).
64. R. Santucci, L. Fiorucci, F. Sinibaldi, F. Polizio, A. Desideri, and F. Ascoli, The heme-containing N-fragment (residues 1-56) of cytochrome *c* is a *bis*-histidine functional system, *Arch. Biochem. Biophys.* **379**(2), 331–336 (2000).
65. K. A. Dill, Dominant forces in protein folding, *Biochemistry* **29**(31), 7133–7155 (1990).
66. C. Tanford, *The Hydrophobic Effect: Formation of Micelles and Biological Membranes*, 2nd ed. (New York: Wiley, 1980).
67. B. Widom, P. Bhimalapuram, and K. Koga, The hydrophobic effect, *Phys. Chem. Chem. Physics* **5**(15), 3085–3093 (2003).
68. N. T. Southall, K. A. Dill, and A. D. J. Haymet, A view of the hydrophobic effect, *J. Phys. Chem. B* **106**(3), 521–533 (2002).
69. F. M. Richards, Areas, volumes, packing, and protein structure, *Annu. Rev. Biophys. Bioeng.* **6**, 151–176 (1977).
70. M. H. Klapper, On the nature of the protein interior, *Biochim. Biophys. Acta* **229**(3), 557–566 (1971).
71. C. Chothia, The nature of the accessible and buried surfaces in proteins, *J. Mol. Biol.* **165**, 1–14 (1976).
72. G. D. Rose, A. R. Geselowitz, G. J. Lesser, R. H. Lee, and M. H. Zehfus, Hydrophobicity of amino acid residues in globular proteins, *Science* **229**(4719), 834–838 (1985).
73. J. R. Livingstone, R. S. Spolar, and M. T. Record, Contribution to the thermodynamics of protein folding from the reduction in water-accessible nonpolar surface area, *Biochemistry* **30**(17), 4237–4244 (1991).
74. R. S. Spolar, J. H. Ha, and M. T. Record, Hydrophobic effect in protein folding and other noncovalent processes involving proteins, *Proc. Natl. Acad. Sci. USA* **86**(21), 8382–8385 (1989).
75. R. L. Baldwin, Temperature dependence of the hydrophobic interaction in protein folding, *Proc. Natl. Acad. Sci. USA* **83**(21), 8069–8072 (1986).
76. R. S. Spolar, J. R. Livingstone, and M. T. Record, Use of liquid-hydrocarbon and amide transfer data to estimate contributions to thermodynamic functions of protein folding from the removal of nonpolar and polar surface from water, *Biochemistry* **31**(16), 3947–3955 (1992).
77. S. D. Pickett and M. J. E. Sternberg, Empirical scale of side-chain conformational entropy in protein folding, *J. Mol. Biol.* **231**(3), 825–839 (1993).
78. K. A. Dill, Theory for the folding and stability of globular proteins, *Biochemistry* **24**(6), 1501–1509 (1985).
79. W. J. Becktel and J. A. Schellman, Protein stability curves, *Biopolymers* **26**(11), 1859–1877 (1987).
80. R. S. Spolar and M. T. Record, Coupling of local folding to site-specific binding of proteins to DNA, *Science* **263**(5148), 777–784 (1994).
81. J. Kyte and R. F. Doolittle, A simple model for displaying the hydrophobic character of a protein, *J. Mol. Biol.* **157**, 105–132 (1982).
82. N. Kurt and S. Cavagnero, The burial of solvent-accessible surface area is a predictor of polypeptide folding and misfolding as a function of chain elongation, *J. Am. Chem. Soc.* **127**(45), 15690–15691 (2005).
83. O. V. Tsodikov, M. T. Record, and Y. V. Sergeev, Novel computer program for fast exact calculation of accessible and molecular surface areas and average surface curvature, *J. Comput. Chem.* **23**(6), 600–609 (2002).

84. N. Kurt and S. Cavagnero, The burial of solvent-accessible surface area is a predictor of polypeptide folding and misfolding as a function of chain elongation, *J. Am. Chem. Soc.* **127**(45), 15690–15691 (2005).
85. H. M. Berman, J. Westbrook, Z. Feng, G. Gilliland, T. N. Bhat, H. Weissig, I. N. Shindyalov, and P. E. Bourne, The Protein Data Bank, *Nucleic Acids Res.* **28**, 235–242 (2000).
86. S. Cammer, A. Tropsha, and C. W. Carter, Protein Structure Workbench, contact order profile calculation, http://mmlsun4.pha.unc.edu/psw/3dworkbench.html.
87. D. B. Sherman, S. X. Zhang, J. B. Pitner, and A. Tropsha, Evaluation of the relative stability of liganded versus ligand-free protein conformations using Simplicial Neighborhood Analysis of Protein Packing (SNAPP) method, *Proteins* **56**, 828–838 (2004).
88. S. B. Zimmerman, and A. P. Minton, Macromolecular crowding: biochemical, biophysical and physiological consequences, *Annu. Rev. Biophys. Bioeng.* **22**, 27–75 (1993).
89. M. T. Record, D. S. Courtenay, D. S. Cayley, and H. J. Guttman, Biophysical compensation mechanisms buffering *E. coli* protein-nucleic acid interactions against changing environments, *Trends Biochem. Sci.* **23**, 190–194 (1998).
90. J. Han and J. Herzfeld, Macromolecular diffusion in crowded solutions, *Biophys. J.* **65**, 1155–1161 (1993).
91. B. van den Berg, R. Wain, C. M. Dobson, and R. J. Ellis, Macromolecular crowding perturbs protein refolding kinetics: implications for folding inside the cell, *EMBO J.* **19**(15), 3870–3875 (2000).
92. A. P. Minton, Implications of molecular crowding on protein assembly, *Curr. Opin. Struct. Biol.* **10**, 34–39 (2000).
93. A. P. Minton, Protein folding: Thickening the broth, *Curr. Biol.* **10**(3), R97–R99 (2000).
94. N. Ban, P. Nissen, J. Hansen, P. B. Moore, and T. A. Steitz, The complete atomic structure of the large ribosomal subunit at 2.4 Angstrom resolution, *Science* **289**, 905–920 (2000).
95. G. Kramer, W. Kudlicki, and B. Hardesty, Cotranslational folding—omnia mea mecum porto? *Int. J. Biochem. Cell. B.* **33**, 541–553 (2001).
96. A. L. Goldberg, Protein degradation and protection against misfolded or damaged proteins, *Nature* **426**, 895–899 (2003).
97. A. L. Goldberg and J. F. Dice, intracellular protein degradation in mammalian and bacterial cells, *Annu. Rev. Biochem.* **43**, 835–869 (1974).
98. U. Schubert, L. C. Anton, J. Gibbs, C. C. Norbury, J. W. Yewdell, and J. R. Bennink, Rapid degradation of a large fraction of newly synthesized proteins by protesomes, *Nature* **404**, 770–774 (2000).
99. F. U. Hartl and M. Hayer-Hartl, Molecular chaperones in the cytosol: from nascent chain to folded protein, *Science* **295**, 1852–1858 (2002).
100. J. Frydman, Folding of newly translated proteins *in vivo*: the role of molecular chaperones, *Annu. Rev. Biochem.* **70**, 603–647 (2001).
101. T. Hesterkamp, S. Hauser, H. Lutcke, and B. Bukau, *Escherichia coli* trigger factor is a prolyl isomerase that associates with nascent polypeptide chains, *Proc. Natl. Acad. Sci. USA* **93**, 4437–4441 (1996).
102. E. A. Craig, H. C. Eisenman, and H. A. Hundley, Ribosome-tethered molecular chaperones: the first line of defense against misfolding? *Curr. Opin. Microbiol.* **6**, 157–162 (2003).
103. S. Rospert, Y. Dubaquie, and M. Gautschi, Nascent-polypeptide-associated complex, *Cell. Mol. Life Sci.* **59**, 1632–1639 (2002).
104. B. Bukau and A. L. Horwich, The Hsp70 and Hsp60 chaperone machines, *Cell* **92**, 351–366 (1998).
105. J. C. Young, J. M. Barral, and F. U. Hartl, More than folding: localized functions of cytosolic chaperones, *Trends Biochem. Sci.* **28**(10), 541–547 (2003).

Determinants of Protein Folding and Aggregation in P22 Tailspike Protein

Matthew J. Gage,[1,2] Brian G. Lefebvre,[1,3] and Anne S. Robinson[1,4]

The P22 tailspike protein is a well-studied model system for understanding multimeric protein folding and aggregation. It is one of the few systems for which both *in vivo* and *in vitro* folding pathways have been well characterized. Aggregation of the P22 tailspike occurs through a multimeric addition pathway in which both monomeric and multimeric protein can associate with existing aggregates promoting aggregate growth. Native tailspike protein can be recovered from aggregated tailspike protein using hydrostatic pressure, though some of the protein remains aggregated. Hydrostatic pressure also has provided insight into the effects of two temperature-sensitive folding (*tsf*) mutants. Pressure-treatment of E196K leads to nativelike trimer recovery, while pressure treatment of G244R does not lead to a similar recovery of trimer, indicating that the two mutations affect different steps on the folding pathway. Tailspike aggregates form both clusters and long fiberlike aggregates, as visualized by both atomic force microscopy and confocal microscope, and coaggregate specifically with truncated tailspike protein but not unrelated proteins. Interestingly, more fiber-like aggregates are formed by tailspike truncations in which the C-terminus has been deleted, implying that the C-terminus might disrupt aggregate formation.

1. INTRODUCTION

The study of folding and aggregation is of great interest due to the increasing number of diseases that involve either protein misfolding or protein aggregation. A number of different model systems have been used to study these processes, including lysozyme,[1]

---

[1] Chemical Engineering Department, University of Delaware, Newark, DE 19716
[2] Current address: Department of Chemistry and Biochemistry, Northern Arizona University, P.O. Box 5498, Flagstaff, AZ 86011
[3] Current address: Department of Chemical Engineering, Rowan University, 201 Mullica Hill Rd., Glassboro, NJ 08028
[4] To whom correspondence should be addressed: Anne S. Robinson, 150 Academy St., Newark, DE 19716; email: robinson@che.udel.edu

barnase,[2] and β_2-microglobulin.[3] The majority of model systems are small (5-20 kDa), monomeric proteins. However, there are natural limits to the amount of information these model systems can provide since the majority of proteins in the cell are larger then 20 kDa and often are involved in multimeric complexes.

Unlike most model systems, the P22 tailspike protein is a large (72 kDa per monomer), trimeric protein whose *in vivo* and *in vitro* folding and aggregation pathways have been well characterized. While study of this protein is necessarily more complex then many other model proteins, the insights on folding and aggregation gathered from this system offer great value. A review of the factors that promote tailspike aggregation and the current models of aggregate growth are presented here.

2. P22 TAILSPIKE PROTEIN

2.1. P22 Tailspike Structure

The P22 tailspike protein is a homotrimeric protein with each monomer containing 666 amino acids. The protein is both protease resistant and SDS-resistant[4] and is thermostable beyond 80° C.[5] The structure of the protein was determined in two parts: the main body in 1994[6] and the head-binding domain in 1997.[7]

There are three main structural elements to the protein; the head-binding domain, the β-helix domain, and the caudal-fin domain (Figure 1A). The head-binding domain is formed by two antiparallel β-sheets, one of five strands, and one of three strands.[7] The N-terminus is the first region of the protein to undergo thermal denaturation[8] and this region does not play a significant role in the thermostability, sodium dodecyl sulfate (SDS) resistance, and folding kinetics of the protein.[9]

The main body of the protein is the β-helix domain. This domain consists of a 13-coil right-handed β-helix.[6] The helix has a triangular cross section, with each side composed of a parallel β-sheet (A, B, and C respectively, Figure 1B). There is also a large 62-residue loop inserted between the third strands of the B and C β-sheets, which is referred to as the dorsal fin.[6] The β-helix has a hydrophilic interface with about 65 residues involved in the intersubunit contacts, but there is no intertwining of the subunits in this region.[6] There are also about 80 ordered water molecules along the interface (Figure 1B), which mediate the majority of the subunit-subunit contacts.[6]

The final structural element is the caudal fin, which can be divided further into three sections; the interdigitated region and two antiparallel β-sheets (D and E, respectively) (Figure 1C). The interdigitated region consists of a 15-residue loop that intertwines with the identical loops from the other two subunits.[6] The D β-sheet is five stranded with each strand having two to six amino acids, while the E β-sheet is three stranded with each strand being six residues long.[6] The D and E β-sheets pack together with their symmetry partners to form a triangular structure, though the D and E sheets lie approximately 90° from each other.[6] The core of this interface is dominated by hydrophobic contacts.

2.2. P22 Tailspike Folding and Aggregation Pathway

The P22 tailspike protein is one of the few systems for which both the *in vivo* and *in vitro* folding and aggregation pathways have been well characterized.[9–17] Folding

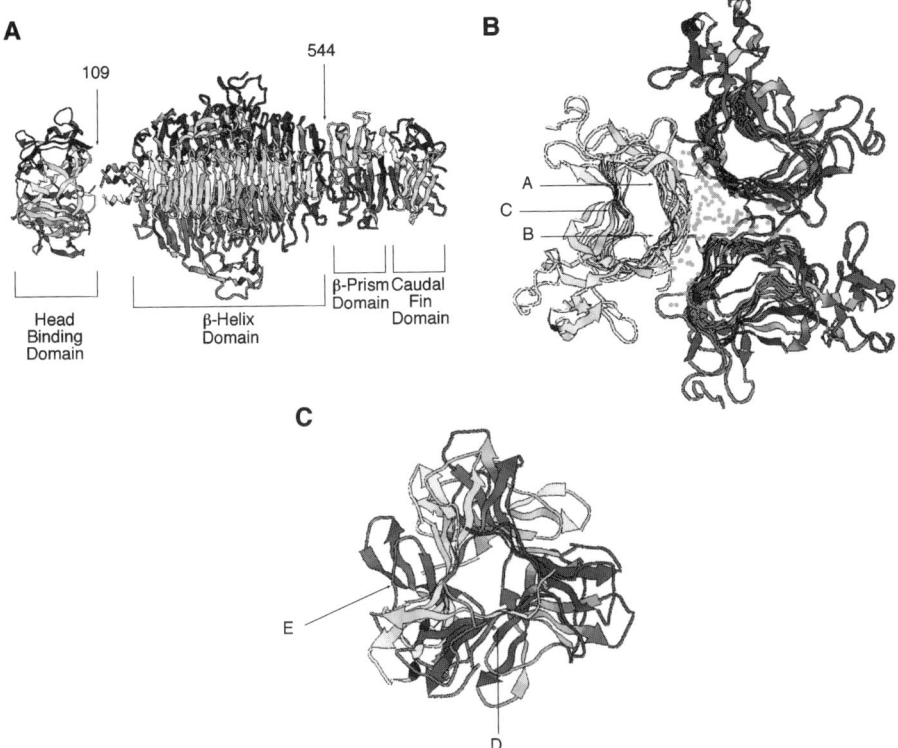

FIGURE 1. Structure of the P22 tailspike protein. (A) Ribbon diagram of the structure of P22 tailspike protein. The three subunits are colored red, blue, and yellow. The four main structural features are indicated on the structure. The arrows highlighting residues 109 and 544 indicate the approximate location of the truncations described in this work. (B) Ribbon diagram of the β-helix domain interface. This view is looking down the interface between the three β-helix domains from the N terminus. The A, B, and C β-sheets that compose the helix are indicated. The three β-helices that compose the helix are colored as in panel A. The light blue balls in the space between the subunits indicate the position of the ordered waters in the structure. Note that the interactions between the domains are predominately mediated by water. (C) Ribbon diagram of the C-terminus, oriented to view down the long axis of the protein from the β-helices to the end of the protein. The D and E β-sheets that compose the C-terminus are indicated and the three subunits are colored as in panels A and B. (A color version of this figure appears between pages 264 and 265.)

occurs on the order of minutes *in vivo* and on the order of minutes to days *in vitro*.[9,10] A current schematic of the folding and aggregation pathways is shown in Figure 2.

The initial step in the folding pathway is formation of a thermolabile monomer species. Two partially folded monomer chains associate together through hydrophobic interactions to form a dimeric species.[17] Association of a second monomer chain leads to a protrimer species, which does not contain the thermal stability or SDS and protease-resistance of the native trimer.[10] The protrimer matures to the native trimer by a currently poorly understood structural rearrangement.

One interesting feature of the folding of P22 tailspike is the formation of a transient disulfide bond during folding.[18–20] There are eight cysteine residues in P22 tailspike. In the final, mature form of the protein, all eight cysteines are reduced.[6,21] However,

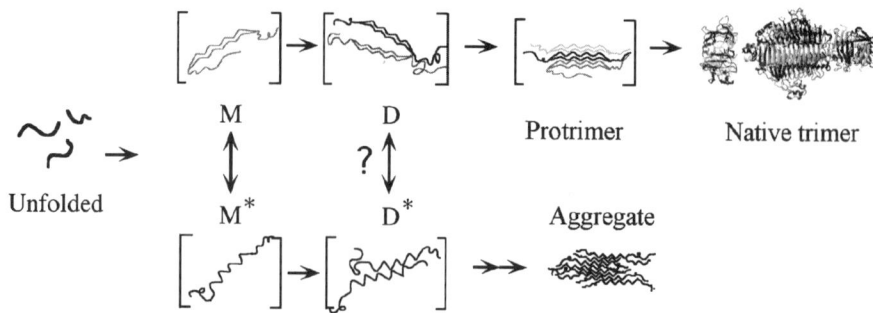

FIGURE 2. Folding and aggregation pathway of P22 tailspike. Unfolded protein begins to fold and enters either the folding pathway (M) or the aggregation pathway (M*). The monomer species may change between folding-competent and aggregate-prone conformations depending on conditions until it dimerizes with another monomer in the same state. Interaction between the pathways in the dimer state also has been proposed. The folding competent dimer adds a monomer to make the protrimer species which then undergoes structural rearrangement to form the final trimer species. Aggregates add monomers sequentially and also as clusters to form large aggregates.

single-mutant substitution of the three C-terminal cysteines has been shown to significantly affect folding and/or assembly kinetics *in vivo*[19] and *in vitro*.[20] A nonnative disulfide bond is formed during the assembly process that is broken during the maturation of the protrimer species.[18–20] While the exact role of the transient disulfide is not completely understood, the presence of cysteine residues at positions 496, 613, and 635 appear to be necessary for efficient trimer formation.

On the aggregation pathway, unfolded or newly synthesized protein forms an aggregation-prone species, which can associate with a second aggregation-prone monomer to form a dimer aggregate. It has been proposed that folding competent dimers can be structurally altered to form aggregate prone dimers, but this has not been convincingly proven. Additional aggregation-prone species can be added to form large aggregate species. A mechanistic description of aggregate growth will be discussed in more detail later.

3. EXPERIMENTAL METHODS TO STUDY P22 TAILSPIKE AGGREGATION

3.1. Circular Dichroism and Fluorescence Spectroscopy

Circular dichorism (CD) is a commonly used technique for probing secondary and tertiary structural changes that accompany aggregation. Many protein aggregates, including P22 tailspike aggregates, are primarily β-sheet in nature. Native P22 tailspike has a large minimum that occurs between 216 and 220 nm in the far-UV, while aggregated tailspike shows a slightly red-shifted minimum (Figure 3A). In addition, the magnitude of the peak increases, suggesting that aggregation promotes a more extended β-sheet conformation. As the native tailspike protein also is largely β-sheet in structure,

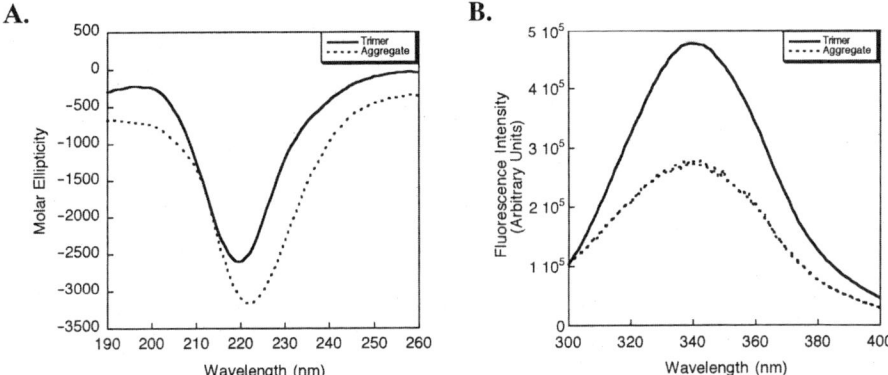

FIGURE 3. (A) CD spectra of 100 μg/ml native P22 tailspike trimer (solid line) and P22 tailspike aggregates (dashed line). Aggregated tailspike protein shows a slightly increased β-sheet signature. (B) Fluorescence spectra of 100 μg/ml native P22 tailspike (solid line) and P22 tailspike aggregate (dashed line). An excitation wavelength of 280 nm was used to excite the tryptophans in the protein. Aggregated tailspike protein shows a decreased fluorescence intensity and a slight shift in the center of mass.

it is difficult to distinguish conformational shifts between native protein and tailspike aggregates using this approach.

Shifts in the fluorescence spectra of tryptophan are indicative of changes in structure and in solvent exposure and provide another spectroscopic tool to study aggregation. The P22 tailspike monomer contains seven tryptophans, six of which lie in a ring around the center of the β-helix and one lies in the C-terminus. Spectra of the native tailspike trimer have strong fluorescent intensities and have a center of mass at 343 nm. Aggregation of the trimer results in a significant decrease in fluorescent intensity and a very slight shift in the center of mass (341 nm for aggregate) (Figure 3B). This indicates that in aggregated tailspike protein, the tryptophans reside in a different environment than in the native protein, resulting in increased quenching (due either to proximity to other fluorophores or due to increased solvent accessibility). This intensity change provides a potentially useful probe to differentiate the transition of unfolded protein to folded protein verses the transition of unfolded protein to aggregate if only one transition is occurring. However, fluorescence has not proven useful with the tailspike protein to study the mixed populations of folded protein and aggregates that occur during refolding reactions.

It therefore has been necessary to develop new techniques to study protein aggregation since some of the more commonly used techniques to study conformational changes are not sufficient. Two techniques that we have successfully employed are nondenaturing gel electrophoresis and size-exclusion chromatography (SEC).

3.2. Nondenaturing Gel Electrophoresis

Gel electrophoresis is one of the most commonly used methods for separation of protein samples. Following separation in an acrylamide matrix, the proteins can be stained for analysis or transferred to nitrocellulose for Western blot analysis. One of the more common types of gel electrophoresis is sodium dodecyl sulfate-polyacrylamide gel electrophoresis (SDS-PAGE), where the protein sample is prepared in a buffer containing the surfactant SDS. This generally disrupts noncovalent interactions, so the protein is viewed in the monomeric state. Under these conditions, tailspike protein chains on the folding pathway and aggregating protein have the same mobility, which is not useful in studying intermediates along the pathways. However, under nondenaturing (native) conditions, which does not disrupt protein-protein interactions, intermediates in the folding pathway can be visualized, making feasible the study of folding competent verses aggregate prone species using Western blot techniques.[22,23] Later in the chapter, we will show examples of using this technique to identify changes in aggregation due to mutation or solution conditions.

3.3. Size Exclusion Chromatography

A second method that we have employed to study folding and aggregation intermediates is size exclusion chromatography (SEC), which separates protein species based on their Stokes radius. Larger proteins generally migrate faster than smaller proteins, allowing analysis of the composition of a sample. We have used a TSK3000 gel filtration column to analyze the species composition of refolding samples at different time points and correlated this with gel electrophoresis analysis of the refolding samples to measure the formation and disappearance rates of the various refolding species (Figure 4).[24] One limitation of SEC is its ability to resolve species with similar apparent molecular weights. This has created difficulties in resolving the dimer and trimer species due to the similar Stokes radius of the species. In addition, the ability to resolve large macromolecular

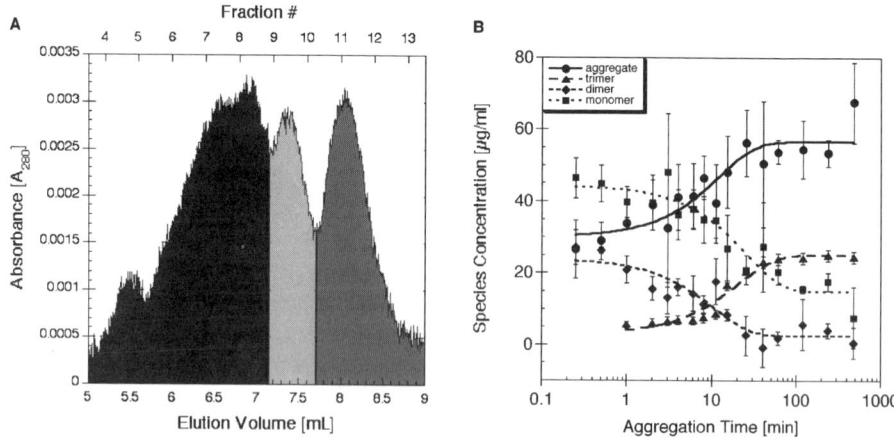

FIGURE 4. (A) Chromatograph of a typical refolding reaction (5 minutes at 25° C). The black portion of the curve represents aggregated protein, the light gray represents trimer and dimer, and the dark gray represents monomer. The data were fit by the trapezoid rule as other fits did not improve the data quality. (B) Species concentrations over time for refolding reactions at 25° C. The rate of each reaction was determined using a first-order fit.

4. FACTORS CONTROLLING AGGREGATION OF P22 TAILSPIKE

4.1. Temperature-Sensitive Folding Mutations

There are two main factors controlling the partitioning between the folding-competent and aggregation-prone monomers; temperature (*in vivo* and *in vitro*) and concentration (*in vitro*). In general, higher temperatures and concentrations promote aggregation, while lower temperatures and concentrations favor proper folding.

One of the most powerful tools in characterizing the partitioning between the folding and aggregation pathways of the P22 tailspike protein has been mutagenesis. There have been a large number of P22 tailspike mutations identified that result in altered partitioning between the folding and aggregation pathways. The largest class of mutants is the temperature-sensitive folding (*tsf*) mutants. The *tsf* mutants are capable of forming nativelike trimer at permissive temperatures (28° C, *in vivo*), but form inclusion bodies at restrictive temperatures (38° C). At the restrictive temperatures, the mutant chains associate in a partially folded state that is protease-sensitive and SDS-labile and that is unreactive with native tailspike antibodies.[11,25] The *tsf* mutations are clustered in the β-helix domain, especially in loop regions.[26,27] Also, a number of the *tsf* mutations have been shown to destabilize the isolated β-helix domain,[5,28] indicating that this domain may resemble the critical thermolabile intermediate that determines partitioning between the folding and aggregation pathways.

Four mutations have been identified that act as global suppressors of the *tsf* phenotype: V331G, V331A, A334V, and A334I.[29–31] Positions 331 and 334 are located in the middle of the β-helix domain, on the sixth coil of the helix.[6] V331 is located in the turn between the B and C β-sheets, while A334 is the first residue in β-sheet C.[6] Thermodynamic studies on isolated β-helix domain proteins have shown that the global suppressors increase the stability of both the native β-helix domain and of β-helix domains containing both suppressor and *tsf* mutations.[32]

Aggregated polypeptide chains of both the native tailspike protein and *tsf* mutants increase with increasing temperature.[10,12,25] The protein found within the tailspike inclusion bodies is both SDS-labile and protease-sensitive, indicating that the inclusion body is not composed of aggregated native tailspike.[12] Pulse-chase experiments with the native tailspike protein and with two *tsf* mutants have shown that there is a soluble intermediate that is involved in the growth of the aggregate.[12] Furthermore, this intermediate becomes more aggregation-prone as the incubation temperature increases. Once the protein becomes aggregated, it is irreversibly lost, as the amount of recoverable protein in temperature shift experiments is decreased with longer incubation times.[12]

In addition to *tsf* mutants, there are also several mutations in the C-terminus of the tailspike protein that stall assembly of the tailspike trimer. Mutations that either inhibit or slow progression through the folding pathway exhibit stable, long-lived monomeric and dimeric species, rather than increase the aggregation propensity. These have been seen for the three C496S, C613S, and C635S mutants used to study the role of the transient disulfide in the folding pathway.[20] In addition, the mutants R563Q and A575T, which inhibit protrimer formation, also exist in a stable long-lived soluble monomeric state.[33] The long-lived nature of monomer-folding intermediates in these proteins, along with the *tsf* data, suggests that proper formation of the β-helical-like folding intermediate may be the critical junction between the folding and aggregation pathways.

4.2. Buffer Conditions

Typical refolding conditions for tailspike are 50 mM Tris-HCl or 50 mM phosphate at pH 7.6 and 1 mM EDTA. Our laboratory also has investigated the effects of a number of folding additives in an attempt to improve the yield of refolded protein. In general, additives such as L-arginine, *bis*-ANS, glycerol, polyethylene glycol (ave. MW = 3350 g/mol), and Tween-20 did not increase the yields of folded protein and in many cases actually increased aggregation.[34] Most successful folding additives work by reducing associations between proteins. These additives also may inhibit association between correctly folded proteins, adversely affecting assembly of the P22 tailspike protein.

Refolding pH is another factor that significantly affects aggregation of the P22 tailspike protein. Tailspike refolds efficiently between pH 6.0 and 10 (Figure 5). However, at pH's greater than 10, assembly of the protein is inhibited, as increased population of monomer and dimer intermediates are observed. In contrast, aggregation is favored at pHs below 6. When refolding is initiated at pHs between 3 and 6, aggregation formation occurs very rapidly. A similar high aggregation propensity is seen with refolding reactions using isolated β-helix domain protein (data not shown). Confocal microscopy demonstrates that the morphology of aggregates formed at pH 3 is similar to that of aggregates formed at pH 7 (data not shown). However, while all the protein becomes aggregated at pH 3,

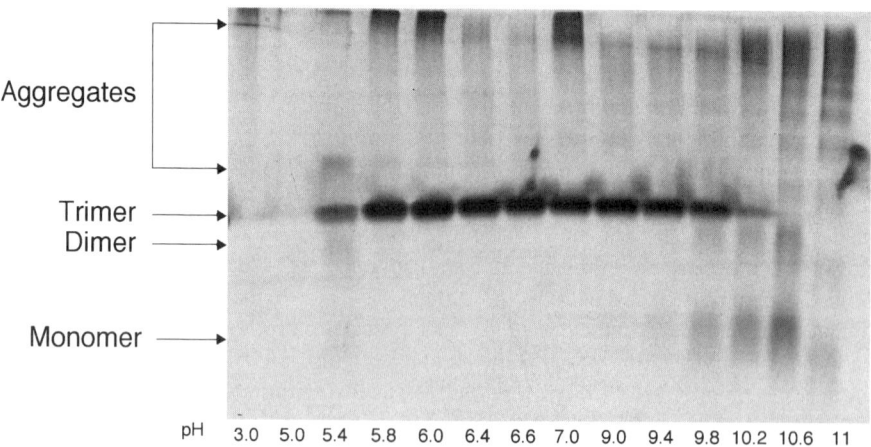

FIGURE 5. Nondenaturing gel of refolded native tailspike protein at 100 μg/ml. Refolding buffers were prepared at the various pH's indicated and protein was refolded at 100 μg/ml at 20° C overnight. Protein was separated using a 7% nondenaturing gel and visualized by silver staining. Protein refolded between pH 9.8 and 11 show decreased ability to assemble properly, while the protein completely aggregates below pH 5.4.

there is a significant amount of tailspike protein that remains soluble, either as small aggregates or as folded trimer at pH 7.0. These results demonstrate two significant features of tailspike folding and aggregation. First, assembly of the tailspike trimer likely requires formation of an ionic association involving either arginine or lysine residues, as assembly intermediates appear stalled around the pKa values of the side chains of these amino acids. In addition, the significant effect of low pH on aggregate formation indicates that low pH destabilizes the monomeric folding intermediate. The mechanism of aggregate growth is likely not affected by low pH, because aggregate morphology appears to be independent of pH.

5. AGGREGATE GROWTH

Intermediates in refolding and aggregation of P22 tailspike can be trapped by mixing folding reaction samples with ice-cold sample buffer and keeping the samples on ice.[35] Figure 6 shows an example of the distribution of species from a typical refolding reaction under conditions that produce both aggregated tailspike protein and productive folding intermediates. Aggregation occurs rapidly, as evidenced by the ladder of multimeric species accumulated during the dead time of mixing. The multimers at time $t = 0$ disappear if refolding is initiated on ice (data not shown), which favors productive folding.[23] Ferguson plot analysis has demonstrated that these bands correspond to aggregation species ranging from monomeric species to pentameric species and beyond.[35] As time progresses, the concentration of the trimeric and tetrameric aggregates increases, with their maximum concentration observed between 4 and 20 minutes. Both species then decrease in concentration in subsequent timepoints, as the trimeric and tetrameric

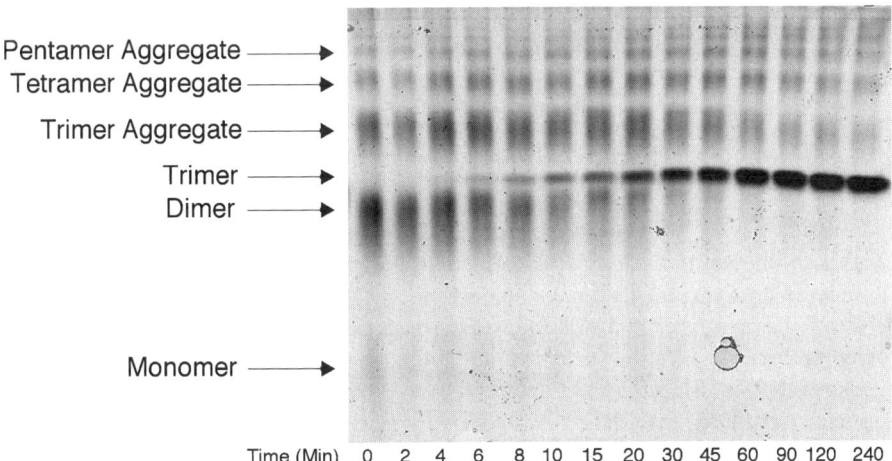

FIGURE 6. Native tailspike protein refolded at 100 μg/ml and 20° C. Samples were collected from the refolding reaction at the indicated timepoints and quenched with ice-cold nondenaturing gel buffer. Samples were separated using a 7% nondenaturing gel and visualized by silver staining. Highlighted are the various aggregate species. Note the progression of monomer to dimer to higher order aggregates as time progresses.

aggregates contribute to the formation of larger aggregate species. This approach works not only to observe the P22 tailspike protein aggregates, but has been shown to also isolate aggregation intermediates in other proteins, such as the P22 coat proteins and carbonic anhydrase II (CAB),[35] indicating that aggregate growth can be suspended by low temperatures, aiding in isolation of aggregation intermediates.

Identification of aggregation products for the P22 coat protein and CAB is easier than with the P22 tailspike, since both proteins are monomeric. While some of the monomeric species also are likely to be misfolded, any species larger than the monomer is clearly an aggregated species. In contrast, the P22 tailspike is trimeric and there are monomer, dimer, and trimeric intermediates in both the folding and aggregation pathways; thus distinguishing between folding-competent and aggregation-prone species is a much more difficult problem. This difficulty has been overcome through the characterization of epitope-specific monoclonal antibodies.[36] Antibodies have been generated that recognize the native tailspike proteins and the protrimer species.[22] In addition, there are a number of antibodies that react only with aggregate prone species.[22] The combination of these antibodies has helped to identify productive-folding species verses aggregation-prone species.[23]

There are two general models for aggregate growth: a sequential addition model and a multimeric addition. In the sequential addition model, monomeric species add to larger aggregate cores, causing growth of the aggregate. In the multimeric addition mechanism, aggregates of various sizes are added together to form larger aggregates. For P22 tailspike, we observed that aggregate growth continues even in the absence of monomeric intermediates, as shown in Figure 5. In addition, two-dimensional electrophoresis demonstrated that multimeric intermediates continue to aggregate in the absence of monomeric protein.[37] This indicates that larger molecular weight species also are able to assemble into higher-order aggregates. This is supported by static light-scattering data that are consistent with a model of multimeric addition and not linear growth.[37] These results indicate that tailspike aggregates form by generating small aggregating units, which can then add together to form larger aggregates.

6. PRESSURE EFFECTS ON AGGREGATES

We have used hydrostatic pressure to study aggregate dissociation. Application of pressure disrupts protein-protein interactions because protein-water bonds are shorter (lower volume) than protein-protein interactions. It has been shown in our laboratory that application of pressure to tailspike aggregates results in reasonable recovery of properly folded protein. Application of hydrostatic pressure to aggregates that were formed at 37° C resulted in dissociation of about 60% of the aggregates.[24] These aggregates are dissociated primarily to folding-competent monomeric and dimeric intermediates, which were found by treating the sample at 240 MPa for between 5 and 90 minutes (~80% yield of trimer from dissociated aggregates).[38]

One interesting feature of pressure dissociation is that 40% of the aggregates are unperturbed by pressure. Under confocal microscopy there is no difference between aggregates prior to and after pressure treatment (unpublished data). Repeated cycles of pressure do not disrupt this pool of aggregates, indicating that there are differences in the morphology of the aggregates. It is possible that pressure is removing the proteins on the surface of the aggregates, but that the core of the aggregates remains untouched.[38] Addition of urea helps to break up this pool of aggregates, but the presence of additional urea presumably disrupts subsequent assembly reactions, which is detrimental to trimer formation.[38]

Pressure has proved to be a useful tool in studying the phenotypes of *tsf* aggregates. We have studied the role of two different *tsf* mutants, G244R and E196K. The E196K mutant shows a near-lethal phenotype *in vitro*, even at 20° C. However, refolding for this mutant is much more efficient if refolding is initiated on ice, indicating that the mutation affects the early stages of refolding.[39] When aggregates of this mutant are pressure treated, nativelike recovery of trimer is observed.

In contrast to the effects seen for the E196K mutation, the G244R mutation shows similar effects as wild-type tailspike during *in vitro* refolding experiments at 20° C and in monomer refolding kinetics.[39] However, pressure treatment of aggregates does not result in significant recovery of properly folded tailspike protein, although it yields similar dissociation of aggregates to monomers and dimers.[39] This implies that the G244R mutation affects the formation of the dimer species. Pressure appears to dissociate the aggregates to the folding-competent monomer state, which would account for the difference in effect for the two different mutants. In the case of E196K, the *tsf* mutation affects the formation of the folding-competent monomer and pressure-generated monomers are past this point in the folding pathway. In contrast, the G244R mutation affects the dimer formation, which is the first step in the folding pathway that the pressure-generated species must pass through on their way to folded trimer. Pressure has helped to distinguish between these two phenotypes.

7. AGGREGATE MORPHOLOGY

The morphology of tailspike aggregates has been studied using both confocal microscopy and atomic force microscopy (AFM). Aggregates of the full-length protein formed two different morphologies, ellipsoid clusters and long, fiberlike aggregates, as seen by AFM (Figure 7). The fibers were generally 510 nm long, 145 nm wide, and 25 nm tall, while clusters seen were generally 110 nm long, 45 nm wide, and 20 nm tall. The size of the two types of aggregates fits a model of aggregate growth by axial addition of the ellipsoid clusters.

A close examination of the clusters reveals small bumps along the surface of the cluster that are spaced about 20 nm apart, consistent with the dimensions of a monomer or dimer species (Figure 7C). The ellipsoid clusters may be formed by addition of monomer and dimer chains and then once formed, the ellipsoids may associate to form the larger aggregate species. This two-stage aggregate growth has been reported for other systems, such as amyloid β peptide[40] and α-synuclein.[41]

We also have studied aggregate structure using confocal microscopy. Native tailspike protein was labeled with a fluorescent probe (Alexa-488, Molecular Probes) and aggregated at 37° C. As can be seen in Figure 8A, native tailspike consistently forms

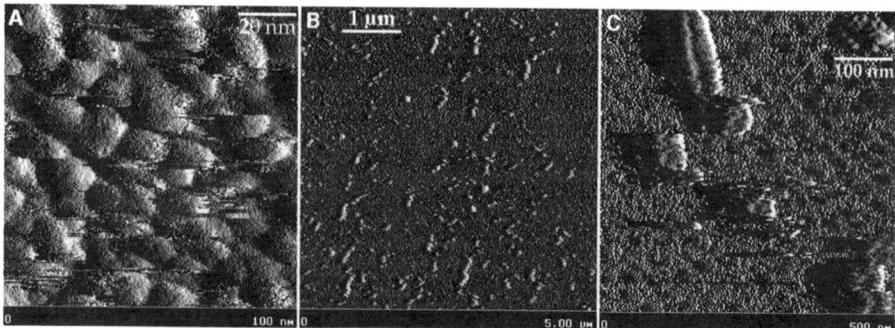

FIGURE 7. (A) AFM image of native P22 tailspike trimer. The dimensions of the trimer (18 nm long and 10 nm wide) are in good agreement with the dimensions determined from the crystal structure. (B) AFM image of aggregated tailspike protein. Two aggregate morphologies, clusters and fibers, can be seen. The clusters are 110 nm long, 45 nm wide, and 20 nm tall, while the fibers are 510 nm long, 145 nm wide, and 25 nm tall. (C) Increased magnification of tailspike aggregates, showing a fiber and a cluster. The arrow highlights the bumps along the edge of the cluster. These bumps are similar in size to the monomeric protein. (A color version of this figure appears between pages 264 and 265.)

small, threadlike aggregates that are 5 μm long and 0.5–1 μm wide. In contrast, aggregates formed by a truncated version of the tailspike protein (amino acids 109-544) generally form much longer and more ribbonlike aggregates under the same conditions. These aggregates also tend to have a much greater variety of dimensions than aggregates of the native tailspike protein.

These differences in appearance were exploited to address one of the most important questions about aggregate formation. Specifically, whether different proteins have the

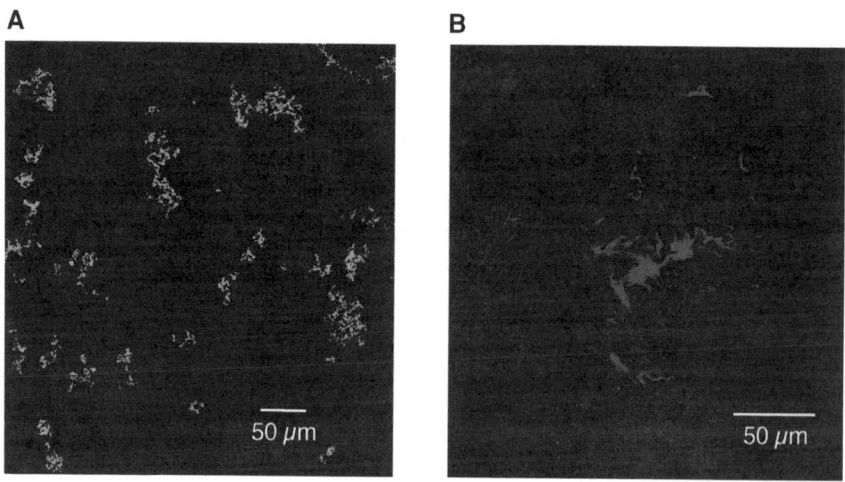

FIGURE 8. Native and truncated P22 tailspike protein visualized by confocal microscopy. (A) Aggregated full-length native tailspike protein labeled with Alexa-488 as a fluorophore. (B) Aggregated truncated tailspike protein (residues 109-544) labeled with Alexa-488 as a fluorophore. Notice the longer, more ribbonlike aggregate structure of the truncated protein. (A color version of this figure appears between pages 264 and 265.)

ability to associate, or coaggregate, during aggregation. This has been investigated using both the P22 tailspike and P22 coat proteins. King and colleagues mixed denatured forms of both proteins together under aggregating conditions and examined the aggregates by Western blotting.[42] Higher-order aggregates reacted to either the tailspike antibodies or to the coat protein antibodies, but not to both, indicating during the initial stages of aggregation the two proteins do not associate.[42] This was a significant result, as it implied that specific contacts are necessary for aggregate growth and that not all of these contacts exist in all proteins.

We have done further studies of coaggregation of larger aggregates using confocal microscopy. We have coaggregated fluorescently labeled full-length tailspike protein with fluorescently labeled tailspike truncations and examined the degree of association. Labeled, denatured protein was mixed under aggregating conditions to form mixtures of aggregates, which were examined using confocal microscopy. In general, the truncated tailspike protein and full-length tailspike protein colocalize (Figure 9). There are places where one label occurs without the other label, indicating that potentially these coaggregates are formed by multimeric association rather than by single monomer units. This is further borne out by similar coaggregation studies using full-length TSP and CAB (Figure 9). In mixed aggregation reactions, there was a significantly lower degree of colocalization of the fluorophores. However, some coaggregation did appear to exist. This is thought to be due to the association of larger multimeric ensembles.

The coaggregation and aggregation studies that we have performed with confocal microscopy also reveal some interesting features of aggregate morphology. Aggregates of the full-length tailspike protein and of the N-terminus truncation have a much shorter morphology than aggregates of the β-helix domain. This is seen also with coaggregates, where aggregates that contain a truncation with the C-terminus deleted have a much longer, filamentous structure then aggregates where both proteins contain the C-terminus. These results imply that the C-terminus may disrupt the association of multimeric aggregates, preventing association of larger, multimeric complexes.

These results provide some valuable insights into aggregate formation and the specificity of aggregation. Confocal microscopy provides us with the ability to visualize the presence of two different proteins in aggregates using different fluorescent dyes. One limitation to this approach is resolution. Confocal microscopy is limited to the level of micrometers. It is not possible to get to the level of the individual proteins to look at association. One question that cannot be answered by these experiments is whether the colocalization seen is due to association of monomeric protein or whether it is due to association of multimers. In contrast to confocal microscopy, electron microscopy and AFM can image at much higher resolution, but do not have the ability of easily distinguishing between two similar proteins associated together. Therefore, even with the resolution limitations, confocal microscopy is the most effective method to image these kinds of systems.

Because of the limited resolution, this technique only provides general information about coaggregation. One important question is whether there are any effects due to labeling. To address this, unlabeled coaggregates have been studied using Congo red to verify that aggregate morphology is not dependent on the presence of the fluorescent

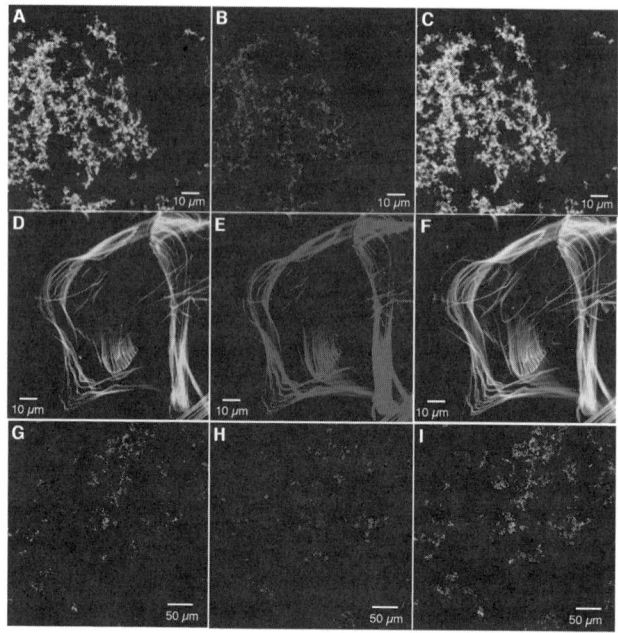

FIGURE 9. Coaggregated tailspike protein: 100 μg/ml of labeled native tailspike protein was mixed with 100 μg/ml of either truncated tailspike protein or CAB at 37° C and a final concentration of 200 μg/ml. (A), (B), (C) Coaggregated full-length tailspike with truncated tailspike protein (residues 109-666). Labeled full-length tailspike is in A, labeled 109-666 in B, and a composite of both is seen in C. (D), (E), (F) Coaggregated full-length tailspike with truncated tailspike protein (residues 109-544). Labeled full-length tailspike is in D, labeled 109-544 in E, and a composite of both is seen in F. (G), (H), (I) Coaggregated full-length tailspike with CAB (10X tailspike concentration). Labeled full-length tailspike is in G, labeled CAB in H, and a composite of both is seen in I. Note that the CAB and tailspike do not colocalize very well, while the full-length tailspike colocalizes with each of the truncations very well. (A color version of this figure appears between pages 264 and 265.)

probe. In addition, the limited colocalization of tailspike with CAB provides a nice control to indicate that there is level of specificity in aggregation.

Tailspike aggregates appear in both AFM and confocal images to form fiberlike aggregates. However, in both cases, the fiber diameters are an order of magnitude larger than typical amyloid fibrils (0.15 μm for AFM and 0.2 to 0.5 μm for confocal). In studies of the isolated β-helix domain, similar-sized fibers have been shown to have the characteristic Congo red birefringence associated with amyloid fibrils.[43] Therefore, this implies that tailspike aggregates may pack in a manner similar to amyloid fibrils.

8. DISCUSSION

Biochemical study of protein aggregates is an extremely challenging undertaking since many biochemical and biophysical techniques require soluble or a well-ordered arrangement of the protein of interest. Most protein aggregates are insoluble or rapidly become so and are only partially ordered, making some standard biochemical and biophysical techniques ineffective. It has been possible to gain some insight into aggregate structure with solid-state NMR, X-ray fiber diffraction, and cryoelectron microscopy. While atomic resolution information about aggregate structure has remained elusive, it has been possible to gain insight into aggregation kinetics and some of the conformational changes associated with aggregation and misfolding through spectroscopy and methods such as size exclusion chromatography and gel electrophoresis.

The P22 tailspike has been a very useful model system for the study of protein aggregation. The large size of the protein allows examination of the role of specific domains in aggregate formation. Also, the large number of known aggregation-promoting mutations has helped to characterize the system. The β-helical nature of the protein makes it an attractive system to study aggregate growth and formation in light of recent modeling studies, suggesting that protofibrils may have a β-helical structure.[44,45]

The nature of P22 tailspike aggregates has been studied for a number of years. Tailspike aggregates show a high degree of β-sheet structure, similar to the native protein, indicating that aggregated protein partially may resemble the native state of the protein. However, since some of the monoclonal antibodies that recognize the native tailspike epitopes are not reactive with tailspike aggregates, the protein does not closely resemble the native structure.

Efficiency of tailspike refolding decreases with increases in temperature, indicating that there are potentially "traps" on the folding landscape that become accessible at higher temperature, resulting in a higher degree of aggregation. This is supported by the large number of mutations localized in the β-helix domain that result in elevated levels of aggregation at higher temperatures. This suggests that a perturbation to early β-helix domain packing results in increases in aggregation.

The use of hydrostatic pressure has shown that the protein exists in different conformations within tailspike aggregates. Application of 240 MPa hydrostatic pressure dissociates ~60% of aggregates into monomer and dimer. The remaining 40% of the protein does not dissociate with higher application of pressure. There is increased dissociation with the addition of urea, but this also results in reduced yields of refolded protein. Hydrostatic pressure also has provided insight into the mechanism of two *tsf* mutants, revealing one destabilized formation of the monomer, while the other destabilized formation of the dimer.

Confocal microscopy and AFM have shown that the tailspike protein forms filamentous aggregates, resembling fibrils. However, tailspike filaments are larger in diameter than typical fibrils and are not as stable, indicating a different packing arrangement. In addition, coaggregation studies of the tailspike protein with truncated tailspike proteins have shown that removal of the C-terminus leads to formation of much longer aggregates, suggesting that different domains of the tailspike protein pack into aggregates differently and that this can affect the overall morphology of the aggregate.

The techniques that we have developed to study aggregation of the P22 tailspike are general techniques that can be applied to both monomer and multimeric aggregating systems. In our laboratory, we have extended these techniques to explore the folding and aggregation of an engineered single-chain antibody, a monomeric protein. Size-exclusion chromatography and native PAGE have revealed some interesting properties of this system.

The advantage of the techniques that we have been developing is their ability to capture early aggregation events under physiological conditions. Some of the techniques used to study aggregation require formation of a certain size of aggregate, making them unsuitable to study the initial nucleation events. As these events are critical to the initiation of aggregation, we have been developing ways to better understand these initial events.

There are now a number of techniques available to follow different stages of aggregation and also some structural information about proteins within aggregates. However, high-resolution structural information about aggregation-prone intermediates has remained elusive. With the growing evidence that small oligomeric aggregates may be the toxic species in aggregation-related diseases, a better understanding of the structural changes that lead to aggregation is important. This is the next frontier in aggregation research as the techniques to study these structural changes do not currently exist.

ACKNOWLEDGMENTS

The authors would like to thank Junghwa Kim, Brenda Danek, Jennifer Zak, Noelle Comolli, and Dana Ungerbuehler for their contributions to this work. This work has been supported in part by NIH P20 RR15588 and R01 GM-60543 and NSF BES 99-84312.

REFERENCES

1. A. Matagne and C. M. Dobson, The folding process of hen lysozyme: a perspective from the "new view," *Cell Mol. Life Sci.* **54**, 363–371 (1998).
2. V. Daggett and A. Fersht, The present view of the mechanism of protein folding, *Nat. Rev. Mol. Cell Biol.* **4**, 497–502 (2003).
3. F. Chiti, E. De Lorenzi, S. Grossi, P. Mangione, S. Giorgetti, G. Caccialanza, C. M. Dobson, G. Merlini, G. Ramponi, and V. Bellotti, A partially structured species of beta 2-microglobulin is significantly populated under physiological conditions and involved in fibrillogenesis, *J. Biol. Chem.* **276**, 46714–46721 (2001).
4. D. P. Goldenberg and J. King, Temperature-sensitive mutants blocked in the folding or subunit of the bacteriophage P22 tail spike protein. II. Active mutant proteins matured at 30 degrees C, *J. Mol. Biol.* **145**, 633–651 (1981).
5. J. M. Sturtevant, M.-H. Yu, C. Haase-Pettingell, and J. King, Thermostability of temperature-sensitive folding mutants of the P22 tailspike protein. *J. Biol. Chem.* **264**, 10693–10698 (1989).
6. S. Steinbacher, R. Seckler, S. Miller, B. Steipe, R. Huber, and P. Reinemer, Crystal structure of P22 tailspike protein: interdigitated subunits in a thermostable trimer, *Science* **265**, 383–385 (1994).
7. S. Steinbacher, S. Miller, U. Baxa, N. Budisa, A. Weintraub, R. Seckler, and R. Huber, Phage P22 tailspike protein: crystal structure of the head-binding domain at 2.3 Å, fully refined structure of the endorhamnosidase at 1.56 Å resolution, and the molecular basis of O-antigen recognition and cleavage, *J. Mol. Biol.* **267**, 865–880 (1997).
8. B. L. Chen and J. King, J. Thermal unfolding pathway for the thermostable P22 tailspike endorhamnosidase, *Biochemistry* **30**, 6260–6269 (1991).
9. M. Danner, A. Fuchs, S. Miller, and R. Seckler, Folding and assembly of phage P22 tailspike protein lacking N-terminal, head-binding domain, *Eur. J. Biochem.* **215**, 653–661 (1993).

10. D. Goldenberg and J. King, Trimeric intermediate in the *in vivo* folding and subunit assembly of the tailspike endorhamnosidase of bacteriophage P22, *Proc. Natl. Acad. Sci. USA* **79**, 3403–3407 (1982).
11. D. Goldenberg, D. H. Smith, and J. King, Genetic analysis of the folding pathway for the tailspike protein of phage P22, *Proc. Natl. Acad. Sci. USA* **80**, 7060–7064 (1983).
12. C. Haase-Pettingell and J. King, Formation of aggregates from a thermolabile *in vivo* folding intermediate in P22 tailspike maturation a model for inclusion body formation, *J. Biol. Chem.* **263**, 4977–4983 (1988).
13. J. King and M.-H. Yu, Mutational analysis of protein folding pathways: The P22 tailspike endorhamnosidase, *Meth. Enzymol.* **131**, 250–266 (1986).
14. M. Danner and R. Seckler, Mechanism of phage P22 tailspike folding mutations, *Prot. Sci.* **2**, 1869–1881 (1993).
15. M. Beibinger, S. C. Lee, S. Steinbacher, P. Reinemer, R. Huber, M.-H. Yu, and R. Seckler, Mutations that stabilize folding intermediates of phage P22 tailspike protein: Folding *in vivo* and *in vitro*, stability, and structural context, *J. Mol. Biol.* **249**, 185–194 (1995).
16. R. Seckler, A. Fuchs, J. King, and R. Jaenicke, Reconstitution of the thermostable trimeric phage P22 tailspike from denatured chains *in vitro*, *J. Biol. Chem.* **264**, 11750–11753 (1989).
17. M. Gage and A. S. Robinson, C-Terminal hydrophobic interactions play a critical role in oligomeric assembly of the P22 tailspike trimer, *Protein Sci.* **12**, 2732–2747 (2003).
18. A. S. Robinson and J. King, Disulphide-bonded intermediate on the folding and assembly pathway of a non-disulphide bonded protein, *Nature Struct. Biol.* **4**, 450–455 (1997).
19. C. Haase-Pettingell, S. D. Betts, S. W. Raso, L. Stuart, A. Robinson, and J. King, Role for cysteine residues in the *in vivo* folding and assembly of the phage P22 tailspike, *Prot. Sci.* **10**, 397–410 (2000).
20. B. L. Danek and A. S. Robinson, Nonnative interactions between cysteines direct productive assembly of P22 tailspike protein, *Biophys. J.* **85**, 3237–3247 (2003).
21. D. Sargent, J. M. Benevides, M.-H. Yu, J. King, and G. J. Thomas, Jr., Secondary structure and thermostability of the phage P22 tailspike. Analysis by Raman spectroscopy of the wild-type protein and a temperature-sensitive folding mutant, *J. Mol. Biol.* **199**, 491–502 (1988).
22. M. A. Speed, T. Morshead, D. I. C. Wang, and J. King, Conformation of P22 tailspike folding and aggregation intermediates probed by monoclonal antibodies, *Prot. Sci.* **6**, 99–108 (1997).
23. S. D. Betts and J. King, Cold rescue of the thermolabile tailspike intermediate at the junction between productive folding and off-pathway aggregation, *Prot. Sci.* **7**, 1516–1523 (1998).
24. B. G. Lefebvre and A. S. Robinson, Pressure treatment of tailspike aggregates rapidly produces on-pathway folding intermediates, *Biotechnol. Bioeng.* **82**, 595–604 (2003).
25. D. H. Smith and J. King, Temperature-sensitive mutants blocked in the folding or subunit assembly of the bacteriophage P22 tailspike protein. III. Intensive polypeptide chains synthesized at 39 degrees C, *J. Mol. Biol.* **145**, 653–676 (1981).
26. B. Fane and J. King, Identification of sites influencing the folding and subunit assembly of the P22 tailspike polypeptide chain using nonsense mutations, *Genetics*, 157–171 (1987).
27. C. Haase-Pettingell and J. King, Prevalence of temperature sensitive folding mutations in the parallel beta coil domain of the phage P22 tailspike endorhamnosidase, *J. Mol. Biol.* **267**, 88–102 (1997).
28. B. Schuler, F. Furst, F. Osterroth, S. Steinbacher, R. Huber, and R. Seckler, Plasticity and steric strain in a parallel β-helix: rational mutations in the P22 tailspike protein, *Proteins: Struct, Funct, Genet.* **39**, 89–101 (2000).
29. R. Villafane, A. Fleming, and C. Haase-Pettingell, Isolation of suppressors of temperature-sensitive folding mutations, *J. Bacteriol.* **176**, 137–142 (1994).
30. B. Fane and J. King, Intragenic suppressors of folding defects in the P22 tailspike protein, *Genetics* **127**, 263–277 (1991).
31. B. Fane, R. Villafane, A. Mitraki, and J. King, Identification of global suppressors for temperature-sensitive folding mutations of the P22 tailspike protein, *J. Biol. Chem.* **266**, 11640–11648 (1991).
32. B. Schuler and R. Seckler, P22 tailspike folding mutants revisited: effects on the thermodynamic stability of the isolated β-helix domain, *J. Mol. Biol.* **281**, 227–234 (1998).

33. J. F. Kreisberg, S. D. Betts, C. Haase-Pettingell, and J. King, The interdigitated β-helix domain of the P22 tailspike protein acts as a molecular clamp in trimer stabilization, *Prot. Sci.* **11**, 820–830 (2002).
34. K. Whitehead, *The Effects of Chemical Additives on the in vitro Refolding of the P22 Tailspike Protein*, Undergraduate Senior Thesis, University of Delaware (2002).
35. M. Speed, D. I. C. Wang, and J. King, Multimeric intermediates in the pathway to the aggregated inclusion body state for P22 tail spike polypeptide chains, *Prot. Sci.* **4**, 900–908 (1995).
36. B. Friguet, l. Djavadi-Ohaniance, C. Haase-Pettingell, J. King, and M. E. Goldberg, Properties of monoclonal antibodies selected for probing the conformation of wild type and mutant forms of the P22 tailspike endorhamnosidase, *J. Biol. Chem.* **265**, 10347–10351 (1990).
37. M. Speed, *Characterization of the Aggregation Pathway Competing with Productive Protein Folding*. Ph.D. thesis, Massachusetts Institute of Technology, Cambridge, MA (1996).
38. B. G. Lefebvre, M. J. Gage, and A. S. Robinson, Maximizing recovery of native protein from aggregates by optimizing pressure treatment, *Biotechnol. Progr.* **20**, 623–629 (2004).
39. B. G. Lefebvre, N. K. Comolli, M. J. Gage, and A. S. Robinson, Pressure dissociation studies provide insight into oligomerization competence of temperature-sensitive folding mutants of P22 tailspike, *Prot. Sci.* **13**, 1538–15346 (2004).
40. J. D. Harper, S. S. Wong, C. Lieber, and P. T. Lansbury, Jr., Assembly of A beta amyloid protofibrils: an *in vitro* model for a possible early event in Alzheimer's disease, *Biochemistry* **38**, 8972–8980 (1999).
41. K. A. Conway, J. D. Harper, and P. T. Lansbury, Jr., Fibrils formed *in vitro* from alpha-synuclein and two mutant forms linked to Parkinson's disease are typical amyloid, *Biochemistry* **39**, 2552–2563 (2000).
42. M. A. Speed, D. I. C. Wang, and J. King, Specific aggregation of partially folded polypeptide chains: The molecular basis of inclusion body composition, *Nature Biotech.* **14**, 1283–1287 (1996).
43. B. Schuler, R. Rachel, and R. Seckler, Formation of fibrous aggregates from a non-native intermediate: the isolated P22 tailspike β-helix domain, *J. Biol. Chem.* **274**, 18589–18596 (1999).
44. D. T. Downing and N. D. Lazo, Molecular modelling indicates that the pathological conformations of prion proteins might be beta-helical, *Biochem. J.* **343** Pt 2, 453–460 (1999).
45. N. D. Lazo and D. T. Downing, Amyloid fibrils may be assembled from beta-helical protofibrils, *Biochemistry* **37**, 1731–1735 (1998).

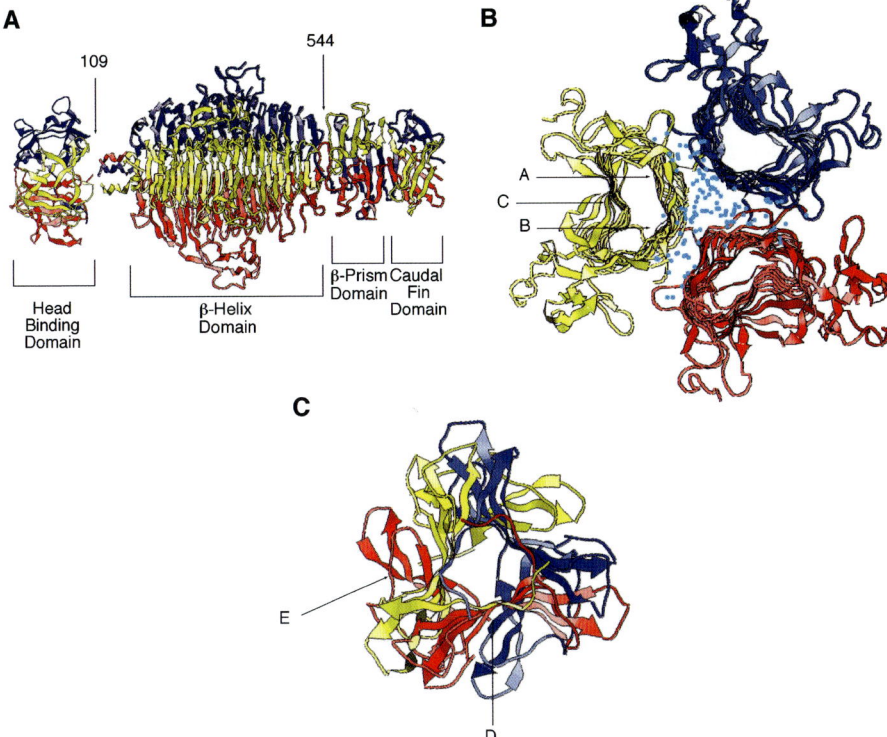

FIGURE 1. Structure of the P22 tailspike protein. (A) Ribbon diagram of the structure of P22 tailspike protein The three subunits are colored red, blue, and yellow. The four main structural features are indicated on the structure. The arrows highlighting residues 109 and 544 indicate the approximate location of the truncations described in this work. (B) Ribbon diagram of the β-helix domain interface. This view is looking down the interface between the three β-helix domains from the N terminus. The A, B, and C β-sheets that compose the helix are indicated. The three β-helices that compose the helix are colored as in panel A. The light blue balls in the space between the subunits indicate the position of the ordered waters in the structure. Note that the interactions between the domains are predominately mediated by water. (C) Ribbon diagram of the C-terminus, oriented to view down the long axis of the protein from the β-helices to the end of the protein. The D and E β-sheets that compose the C-terminus are indicated and the three subunits are colored as in panels A and B.

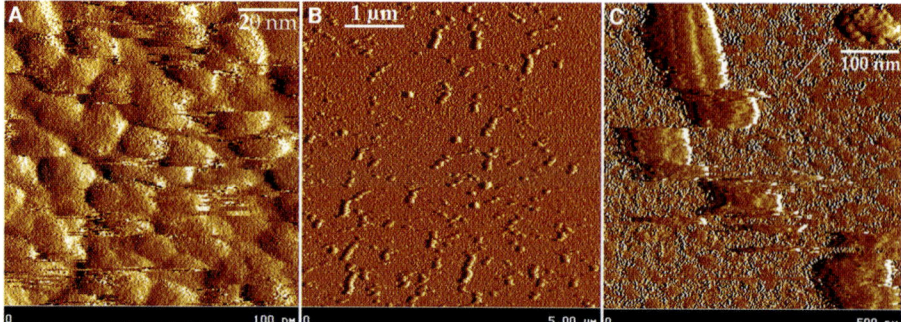

FIGURE 7. (A) AFM image of native P22 tailspike trimer. The dimensions of the trimer (18 nm long and 10 nm wide) are in good agreement with the dimensions determined from the crystal structure. (B) AFM image of aggregated tailspike protein. Two aggregate morphologies, clusters and fibers, can be seen. The clusters are 110 nm long, 45 nm wide, and 20 nm tall, while the fibers are 510 nm long, 145 nm wide, and 25 nm tall. (C) Increased magnification of tailspike aggregates, showing a fiber and a cluster. The arrow highlights the bumps along the edge of the cluster. These bumps are similar in size to the monomeric protein.

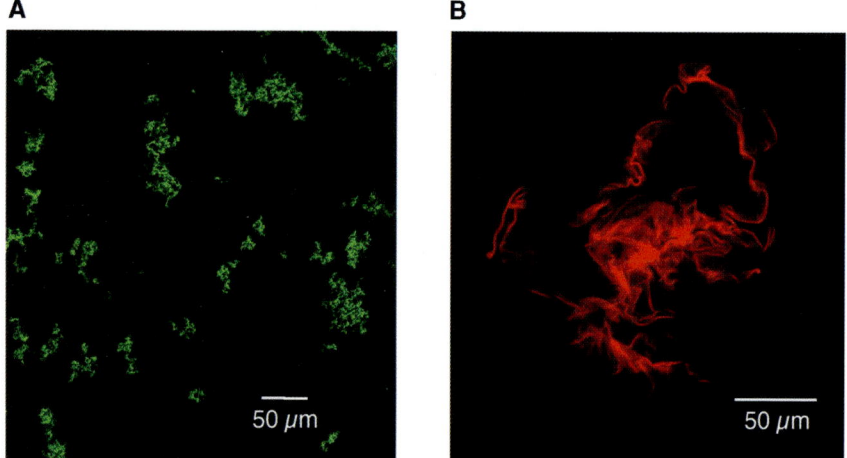

FIGURE 8. Native and truncated P22 tailspike protein visualized by confocal microscopy. (A) Aggregated full-length native tailspike protein labeled with Alexa-488 as a fluorophore. (B) Aggregated truncated tailspike protein (residues 109-544) labeled with Alexa-488 as a fluorophore. Notice the longer, more ribbonlike aggregate structure of the truncated protein.

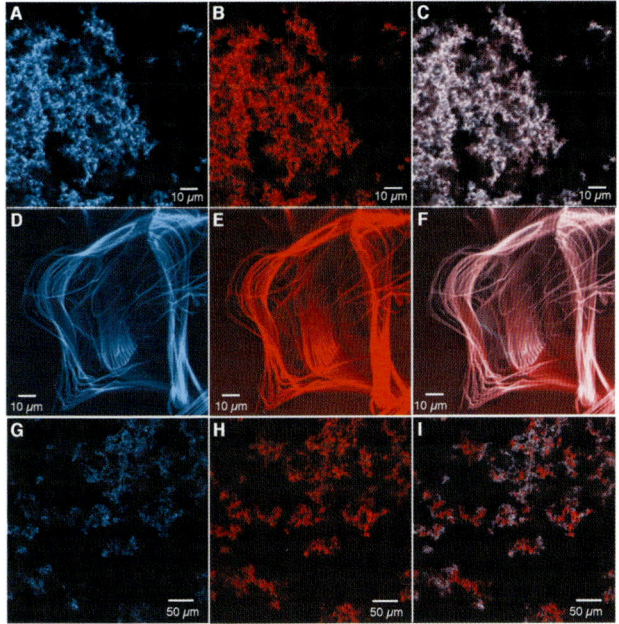

FIGURE 9. Coaggregated tailspike protein: 100 μg/ml of labeled native tailspike protein was mixed with 100 μg/ml of either truncated tailspike protein or CAB at 37° C and a final concentration of 200 μg/ml. (A), (B), (C) Coaggregated full-length tailspike with truncated tailspike protein (residues 109-666). Labeled full-length tailspike is in A, labeled 109-666 in B, and a composite of both is seen in C. (D), (E), (F) Coaggregated full-length tailspike with truncated tailspike protein (residues 109-544). Labeled full-length tailspike is in D, labeled 109-544 in E, and a composite of both is seen in F. (G), (H), (I) Coaggregated full-length tailspike with CAB (10X tailspike concentration). Labeled full-length tailspike is in G, labeled CAB in H, and a composite of both is seen in I. Note that the CAB and tailspike do not colocalize very well, while the full-length tailspike colocalizes with each of the truncations very well.

Factors Affecting the Fibrillation of α-Synuclein, a Natively Unfolded Protein

Anthony L. Fink[1]

Substantial evidence has accumulated in the last few years that the aggregation/fibrillation of α-synuclein is a critical factor in Parkinson's disease (PD). α-Synuclein is a major component of intracellular inclusions known as Lewy bodies which are the pathological hallmark of PD. Mutations of α-synuclein have been associated with rare cases of familial PD. α-Synuclein is a natively unfolded protein of unknown function. A partially folded intermediate has been shown to be critical for α-synuclein fibrillation. A variety of endogenous and exogenous factors affect the kinetics of α-synuclein fibrillation, leading to both acceleration and inhibition. This chapter reviews some of these factors and the potential mechanisms that underlie their effects. Among the factors discussed are pesticides; metals; polycations, including histones; glycosaminoglycans (GAGs), such as heparin; lipids and membranes; macromolecular crowding; and oxidation. Acceleration typically arises from conditions that increase the concentration of the critical amyloidogenic intermediate. Inhibition typically arises from situations in which either the monomer or nonfibrillogenic oligomers are stabilized. The wide range of factors that cause acceleration of fibrillation suggests that there may be cellular chaperones or other cellular agents that normally prevent the aggregation of α-synuclein. Many of the findings concerning the aggregation of α-synuclein are applicable to other proteins that are prone to aggregate.

1. INTRODUCTION

α-Synuclein is a relatively abundant brain protein of 140 amino acids and of unknown function, which belongs to the class of proteins known as natively unfolded, i.e., the

---

Department of Chemistry and Biochemistry, University of California at Santa Cruz, Santa Cruz, CA 95064
[1] To whom correspondence should be addressed: Anthony Fink, Department of Chemistry and Biochemistry, University of California at Santa Cruz, Santa Cruz, CA 95064; email: enzyme@ucsc.edu

purified protein at neutral pH is disordered and lacks detectable secondary or tertiary structure.[1,2] No information currently is available regarding the conformation of α-synuclein *in vivo*; however, it is likely that it is folded (or at least partially folded) when it interacts with its binding partners and membranes. A large fraction of eukaryotic proteins, especially those involved in cell cycle regulation, are believed to be intrinsically disordered until they bind their target ligands, such as DNA. Substantial evidence implicates the aggregation of α-synuclein as a key step in the development of Parkinson's disease.[3] α-Synuclein has been shown to be present in high concentration at presynaptic terminals[4] and to bind to membrane vesicles.[5-13]

The cause of Parkinson's disease (PD) is unknown, but considerable evidence suggests a multifactorial etiology involving genetic and environmental factors. PD is the second most common neurodegenerative disorder after Alzheimer's disease, affecting an estimated 1.5 million people in the United States. The two pathological hallmarks of PD are the loss of dopaminergic neurons in the substantia nigra region of the brain and the presence of intracellular inclusions, Lewy bodies (LBs), and Lewy neurites in the surviving neurons. The main symptoms of PD arise from neuromuscular dysfunction, due to the low levels of the neurotransmitter dopamine in the striatum. Recent work has shown that, except in extremely rare cases, there appears to be no direct genetic basis.[14] However, several studies have implicated environmental factors, especially pesticides and metals,[15-24] as well as oxidative stress.[25-28]

Extensive data support the hypothesis that "pathological" or abnormal aggregation[29] arises from a key partially folded intermediate precursor. Such intermediates have sizable nonpolar patches (i.e., contiguous hydrophobic side chains) on their surface, which lead to hydrophobic interactions between molecules, resulting in specific intermolecular interactions and aggregation. These hydrophobic patches are absent in the fully unfolded state. Factors that increase the concentration of such intermediates will favor aggregation. Protein aggregation is a very complex process, typically with multiple soluble/insoluble states, multiple conformations, multiple intermediates, kinetic competition between multiple pathways, competition between fibrillar and amorphous aggregates, and multiple filamentous states (e.g., protofilaments, protofibrils, fibrils). α-Synuclein forms a partially folded intermediate whose population correlates with increased formation of fibrils.[30] Both fibrillar and amorphous aggregates may be formed from α-synuclein.

Although the cause of sporadic PD is unknown, a positive correlation between the prevalence of PD and industrialization has been recognized[15] and this disorder is now considered likely to be an "environmental" disease. In fact, as noted above, several studies have implicated such environmental factors as pesticides, herbicides, and heavy metals in the origin of PD.[15-24,31-34] It appears that the multifactorial basis of PD is due to a combination of genetic susceptibility and environmental factors and arises from the common pathway of α-synuclein aggregation (Figure 1).[35]

This chapter summarizes some of the evidence that indicates that a wide variety of factors affect the fibrillation of α-synuclein *in vitro* and also may have similar effects *in vivo*, and thus are potential contributors to the disease. Knowledge of how such factors affect the rate of α-synuclein fibrillation also should shed light about the underlying molecular mechanism of α-synuclein aggregation.

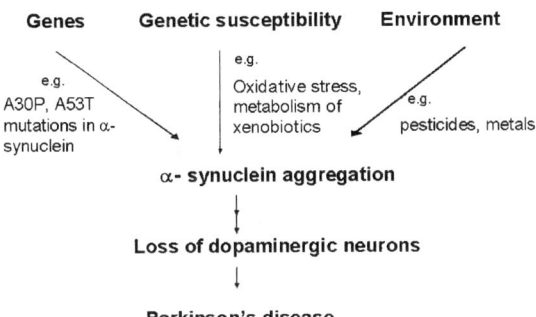

FIGURE 1. Proposed mechanism for environmental factors and genetic susceptibility leading to α-synuclein aggregation and Parkinson's disease.

2. MULTIPLE FACTORS AFFECT THE RATE OF α-SYNUCLEIN FIBRILLATION

Recent investigations suggest that many factors can significantly accelerate the rate of α-synuclein fibrillation.[13,30,36–43] Conversely, many factors can inhibit the rate of fibrillation.[12,16,37,44,45] The large number of factors that accelerate α-synuclein fibrillation suggests that there might be an intrinsic mechanism to prevent α-synuclein from aggregating under normal conditions and that it is possible that it is the loss of some inhibitor of aggregation (such as a chaperone) that may go awry in Parkinson's disease.

Before going into details, a few definitions are in order. We define aggregation as the abnormal self-association of proteins, leading either to soluble oligomers or insoluble deposits. We define fibrillation or fibril formation as the formation of filamentous, ordered polymers, usually referred to as amyloid fibrils. Protein fibrils contain a core of cross-β-sheet structure in which β-strands run perpendicular to the axis of the fibril. Fibrils are usually 7–12 nm in diameter and up to a few microns in length. The thinnest forms are smooth, often designated as protofilaments, whereas wider fibrils often show a twisted or ropelike morphology, reflecting several (usually 2–6) protofilaments wrapped around each other. By morphology we mean the overall macroscopic appearance of the deposits, as evidenced by electron microscopy, for example. Acceleration of fibrillation is used in the sense of an increased rate of formation of fibrils. Inhibition of fibrillation means both slower kinetics of fibril formation as well as fewer fibrils (i.e., less protein in the form of fibrils) being formed (usually the two go together). A feature of protein fibrillation is that the length of the fibrils is normally very heterogeneous.

2.1. Properties of α-Synuclein

α-Synuclein is an intrinsically unstructured (natively unfolded) protein. Analysis of the amino acid sequences of small globular folded proteins and "natively unfolded" proteins indicate that "natively unfolded" proteins result from a combination of low overall hydrophobicity and large net charge.[2] The N-terminal domain of α-synuclein consists of seven repeats of 11 residues predicted to form amphiphilic helices and a C-terminal

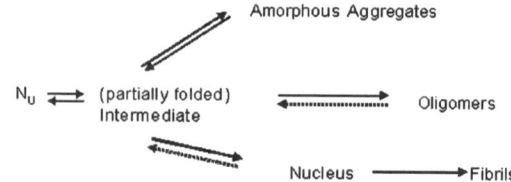

SCHEME 1.

tail enriched in acidic residues and proline and predicted to remain disordered. The pI is 4.7, yielding a protein with a large net negative charge at neutral pH. On binding to membranes composed of phospholipids with acidic head groups the N-terminal domain of α-synuclein adopts a helical conformation.

The molecular mechanisms underlying α-synuclein aggregation remain unknown; however, extant data suggest a minimum kinetics scheme of the sort shown in Scheme 1, where N_U is the monomeric natively unfolded state. Interestingly, as noted, secondary structure prediction schemes and experimental data show that the N-terminal domain is potentially helical and the C-terminal domain is disordered. Thus, the transformation to β-structure in the fibrils is perhaps surprising and is presumably driven by the thermodynamic stability of the fibrils.

2.2. Factors that Accelerate α-Synuclein Fibrillation

A number of factors have been identified that lead to significant acceleration in the rate of α-synuclein fibrillation *in vitro*. Some, such as increasing α-synuclein concentration, are readily explicable, in this case since aggregation involves self-association with at least a bimolecular kinetic step. That elevated levels of α-synuclein may be a cause of PD is bolstered by the recent reports that one form of familial early onset PD arises from triplication of the gene for α-synuclein.[46,47]

The kinetics of fibrillation, as well as aggregation in general, are usually attributed to a nucleated polymerization type of process and are manifested as sigmoidal kinetic curves, characterized by an initial lag followed by exponential growth and final leveling off (see, for example, Figure 2). The initial lag is attributed to the formation of a critical amyloidogenic nucleus, which once formed is followed by rapid fibril elongation/growth. Formation of the nucleus is associated with the lag seen in Figure 2 and the fibril elongation phase is manifested by the subsequent exponential increase in thioflavin T (ThT) signal in Figure 2. Thus, acceleration of fibrillation may be manifested by a decrease in lag time, an increase in elongation rate, or both. With increasing α-synuclein concentration the lag time decreases and the rate of elongation increases, both showing first-order dependence on the protein concentration.[36] The ThT fluorescence signal reflects the total amount of fibrillar material. ThT fluorescence is dramatically increased by ThT binding to fibrils. Very much smaller increases are sometimes observed on binding to oligomeric or partially folded intermediates. In many, but not all cases, the intensity of the ThT fluorescence signal is proportional to the amount of fibrils.

Some of the observed factors that accelerate α-synuclein fibrillation have the potential to be important in the pathogenesis of the disease, whereas others do not. Examples

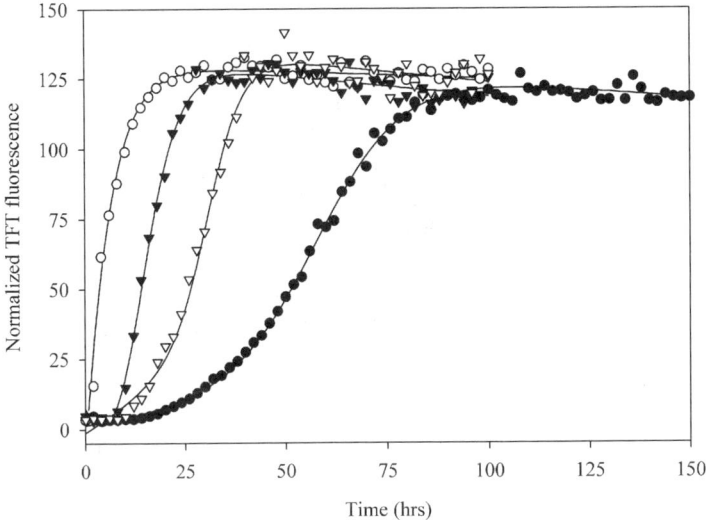

FIGURE 2. Effect of metals on the kinetics of fibrillation of α-synuclein. Fibrillation monitored by thioflavin T fluorescence; 0.5 mg/ml α-synuclein was incubated at pH 7.5, 37° C. Solid circles = control; open triangles = $CuCl_2$; close triangles = $MgCl_2$; open circles = $PbCl_2$. All metals at 500 μM.

of the latter include agitation, low pH, and temperatures >40° C.[30] Agitation of solutions of α-synuclein incubated at 37° C leads to dramatic increases in the rate of fibrillation, decreasing the time for observing fibrils from many weeks/months to hours in in vitro systems. The underlying molecular basis for this effect is unclear, but most likely involves effects of the air-water interface leading to partial folding of α-synuclein and formation of the critical amyloidogenic intermediate. If this is the case, then it is likely that comparable intracellular environments, such as hydrophobic surfaces, would have similar effects. As discussed subsequently, membrane surfaces can lead to acceleration of α-synuclein fibrillation under some circumstances.[13]

2.2.1. Metal ions

Occupational exposure to specific metals, especially manganese, copper, lead, iron, mercury, zinc, and aluminum, appears to be a risk factor for Parkinson's disease based on clinical and epidemiological studies.[21,22,31,32,48–56] Elevated levels of several of these metals have been reported in the substantia nigra of PD subjects. A correlation with increased incidence of PD has been reported for individuals with mercury amalgam fillings[55] and in welders.[57,58] There are a large number of possible ways in which toxic metal exposure could lead to increased risk of PD. Perhaps the simplest would be a direct effect of the metal on the aggregation of α-synuclein. Support for this hypothesis comes from investigations of the effect of various metals on inducing conformational changes and affecting the kinetics of fibrillation of α-synuclein in vitro, which demonstrated that metals not only can induce partial folding of α-synuclein, but also substantially can increase the rate of fibrillation.[36,59]

For example, di- and trivalent metal ions caused significant accelerations in the rate of α-synuclein fibril formation. Aluminum was the most effective, along with copper (II), iron (III), cobalt (III), mercury (II), lead (II), and manganese (II).[36,59] Some typical data are shown in Figure 2. The effectiveness correlated with increasing ion charge density. A correlation was noted between efficiency in stimulating fibrillation and inducing a conformational change, ascribed to formation of the key amyloidogenic partially folded intermediate. The mechanism for this presumably involves the metal ions binding to the negatively charged carboxylates, thus masking the electrostatic repulsion and facilitating collapse to the partially folded conformation. Studies on the effect of increasing concentration of various salts suggests that the Debye-Hückel effect occurs at low concentrations and that ion effects at higher concentrations probably reflect effects involving competition between interactions involving water, ions, and protein.[60]

The potential for ligand bridging by polyvalent metal ions was proposed to be an important factor in the metal-induced conformational changes of α-synuclein. The results indicate that low concentrations of some metals can directly induce α-synuclein fibril formation.[36,59] For the interaction of α-synuclein with copper II binding sites were found with dissociation constants of ≤5 μM and ~ 100 μM. The findings also revealed that much lower metal concentrations are required for fibrillation at higher α-synuclein concentrations: since the cellular levels of α-synuclein could be hundreds of micromolar, it is probable that the *in vitro* observations are germane to the *in vivo* situation. Calcium has been reported to bind to α-synuclein and be important in its interaction with lipids.[61]

2.2.2. *Pesticides* Parkinsonism has been associated with long-term occupational exposure to pesticides as well as farming, rural living, well-water drinking, and exposure to agricultural chemicals. A large number of epidemiological studies have shown increased risk of PD in populations exposed to pesticides.[62–65] Specific pesticides that have been implicated include paraquat, organochlorine compounds, dieldrin, 1,1'-(2,2-dichloroethenyl diene)-*bis*(4-chlorobenzene) (p,p-DDE), hexachlorocyclohexane (lindane), and rotenone. *In vitro* studies have shown that several commonly used pesticides, including rotenone, dieldrin, and paraquat, induce a conformational change in α-synuclein and significantly accelerate the rate of formation of α-synuclein fibrils.[39] The simplest hypotheses for the association of pesticides with PD include that pesticides are directly toxic to mitochondria, or modulate xenobiotic metabolism, or adversely affect proteasomal function, or directly affect α-synuclein levels and fibrillation.

Support for the latter hypothesis comes from the observation that a number of pesticides, representing different chemical classes, directly stimulate the rate of formation of fibrils when incubated with α-synuclein at pH 7.5, 37° C.[39] Of the pesticides examined, those having the most significant accelerating effects were rotenone, DDT, 2,4-D, dieldrin, diethyldithiocarbamate, paraquat, Maneb, Trifluralin, parathion, and imidazoldinethione. Some results are shown in Figure 3 for two other pesticides. A number of the pesticides examined had no significant effect on the kinetics of α-synuclein fibrillation. The effect of varying the concentration of pesticide was investigated for paraquat[66] and rotenone. In both cases increasing concentration of pesticide led to increased rates of fibrillation. For rotenone the lag times decreased from 75 hr for the control to 42 hr for 10 μM rotenone. The underlying mechanism is probably preferential binding of the relatively hydrophobic pesticides to the amyloidogenic partially folded intermediate of α-synuclein, accounting for the observed conformational changes, thereby increasing its population and leading to fibrillation.[39]

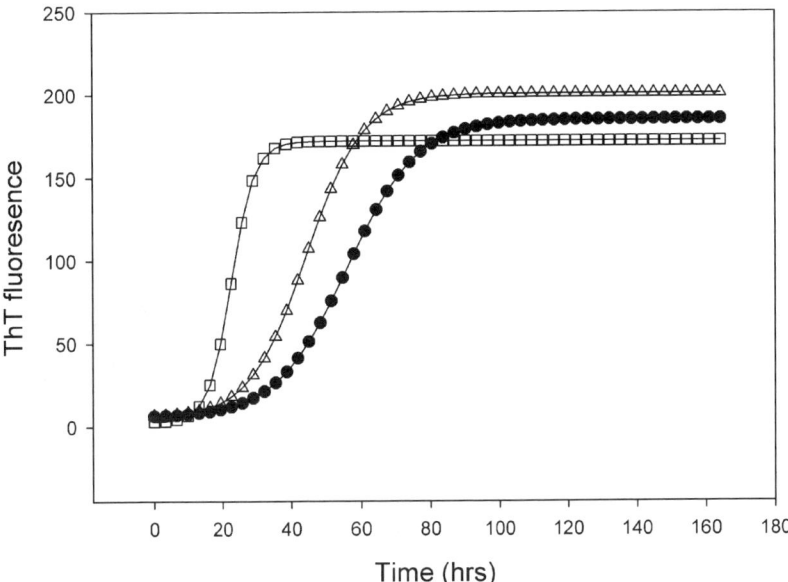

FIGURE 3. Effect of pesticides on the rate of fibrillation of α-synuclein. Experimental conditions were identical to those described in Figure 2. Solid circles = control; triangles = 100 μm dibrom; squares = 100 μM simazine. The data have been smoothed.

Since many individuals at risk owing to pesticide and/or metal exposure are expected to accumulate either both metals and pesticides, or multiple forms of metals or pesticides, in their brains, it is important to know whether the effects are additive. In fact, the simultaneous presence of Al(III) and the pesticide DDC led to conformational changes and fibril formation kinetics that were greater than expected from noncooperative effectors. These results suggest that it is the total neuron load of metals and xenobiotic chemicals, such as pesticides that is critical in determining the rate of physiological fibrillation, supporting the notion that environmental factors, in conjunction with genetic susceptibility, form the underlying molecular basis for idiopathic PD.[2]

Mice exposed to the herbicide paraquat developed α-synuclein deposits, indicating that the abovenoted *in vitro* results can be extrapolated to physiological conditions.[66] Thioflavin S-positive structures accumulated in neurons in the substantia nigra, and dual labeling with anti-α-synuclein antibodies and confocal imaging confirmed that these aggregates contained α-synuclein. In addition, transient up-regulation of α-synuclein was observed on exposure to paraquat, suggesting that both up-regulation of α-synuclein as a consequence of toxicant insult and direct interactions between the protein and environmental factors are potential mechanisms leading to α-synuclein pathology in neurodegenerative diseases.

2.2.3. *Glycosaminoglycans* Glycosaminoglycans (GAGs) routinely are found associated with amyloid deposits in most amyloidosis diseases, and there is evidence to support an active role of GAGs in amyloid fibril formation in some cases. In contrast to the extracellular amyloid deposits, the α-synuclein deposits in Lewy body diseases are

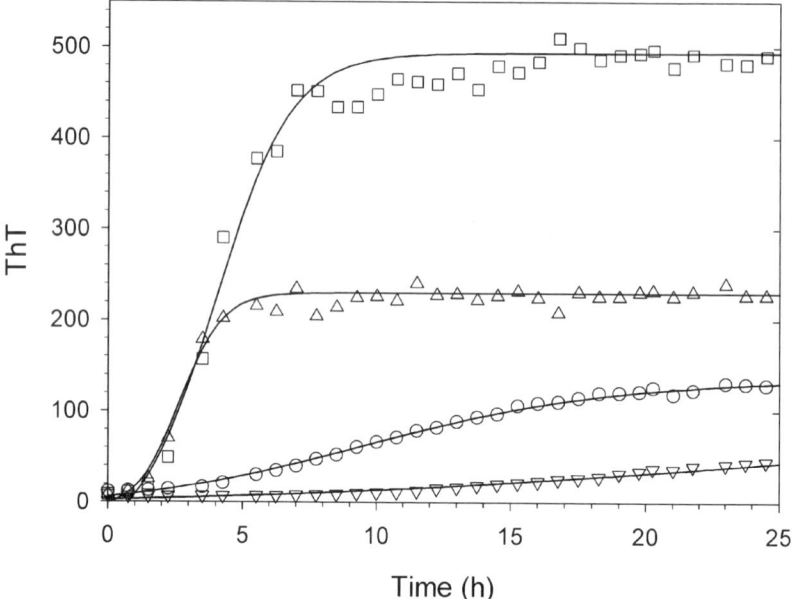

FIGURE 4. Effect of glycosaminoglycans on fibrillation of α-synuclein. Experimental conditions are identical to those described in Figure 2 except the protein concentration was 1 mg/ml. Inverted triangles = control; circles = keratin sulfate; triangles = heparin sulfate; squares = heparin.

intracellular, and thus it is less clear whether GAGs may be involved. *In vitro* studies show that certain GAGs (heparin, heparan sulfate) and other highly sulfated polymers (dextran sulfate) significantly stimulate the formation of α-synuclein fibrils (Figure 4).[40] The different final signals in ThT intensity may result from a number of possible sources, including competition between the GAG and ThT or a mixture of fibrils and nonfibrillar aggregates.

Heparin binding sites typically contain clusters of basic amino acid residues capable of binding to the negatively charged heparin polymer.[67] The N-terminal region of α-synuclein contains multiple repeats of the consensus sequence KTKEGV, so this region of α-synuclein presumably constitutes the GAG binding motif.[40] Interestingly, the interaction of GAGs with α-synuclein is quite specific, since some GAGs, e.g., keratan sulfate, had negligible effect. The molar ratio of heparin to α-synuclein and the incorporation of fluorescein-labeled heparin into the fibrils demonstrate that the heparin is integrated into the fibrils and is not just a catalyst for fibrillation. The apparent dissociation constant for heparin in stimulating α-synuclein fibrillation was 0.19 μM, indicating a strong affinity. Similar effects of heparin were observed with the A53T and A30P mutants of α-synuclein. Since there is some evidence that Lewy bodies may contain glycosaminoglycans,[68] these observations may be very relevant in the context of the etiology of PD.

2.2.4. Macromolecular crowding Intracellular proteins exist in a very crowded milieu, very different from that typically used to study them *in vitro*; for example, the protein concentration in cells is typically in the 150–300 mg/ml range. Macromolecular crowding is expected to have several significant effects on protein aggregation, with the major effects arising from excluded volume and increased viscosity. Consequently, macromolecular crowding could lead to a significant acceleration in the rate of protein aggregation and formation of amyloid fibrils. Various types of polymers, from neutral polyethylene glycols and polysaccharides (Ficolls, dextrans) to inert proteins, have been observed to dramatically accelerate the rate of α-synuclein fibrillation.[42,69] The stimulation of fibrillation increases with increasing length of polymer, as well as increasing polymer concentration. At lower polymer concentrations (typically up to ∼100 mg/ml) the major effect is ascribed to excluded volume, whereas at higher polymer concentrations evidence of opposing viscosity effects become apparent.[42]

In general, the rate of fibrillation in the presence of crowding agents was increased very substantially relative to the rate of fibrillation in the absence of such agents (Figure 5A). When fibrillation was studied as a function of the concentration of crowding agent added, the lag time initially decreased at low concentrations of crowding agent due to the excluded volume effect, then increased or remained relatively constant at high concentrations of crowding agent, reflecting the effect of the high viscosity of the crowding agent solution. Varying molecular weight PEGs were used to study the effects of different sizes of crowding agents on fibrillation. For similar concentrations of PEG, the fibrillation lag time appeared to follow similar kinetics to those found by varying the concentration of crowding agent, an initial decrease with low molecular weight PEG followed by an increase with high molecular weight PEG.[42]

As discussed previously, many metals can cause a significant acceleration of α-synuclein fibrillation under dilute solution conditions. Similarly, the presence of certain metals leads to acceleration of α-synuclein fibrillation in crowded milieu, such as high concentrations of BSA, Ficolls, and polyethyleneglycols (Figure 5B). The faster fibrillation in crowded environments in the presence of heavy metals suggests a simple molecular basis for the observed elevated risk of PD due to environmental exposure to metals.[70]

The very dramatic increases in rate of α-synuclein fibrillation in the presence of macromolecular crowding (up to three to four orders of magnitude decrease in lag time) suggests that in the cell there must be mechanisms to prevent α-synuclein aggregation from occurring.

2.2.5. *Polycations* A variety of polyamines (e.g., spermine, polylysine, polyarginine, and polyethylenimine) accelerate the fibrillation of α-synuclein.[41,71] Long-chain polyethylenimines (PEIs) and histones are effective in the nanogram/milliliter range, whereas shorter versions, e.g., spermidine, are less effective at such low concentrations.[41] The formation of α-synuclein-polycation complexes was not accompanied by significant structural changes in α-synuclein. However, α-synuclein fibrillation was dramatically accelerated in the presence of polycations. The magnitude of the accelerating effect depended on the nature of the polymer, its length and concentration, and suggests a potential critical role of electrostatic interactions in which neutralization of the net negative charge on the protein leads to collapse to the partially folded amyloidogenic intermediate and thence fibrillation.[41]

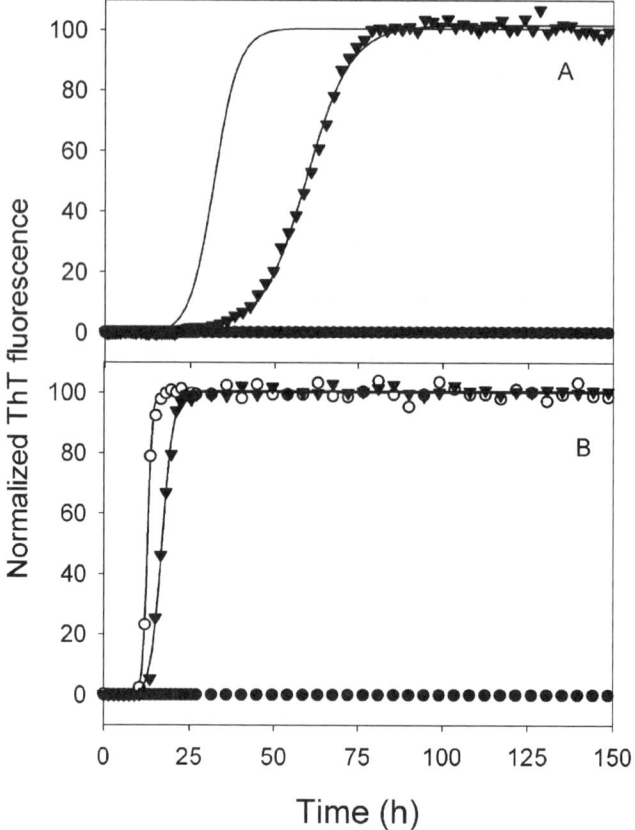

FIGURE 5. Effect of macromolecular crowding on α-synuclein fibrillation. (A) PEG-3350 (solid line) and Ficoll-400000 (filled triangles), at a concentration of 100 mg/ml dramatically accelerates the rate of α-synuclein fibrillation; filled circles represent fibrillation of α-synuclein in the absence of crowding agents. These experiments were done with limited agitation, hence the long lag time in the absence of crowding agents. (B). Even in the presence of crowding agents (100 mg/ml PEG-3350), metals [500 μM Zn^{2+} (open circles), 500 μM Pb^{2+} (black triangles)] still cause an additional acceleration of α-synuclein fibrillation. Black circles represent control experiments.

Histones form a specific complex with α-synuclein and are found in the fibrils in the same (1:2) stoichiometry.[72] α-Synuclein also aggregates and colocalizes with histones in the nuclei of nigral neurons from mice exposed to a toxic insult (i.e., injections of the herbicide paraquat).[72] The significance of these observations remains to be determined, but the interaction with histones suggests a possible role in gene regulation: the "nuclein" part of α-synuclein comes from the fact that it was initially detected in the nucleus. These observations indicate that translocation of α-synuclein into the nucleus and binding with histones could play a role in α-synuclein pathophysiology. The effects of short polyamines such as spermidine in accelerating α-synuclein fibrillation also may be physiologically important, since significant concentrations of these species are present in certain regions of the cell.

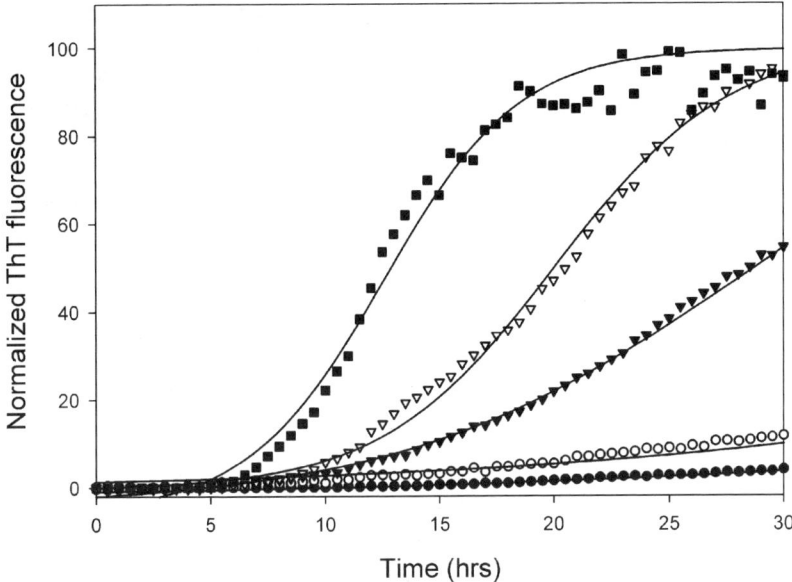

FIGURE 6. Effect of salt concentration on the rate of α-synuclein fibrillation. Conditions were pH 7.4 (20 mM buffer), 37° C, with 0 mM (filled circles), 30 mM (open circles), 100 mM (filled triangles), 200 mM (open triangles), and 500 mM (filled squares) NaCl added.

2.2.6. High ionic strength and specific anions The rate of *in vitro* α-synuclein fibril formation increases dramatically with increasing ionic strength, reaching a maximal effect by 300 mM for NaCl (Figure 6).[37] The accelerating effect is attributed to increased population of the amyloidogenic intermediate, as anions induce partial folding of α-synuclein at neutral pH. Circular dichroism spectra show that at a concentration of 200 mM a variety of different salts induce comparable partial folding of α-synuclein. However, the kinetics of fibrillation varied significantly in the presence of different anions, indicating that specific anions differentially affect the nucleation and/or elongation stages of the fibrillation process. The major effect of anions on the conformation of α-synuclein occurred at quite low anion concentrations (<10 mM). This suggests two different types of effects of the anions. The magnitude of the accelerating effect of the anions generally follows their position in the Hofmeister series, indicating a major role of protein-water-anion interactions in the process at salt concentrations above 10 mM. Below this concentration, electrostatic effects dominated in the mechanism of anion-induced fibrillation. The acceleration of fibrillation by anions also is dependent on the cation. Since the concentration of certain anions may be quite high in selected spatial and temporal regions of cells, these changes could have significant physiological effects on the rate of α-synuclein fibrillation.

2.2.7. Low pH and elevated temperature Either a decrease in pH or an increase in temperature transforms α-synuclein into a partially folded conformation.[30] The presence of this intermediate is correlated strongly with the enhanced formation of α-synuclein fibrils. Plots of the rate of fibrillation against pH show a maximum at pH 3, which corresponds to the maximum population of the partially folded intermediate. The hydrodynamic size of α-synuclein is also a minimum at pH 3, corresponding to the increased packing density of the intermediate.[30] The effect of low pH is to protonate carboxylates, leading to a decrease in the net charge and allowing the intrinsic hydrophobicity to drive partial folding to the amyloidogenic intermediate. This is the same mechanism believed responsible for the effects of metals in accelerating α-synuclein fibrillation.

Elevated temperatures also lead to partial folding for the related reason that the hydrophobic interactions increase with increasing temperature, and thus favor collapse of the natively unfolded protein into the partially folded intermediate. These findings are in accord with the hypothesis that α-synuclein is natively unfolded at neutral pH due to its high net charge and low intrinsic hydrophobicity. Most importantly, however, the correlation between population of the partially folded intermediate and fibrillation demonstrates that the intermediate is a critical species for fibrillation. In fact, all factors or conditions that lead to accelerated α-synuclein fibrillation also lead to population of this partially folded intermediate.[30]

2.3. Factors that may Accelerate or Inhibit α-Synuclein Fibrillation

2.3.1. Membranes and lipids
Membranes and lipid oxidation are likely to be very important in the "life cycle" of α-synuclein. A lipid-binding motif is found in α-synuclein.[73] The reported interactions of α-synuclein with membranes and ensuing effects are somewhat contradictory, probably reflecting the different conditions used in the experiments.[5-13,74,75] Membranes have been reported to accelerate the fibrillation of α-synuclein and a recent report suggests that α-synuclein aggregation may occur on membrane surfaces and that membranes preferentially induce α-synuclein oligomers.[10] However, it also has been reported that α-synuclein binds tightly to neutral and anionic membranes and that the membranes inhibit fibrillation.[11] Recent studies suggest that the outcome of the interaction of α-synuclein with membranes is very dependent on factors such as the relative concentrations of α-synuclein and lipids, nature of the vesicles, nature of the head groups, and so on, at least partly explaining the contradictory reports.[12,13]

In well-defined *in vitro* experiments it has been shown that the presence of vesicles may lead to either inhibition or acceleration of α-synuclein fibrillation (Figure 7), depending on a variety of factors such as the ratio of α-synuclein/lipid, the nature of the lipid, and the nature of the vesicles (LUV vs. SUV, entrapped vs. nonentrapped).[12,13] Significant changes in the conformation of α-synuclein were noted on interaction with lipids. Conditions that lead to substantial helical structure, such as an excess ratio of lipid to protein by mass, result in inhibition of fibrillation. There also is evidence that intermediates in α-synuclein fibrillation can make vesicles leaky.[13,76]

The extant data from *in vivo* and *in vitro* studies indicate that α-synuclein can interact with membranes, leading to either acceleration or inhibition of fibrillation and that oligomers formed during aggregation can cause membrane permeability. The latter observation suggests a simple explanation for the potential cytotoxicity of α-synuclein.

2.3.2. Alcohols
Although it is unlikely that alcohol levels will reach sufficient levels in neurons to affect α-synuclein conformation and aggregation *in vivo*, studies on the effects

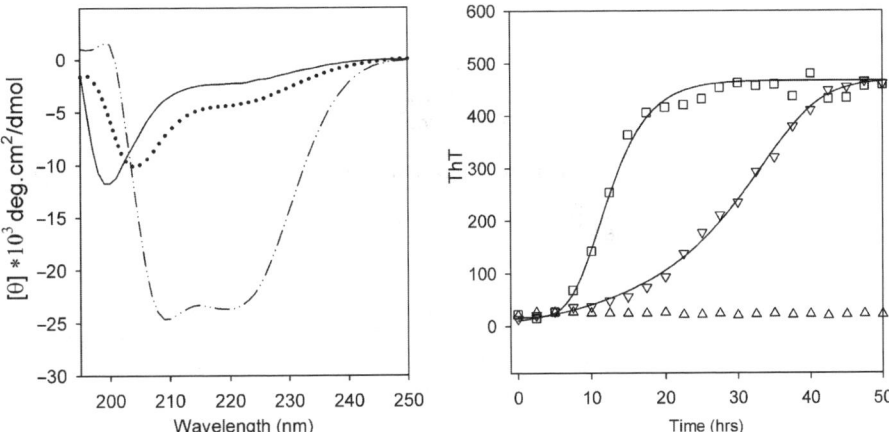

FIGURE 7. Effect of membrane vesicles on α-synuclein conformation (left) and fibrillation (right). Small unilamellar vesicles of phosphatidyl choline lipids were incubated with α-synuclein at a 1:5 mass ratio (dotted line, open squares) or a 5:1 ratio (dot-dash line, triangles). The controls (no vesicles) are shown by the solid line and inverted triangles. At low lipid:protein ratios, α-synuclein forms the amyloidogenic partially folded intermediate leading to accelerated fibrillation, whereas at high lipid levels α-synuclein forms a helical conformation that does not fibrillate.

of alcohols on α-synuclein have been useful in establishing the existence of previously unknown relatively stable conformations.

An investigation of the effects of simple and fluorinated alcohols on the structural and fibrillation properties of α-synuclein revealed that both acceleration and inhibition of fibrillation could occur, depending on the nature of the alcohol and its concentration.[37] All the solvents studied induced folding of α-synuclein, with the common first stage being formation of the amyloidogenic partially folded intermediate, leading to accelerated fibrillation, in some cases at <1% alcohol. However, fibrillation was completely inhibited at high concentrations of organic solvents due to formation of β-structure-enriched oligomers with methanol, ethanol, propanol, and moderate concentrations of trifluoroethanol (TFE), or because of the appearance of a highly α-helical conformation at high TFE and hexafluoroisopropanol concentrations. To some extent, these conformational effects mimic those observed in the presence of phospholipid vesicles and can explain some of the observed effects of membranes on α-synuclein fibrillation. In addition, these studies mapped out the conformational space accessible to α-synuclein.[37]

2.3.3. *Osmolytes* Osmolytes increase the stability of proteins by increasing the free energy of the unfolded state. Thus the presence of an osmolyte would be expected to increase the population of the α-synuclein partially folded intermediate, and thus increase the rate of fibril formation. This was observed to be the case at low to moderate concentrations of osmolytes such as glycerol and trimethylamine-N-oxide (TMAO).[43] For example, concentrations of TMAO in the vicinity of 1 M induced partial folding of α-synuclein and significantly increased fibrillation. At higher concentrations (≥3 M) of TMAO, α-synuclein became tightly folded, with substantial helical character and formed oligomers. Fibrillation was inhibited at these high concentrations of TMAO.[43]

Thus, it is likely that the presence of osmolytes or other kosmotropes (protein stabilizers) in dopaminergic neurons will increase the rate of α-synuclein fibrillation.

2.4. Factors that Inhibit α-Synuclein Fibrillation

2.4.1. β- and γ-synuclein Two homologous proteins, β- and γ-synuclein, also are abundant in the brain. These molecules are highly homologous to α-synuclein, the most notable difference being the absence of a stretch of 11 residues near the middle of the N-terminal domain in β-synuclein. These synucleins also are natively unfolded proteins. β-Synuclein, in spite of its similarities to α-synuclein, does not form fibrils, even under forcing conditions such as at low pH.[45] γ-Synuclein, unlike its homologs, forms a soluble oligomer at relatively low concentrations, which appears to be an off-fibrillation pathway species, and probably accounts for the very much slower fibrillation than with α-synuclein.[45] Interestingly, even though they have similar biophysical properties to α-synuclein, β- and γ-synucleins inhibit α-synuclein fibril formation. Complete inhibition of α-synuclein fibrillation was observed at 4:1 molar excess of β- and γ-synucleins.[45] No significant incorporation of β-synuclein into the fibrils was detected. The lack of fibrils formed by β-synuclein is most readily explained by the absence of a stretch of hydrophobic residues from the middle region of the protein, which leads to preferential stabilization of a soluble oligomer. The inhibition of α-synuclein fibrillation by β-synuclein has been confirmed in *in vivo* experiments.[77] Thus, a decrease in the levels of β-synuclein could be a factor in elevated levels of α-synuclein fibrillation *in vivo*, and thus a factor in the etiology of PD.

2.4.2. Catecholamines Interestingly, dopamine and related catecholamines have been shown to inhibit α-synuclein aggregation.[78,79] The inhibitory activity of dopamine and L-dopa is attributed predominantly to their oxidation to quinone derivatives that covalently modify α-synuclein, leading to the accumulation of soluble oligomers (protofibrils). Since the nigral neurons that die in PD have high levels of dopamine, it is likely that any situation leading to a decrease in the antioxidant level of the neuron will lead to elevated levels of dopamine and other catecholamine oxidation products, and thus increased aggregation of α-synuclein.

2.4.3. Oxidation of methionine Oxidative stress has been implicated in the pathogenesis of PD.[80,81] Usually, reactive oxygen species (ROS) are rendered harmless by a combination of antioxidants, metal chelators, or enzymatic reduction. However, when the level of ROS exceeds some threshold, these defense mechanisms may fail and the cell may suffer from oxidative stress. All amino acids are susceptible to oxidation, although their susceptibilities vary greatly, with methionine being one of the most readily oxidized amino acid constituents of proteins.[82] Methionine is easily oxidized to methionine sulfoxide.

The effect of oxidation of the four methionine residues in α-synuclein to the sulfoxides has been investigated. Rather surprisingly, the fibrillation of α-synuclein at neutral pH was completely inhibited by methionine oxidation.[44] This appears to be due to the formation of a stable, off-pathway oligomer. Very significantly, the addition of small molar excesses of Met-oxidized α-synuclein to α-synuclein led to inhibition of fibrillation.

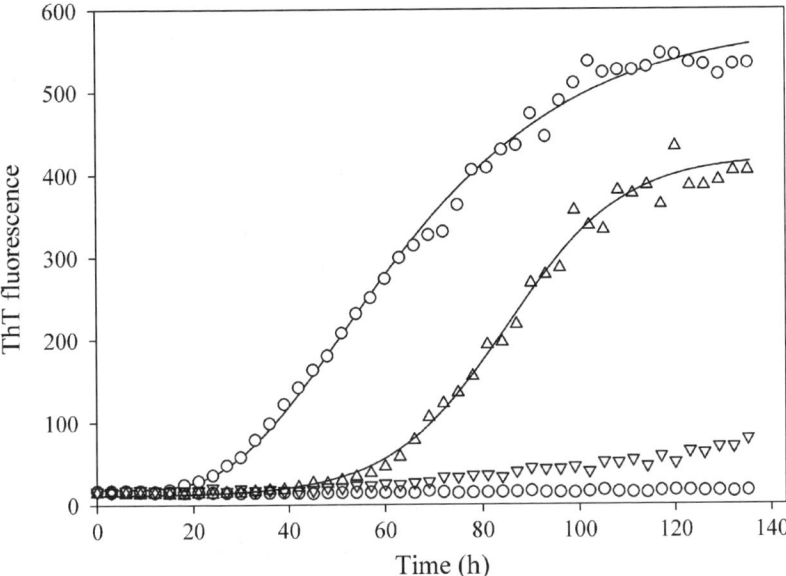

FIGURE 8. Effect of methionine-oxidized α-synuclein on fibrillation of unmodified α-synuclein. Control (unmodified, top curve, open circles), 1:2 with oxidized protein (triangles), 1:4 with oxidized protein (inverted triangles), and Met-oxidized α-synuclein (bottom curve, open circles).

A fourfold molar excess of the oxidized protein was sufficient to completely inhibit α-synuclein fibril formation (Figure 8).[44] Furthermore, in the mixtures of nonoxidized and oxidized proteins, methionine-oxidized α-synuclein affected primarily the duration of lag-time and did not affect the elongation time. Thus, interaction of methionine-oxidized α-synuclein with the nonoxidized protein inhibited the nucleation process but not the elongation of fibrils.[44] Both oxidized and nonoxidized α-synucleins were natively unfolded under conditions of neutral pH, with the oxidized protein being slightly more disordered. Both proteins adopted identical partially folded conformations under conditions of acidic pH.

The degree of inhibition of fibrillation by MetO α-synuclein has been shown to be proportional to the number of oxidized methionines. This was demonstrated by selectively converting Met residues into Leu, prior to Met oxidation. The results showed that with one oxidized Met the kinetics of fibrillation were comparable to those for the control (nonoxidized), and with increasing numbers of methionine sulfoxides the kinetics of fibrillation became progressively slower.

The effect of several metals on the structural properties of methionine-oxidized human α-synuclein and its propensity to fibrillate also has been investigated.[59] The presence of the metals induced partial folding of both oxidized and nonoxidized α-synucleins. Although the fibrillation of α-synuclein was inhibited completely by methionine-oxidation, the presence of certain specific metals (Ti^{3+}, Zn^{2+}, Al^{3+}, and Pb^{2+}) overcame this inhibition and led to accelerated fibrillation. These findings indicate that a combination of oxidative stress and environmental metal pollution could play an important role in triggering the fibrillation of α-synuclein, and hence possibly PD.

2.4.4. Nitrated α-synuclein A variety of evidence also implicates nitrative stress in PD, and nitrated α-synuclein has been identified in Lewy bodies.[26,80,83–86] *In vivo* nitration of many proteins occurs due to the presence of peroxynitrite formed from the oxidation of nitric oxide. An examination of the effect of nitration on the propensity of α-synuclein to fibrillate *in vitro,* however, showed that fibril formation of α-synuclein at neutral pH was completely inhibited by nitration due to the formation of stable soluble oligomers (apparently octamers).[87] The nitrated α-synuclein also inhibits fibrillation of unmodified α-synuclein *in vitro,* suggesting that the stable oligomers are located off the fibrillation pathway. This inhibition of fibrillation was observed even in the presence of significant substoichiometric concentrations of nitrated α-synuclein. These observations suggest that nitration of soluble α-synuclein may be a protective factor in PD rather than a causative one and that the observation of nitrated α-synuclein in Lewy bodies may reflect nitration of the protein already in the fibrillar form.

2.5. Generalization of α-Synuclein Findings to other Proteins

In general, it is likely that many of the factors that affect the aggregation of α-synuclein apply to other proteins. For example, agitation, low pH, high temperature, macromolecular crowding, and destabilizing mutations appear to be very general factors that accelerate the aggregation or fibrillation of most proteins. Other agents such as GAGs and low concentrations of alcohols also stimulate the aggregation of most proteins.

3. CONCLUSIONS

The focus of this chapter has been on studies, especially *in vitro*, that show that a significant number of factors, both those found naturally in neurons and others, such as environmental contaminants, can affect substantially the rate of α-synuclein fibrillation. In particular, the apparent large number of diverse factors that can accelerate the rate of α-synuclein aggregation/fibrillation, in some cases very dramatically, is a striking finding, which raises the critical question of how dopaminergic neurons are able to minimize α-synuclein aggregation under normal conditions. It would seem that there must be a combination of factors that minimize its aggregation, ranging from identified agents such as β-synuclein, whose concentration is believed to be higher than that of α-synuclein[88,89] to as yet unidentified putative chaperones. A limited amount of data suggests that many compounds may affect α-synuclein aggregation in an additive or synergistic fashion. Thus, the presence of several compounds at low concentrations may have a major effect on α-synuclein aggregation, far greater than that of each individual component. The factors that accelerate and inhibit α-synuclein fibrillation are listed in Table 1. Many of the factors that cause inhibition of α-synuclein fibrillation *in vitro* do so by leading to the accumulation of the off-pathway oligomers. On the other hand,

TABLE 1. Factors that affect the rate of α-synuclein *in vitro*. An asterisk indicates that both acceleration and inhibition are observed, depending on the specific conditions

Factors that accelerate fibrillation	Factors that inhibit fibrillation
Agitation	Alchohols*
Alcohols*	Lipids*
Certain metals	Methionine oxidation
Certain pesticides	Miscellaneous inhibitory compounds
GAGs	Tyrosine nitration
High ionic strength	Osmolytes*
High temperature	β-synuclein, γ-synuclein
Histones	
Lipids and membranes*	
Low pH	
Molecular crowding	
Osmolytes*	
Other soluble, nonpolar compounds	
Polyamines	
A53T mutation	
High α-synuclein concentration	

factors that accelerate α-synuclein fibrillation tend to lead to increased population of the amyloidogenic intermediate.

REFERENCES

1. P. H. Weinreb, W. Zhen, A. W. Poon, K. A. Conway, and P. T. Lansbury, Jr., NACP, a protein implicated in Alzheimer's disease and learning, is natively unfolded, *Biochemistry* **35**, 13709–13715 (1996).
2. V. N. Uversky, J. R. Gillespie, and A. L. Fink, Why are "natively unfolded" proteins unstructured under physiologic conditions? *Proteins* **41**, 415–427 (2000).
3. J. Q. Trojanowski and V. M. Lee, Parkinson's disease and related alpha-synucleinopathies are brain amyloidoses, *Ann. N. Y. Acad. Sci.* **991**, 107–110 (2003).
4. M. C. Irizarry, T. W. Kim, M. McNamara, R. E. Tanzi, J. M. George, D. F. Clayton, and B. T. Hyman, Characterization of the precursor protein of the non-A beta component of senile plaques (NACP) in the human central nervous system, *J. Neuropathol. Exp. Neurol.* **55**, 889–895 (1996).
5. L. Maroteaux, J. T. Campanelli, and R. H. Scheller, Synuclein: a neuron-specific protein localized to the nucleus and presynaptic nerve terminal, *J Neurosci* **8**, 2804–2815 (1988).
6. A. Iwai, E. Masliah, M. Yoshimoto, N. Ge, L. Flanagan, H. A. de Silva, A. Kittel, and T. Saitoh, The precursor protein of non-A beta component of Alzheimer's disease amyloid is a presynaptic protein of the central nervous system, *Neuron* **14**, 467–475 (1995).
7. N. B. Cole, D. D. Murphy, T. Grider, S. Rueter, D. Brasaemle, and R. L. Nussbaum, Lipid droplet binding and oligomerization properties of the Parkinson's disease protein alpha-synuclein, *J. Biol. Chem.* **277**, 6344–6352 (2002).
8. W. S. Davidson, A. Jonas, D. F. Clayton, and J. M. George, Stabilization of alpha-synuclein secondary structure upon binding to synthetic membranes, *J. Biol. Chem.* **273**, 9443–9449 (1998).
9. E. Jo, J. McLaurin, C. M. Yip, P. George-Hyslop, and P. E. Fraser, alpha-Synuclein membrane interactions and lipid specificity, *J. Biol. Chem* **275**, 34328–34334 (2000).

10. H. J. Lee, C. Choi, and S. J. Lee, Membrane-bound alpha-synuclein has a high aggregation propensity and the ability to seed the aggregation of the cytosolic form, *J. Biol. Chem.* **277**, 671–678 (2002).
11. V. Narayanan and S. Scarlata, Membrane binding and self-association of alpha-synucleins, *Biochemistry* **40**, 9927–9934 (2001).
12. M. Zhu and A. L. Fink, Lipid binding inhibits alpha-synuclein fibril formation, *J. Biol. Chem.* **278**, 16873–16877 (2003).
13. M. Zhu, J. Li, and A. L. Fink, The association of alpha-synuclein with membranes affects bilayer structure, stability, and fibril formation, *J. Biol. Chem.* **278**, 40186–40197 (2003).
14. C. M. Tanner, R. Ottman, S. M. Goldman, J. Ellenberg, P. Chan, R. Mayeux, and J. W. Langston, Parkinson disease in twins: an etiologic study, *J. Am. Assoc. Med.* **281**, 341–346 (1999).
15. C. M. Tanner, The role of environmental toxins in the etiology of Parkinson's disease, *Trends Neurosci.* **12**, 49–54 (1989).
16. M. Hashimoto, L. J. Hsu, Y. Xia, A. Takeda, A. Sisk, M. Sundsmo, and E. Masliah, Oxidative stress induces amyloid-like aggregate formation of NACP/alpha-synuclein *in vitro*, *Neuroreport* **10**, 717–721 (1999).
17. C. Hertzman, M. Wiens, B. Snow, S. Kelly, and D. Calne, A case-control study of Parkinson's disease in a horticultural region of British Columbia, *Mov. Disord.* **9**, 69–75 (1994).
18. E. C. Hirsch, J. P. Brandel, P. Galle, F. Javoy-Agid, and Y. Agid, Iron and aluminum increase in the substantia nigra of patients with Parkinson's disease: an X-ray microanalysis, *J. Neurochem.* **56**, 446–451 (1991).
19. P. F. Good, C. W. Olanow, and D. P. Perl, Neuromelanin-containing neurons of the substantia nigra accumulate iron and aluminum in Parkinson's disease: a LAMMA study, *Brain Res.* **593**, 343–346 (1992).
20. M. Yasui, T. Kihira, and K. Ota, Calcium, magnesium and aluminum concentrations in Parkinson's disease, *Neurotoxicology* **13**, 593–600 (1992).
21. J. M. Gorell, B. A. Rybicki, J. C. Cole, and E. L. Peterson, E. L. Occupational metal exposures and the risk of Parkinson's disease, *Neuroepidemiology* **18**, 303–308 (1999).
22. J. M. Gorell, C. C. Johnson, B. A. Rybicki, E. L. Peterson, G. X. Kortsha, G. G. Brown, and R. J. Richardson, Occupational exposure to manganese, copper, lead, iron, mercury and zinc and the risk of Parkinson's disease, *Neurotoxicology* **20**, 239–247 (1999).
23. W. Hellenbrand, H. Boeing, B. P. Robra, A. Seidler, P. Vieregge, P. Nischan, J. Joerg, W. H. Oertel, E. Schneider, and G. Ulm, Diet and Parkinson's disease. II: A possible role for the past intake of specific nutrients. Results from a self-administered food-frequency questionnaire in a case-control study, *Neurology* **47**, 644–650 (1996).
24. E. Altschuler, Aluminum-containing antacids as a cause of idiopathic Parkinson's disease, *Med. Hypotheses* **53**, 22–23 (1999).
25. D. A. Butterfield and J. Kanski, Brain protein oxidation in age-related neurodegenerative disorders that are associated with aggregated proteins, *Mech. Ageing Dev.* **122**, 945–962 (2001).
26. H. Ischiropoulos and J. S. Beckman, Oxidative stress and nitration in neurodegeneration: cause, effect, or association? *J. Clin. Invest.* **111**, 163–169 (2003).
27. S. Kanda, J. F. Bishop, M. A. Eglitis, Y. Yang, and M. M. Mouradian, Enhanced vulnerability to oxidative stress by alpha-synuclein mutations and C-terminal truncation, *Neuroscience* **97**, 279–284 (2000).
28. Y. Zhang, V. L. Dawson, and T. M. Dawson, Oxidative stress and genetics in the pathogenesis of Parkinson's disease, *Neurobiol. Dis.* **7**, 240–250 (2000).
29. A. L. Fink, Protein aggregation: folding aggregates, inclusion bodies and amyloid, *Folding & Design* **3**, 9–15 (1998).
30. V. N. Uversky, J. Li, and A. L. Fink, Evidence for a partially folded intermediate in alpha-synuclein fibril formation, *J. Biol. Chem.* **276**, 10737–10744 (2001).
31. B. A. Rybicki, C. C. Johnson, J. Uman, and J. M. Gorell, Parkinson's disease mortality and the industrial use of heavy metals in Michigan, *Mov. Disord.* **8**, 87–92 (1993).

32. J. M. Gorell, C. C. Johnson, B. A.. Rybicki, E. L. Peterson, G. X. Kortsha, G. G. Brown, and R. J. Richardson, Occupational exposures to metals as risk factors for Parkinson's disease, *Neurology* **48**, 650–658 (1997).
33. K. L. Kirkey, C. C. Johnson, B. A. Rybicki, E. L. Peterson, G. X. Kortsha, and J. M. Gorell, Occupational categories at risk for Parkinson's disease, *Am. J. Ind. Med.* **39**, 564–571 (2001).
34. D. G. Le Couteur, A. J. McLean, M. C. Taylor, B. L. Woodham, and P. G. Board, Pesticides and Parkinson's disease, *Biomed. Pharmacother.* **53**, 122–130 (1999).
35. V. N. Uversky, J. Li, M. Zhu, K. Bower, and A. L. Fink, Synergistic effects of pesticides and metals on the fibrillation of alpha-synuclein: Implications for Parkinson's disease, *Neurotoxicology* **23**, 527–536 (2002).
36. V. N. Uversky, J. Li, and A. L. Fink, Metal-triggered structural transformations, aggregation, and fibrillation of human alpha-synuclein. A possible molecular link between Parkinson's disease and heavy metal exposure, *J. Biol. Chem.* **276**, 44284–44296.
37. L. A. Munishkina, C. Phelan, V. N. Uversky, and A. L. Fink, Conformational behavior and aggregation of alpha-synuclein in organic solvents: modeling the effects of membranes, *Biochemistry* **42**, 2720–2730 (2003).
38. K. A. Conway, J. D. Harper, and P. T. Lansbury, Accelerated *in vitro* fibril formation by a mutant α-synuclein linked to early-onset Parkinson disease, *Nature Med.* **4**, 1318–1320 (1998).
39. V. N. Uversky, J. Li, and A. L. Fink, Pesticides directly accelerate the rate of alpha-synuclein fibril formation: a possible factor in Parkinson's disease, *FEBS Lett.* **500**, 105–108 (2001).
40. J. A. Cohlberg, J. Li, V. N. Uversky, and A. L. Fink, Heparin and other glycosaminoglycans stimulate the formation of amyloid fibrils from alpha-synuclein *in vitro*, *Biochemistry* **41**, 1502–1511 (2002).
41. J. Goers, V. N. Uversky, and A. L. Fink, Polycation-induced oligomerization and accelerated fibrillation of human alpha-synuclein *in vitro*, *Prot. Sci.* **12**, 702–707 (2003).
42. V. N. Uversky, M. Cooper, K. S. Bower, J. Li, and A. L. Fink, Accelerated alpha-synuclein fibrillation in crowded milieu, *FEBS Lett* **515**, 99–103 (2002).
43. V. N. Uversky, J. Li, and A. L. Fink, Trimethylamine-N-oxide-induced folding of alpha-synuclein, *FEBS Lett.* **509**, 31–35 (2001).
44. V. N. Uversky, G. Yamin, P. O. Souillac, J. Goers, C. B. Glaser, and A. L. Fink, Methionine oxidation inhibits fibrillation of human alpha-synuclein *in vitro*, *FEBS Lett.* **517**, 239–244 (2002).
45. V. N.Uversky, J. Li, P. O. Souillac, I. S. Millett, S. Doniach, R. Jakes, M. Goedert, and A. L. Fink, Biophysical properties of the synucleins and their propensities to fibrillate: inhibition of alpha-synuclein assembly by beta- and gamma-synucleins, *J. Biol. Chem.* **277**, 11970–11978 (2002).
46. J. Bradbury, Alpha-synuclein gene triplication discovered in Parkinson's disease, *Lancet Neurol.* **2**, 715 (2003).
47. A. B. Singleton, M. Farrer, J. Johnson, A. Singleton, S. Hague, J. Kachergus, M. Hulihan, T. Peuralinna, A. Dutra, R. Nussbaum, S. Lincoln, A. Crawley, M. Hanson, D. Maraganore, C. Adler, M. R. Cookson, M. Muenter, M. Baptista, D. Miller, J. Blancato, J. Hardy, and K. Gwinn-Hardy, alpha-Synuclein locus triplication causes Parkinson's disease, *Science* **302**, 841 (2003).
48. D. T. Dexter, P. Jenner, A. H. Schapira, and C. D. Marsden, Alterations in levels of iron, ferritin, and other trace metals in neurodegenerative diseases affecting the basal ganglia. The Royal Kings and Queens Parkinson's Disease Research Group, *Ann. Neurol.* **32 Suppl**, S94–100 (1992).
49. G. C. Gazzaniga, B. Ferraro, M. Camerlingo, L. Casto, M. Viscardi, and A. Mamoli, A case control study of CSF copper, iron and manganese in Parkinson disease, *Ital. J Neurol. Sci.* **13**, 239–243 (1992).
50. F. J. Jimenez-Jimenez, J. A. Molina, M. V. Aguilar, I. Meseguer, C. J. Mateos-Vega, M. J. Gonzalez-Munoz, F. De Bustos, A. Martinez-Salio, M. Orti-Pareja, M. Zurdo, and M. C. Martinez-Para, Cerebrospinal fluid levels of transition metals in patients with Parkinson's disease, *J. Neural Transm.* **105**, 497–505 (1998).
51. E. Kienzl, L. Puchinger, K. Jellinger, W. Linert, H. Stachelberger, and R. F. Jameson, The role of transition metals in the pathogenesis of Parkinson's disease, *J. Neurol. Sci.* **134 Suppl**, 69–78 (1995).

52. H. H. Liou, M. C. Tsai, C. J. Chen, J. S. Jeng, Y. C. Chang, S. Y. Chen, and R. C. Chen, Environmental risk factors and Parkinson's disease: a case-control study in Taiwan, *Neurology* **48**, 1583–1588 (1997).
53. E. B. Montgomery, Jr., Heavy metals and the etiology of Parkinson's disease and other movement disorders, *Toxicology* **97**, 3–9 (1995).
54. L. M. Sayre, G. Perry, C. S. Atwood, and M. A. Smith, The role of metals in neurodegenerative diseases, *Cell Mol. Biol.* **46**, 731–741 (2000).
55. A. Seidler, W. Hellenbrand, B. P. Robra, P. Vieregge, P. Nischan, J. Joerg, W. H. Oertel, G. Ulm, and E. Schneider, Possible environmental, occupational, and other etiologic factors for Parkinson's disease: a case-control study in Germany, *Neurology* **46**, 1275–1284 (1996).
56. L. S. Wechsler, H. Checkoway, G. M. Franklin, and L. G. Costa, A pilot study of occupational and environmental risk factors for Parkinson's disease, *Neurotoxicology* **12**, 387–392 (1991).
57. A. H. Sadek, R. Rauch, and P. E. Schulz, (2003). Parkinsonism due to manganism in a welder, *Int. J Toxicol.* **22**, 393–401 (2003).
58. N. Hakansson, P. Gustavsson, C. Johansen, and B. Floderus, Neurodegenerative diseases in welders and other workers exposed to high levels of magnetic fields, *Epidemiology* **14**, 420–426 (2003).
59. G. Yamin, C. B. Glaser, V. N. Uversky, and A. L. Fink, Certain metals trigger fibrillation of methionine-oxidized α-synuclein, *J. Biol. Chem.* **278**, 27630–27635 (2003).
60. L. A. Munishkina, J. Henriques, V. N. Uversky, and A. L. Fink, Role of protein-water interactions and electrostatics in alpha-synuclein fibril formation, *Biochemistry* **43**, 3289–3300 (2004).
61. M. S. Nielsen, H. Vorum, E. Lindersson, and P. H. Jensen, Ca^{2+} binding to alpha-synuclein regulates ligand binding and oligomerization, *J. Biol. Chem.* **276**, 22680–22684 (2001).
62. R. Betarbet, T. B. Sherer, G. MacKenzie, M. Garcia-Osuna, A. V. Panov, and J. T. Greenamyre, Chronic systemic pesticide exposure reproduces features of Parkinson's disease, *Nat. Neurosci.* **3**, 1301–1306 (2000).
63. B. I. Giasson and V. M. Lee, A new link between pesticides and Parkinson's disease, *Nat. Neurosci.* **3**, 1227–1228 (2000).
64. T. B. Sherer, R. Betarbet, and J. T. Greenamyre, Pathogenesis of Parkinson's disease, *Curr. Opin. Investig. Drugs* **2**, 657–662 (2001).
65. D. A. Di Monte, M. Lavasani, and A. B. Manning-Bog, Environmental factors in Parkinson's disease, *Neurotoxicology* **23**, 487–502 (2002).
66. A. B. Manning-Bog, A. L. McCormack, J. Li, V. N. Uversky, A. L. Fink, and D. A. Di Monte, The herbicide paraquat causes up-regulation and aggregation of alpha-synuclein in mice, *J. Biol. Chem.* **277**, 1641–1644 (2002).
67. A. D. Cardin and H. J. Weintraub, Molecular modeling of protein-glycosaminoglycan interactions, *Arteriosclerosis* **9**, 21–32 (1989).
68. G. Perry, P. Richey, S. L. Siedlak, P. Galloway, M. Kawai, and P. Cras, Basic fibroblast growth factor binds to filamentous inclusions of neurodegenerative diseases, *Brain Res.* **579**, 350–352 (1992).
69. M. D. Shtilerman, T. T. Ding, and P. T. Lansbury, P. T., Jr., Molecular crowding accelerates fibrillization of alpha-synuclein: could an increase in the cytoplasmic protein concentration induce Parkinson's disease? *Biochemistry* **41**, 3855–3860 (2002).
70. L. A. Munishkina, E. M. Cooper, V. N. Uversky, and A. L. Fink, The effect of macromolecular crowding on protein aggregation and amyloid fibril formation, *J. Mol. Recognit.* **17**, 456–464 (2004).
71. T. Antony, W. Hoyer, D. Cherny, G. Heim, T. M. Jovin, and V. Subramaniam, Cellular polyamines promote the aggregation of alpha-synuclein, *J. Biol. Chem.* **278**, 3235–3240 (2003).
72. J. Goers, A. B. Manning-Bog, A. L. McCormack, I. S. Millett, S. Doniach, D. A. Di Monte, V. N. Uversky, and A. L. Fink, Nuclear localization of alpha-synuclein and its interaction with histones, *Biochemistry* **42**, 8465–8471 (2003).
73. R. Sharon, M. S. Goldberg, I. Bar-Josef, R. A. Betensky, J. Shen, and D. J. Selkoe, Alpha-synuclein occurs in lipid-rich high molecular weight complexes, binds fatty acids, and shows homology to the fatty acid-binding proteins, *Proc. Natl. Acad. Sci. USA* **98**, 9110–9115 (2001).
74. R. N. Cole and G. W. Hart, Cytosolic O-glycosylation is abundant in nerve terminals, *J. Neurochem.* **79**, 1080–1089 (2001).

75. P. J. McLean, H. Kawamata, S. Ribich, and B. T. Hyman, Membrane association and protein conformation of alpha-synuclein in intact neurons. Effect of Parkinson's disease-linked mutations, *J. Biol. Chem.* **275**, 8812–8816 (2000).
76. M. J. Volles and P. T. Lansbury, Jr., Vesicle permeabilization by protofibrillar alpha-synuclein is sensitive to Parkinson's disease-linked mutations and occurs by a pore-like mechanism, *Biochemistry* **41**, 4595–4602 (2002).
77. M. Hashimoto, E. Rockenstein, M. Mante, M. Mallory, and E. Masliah, Beta-synuclein inhibits alpha-synuclein aggregation: a possible role as an anti-parkinsonian factor, *Neuron* **32**, 213–223 (2001).
78. K. A. Conway, J. C. Rochet, R. M. Bieganski, and P. T. Lansbury, Jr., Kinetic stabilization of the alpha-synuclein protofibril by a dopamine- alpha-synuclein adduct, *Science* **294**, 1346–1349 (2001).
79. J. Li, M. Zhu, A. B. Manning-Bog, D. A. Di Monte, and A. L. Fink, Dopamine and L-dopa disaggregate amyloid fibrils: implications for Parkinson's and Alzheimer's disease, *FASEB J.* **18**, 962–964 (2004).
80. J. E. Duda, B. I. Giasson, Q. Chen, T. L. Gur, H. I. Hurtig, M. B. Stern, S. M. Gollomp, H. Ischiropoulos, V. M. Lee, and J. Q. Trojanowski, Widespread nitration of pathological inclusions in neurodegenerative synucleinopathies, *Am. J. Pathol.* **157**, 1439–1445 (2000).
81. P. Jenner, D. T. Dexter, J. Sian, A. H. Schapira, and C. D. Marsden, Oxidative stress as a cause of nigral cell death in Parkinson's disease and incidental Lewy body disease, The Royal Kings and Queens Parkinson's Disease Research Group, *Ann. Neurol.* **32 Suppl**, S82–S87 (1992).
82. W. Vogt, Oxidation of methionyl residues in proteins: tools, targets, and reversal, *Free Radic. Biol. Med.* **18**, 93–105 (1995).
83. B. I. Giasson, J. E. Duda, I. V. Murray, Q. Chen, J. M. Souza, H. I. Hurtig, H. Ischiropoulos, J. Q. Trojanowski, and V. M. Lee, Oxidative damage linked to neurodegeneration by selective alpha-synuclein nitration in synucleinopathy lesions, *Science* **290**, 985–989 (2000).
84. E. H. Norris, B. I. Giasson, H. Ischiropoulos, and V. M. Lee, Effects of oxidative and nitrative challenges on alpha-synuclein fibrillogenesis involve distinct mechanisms of protein modifications, *J. Biol. Chem.* **278**, 27230–27240 (2003).
85. E. Paxinou, Q. Chen, M. Weisse, B. I. Giasson, E. H. Norris, S. M. Rueter, J. Q. Trojanowski, V. M. Lee, and H. Ischiropoulos, Induction of alpha-synuclein aggregation by intracellular nitrative insult, *J. Neurosci.* **21**, 8053–8061 (2001).
86. J. M. Souza, B. I. Giasson, Q. Chen, V. M. Lee, and H. Ischiropoulos, Dityrosine cross-linking promotes formation of stable alpha-synuclein polymers. Implication of nitrative and oxidative stress in the pathogenesis of neurodegenerative synucleinopathies, *J. Biol. Chem.* **275**, 18344–18349 (2000).
87. G. Yamin, V. N. Uversky, and A. L. Fink, Nitration inhibits fibrillation of human alpha-synuclein in vitro by formation of soluble oligomers, *FEBS Lett.* **542**, 147–152 (2003).
88. J. E. Duda, U. Shah, S. E. Arnold, V. M. Lee, and J. Q. Trojanowski, The expression of alpha-, beta-, and gamma-synucleins in olfactory mucosa from patients with and without neurodegenerative diseases, *Exp. Neurol.* **160**, 515–522 (1999).
89. J. E. Galvin, T. M. Schuck, V. M. Lee, and J. Q. Trojanowski, Differential expression and distribution of alpha-, beta-, and gamma- synuclein in the developing human substantia nigra, *Exp. Neurol.* **168**, 347–355 (2001).

Molten Globule–Lipid Bilayer Interactions and Their Implications for Protein Transport and Aggregation

Lisa A. Kueltzo[1] and C. Russell Middaugh[2,3]

The molten globule (MG) state of proteins is characterized by a reduction in tertiary but a retention of secondary structure in comparison to the native state, accompanied by an increase in hydrodynamic radius. It is currently accepted as a common intermediate in protein-folding pathways, and although often transient it has been found to be the predominant collection of states in some biological systems. The exposure of internal hydrophobic surfaces, along with the increased flexibility of the molten globule state, also increases its tendency to interact with amphipathic lipid bilayers. It therefore is not surprising that MG states have been repeatedly suggested as key elements in the transport of proteins across cell membranes, most recently in the context of non-classical protein transport. Unfortunately, this structurally labile conformation is highly prone to self-association, aggregation, and precipitation. This presents a significant problem when analyzing protein-membrane interactions, since the conformational state most suited to membrane interactions may be one of much lower stability. When aggregation occurs simultaneously with membrane interaction, questions can arise regarding both the quality of the sample as well as the validity of any mechanistic conclusions. In this chapter, we discuss the molten globule-lipid bilayer interaction in the context of nonclassically transported proteins, focusing on the potential role MG states may play in the mechanism of transport as well as the difficulties that the structural lability of MG states can induce in both data acquisition and interpretation.

---

[1] Globe Immune, Inc., 1450 Infinite Drive, Louisville, CO 80027
[2] Department of Pharmaceutical Chemistry, University of Kansas, 2095 Constant Ave., Lawrence, KS 66047
[3] To whom correspondence should be addressed: C. Russell Middaugh, Department of Pharmaceutical Sciences, University of Colorado Health Sciences Center, 4200 E. 9th Ave., Denver, C0 80262; email: middaugh@ku.edu

1. INTRODUCTION

The molten globule (MG) state of proteins is a partially unfolded conformation characterized by an absence of tertiary structure but maintenance of most of the macromolecule's native α-helices and/or β-sheets. This is typically accompanied by an overall decrease in compactness (increase in hydrodynamic radius of the protein) and exposure of interior apolar surfaces to the bulk solvent. The MG state is accepted as a common intermediate in many protein folding pathways (Figure 1), and although often transient it has been found to be a predominant collection of conformational states in some biological systems. Such states are often unstable, however, being prone to self-association, aggregation, and precipitation. This can have major consequences for the phenomenon of protein-lipid bilayer interactions. It has been shown repeatedly that the more expanded structure of the

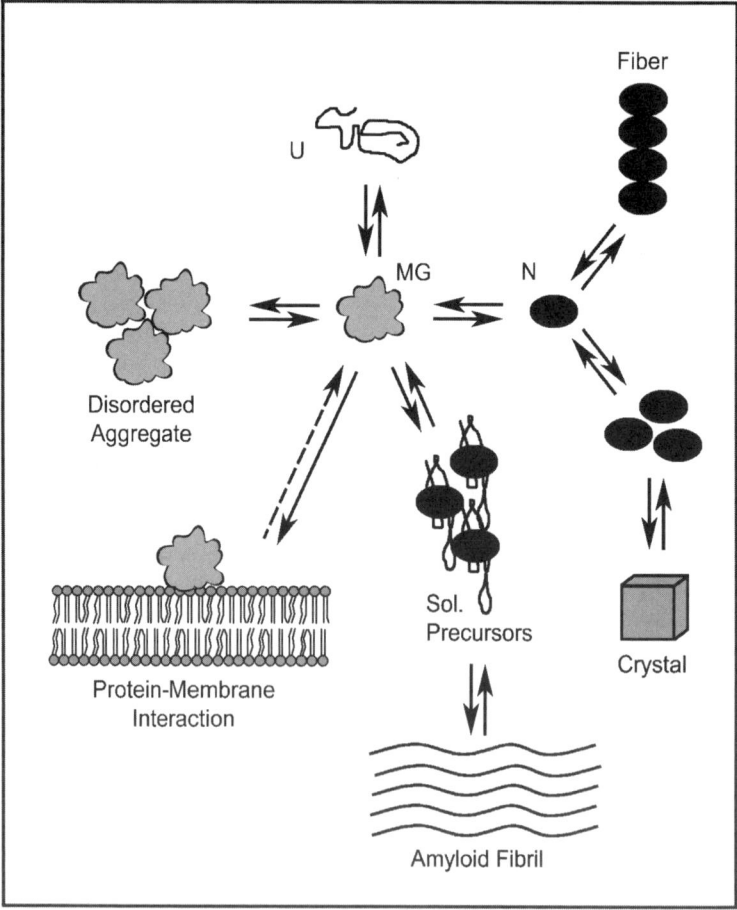

FIGURE 1. Potential roles of molten globule states in protein folding. N: Native, MG: Molten globule, U: unfolded. Adapted from Dobson and Karplus.[6]

MG state is capable of interacting with amphipathic lipid bilayers.[1-3] Even if a MG state is not present under bulk solution conditions, such forms of proteins may be induced by conditions at the cell surface, such as the lower pH present at membrane interfaces[4] or by components of the lipid bilayer itself.[5]

Thus we are presented with a situation in which conformational states involved in membrane interactions also may be of low stability. Not surprisingly, this can produce a number of difficulties in both data acquisition and interpretation. Not only does protein aggregation produce questions of sample quality (e.g., the heterogeneity of the sample increases), but it interferes in a number of common spectroscopic approaches (fluorescence, ultraviolet absorbance, circular dichroism, etc.). When investigating the mechanism of protein-membrane interactions, how can one be certain that concomitant aggregation processes are not occurring, significantly complicating data analysis? Furthermore, the presence of MG states may be of functional importance. Does membrane interaction depend on the presence of the MG state? Is the MG state required for proper biological function of the protein during and following membrane insertion or passage? In such cases, attempting to simplify the situation by shifting the conformational equilibrium of the protein away from the MG state may be ill-advised. These are all questions that arise from the presence of MG states in protein-membrane systems.

In this chapter, we examine molten globule-lipid bilayer interactions in the context of nonclassic protein transport. A number of proteins and peptides, including the herpesvirus protein VP22, HIV Tat, various homeoproteins as well as a variety of cationic peptides, exhibit what has become known as nonclassical transport activity. Nonclassical transport is generally defined as entry into and/or exit from cells in the absence of previously known energy-dependent mechanisms. The majority of nonclassical peptides and proteins appear to be able to mediate macromolecule delivery.[7] By creating fusion constructs through covalent or noncovalent interactions, these vectors have been shown to transport associated (macro)molecules through lipid bilayers with no loss of biological function of the attached molecules *in vitro* and *in vivo*.[8-10] To date, nonclassical transport-effected delivery has been employed to deliver proteins,[12-14] peptides,[15] peptide nucleic acids,[16] and oligonucleotides,[17] among other macromolecules into a wide variety of cell types.

Although a number (>20) of nonclassic delivery vectors are in early stages of development, no general consensus has been reached explaining their mechanism of transport. A variety of hypotheses, however, have been postulated. The highly cationic nature of the vast majority of these vectors, many of which are arginine-rich peptides or proteins possessing clustered basic sequences, has suggested that direct interactions between the vectors and cell-surface polyanions mediate at least the initial stages of transport.[18-20] A growing body of evidence suggests that some form of endocytosis may be involved in this process. One nonmutually exclusive idea proposes a role for molten globule-like states of proteins in transport[21]; this is supported by evidence that many nonclassically transported proteins (e.g., VP22 and FGF-1) appear to adopt MG-like states under physiological solution conditions (neutral pH, moderate temperature).[22,23] Additionally, the majority of peptide based vectors, although in some cases able to adopt helical conformations, appear to have no obvious requirements for ordered structure for transport.[24-26] These observations are further supported by arguments that MG-like

TABLE 1. Solution conditions under which study proteins exist in a molten globule or molten globule-like state.

Protein	pH	Temperature (°C)	Conformational state
VP22.C1[22]	7.4	25–40°C	MG-like
Fibroblast growth factor-1[28]	7.0	30–40°C	MG
Bovine granulocyte colony stimulating factor[29]	2–3 4	~30–50°C 45–55°C	MG Highly ordered MG
Keratinocyte growth factor –2	6.2	37°C	MG-like
Porcine somatotropin[30]	4	25°C	MG

states have the potential for bilayer transport in other systems.[27] The specific mechanism by which MG-effected transport is accomplished, however, yet has to be convincingly demonstrated. This may result in part from the fact that although a plethora of evidence exists supporting a role for polycations in nonclassical transport, less information is available supporting a role for MG states and the expanded conformation of peptides. Therefore, to further explore the potential role MG-like states may play in nonclassic protein transport, the interactions of a series of proteins with two lipid bilayer models are examined here.

The proteins studied here include VP22.C1 (the transport-competent, C-terminal half of the nonclassical transport herpes virus protein VP22), fibroblast growth factor-1 (FGF-1), keratinocyte growth factor-2 (KGF-2), bovine granulocyte colony stimulating factor (bGCSF), and porcine somatotropin (pST). These proteins are basic in nature (net positive charge at neutral pH), of similar size (16 kDa to 23 kDa), and represent a diverse group of structural families and biological functions. All have been shown to adopt molten globulelike states under various conditions of temperature and pH [22,28–30] (Table 1).

The herpesvirus protein VP22 possesses both nonclassical import and export activity[8] and is a candidate for nonclassical transport-based macromolecule delivery.[8,13] FGF-1 has demonstrated nonclassical secretory activity[31] as well as extensive nonspecific polyanion-binding ability.[32] Lipid bilayer interactions have been demonstrated previously for FGF-1 and a dependence on protein conformation and lipid bilayer surface charge has been established.[33] Although less information is available for KGF-2, polyanion binding and MG-like conformations have been demonstrated in this laboratory (unpublished results). bGCSF and pST are four-helix bundle proteins with potential veterinary therapeutic applications.[34,35] Neither protein has demonstrated nonclassical transport activity to date, although no evidence against potential transport activity is available either. These proteins are included to explore the possibility that lipid interaction, to a varying degree, may be a general property of MG-like states.

Two model lipid systems are employed. A fluorescence-based approach analyzes the release of fluorescent dye encapsulated in liposomes in the presence of the various proteins. The second assay employs absorbance spectroscopy to monitor a chromophore consisting of polymerized diacetylene lipids. The absorbance spectrum of the chromophore is environmentally sensitive, and when incorporated into phospholipid bilayers can serve as a sensor for phospholipid-ligand interactions.[36] The assays are conducted under varying conditions of temperature, pH, and liposome composition. Additionally, the role that the lipid bilayer may play in stabilizing partially unfolded structures during transport is examined. Work by Bogdanov and colleagues has shown that the lipid bilayer is a vital component in the refolding of certain integral membrane proteins, acting as a putative molecular chaperone.[37,38] Some evidence also suggests that lipids may play a role in the folding of soluble proteins, especially those with membrane insertion domains.[4,39] We therefore also explore whether the lipid bilayer can indeed act as a molecular chaperone for globular proteins by examining the effect of liposomes of varying compositions on the aggregation and refolding of FGF-1.

2. MATERIALS AND METHODS

Although it is not the purpose of this work to review the various experimental techniques employed in lipid bilayer interaction studies, a brief overview of the methods employed for this particular study is warranted. Protein was either purified as previously described (VP22.C1[22]) or received in a highly purified form from the following collaborators: recombinant bGCSF and pST were provided by the Bioprocess Research and Development department of Pfizer, Inc., recombinant KGF-2 by Human Genome Sciences, and recombinant human FGF-1 by Merck & Co. All lipids, fluorescent probes, and other reagents were obtained from commercial sources.

The methods presented here have been described previously. Model lipid bilayers were prepared with dioleoylphosphatidylcholine (DOPC) and/or dioleolyphosphatidylglycerol (DOPG) in various ratios, producing zwitterionic (neutral) or anionic vesicles, respectively. Liposomes were prepared by a thin film method[40] and were hydrated in 6.2 mM sodium phosphate and 120 mM NaCl, pH 7.4 (blank), or with one of two fluorescent probe solutions: 100 mM 5,6-carboxyfluorescein (Fln) in water, pH 7.0, or 50 mM sulforhodamine B (rhoB), pH 7.0. The RhoB-containing liposomes were used in place of carboxyfluorescein for low pH samples, since carboxyfluorescein fluorescence is extensively quenched below pH 5. For dye-loaded liposomes, excess dye was separated from the loaded liposomes by size-exclusion chromatography. The hydrodynamic diameter of lipid preparations was determined by dynamic light scattering, with an average size of $\sim$70 nm and $\sim$150 nm for blank and dye-containing vesicles, respectively. Kinetic dye release studies were conducted based on modifications of the work of Mach and Middaugh.[33] Maximum dye release was defined as the signal observed after complete disruption of the vesicles in the presence of 0.1% Triton. During fluorescence experiments, a scattering signal was simultaneously monitored at 180° from the emission signal employing a second detector. Sample pH and temperature were customized for each protein to induce MG-like states.

Polydiacetylene/phospholipid vesicles were prepared as previously described[36,41] (Figure 3). An average liposome size of 400 nm was determined by light-scattering methods. As shown by Kolusheva and colleagues,[36] changes in the polydiacetylene absorbance spectrum can be quantified by determining the colorimetric response (%CR), defined by:

$$\%CR = \left(\frac{PB_0 - PB_1}{PB_0}\right) \times 100\%, \qquad (1)$$

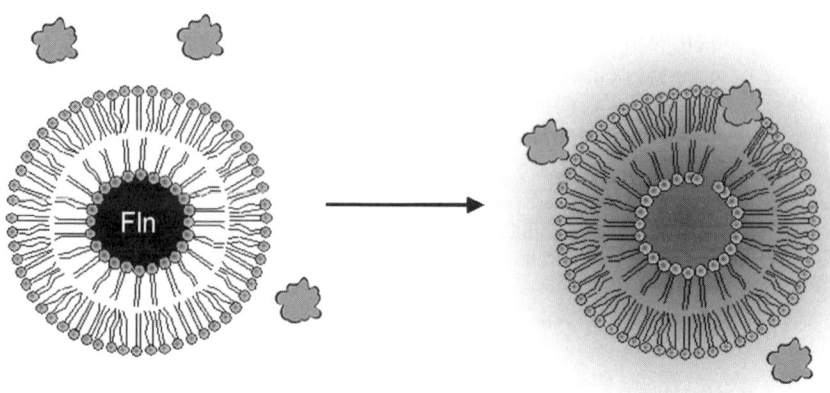

FIGURE 2. Fluorescent dye release assay. Concentrated dye (e.g., 100 mM 5,6-carboxyfluorescein) is encapsulated within a liposome at such concentration as to quench the fluorescence of the dye. Upon interaction with a protein or other ligand that perturbs the bilayer, the dye is released into the surrounding solution. The dilution occurring upon release relieves the self-quenching effect, causing an increase in the fluorescence signal. The extent of the signal is proportional to the degree of bilayer perturbation.

where

$$PB = \frac{A_{660nm}}{A_{660nm} + A_{550nm}}, \qquad (2)$$

In this approach, an increase in %CR indicates an interaction with the lipid component of the vesicle. The PB_0 and PB_1 represent the PB values in the absence (control) and presence of a ligand (e.g., protein), respectively. (Binding to charged head groups of diacetylene or phospholipid components without bilayer perturbation provides little to no change in %CR.) Changes in the %CR are difficult to compare across studies, since they depend on the molar ratio of the PDA to the phospholipid components as well as the protein to phospholipid ratio. An example of an extremely high increase in CR is presented by Kolusheva et al.,[36] who saw an increase of ~80% with the addition of high concentrations of melittin, a membrane-disrupting peptide, to vesicles similar to those employed in this study. Since the aggregation present in some samples examined here precluded a strict application of this method due to inflation of apparent absorbance values by extensive light scattering, the above equation was applied to second-derivative intensities at similar wavelengths. This approach minimizes scattering contributions since derivative analysis is insensitive to the broad light-scattering background signal.

The kinetics of FGF-1 aggregation were determined by examining turbidity at 350 nm as a function of time as described previously.[42] Refolding studies were conducted employing methods based on the work of Edwards and colleagues[42] and the procedures of Dabora et al.,[43] which are based on the quenching of the fluorescence of the protein's single Trp residue when the native state is formed during refolding from a 2 M guanidine hydrochloride solution.

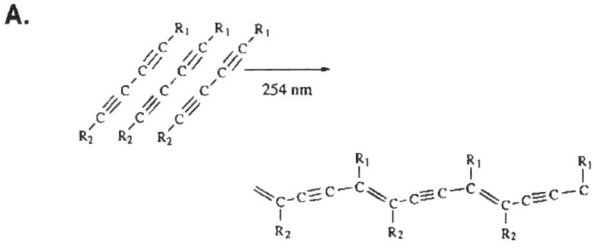

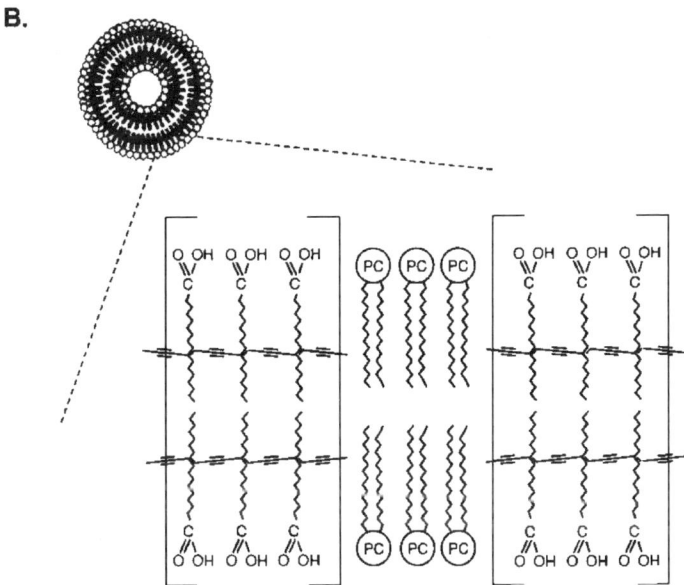

FIGURE 3. Formation of polydiacetylene vesicles. (A) Upon irradiation at 254 nm, diacetylene monomers undergo a 1,4 addition reaction to form a polymer (polydiacetylene, or PDA) containing a blue-colored chromophore. (B) Cross section of a PDA/phospholipid vesicle, showing interspersed regions of phospholipid and polymerized diacetylene. Interaction of a ligand with the phospholipid regions causes a perturbation of the adjacent PDA network, resulting in a visible color change in the sample. Adapted from refs. 36 and 44.

3. INTERACTION OF MODEL PROTEINS WITH LIPID BILAYERS

3.1. Dye Release Assay

One approach to determining the effect of MG-state formation on protein-lipid interactions is to employ a dye release model.[33] This is based on the self-quenching property of fluorescent dyes such as 5,6-carboxyfluorescein (Fln) and sulforhodamine B (rhoB) when encapsulated at high concentrations within liposomes. Significant perturbation of the liposome bilayer results in release of the dye into the surrounding medium, resulting in dilution and an increase in fluorescence intensity as the self-quenching is relieved.

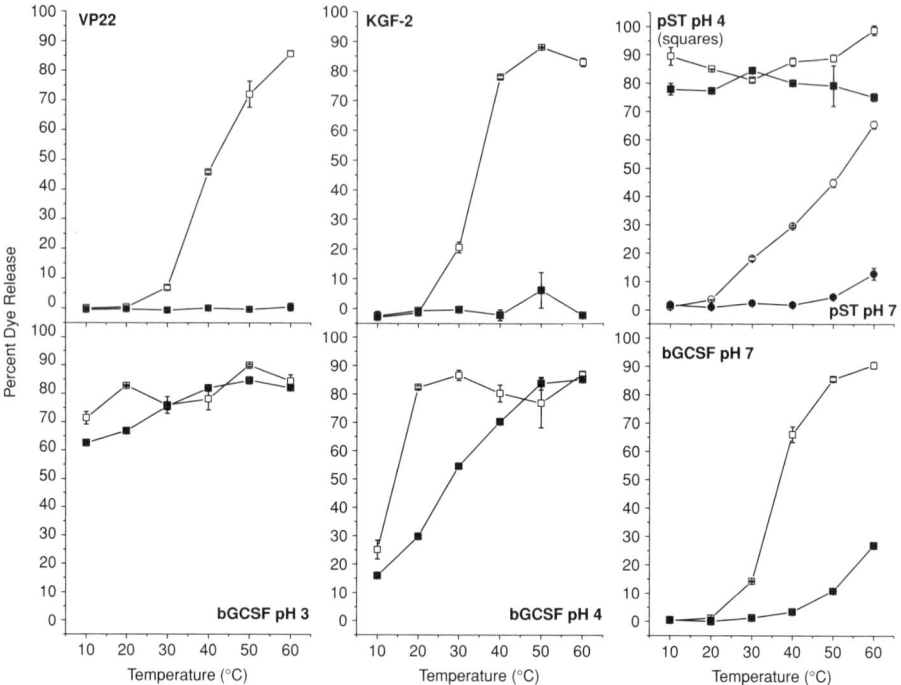

FIGURE 4. Release of fluorescent dye from liposomes in the presence of protein. Extent of release was determined as percent of release observed in the presence of 0.1% Triton. The DOPC liposomes are represented by closed symbols and DOPC:DOPG vesicles at a 1:1 molar ratio are represented by open symbols. The pST data are represented by squares at pH 4 and circles at pH 7. The protein concentration used was 1 mM while the total lipid concentration was 100 mM. The buffers used were 10 mM citrate and 150 mM NaCl at pH 3, 4, or 7 (bGCSF); 10 μM citrate and 150 mM NaCl at pH 4 or 10 μM phosphate and 150 mM NaCl at pH 7 (pST); 10 mM phosphate and 200 mM NaCl at pH 7.4 (VP22.C1); and 20 mM Na acetate with 125 mM NaCl and 1 mM EDTA at pH 6.2 for KGF-2.

As mentioned previously, VP22.C1, KGF-2, and FGF-1 all adopt MG-like conformations under near-neutral pH (7.4, 6.2, and 7.4 respectively) and moderate temperature (∼30°C to 45°C) conditions.[22,23] The extent of dye release induced by VP22 and KGF-2 as a function of both vesicle charge and sample temperature is illustrated in Figure 2. Here it is shown that the extent of dye release, and therefore the interaction of these proteins with the lipid bilayer, increases dramatically under MG-promoting solution conditions. A specific requirement for negatively charged lipid surfaces also is demonstrated for both proteins, since no interactions are observed in the presence of neutral zwitterionic vesicles.

These results are very similar to those observed by Mach and Middaugh in FGF-1 lipid-binding experiments.[33] This similarity does not extend, however, to the rate of dye release. The slightly faster rate and the biphasic nature of the kinetics of the representative release profiles shown in Figure 5D suggest subtle differences in the way the proteins interact with lipid bilayers. For example, the penetration depth of the proteins may differ

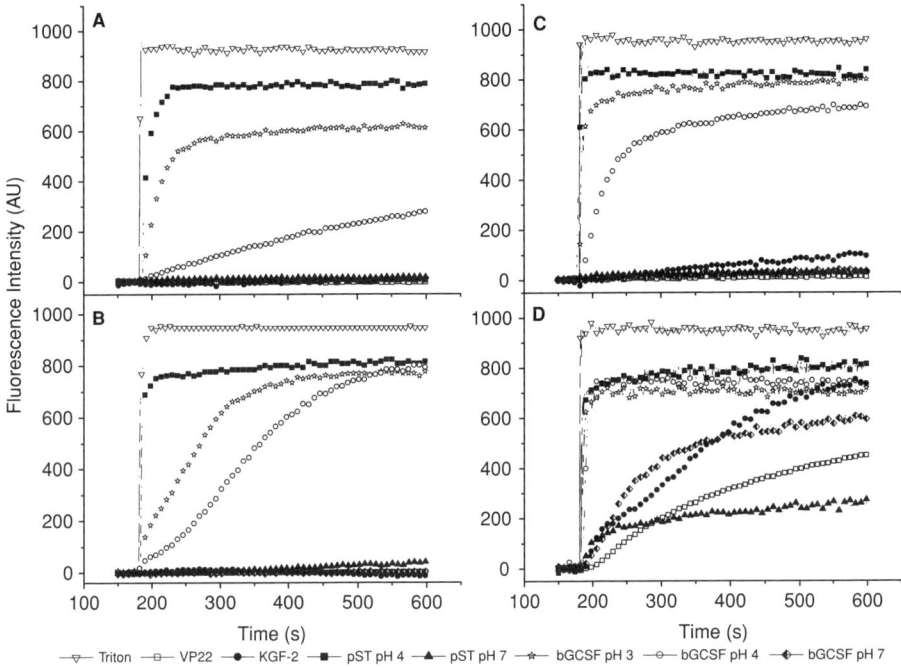

FIGURE 5. Representative kinetic traces of fluorescent dye release induced by protein at 20°C (A,B) and 40°C (C,D). Liposomes employed were 100% DOPC (A,C) and DOPC:DOPG vesicles at a 1:1 molar ratio (B,D). A protein concentration of 1 μM was employed with a total lipid concentration of 100 μM. The buffers used are described in Figure 4.

significantly, or one protein may interact in a 1:1 fashion with the liposome, while others may come together to form higher-order complexes (e.g., dimers, trimers, etc.) which then interact with the liposome. At this time, however, there are no data available to directly confirm such speculations.

In an effort to determine if lipid interaction is a general property of MG-forming proteins, the interactions of pST and bGCSF with model lipid systems also were examined. Under acidic conditions (pH 3 and 4 for bGCSF and pST, respectively), both proteins adopt MG-like conformations.[29,30] Thus, it is not surprising that both proteins induce significant release of fluorescent dye under these conditions. In contrast to the MG-induced dye release of VP22.C1 and KGF-2, however, the release induced by pST at pH4 has no significant dependence on either temperature or liposome surface charge (Figure 2). Similar results were observed for bGCSF at pH 3. Interestingly, since the MG-like conformation of bGCSF adopted at pH 3 is formed between approximately 25 and 40°C.,[29] it is likely that the MG state of bGCSF is not solely responsible for the interaction observed at this pH. The only evidence that conformation is playing any role in the lipid interaction of these two proteins under these conditions is observed in the dye release kinetic profiles (Figure 5). In contrast to the rapid rate observed in all samples for pST at pH 4, a slightly slower rate of release is observed for bGCSF at pH 3 at 20°C than 40°C, with definite biphasic character observed in the presence of DOPG containing vesicles at 20°C.

Release of the dye also is induced by bGCSF at pH 4 (Figure 2, bottom center). At this pH, bGCSF adopts a non-MG-like but temperature-labile conformation.[29] In this case, the extent of dye release is seen to be dependent on both temperature and vesicle charge. In contrast to pH 3, the rate of release appears to increase dramatically with increasing temperature, although the profile is biphasic under some conditions (not illustrated), once again suggesting subtleties in the interaction of the protein with the bilayer.

Under neutral pH conditions, both pST and bGCSF are expected to adopt less flexible conformations. Despite this, a moderate level of release at pH 7 is observed for both proteins, once more dependent on both temperature and lipid vesicle composition. Release is enhanced in the presence of negatively charged vesicles, as seen for VP22.C1 and KGF-2. Overall, a slower rate of release is observed compared to more acidic conditions.

Due to the tendency of the MG state to self-associate, it is quite reasonable to assume that protein aggregation is occurring in these studies. This will increase the difficulty of both data collection and interpretation since this can occur simultaneously with binding events. This also may alter binding/interaction phenomena. Aggregation of liposomes also may potentially affect dye release through liposome fusion; light scattering, however, can be monitored simultaneously to detect changes in particle size. (Aggregation of the protein itself can also occur, but the size of such aggregates in comparison to cross-linked protein-liposome complexes is expected to be quite small.) In this study, a variety of scattering profiles are observed, as shown in Figure 6. As a point of reference, in Panel T it is shown that disruption of the liposomes by Triton results in a decrease in scattering intensity independent of sample conditions.

In the presence of neutral liposomes, the majority of samples demonstrate an initial increase in scattering immediately after protein addition, followed by a return to near pre-protein addition levels (Figure 6A). (Note: the initial spike in scattering intensity observed in all profiles results from the introduction of the needle into the sample for protein injection.) This suggests that for the majority of samples, the release of dye and the sustained increased light scattering are not functionally correlated. The only notable exception to this is KGF-2. In the presence of neutral liposomes KGF-2 exhibits type B scattering (Figure 6B; scattering increases sharply, rapidly returning to almost initial levels and plateaus with a moderate overall increase in scattering at 600 sec) at 40°C and type C, in which an initial jump in scattering upon addition of the protein is sustained throughout the experiment at higher temperatures (Figure 6C). Again, the aggregation in these samples does not appear to correlate with dye release.

Samples containing negatively charged liposomes generally exhibit higher scattering intensities after protein addition. The VP22.C1 samples follow a type C profile, although at higher temperatures final scattering intensities decrease. In contrast, KGF-2 shows a type D scattering profile at all temperatures, with an abrupt increase in scattering followed by a slow decrease to initial levels. At both pH values examined, pST again exhibits type A scattering, with the exception of type B observed at low temperature ($<40°C$) and pH 4. Type A scattering profiles are observed for bGCSF at pH

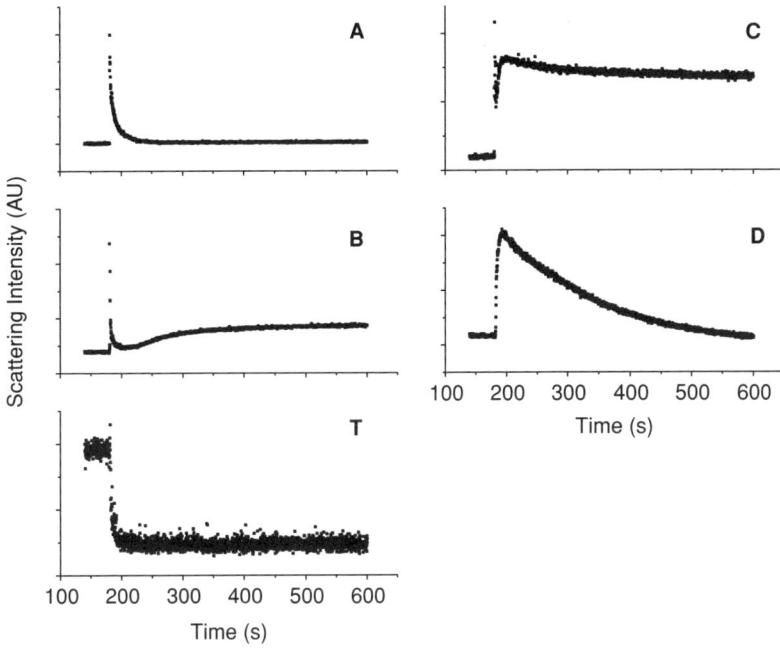

FIGURE 6. Light-scattering profiles observed during dye release studies. Light scattering was monitored at 90° simultaneously with fluorescence emission during dye release studies. The curves illustrated are representative of several scattering trace types that were consistently observed. A type T profile was observed upon addition of Triton to samples; all other curves represent protein induced scattering changes. Experimental conditions are described in Figure 4.

7 and 4 at low temperatures (<50°C), while type C is observed for all other bGCSF samples.

Why is it important to monitor scattering in these experiments? As mentioned previously, any perturbation of the lipid bilayer of the liposome may result in a release of dye, with the extent of release directly proportional to the degree of bilayer disturbance. If the liposomes self-associate in these sample, presumably through cross-linking, bilayer perturbation could occur through either self-association or liposome fusion. This could easily produce a false-positive result for the assay. Monitoring the optical density of the solution and correlating it to the extent of dye release permits the exclusion of such false positives from consideration.

3.2. Polydiacetylene Assay

An alternative approach to examining the interaction of a variety of substances with lipid bilayers has been developed by Kolusheva and colleagues. Following ultraviolet irradiation, diacetylene-based lipids polymerize, resulting in an intensely colored polymer.[44] Perturbation of the polymer backbone results in a color shift (e.g., from blue to red). These spectral changes then can be used to analyze binding interactions.[41]

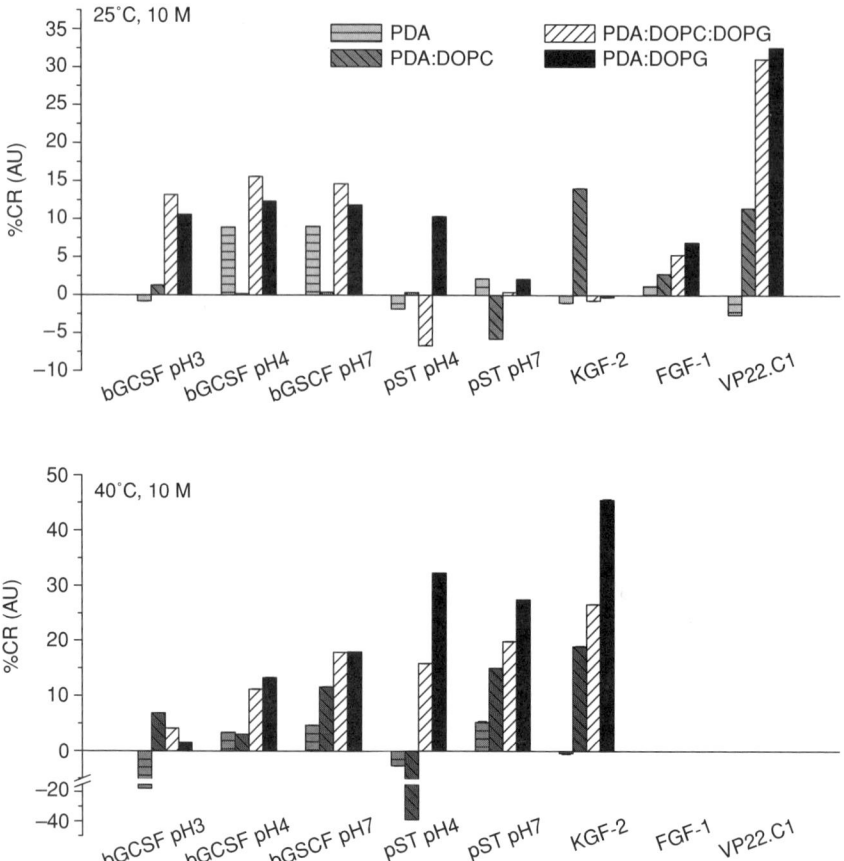

FIGURE 7. Colorimetric response induced by various proteins. The colorimetric response of vesicles composed of PDA, 3:2 [PDA]:[DOPC], 3:2 [PDA]:[DOPG], and 3:1:1 [PDA]:[DOPC]:[DOPG] was determined in the presence of each protein as described in the text. Samples were incubated for 30 min at 25 or 40°C before measurements were collected. A protein concentration of 10 μM was used with a total lipid concentration of 100 μM. The buffers employed are described in Figure 4.

Further studies of the PDA system have shown incorporation of diacetylene lipids into phospholipid vesicles permits monitoring of peptide-phospholipid interactions using a simple colorimetric assay.[36] Here the PDA approach has been extended to protein/bilayer interactions.

In these studies, the colorimetric response (%CR) of the PDA chromophore alone and in combination with DOPC and/or DOPG lipids (creating primarily neutral or anionic vesicles, respectively) was measured at 25°C and 40°C in the presence of each protein (Figure 7). Although little interaction is observed for KGF-2 at 25°C, with the exception of a modest interaction with DOPC containing vesicles (10–15%) at 40°C, strong interactions are seen with negatively charged vesicles. FGF-1 and VP22.C1 demonstrate small and moderate, respectively, charge-dependent interactions with the liposome samples at 25°C, conditions under which both proteins are expected to be in MG-like states. Unfortunately, %CR values at 40°C for VP22.C1- and FGF-1-containing samples were not obtainable. This is due to a strong baseline red shift resulting from the buffer itself, limiting the temperature window in which color changes could be detected.

The interaction detected by this method between bGCSF, pST, and the PDA vesicles is significantly less than that observed in the corresponding dye-release studies. At low temperature (25°C), little to no interaction is seen for pST at either pH, while bGCSF exhibits a low level of interaction under all pH conditions. The interaction of bGCSF appears to be dependent on the composition of the vesicle, with increased binding favored by a negatively charged surface. At increased temperatures, small increases and decreases in the %CR are observed for bGCSF, although no pattern is apparent. In contrast, a significant increase in %CR is observed for pST at both pH 4 and 7. Qualitatively, these results suggest an increase in either surface binding and/or penetration of pST with increasing temperature, with a preference for a higher negative charge density on the vesicle surface. No similar result is observed for bGCSF.

Since the PDA assay employs absorbance spectroscopy, it was possible to collect optical density (OD) data simultaneously during the experiment as a measure of turbidity. Sample aggregation, monitored by an increase in OD at 400 nm in the presence of protein compared to that of vesicles in buffer alone, is shown in Figure 8. In these systems, the aggregation again presumably arises from protein-liposome interactions. At 25°C, little to no aggregation is observed in samples containing PDA or PDA:DOPC liposomes, regardless of protein composition. The DOPG-containing liposome samples, however, are aggregated, generally having an OD of ~0.8 to 1.0. Exceptions to these observations occur with KGF-2 and pST samples. All phospholipid-containing KGF-2 samples have an OD of ~1.0. At pH 4 extensive aggregation (~1.5 to 2.5 OD) is observed for pST samples, especially those containing combination PDA:DOPG liposomes. Furthermore, pST at pH7 induces no detectable aggregation in any of the samples examined. In general, however, aggregation of samples with phospholipid-containing vesicles appears to follow a surface charge-dependent trend with samples containing vesicles with greater amounts of anionic phospholipid exhibiting higher OD after addition of protein.

Incubation at 40°C generally showed small (10–20%) increases in OD at 400 nm compared to 25°C samples, with some exceptions. Aggregation of samples containing PDA:DOPC vesicles and KGF-2 decreased compared to values obtained at 25°C. A similar decrease was observed in samples containing PDA:DOPC vesicles and pST at pH 4. Additionally, the aggregation observed in FGF-1 samples increased dramatically for all liposome compositions examined.

In summary, VP22.C1, FGF-1, and KGF-2 all exhibited a strong interaction with the PDA phospholipid vesicles in the temperature range over which they adopt MG-like conformations. For all three the extent of interaction increased with increasing negative charge of the surface of the vesicles. The temperature and charge dependencies of the interaction of bGCSF and pST also were well defined in this study. In all cases when aggregation was observed (all samples except bGCSF at pH 3 and pST at pH 7) the extent of aggregation increased with increasing negative charge on the liposomes.

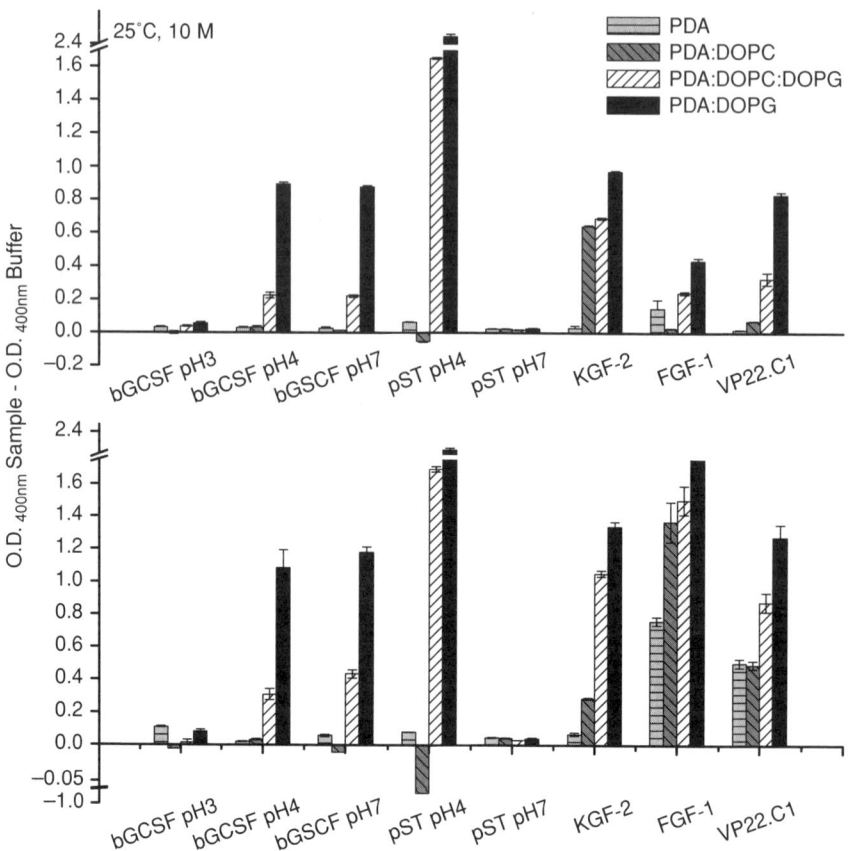

FIGURE 8. Optical density of protein:PDA vesicle samples. Optical density measurements at 400 nm were collected simultaneously with measurement of colorimetric response. Optical density was calculated as the difference between samples containing protein and liposomes and those containing liposomes alone. The vesicle compositions were 100% PDA, 3:2 [PDA]:[DOPC], 3:2 [PDA]:[DOPG], and 3:1:1 [PDA]:[DOPC]:[DOPG]. Samples were incubated for 30 min at 25 or 40°C before measurements were collected. A protein concentration of 10 μM was used with a total lipid concentration of 100 μM. The buffers employed are described in Figure 4.

3.3. Comparison of Dye Release and Polydiacetylene Assays

The two assays employed here produce significantly different results. It is possible that differences in interactions between proteins in MG-like conformations with the two distinct types of lipid bilayers presented here are seen because of the intrinsic heterogeneity of these states in terms of their hydrodynamic behavior, polarity, and detailed intramolecular interactions. The majority of the other differences most likely lie with the assays themselves.

For example, the dye release assay is a kinetic method, while the PDA techniques measure samples in an equilibrium state. Therefore, only the final extent of dye release can be compared to the %CR values. The assays also may possess differences in sensitivity due to the intrinsic differences in their lipid bilayer compositions. The incorporation of the PDA moiety into the phospholipid vesicles produces significantly different liposomes that vary in size as well as overall charge from conventional liposomes. Since the diacetylene lipid employed here is negatively charged,[43] none of the PDA vesicles are truly neutral, although the much smaller head group of the PDA lipid compared to the phospholipids present may not directly interact with the protein. Additionally, no data are available on the dynamics of the lipid bilayer after incorporation of the diacetylene polymer, but it is reasonable to assume that they are quite different from pure phospholipid liposomes. A potential restriction of mobility within the bilayer by the polymerized diacetylene species may significantly alter the effective penetration of the protein molecules. Finally, although NMR studies have shown that peptides interact preferentially with phospholipid head groups when binding to the PDA vesicles employed here,[36] similar studies have not been performed yet with proteins. Therefore, direct protein:PDA interactions cannot be eliminated. Altogether, these conditions suggest that caution should be employed in attempting any direct quantitative comparison between assay results, a course that is followed throughout the later discussion.

In our experience, using multiple approaches to evaluate interactions such as these is extremely valuable, since it provides a multidimensional "picture" of the target phenomenon. Choosing the methods to be included in an overall experimental design, of course, is crucial. The two assays employed here are both theoretically sound and have been successfully used elsewhere. In practice, the PDA assay is accompanied by a greater number of caveats and experimental difficulties. The effect of buffer salts and pH on the %CR appears to be particularly important. Although the dye-release study has its own difficulties (i.e., the large number of repetitions necessary to achieve reproducible kinetics), it may be a better choice as a first approach to studying these types of interactions.

4. CAN LIPID BILAYERS ACT AS MOLECULAR CHAPERONES? EFFECTS ON FGF-1 AGGREGATION AND REFOLDING

To determine if the lipid bilayer could act as a chaperone for globular proteins, the extent of aggregation of FGF-1 was further examined in the presence of liposomes composed of DOPC and DOPG as a function of protein to lipid ratio. As illustrated in Figure 9, no significant effect is observed in the presence of liposomes consisting solely of DOPC. Addition of liposomes containing DOPG produces a variety of effects. Maximum decreases in the extent of aggregation of 40, 70, and 50% are observed upon addition of increasing amounts of liposomes containing 25, 75, and 100% DOPG, respectively. No inhibition is observed in the presence of 50:50 DOPC:DOPG liposomes.

Additionally, the time-dependant turbidity profiles for DOPG samples do not exhibit the initial lag time observed previously for FGF-1 alone[42] or DOPC liposomes (results not illustrated). Immediate aggregation is observed upon mixing of FGF-1 and liposome solutions containing 50% or more anionic lipid, the extent proportional to the amount of DOPG present. This results in an increased OD (measured at 350 nm) at $t = 0$ and prevents analysis of those samples containing large amounts of DOPG. Additionally, the decreases in the extent of aggregation (Figure 9) at ratios near 1:1 w:w liposome:FGF-1

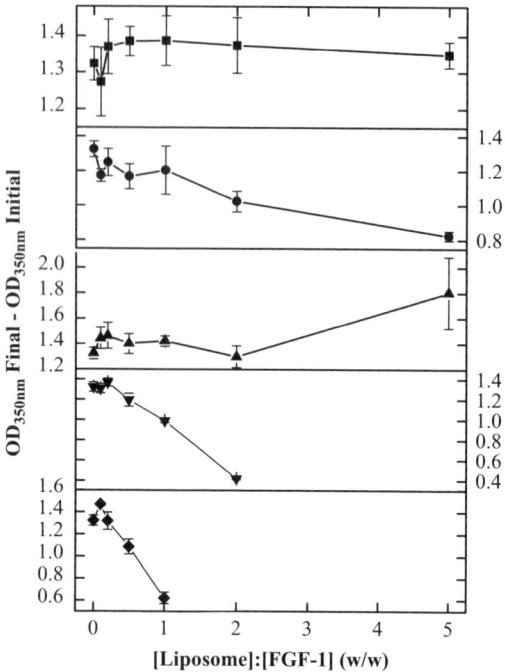

FIGURE 9. Extent of aggregation of FGF-1 at 45°C in the presence of liposomes. The extent of aggregation is defined as the difference between the final and initial asymptote of a sigmoidal curve fit to the raw data (OD$_{350nm}$ Final − OD$_{350nm}$ Initial). (■) FGF-1 and DOPC, (●) FGF-1 and 75% DOPC/25% DOPG, (▲) FGF-1 and 50% DOPC/50% DOPG, (▼) FGF-1 and 25% DOPC/75% DOPG, (♦) FGF-1 and DOPG.

may reflect precipitation of the sample, representing a more complex process than simple thermally induced aggregation of the protein.

Initial rates of aggregation were determined from fits to a sigmoidal function (results not illustrated). Statistical significance was determined from the Student's t-test ($p < 0.05$). A significant decrease (up to ∼30%) in the aggregation rate is observed for samples containing larger amounts of DOPC liposomes (2:1 and 5:1 liposome:protein by weight). In contrast, an increase (∼40%) was observed with a 0.2:1 liposome:protein ratio in the presence of 25% DOPG liposomes, although this is not seen for 25% DOPG samples at other weight ratios. No other significant effects are detected. Rates for samples containing 50% or more DOPG were not determined owing to the initial aggregation of the samples upon mixing.

In addition to determining if liposomes could inhibit the aggregation of FGF-1, the effect of liposomes on the rate of refolding of unfolded FGF-1 also was examined, based on changes in the protein's fluorescence upon dilution from concentrated guanidine hydrochloride solutions.[43] As illustrated in Figure 10, addition of liposomes to FGF-1 solutions does result in a slight increase in the rate of FGF-1 refolding. At 25°C, small but statistically significant increases ($p < 0.05$) are observed in the presence of 50% and 75% DOPG liposomes as well as with larger amounts of 25% DOPG vesicles (1:1 weight

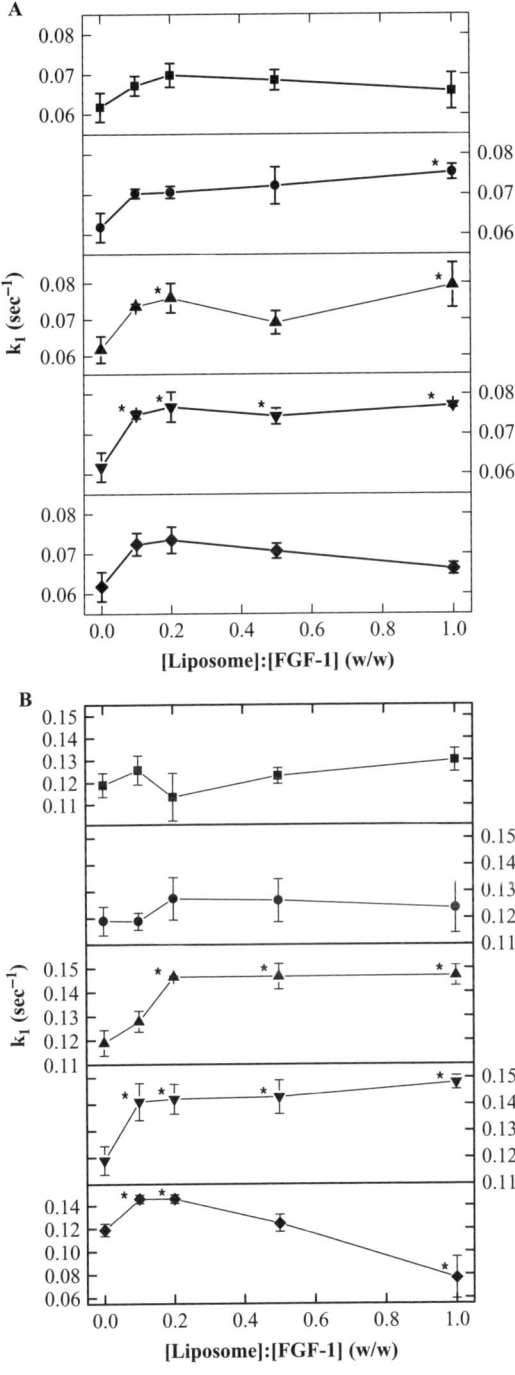

FIGURE 10. Refolding kinetics of FGF-1 in the absence and presence of liposomes. Unfolded FGF-1 was diluted 20-fold into refolding buffer to a final concentration of 5 μg/ml and rates were determined from initial rates of intrinsic fluorescence changes. (■) FGF-1 and DOPC, (●) FGF-1 and 75% DOPC/25% DOPG, (▲) FGF-1 and 50% DOPC/50% DOPG, (▼) FGF-1 and 25% DOPC/75% DOPG, (♦) FGF-1 and DOPG. Studies were conducted at (A) 25°C and (B) 37°C. Statistically significant differences compared to FGF-1 alone (Student's t-test, $p > 0.05$) are starred (*).

ratio). The rate increases exhibited upon addition of 50% and 75% DOPG liposomes are not dependent on liposome concentration. No significant effect is observed in samples containing 100% DOPC or 100% DOPG.

At 37°C, no effect is seen with 100% DOPC or 25% DOPG liposomes, while significant increases occur in the rate of refolding for samples containing 50% and 75% DOPG. In contrast to changes observed at 25°C, the increase in rate seen with the addition of 50% DOPG vesicles does not occur immediately upon introduction of a small quantity of liposome, but rather at a 0.2:1 liposome:FGF-1 ratio. Pure DOPG vesicles do perturb refolding at 37°C. At low liposome to protein ratios (0.1:1 and 0.2:1), increases in rate are observed while at a ratio of 1:1 a significant decrease in the rate of refolding is seen. It should be noted that the aggregation that occurred on mixing in the turbidity studies did not occur here owing to differences in the assay protocols. This allowed higher DOPG content liposome data to be obtained and analyzed.

5. CONCLUSIONS

Studies of multicomponent systems such as protein-membrane interactions frequently involve significant interpretative complications when aggregation is present. In this chapter, we demonstrate some of these difficulties through a mechanistic exploration of nonclassical transport vectors, examining the potential role of the appearance of molten globule states during early stages of protein transport, employing two model lipid systems. The potential value of simple, efficient, and versatile delivery systems based on nonclassical transport is high. Before optimal application of such vectors can be accomplished, however, a thorough understanding of the mechanism of transport must be established.

The majority of studies that endeavor to establish the interaction of proteins in compact or MG-like conformations with lipid bilayers do so employing a single assay with a single protein, making the extrapolation of individual results to general principles problematic at best. Here, the lipid interactions of multiple proteins with a variety of functions and properties are described. From these results it seems plausible to propose that a general connection exists between conformational lability and lipid bilayer penetration. Specifically, those proteins that are thought to possess nonclassic transport activity (e.g., VP22.C1, FGF-1, and KGF-2) display a definite increase in lipid bilayer interactions under MG-inducing solution conditions. In all three cases, this interaction was strongly dependent on the phospholipid composition of the model lipid bilayer. The unexpected results observed for both pST and bGCSF, however, remain to be explained. One potential reason for the enhanced interaction of these two proteins under non-MG conditions compared to the others may be related to structural differences. While KGF-2 and FGF-1 are composed primarily of β-sheet[23] (unpublished results), and VP22.C1 possesses both β-sheet and α-helical structure,[22] both pST and bGCSF are primarily helical proteins.[35,45] Considering the propensity for some helical peptides and proteins to directly interact with lipid membranes, a similar mechanism could be involved here. This may then be enhanced by the conformational lability induced at lower pH values.

Evidence that this type of bilayer interaction is a general property of basic helical proteins, however, is not yet available.

Once contact between a protein and lipid bilayer has been established, one must consider how the system will remain stable as the protein potentially penetrates and passes through the bilayer. One hypothesis is that the lipid molecules may stabilize the protein in a chaperone-like manner. Similar activity has been demonstrated previously for the folding of soluble proteins.[46] During FGF-1 refolding, we observe that the aggregation of FGF-1 depends on both the amount of lipid present as well as the composition of the liposome. Although a quantitative correlation between the amount of anionic phospholipid (DOPG) present and FGF-1 aggregation cannot be clearly seen, it is apparent that increased amounts of DOPG cause both some immediate aggregation and increase the rate of thermally induced protein-lipid bilayer association, similar to results observed with PDA vesicles. There is no effect of pure DOPC liposomes on the refolding rates of FGF-1 at 37°C, but stimulatory effects do appear on an increase in the anionic content of the phospholipid bilayer. At lower temperatures (25°C), a smaller but significant interaction between FGF-1 and vesicles containing anionic lipids also is observed. Thus, it seems that the addition of DOPG may be both advantageous and detrimental, stimulating aggregation as well as the rate of refolding. This suggests a more complex interaction between liposomes and FGF-1 than previously hypothesized.

The secondary purpose of this chapter is to provide a sampling of experimental techniques to study protein-membrane interactions, as well as demonstrate the difficulties one may encounter in the use of these methods. These difficulties arise primarily from the presence of processes occurring simultaneously with binding itself. One such process observed in this study is aggregation (Figures 6 and 8). In general, aggregation increases as the negative charge density of the vesicle surface is increased. This is not surprising, since the excess of negatively charged vesicles present could potentially noncovalently cross link basic protein molecules. Additionally, it is observed with the PDA assay and KGF-2, VP22.C1, and FGF-1 that an increase in temperature produces an overall increase in optical density. These trends do not appear to correlate with changes in %CR, however, suggesting that although aggregation may minimally contribute to the observed %CR, the majority of the signal results from the protein-lipid bilayer interaction. This is supported by an examination of changes in light scattering over time during fluorescence release studies. In the FGF refolding and aggregation inhibition studies, however, aggregation does play a major role in the observations, inhibiting definitive conclusions.

Therefore, when examining unstable protein states such as the molten globule, especially in systems where such states may be even transiently present such as lipid bilayer systems, it is apparent that extreme care must be taken in experimental design and interpretation to minimize or account for the contribution of aggregation to the final results. One approach to this problem is prescreening studies involving optical density measurements to identify aggregation-prone conditions. When designing such experiments, the order of component addition also may be critical. For example, incubating a premixed system at the desired temperature before initiating data collection may result in less ambiguous results than preincubating the components separately and mixing them immediately prior to data collection. These types of considerations are system dependent and must be evaluated for each situation.

A secondary consideration is the choice of assays. Of those presented here, the assay that provided the most clearly interpretable results and was the most reliable experimentally was the dye-release assay. The primary factor in this assay is the choice of dye; in this study, two different dyes were employed owing to the pH limitations of the carboxyfluorescein. This assay has proved useful in other situations as well.[47] The PDA assay, although a novel method for screening ligand-bilayer interactions, is still under development. This study presents one of the first applications of this model to molecules as large as proteins; initial studies were conducted on peptides and small molecule ligands. The main drawback of the PDA model in our hands was the extreme sensitivity of the liposome signal-to-buffer components and temperature; this greatly reduced the overall sensitivity of the method. One important item to note for the refolding study is that this particular study is developed specifically on the principle that the fluorescence of FGF-1 is quenched in the folded state. For other proteins with similar fluorescence characteristics, this approach also should work well to monitor unfolding and refolding. For other proteins, the sensitivity may not be as great, since the changes in signal could be much smaller. The use of extrinsic dyes may have to be considered in these systems.

A final consideration is the nature of the liposome itself. Independent of the choice of assay, liposome composition would appear to be the critical variable. In designing individual experiments, the nature of the hypothesis might dictate the nature of the liposomes. Does one wish to mimic the composition of the cell? Is there a specific property (as in this study where surface charge was the key property) that needs to be varied? Bilayer thickness, which can be altered by employing lipids of varying chain lengths, may be a factor in the study design. All these factors will dictate the final composition of the liposome model.

In conclusion, this work provides evidence that MG-like states of proteins may play a role in nonclassical transport. This function may be enhanced by the potentially stabilizing effect of the lipid bilayer, as indicated by the complex effects of liposomes on the aggregation and refolding of FGF-1. It is evident that further studies and more importantly new, transport-sensitive *in vitro* assays are required before a definite MG-effected mechanism can be established. It is important to note, however, that the role these labile states play may not involve the entry process itself, but could reflect participation in endosomal release processes. This would be consistent with recent suggestions that some form of endocytosis in fact is involved in these unusual transport pathways. This work also demonstrates some of the difficulties inherent in studying protein-lipid bilayer interactions, especially when dealing with the aggregation-prone MG state. By exercising care and good judgment in assay choice and design, processes such as aggregation may be minimized.

REFERENCES

1. I. Shin, I. Silman, and L. M. Weiner, Interaction of partially unfolded forms of Torpedo acetylcholinesterase with liposomes, *Protein Sci.* **5**, 42–51 (1996).

2. S. Banuelos and A. Muga, Binding of molten globule-like conformations to lipid bilayers. Structure of native and partially folded alpha-lactalbumin bound to model membranes, *J. Biol. Chem.* **270**, 29910–29915 (1995).
3. F. G. van der Goot, J. H. Lakey, and F. Pattus, The molten globule intermediate for protein insertion or translocation through membranes, *Trends Cell Biol.* **2**, 343–348 (1992).
4. F. G. van der Goot, J. M. Gonzalez-Manas, J. H. Lakey, and F. Pattus, A "molten-globule" membrane-insertion intermediate of the pore-forming domain of colicin A, *Nature* **354**, 408–410 (1991).
5. I. Shin, D. Kreimer, I. Silman, and L. Weiner, Membrane-promoted unfolding of acetylcholinesterase: a possible mechanism for insertion into the lipid bilayer, *Proc. Natl. Acad. Sci. USA* **94**, 2848–2852 (1997).
6. C. M. Dobson and M. Karplus, The fundamentals of protein folding: bringing together theory and experiment, *Curr. Opin. Struct. Biol.* **9**, 92–101 (1999).
7. J. Hawiger, Noninvasive intracellular delivery of functional peptides and proteins, *Curr. Opin. Chem. Biol.* **3**, 89–94 (1999).
8. G. Elliott and P. O'Hare, Intercellular trafficking and protein delivery by a herpesvirus structural protein, *Cell* **88**, 223–233 (1997).
9. A. D. Frankel and C. O. Pabo, Cellular uptake of the tat protein from human immunodeficiency virus, *Cell* **55**, 1189–1193 (1988).
10. S. R. Schwarze, A. Ho, A. Vocero-Akbani, and S. F. Dowdy, *In vivo* protein transduction: delivery of a biologically active protein into the mouse, *Science* **285**, 1569–1572 (1999).
11. A. Prochiantz, Getting hydrophilic compounds into cells: lessons from homeopeptides, *Curr. Opin. Neurobiol.* **6**, 629–634 (1996).
12. W. F. Cheng, C. F. Hung, K. F. Hsu, C. Y. Chai, L. He, J. M. Polo, L. A. Slater, M. Ling, and T. C. Wu, Cancer immunotherapy using Sindbis virus replicon particles encoding a VP22-antigen fusion, *Hum. Gene Ther.* **13**, 553–568 (2002).
13. M. S. Dilber, A. Phelan, A. Aints, A. J. Mohamed, G. Elliott, C. I. Edvard Smith, and P. O'Hare, Intercellular delivery of thymidine kinase prodrug activating enzyme by the herpes simplex virus protein, VP22, *Gene Ther.* **6**, 12–21 (1999).
14. M. C. Morris, J. Depollier, J. Mery, F. Heitz, and G. Divita, A peptide carrier for the delivery of biologically active proteins into mammalian cells, *Nat. Biotech.* **19**, 1173–1176 (2001).
15. L. Chen, L. R. Wright, C. H. Chen, S. F. Oliver, P. A. Wender, and D. Mochly-Rosen, Molecular transporters for peptides: delivery of a cardioprotective epsilonPKC agonist peptide into cells and intact ischemic heart using a transport system, R(7), *Chem. Biol.* **8**, 1123–1129 (2001).
16. K. Braun, P. Peschke, R. Pipkorn, S. Lampel, M. Wachsmuth, W. Waldeck, E. Friedrich, and J. Debus, A biological transporter for the delivery of peptide nucleic acids (PNAs) to the nuclear compartment of living cells, *J. Mol. Biol.* **318**, 237–243 (2002).
17. A. Astriab-Fisher, D. Sergueev, M. Fisher, B. R. Shaw, and R. L. Juliano, Conjugates of antisense oligonucleotides with the Tat and antennapedia cell-penetrating peptides: effects on cellular uptake, binding to target sequences, and biologic actions, *Pharm. Res.* **19**, 744–754 (2002).
18. D. J. Stauber, A. D. DiGabriele, and W. A. Hendrickson, Structural interactions of fibroblast growth factor receptor with its ligands, *Proc. Natl. Acad. Sci. USA* **97**, 49–54 (2000).
19. M. Rusnati, G. Tulipano, D. Spillmann, E. Tanghetti, P. Oreste, G. Zoppetti, M. Giacca, and M. Presta, Multiple interactions of HIV-1 Tat protein with size-defined heparin oligosaccharides, *J. Biol. Chem.* **274**, 28198–28205 (1999).
20. D. Derossi, G. Chassaing, and A. Prochiantz, Trojan peptides: the penetratin system for intracellular delivery, *Trends Cell Biol.* **8**, 84–87 (1998).
21. L. A. Kueltzo and C. R. Middaugh, Nonclassical transport proteins and peptides: an alternative to classical macromolecule delivery systems, *J. Pharm. Sci.* **92**, 1754–1772 (2003).
22. L. A. Kueltzo, N. Normand, P. O'Hare, and C. R. Middaugh, Conformational lability of herpesvirus protein VP22, *J. Biol. Chem.* **275**, 33213–33221 (2000).
23. R. A. Copeland, H. Ji, A. J. Halfpenny, R. W. Williams, K. C. Thompson, W. K. Herber, K. A. Thomas, M. W. Bruner, J. A. Ryan, D. Marquis-Omer, G. Sanyal, R. D. Sitrin, S. Yamazaki, and

C. R. Middaugh, The structure of human acidic fibroblast growth factor and its interaction with heparin, *Arch. Biochem. Biophys.* **289**, 53–61 (1991).
24. S. Futaki, T. Suzuki, W. Ohashi, T. Yagami, S. Tanaka, K. Ueda, and Y. Sugiura, Arginine-rich peptides: an abundant source of membrane-permeable peptides having potential as carriers for intracellular protein delivery, *J. Biol. Chem.* **276**, 5836–5840 (2001).
25. G. Drin, H. Demene, J. Temsamani, and R. Brasseur, Translocation of the pAntp peptide and its amphipathic analogue AP-2AL, *Biochemistry* **40**, 1824–1834 (2001).
26. D. Derossi, S. Calvet, A. Trembleau, A. Brunissen, G. Chassaing, and A. Prochiantz, Cell internalization of the third helix of the antennapedia homeodomain is receptor-independent, *J. Biol. Chem.* **271**, 18188–18193 (1996).
27. V. E. Bychkova, R. H. Pain, and O. B. Ptitsyn, The "molten globule" state is involved in the translocation of proteins across membranes? *FEBS Lett.* **238**, 231–234 (1988).
28. H. Mach, J. A. Ryan, C. J. Burke, D. B. Volkin, and C. R. Middaugh, Partially structured self-associating states of acidic fibroblast growth factor, *Biochemistry* **32**, 7703–7711 (1993).
29. L. A. Kueltzo and C. R. Middaugh, Structural characterization of bovine granulocyte colony stimulating factor: effect of temperature and pH, *J. Pharm. Sci.* **92**, 1793–1804 (2003).
30. E. J. Parkinson, M. B. Morris, and S. Bastiras, Acid denaturation of recombinant porcine growth hormone: formation and self-association of folding intermediates, *Biochemistry* **39**, 12345–12354 (2000).
31. G. Christofori and S. Luef, Novel forms of acidic fibroblast growth factor-1 are constitutively exported by beta tumor cell lines independent from conventional secretion and apoptosis, *Angiogenesis* **1**, 55–70 (1997).
32. D. B. Volkin, P. K. Tsai, J. M. Dabora, J. O. Gress, C. J. Burke, R. J. Linhardt, R. J., and C. R. Middaugh, Physical stabilization of acidic fibroblast growth factor by polyanions, *Arch. Biochem. Biophys.* **300**, 30–41 (1993).
33. H. Mach and C. R. Middaugh, Interaction of partially structured states of acidic fibroblast growth factor with phospholipid membranes, *Biochemistry* **34**, 9913–9920 (1995).
34. K. Kasraian, A. Kuzniar, D. Earley, B. J. Kamicker, G. Wilson, T. Manion, J. Hong, C. Reiber, and P. Canning, Sustained *in vivo* activity of recombinant bovine granulocyte colony stimulating factor, (rbCSF) using HEPES buffer, *Pharm. Dev. Technol.* **6**, 441–447 (2001).
35. K. Fan, M. Sevoian, and D. Gonzales, Instability studies of porcine somatotropin in aqueous solutions and the possible reagents for its stabilization, *J. Agric. Food Chem.* **48**, 5685–5691 (2000).
36. S. Kolusheva, T. Shahal, and R. Jelinek, Peptide-membrane interactions studied by a new phospholipid/polydiacetylene colorimetric vesicle assay, *Biochemistry* **39**, 15851–15859 (2000).
37. M. Bogdanov and W. Dowhan, Lipid-assisted protein folding, *J. Biol. Chem.* **274**, 36827–36830 (1999).
38. M. Bogdanov, M. Umeda, and W. Dowhan, Phospholipid-assisted refolding of an integral membrane protein. Minimum structural features for phosphatidylethanolamine to act as a molecular chaperone, *J. Biol. Chem.* **274**, 12339–12345 (1999).
39. E. A. Bryson, S. E. Rankin, M. Carey, A. Watts, and T. J. T. Pinhiero, Folding of apocytochrome *c* in lipid micelles: formation of alpha-helix precedes membrane insertion, *Biochemistry* **38**, 9758–9767 (1999).
40. D. D. Lasic, *Liposomes in Gene Delivery* (Boca Raton, New York: CRC Press, 1997).
41. S. Kolusheva, L. Boyer, and R. Jelinek, A calorimetric assay for rapid screening of antimicrobial peptides, *Nat. Biotech.* **18**, 225–227 (2000).
42. K. Edwards, L. Kueltzo, M. Fisher, and C. Middaugh, Complex effects of molecular chaperones on the aggregation and refolding of fibroblast growth factor-1, *Arch. Biochem. Biophys.* **393**, 14–21 (2001).
43. J. M. Dabora, G. Sanyal, and C. R. Middaugh, Effect of polyanions on the refolding of human acidic fibroblast growth factor, *J. Biol. Chem.* **266**, 23637–23640 (1991).
44. S. Okada, S. Peng, W. Spevak, and D. Charych, Color and chromism of polydiacetylene vesicles, *Acc. Chem. Res.* **31**, 229–239 (1998).

45. B. Lovejoy, D. Cascio, and D. Eisenberg, Crystal structure of canine and bovine granulocyte-colony stimulating factore (G-CSF), *J. Mol. Biol.* **234**, 640–653 (1993).
46. G. Zardeneta and P. M. Horowitz, Micelle-assisted protein folding. Denatured rhodanese binding to cardiolipin-containing lauryl maltoside micelles results in slower refolding kinetics but greater enzyme reactivation, *J. Biol. Chem.* **267**, 5811–5816 (1992).
47. L. A. Kueltzo, J. Osiecki, J. Barker, W. L. Picking, B. Ersoy, W. D. Picking, and C. R. Middaugh, Structure-function analysis of invasion plasmid antigen C (IpaC) from *Shigella flexneri*, *J. Biol. Chem.* **278**, 2792–2798 (2003).

Part V
Protein Product Development

Self-Association of Therapeutic Proteins

Implications for Product Development

Mary E. M. Cromwell,[1,2] Chantal Felten,[1] Heather Flores,[1] Jun Liu,[1] and Steven J. Shire[1]

Protein aggregation is typically thought of as a denaturation process that results in the formation of precipitate. However, native proteins may reversibly self-associate to form discrete aggregates of dimer, trimer, and higher molecular weight forms. The rates of reversible aggregate association and dissociation vary dramatically from protein to protein and can lead to misconceptions regarding aggregate content if the incorrect assay is used to determine the extent of aggregation. The implications of the formation of these native aggregate states during pharmaceutical development of therapeutic proteins are numerous. For example, self-association of insulin has been exploited to develop a stable protein formulation. Often, however, the formation of these native-type aggregates is undesired and results in several challenges during product development. These challenges may involve the selection of an appropriate method to accurately monitor the level of aggregation, the development of a robust manufacturing process to minimize the level of aggregate created during production, and the selection of appropriate formulation components to minimize the aggregate content of the final product that is delivered to patients. This chapter will provide case studies demonstrating how self-association of therapeutic proteins resulted in challenges to develop robust methods for measurement of the aggregated state, to control physical properties of the pharmaceutical, and to implement a reasonable manufacturing process to limit the amount of these aggregates. In each case, understanding the thermodynamics and kinetics of aggregate association and dissociation played an essential role in the actions taken.

---

[1] Genentech, Inc., 1 DNA Way, South San Francisco, CA 94080
[2] To whom correspondence should be addressed: Mary Cromwell, Genentech, Inc., 1 DNA Way, So. San Francisco, CA 94080; email: Cromwell.mary@gene.com

1. INTRODUCTION

Protein aggregation may result from protein-protein interactions in either the native or the denatured states. Protein aggregation may result in covalent (e.g., disulfide-linked) association or noncovalent (reversible or irreversible) association. Irreversible aggregation by noncovalent association often is recognized as a major degradation pathway for proteins, especially when formulated at high protein concentrations. Irreversible protein aggregation may result from hydrophobic regions exposed by thermal, mechanical, or chemical processes that alter a protein's native conformation,[1] and often results in the formation of insoluble particulates. When the protein-unfolding step is rate limiting, the kinetics of protein aggregation often show pseudo-first-order behavior[1] rather than the expected higher-order concentration dependency. On the other hand, reversible protein self-association generally has a higher-order concentration dependency, and is often overlooked because of the use of analytical methodologies involving significant dilution of protein, which results in dissociation of weakly associated protein complexes. In addition, there is a perception that reversibility of the protein association will have little impact on stability, administration, or properties of the protein formulation.

Reversible protein self-association may impact protein activity, pharmacokinetics, and safety, for example, due to immunogenicity[2-4] especially under conditions where the rate of dissociation is slow. Even where dissociation is rapid, at high protein concentrations the equilibrium may be shifted toward a greater amount of aggregate due to molecular crowding effects[5-7] and an increase of short-range molecular interactions, such as electrostatic and van der Waals' interactions. This increase in the apparent thermodynamic association constant with increasing protein concentration may be shelf life limiting, especially when during storage covalent linkages such as intermolecular disulfides are generated resulting in irreversible aggregation. After *in vivo* administration, especially for subcutaneous injection where dilution is slow, the protein therapeutic may be placed into a crowded macromolecular environment, which may shift the equilibrium to the associated state even at lower concentrations that result from dilution.

This chapter will present case studies of aggregation observed in four proteins generated as therapeutic entities. In each case, aggregation was observed that led to challenges in the development of each product for pharmaceutical use.

2. ANALYTICAL CONSIDERATIONS

Most analytical technologies that are used to characterize proteins often require dilution to lower concentrations or exposure of the protein to solvent conditions that are very different than the initial formulation composition. This problem is especially important in the analysis of molecular weight/size distribution of proteins undergoing reversible self-association since upon dilution higher molecular weight forms may dissociate and then are undetectable by the assay. Sodium dodecyl sulfate-polyacrylamide gel electrophoresis (SDS-PAGE), non-gel-sieving SDS-capillary electrophoresis, and matrix-assisted laser desorption ionization time-of-flight mass spectrometry (MALDI-TOF MS) often are used to obtain molecular weight information. These techniques are very useful for detecting covalently linked aggregates or SDS nondissociable aggregates but cannot be used to determine noncovalent protein-association states. On the other hand, gel permeation or size-exclusion chromatography (SEC) provides such information but has several problems. Elution times during the chromatography may be altered due to the protein's interaction with the chromatographic resin. In addition SEC is dependent on hydrodynamic volume rather than mass, and thus the use of globular protein standards to estimate the molecular weight may be erroneous. The impact of hydrodynamic volume leading to apparently larger proteins than actual size is especially noted for highly glycosylated proteins that have shapes different from the typical globular protein standards.[8,9] The use of on-line low-angle or multiangle static light-scattering detectors coupled with sizing chromatography (LC/LS)[10] allows for the absolute determination of molecular mass of a protein and its higher-order aggregates and fragments during separation by gel sieving. Since measured elution times are no longer used to estimate molecular size, the problem of protein-resin interaction and molecular conformation are no longer important, provided the interactions do not prevent elution of the protein from the chromatographic column. Although LS detectors allow for more accurate mass determination of eluting SEC peaks, erroneous results and conclusions are still possible due to the dilution and limited resolution that occurs during the chromatography. Analytical ultracentrifugation (AUC) may be used to study protein self-association under native conditions and may yield information regarding molecular mass, binding affinities, and the sedimentation coefficient.

Several of the techniques commonly used for studying protein aggregation are briefly described in Table 1. Each technique offers distinct advantages and disadvantages. No single method provides all information regarding protein aggregation. For this reason, it is common to study aggregation using a combination of techniques.

3. RECOMBINANT HUMAN RELAXIN

Relaxin is a naturally occurring pregnancy hormone related in structure to insulin and insulin-like growth factor. Relaxin plays a role in partuition and has been investigated in several clinical trials as a cervical ripening agent prior to labor induction.[11-14] More recent studies have focused on the use of relaxin in treating schleraderma (systemic schlerosis).[15,16] During clinical development of relaxin, a reversible self-association was observed that led to a critical analysis of which analytical techniques were useful in assessing aggregation states.

3.1. Characterization of Size-Variants of Relaxin

The determination of the aggregation state of human relaxin provides an example of a case in which the results from SEC were misleading. When loaded at a concentration of 0.4 mg/ml (67 μM), the protein elutes in 10 mM sodium citrate, 250 mM NaCl at pH 5.0 as a single peak with an apparent molecular weight of 5600, which is comparable to the theoretical molecular weight of 5963 obtained from the amino acid sequence.[17] These data initially led to the conclusion that human relaxin existed as a

TABLE 1. Comparison of common experimental techniques used to study protein aggregation

Technique	Advantages	Disadvantages	Min/max sample conc. (mg/ml)	Cost	Type of data obtained	Time to perform analysis	Other
SEC (UV detection)	Robust; Easy to validate; High throughput; High sensitivity; Determine hydrodynamic information about protein	Sample dilution on injection; Filtration of sample through matrix; Interactions of protein with matrix; Change of solution conditions on injection; Limited buffer conditions; Limited size range; Shear stress on protein; Requires standards to obtain molecular weight	0.1–150	Moderate; Instrument + data collection and analysis: $50–70k	Aggregate amount (quantitative); Molecular weight (qualitative)	<0.5 hr/sample	Standard industry technique for monitoring aggregation content; HPLC may be used for other techniques
AUC (Sedimentation velocity)	Can use formulation buffer and wide range of other solution conditions; Detect high molecular weight aggregates that may be filtered by SEC; Determines a physical parameter, sedimentation coefficient, related to shape and molecular weight; No molecular weight standards required	Extensive training required; Dilution of sample typically necessary; Difficult to validate; Low throughput	0.05–20	Expensive; Instrument + data collection and analysis: $250k	Sedimentation coefficient and aggregate amount (quantitative); Molecular weight (qualitative)	2–3 hr/run 3 samples /run	Can be used to determine small difference (<1%) in sedimentation coefficient as a result of change in conformation or change in molecular mass; Sensitive to change of aggregate distribution

SELF-ASSOCIATION OF THERAPEUTIC PROTEINS

Method	Advantages	Disadvantages	Range	Cost	Information	Time	Notes
AUC (Sedimentation equilibrium)	Can use formulation buffer and wide range of other solution conditions No dilution required Obtain information on binding affinity, molecular weight, and virial coefficient No molecular weight standards required	Extensive training required	0.1–300	Expensive Instrument + data collection and analysis: $250k	Molecular weight, binding constants and virial coefficient (quantitative) Aggregate amount (qualitative)	24–72 hr/run Up to 28 samples/run	Can be used to evaluate reversible association
Multiangle light scattering	Sensitive detection of large particles Used as a detection system with a separating method (e.g., SEC), able to determine absolute molecular weight and radius of gyration	As a detection system with LC, problems of sample dilution, filtration, potential interactions with resin Limited size ranges Difficult to validate Interference from shedding particles from resin	0.1–150	Moderate Instrument + refraction detector: $50k Coupled with HPLC: $100–120k	Molecular weight and radius of gyration (quantitative) Aggregate amount (qualitative)	0.5 hr/sample	Can be used to assess molecular weight homogeneity of sample within an eluting sizing chromatography peak
Dynamic light scattering	Sensitive detection of large aggregates Performed under native conditions	No discrimination between monomer and smaller aggregates Interference from large particles Difficult to validate	1–150	Moderate Instrument: $50–70k	Hydrodynamic radius and diffusion coefficient (quantitative) Molecular weight and aggregate amount (qualitative)	<0.5 hr/sample	Useful to detect trace amounts of very large aggregates

(*cont.*)

TABLE 1. (Cont.)

Technique	Advantages	Disadvantages	Min/max sample conc. (mg/ml)	Cost	Type of data obtained	Time to perform analysis	Other
SDS-PAGE	Inexpensive High throughput Little training required Separation of covalent aggregates Very useful for rapid determination of disulfide cross-linked aggregates	Dissociates noncovalent aggregates Sample preparation may result in aggregate formation Limited resolution, especially if for a glycosylated protein Molecular mass determination is indirect		Inexpensive Equipment: <$20k	Covalent linked aggregate amount (quantitative) Molecular weight (qualitative)	0.5 days/run 10 samples/run	Notoriously inaccurate with hydrophobic and highly glycosylated proteins Different sensitivities may be achieved through selection of detection dye
CE-SDS	High throughput Separation of covalent aggregates Rapid determination of disulfide cross-linked aggregates	Dissociates nonconvalent aggregates Sample preparation may result in aggregate formation Poor resolution with glycosylated protein		Moderate Instrument: $30–50k	Covalently linked aggregate amount (quantitative) Molecular weight (qualitative)	0.5 hr/sample	Choice of UV or laser-induced fluroscence (LIF) detection gives broad range in sensitivity

monomer in solution. Subsequent sedimentation equilibrium studies clearly showed that human relaxin in fact undergoes reversible monomer-dimer self-association[18,19] with a dissociation equilibrium constant, K_d, of 1.6 µM in 10 mM citrate, 0.13 M NaCl, pH 5.0. Typical dilutions during SEC range from 10- to 20-fold, and these dilutions coupled with limited resolution of dimer from monomer, different solution ionic strengths, and the sensitivity of the detector may explain the SEC result. This is a likely explanation since an increase of the loading concentration to 2 mg/ml (308 µM) did result in detection of the dimer.[17] Despite the limitations of SEC, it is possible to investigate weakly associating systems such as relaxin (µM K_d) using short SEC columns and rapid flow rates if the aggregate can be resolved. The K_d and dissociation rate constant can be determined using computer simulation for moving boundary analysis as demonstrated by analysis of monomer-dimer equilibrium of human growth hormone.[20]

Dilution of relaxin resulting in an increase in weight fraction of monomer showed that the dissociation of dimer did not result in a significant change in the far UV circular dichroism (CD) spectrum, suggesting that there is little conformational change upon dissociation of the dimer. However, the near UV CD showed a significant change in the region usually assigned to tyrosine residues. This suggests that a tyrosine residue becomes more exposed upon dissociation of the relaxin dimer. Relaxin crystallizes as a dimer and the lone tyrosine is centered at the dimer interface, supporting the results of the CD concentration difference spectra.[21]

4. RhuMAb VEGF

RhuMAb VEGF is an IgG type I antibody to vascular endothelial growth factor and is the active ingredient in Avastin™. Developed to inhibit angiogenesis, this protein has been shown to be efficacious in phase III clinical trials for the treatment of colorectal cancer.[22] The propensity of this protein to form reversible aggregates was determined during clinical development of rhuMAb VEGF. Characterization of the kinetics and thermodynamics of dimer formation and dissociation showed that the reversible aggregation was controllable by attention to solution conditions.[23]

4.1. Determination of Reversible Aggregation

During evaluation of in-process samples, it was observed that the amount of aggregate in rhuMAb VEGF solutions varied widely, with a mere buffer exchange resulting in a reduction of aggregate levels by several percent. Since no precipitate was observed that could account for the decrease in aggregate, it was suspected that the aggregate forms might be reversible and that they might be in equilibrium with monomer. To test this hypothesis, a rhuMAb VEGF solution was prepared at a moderately high protein concentration (30 mg/ml) and incubated at 40°C to allow aggregate to form. This resulting solution was diluted with the same buffer to 1 mg/ml while maintaining the 40°C incubation temperature. Several time points were analyzed by SEC to determine the content of aggregates in the solution. As displayed in Figure 1, the aggregate content decreased over time. This indicated that the aggregate was indeed reversible.

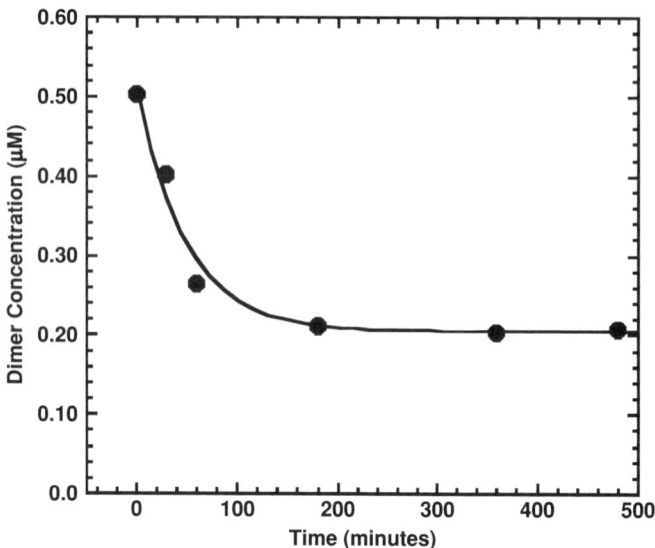

FIGURE 1. The dissociation of rhuMAb VEGF dimer upon dilution from 30 mg/ml to 1 mg/ml. The dissociation was carried out at 40°C with the dimer concentration assessed by SEC. The line through the data points represents a first-order fit to the data.

The equilibrium of rhuMAb VEGF aggregate formation shows temperature dependence. Changing the storage temperature from 5°C to 30°C results in an increase in aggregate from approximately 3% to almost 8% in 10 hours for a 25 mg/ml sample. Returning the sample to 5°C results in a decrease in the aggregate content as displayed in Figure 2.

There is a strong pH dependence of the equilibrium constant for dissociation, K_D. As shown in Table 2, K_D for dimer dissociation is lowest between pH 7.5 and pH 8.0. The K_D at pH 5.5 and lower were not determined because it was not possible to accurately quantify the small amounts of aggregate that formed. The pH dependence implies that there may be an electrostatic interaction that leads to aggregation. However, the isoelectric point, pI, of this protein is approximately 8.2, and thus, the protein would be minimally charged at pH 8.0. Therefore, the observed self-association at near-neutral pH may result simply from charge shielding, in which the repulsive forces between monomers are decreased.

From the results described so far, it is evident that the amount of reversible aggregate forms in rhuMAb VEGF depends on the protein concentration, pH, and temperature. The kinetics of association and dissociation reveal that these changes in aggregate content begin immediately with a change in conditions, and equilibrium is reached sometime between 2 and 10 hours.[23]

4.2. Analytical Methodologies

One of the key quality attributes of pharmaceutical products is the purity. SEC is one of the workhorses in assessing size heterogeneity in biotechnology-derived products.

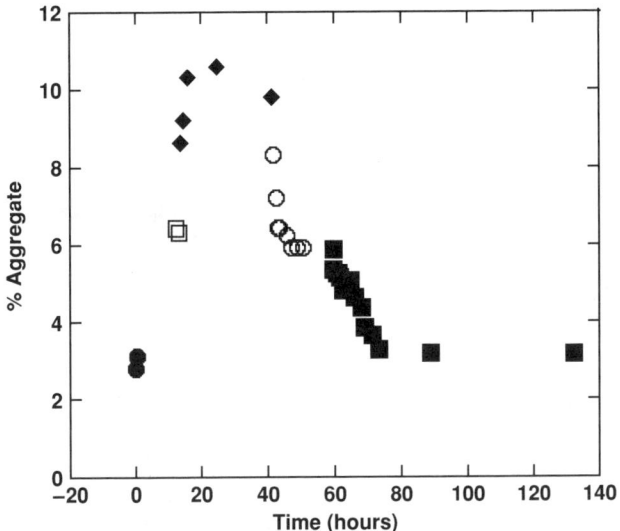

FIGURE 2. The equilibrium concentration of aggregate (predominantly dimer) depends highly on the storage temperature. The results shown are from a single sample stored in a sample chamber in which the chamber temperature was changed as follows: 5°C (●); samples moved from 5°C to 30°C (□); samples moved from 30°C to 40°C (◆); samples moved from 40°C to 30°C (○); and samples moved from 30°C to 5°C (■). The aggregate content was assessed by SEC over time. Once it was determined that equilibrium was achieved, the temperature of the sample chamber was changed, and the reequilibration was followed.

Assessment of the size heterogeneity in Avastin™ proved to be nonroutine due to the reversible nature of the aggregates. The kinetics of association and dissociation of the dimer are in a time frame in which the measured values of aggregates may be affected by sample handling. Additionally, other aggregate forms coelute with the reversible aggregates. These aggregates persist after dilution and the amounts are invariant after incubation for several days in the diluted condition. Due to their persistence, it was deemed necessary that these aggregates should be measured. While it could be argued that the reversible aggregates may not be biologically relevant if the aggregates dissociate upon dilution during IV administration, the persistent aggregates most likely would

TABLE 2. Equilibrium dissociation constant K_D for rhuMAb VEGF as a function of pH.[23] Errors range from 2–6% of the value shown. Dissociation constants were determined at 40°C. All samples were formulated in 10 mM phosphate at the listed pH

pH	K_D (M)
6.5	3.5×10^{-4}
7.0	2.3×10^{-4}
7.5	1.8×10^{-4}
8.0	1.8×10^{-4}
8.5	2.0×10^{-4}

circulate on administration and therefore should be carefully controlled. Thus, a condition was needed in which the persistent aggregates could be measured in the absence of the reversible aggregates. This was achieved by dilution of the product to a protein concentration in which the contribution of the reversible aggregates was minimal (although not zero).

For this product, it is necessary to achieve equilibrium prior to measurement to obtain consistent and meaningful results. Any exposure of the sample to temperatures other than the temperature used for analysis results in a change in the aggregate content. One can imagine taking a sample from a refrigerator, performing the necessary dilution on a lab bench at room temperature (which may vary by a few degrees depending on time of day, location of work area relative to vents/windows) with buffer that may not be at the same temperature as the sample. With multiple samples, an increased exposure time to ambient temperatures would be expected. Additionally, placement of the samples in the HPLC sample chamber would result in having some samples at the chamber temperature for periods of time longer than other samples. All these steps can affect the equilibrium content. Even if great care is taken to perform all steps under refrigerated conditions or on ice, the day-to-day variability in handling time at ambient temperatures could cause significant differences in the resulting aggregate content measured. For these reasons, it is imperative to perform the analysis on samples that have reached equilibrium to avoid assessment of "a moving target."

As described earlier, the equilibrium concentration depends on storage temperature. At what temperature would it be best to test the pharmaceutical product? The recommended storage temperature for the product is $2°C$-$8°C$ for the shelf life of the product. However, the product may experience excursions up to $30°C$ for up to 3 days during manufacturing, shipping, and handling of the pharmaceutical at the site of use. Knowing that the aggregate content increases with increasing temperature, the conservative approach taken for assessment of aggregation was to perform the experiments on samples held at $30°C$. This provides a worst-case scenario about the aggregate content in the drug.

4.3. Implications for Product Development

The presence of reversible aggregates in rhuMAb VEGF resulted in the development of a significantly more rigorous SEC method than typically required for biotechnology derived products. To accurately assess aggregate content, samples are prepared at two protein concentrations to gain information on the total aggregate and the persistent aggregates. Moreover, samples are allowed to come to equilibrium at each concentration to avoid measurement of a moving target.

5. Apo2L/TRAIL

Apo2L/TRAIL is a noncovalent homotrimeric cytokine of the tumor necrosis factor (TNF) superfamily that induces apoptosis in a variety of cancer cells, both *in vitro* and in mouse xenograft tumor models, with little effect on most normal cells. Recombinant soluble Apo2L/TRAIL (amino acids 114-281) consists of three 19.5-kDa monomers

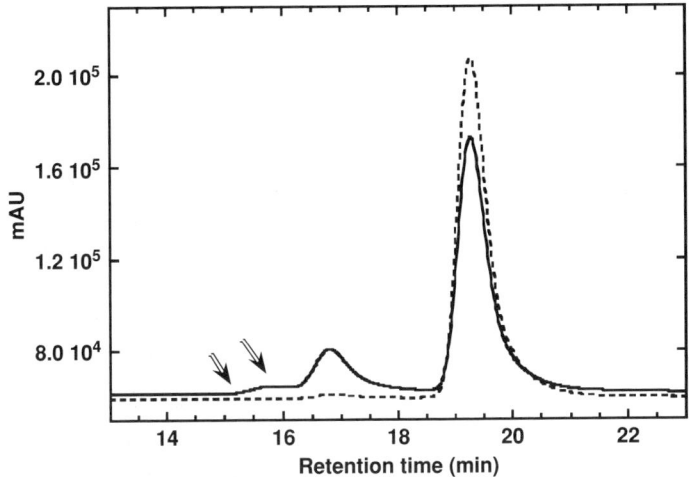

FIGURE 3. Size exclusion chromatogram of a 100 mg/ml Apo2L/TRAIL solution stored for 2 months at 2–8°C (solid line) and a control sample stored at −70°C (dashed line). Upon storage, there is an increase in the amount the hexamer eluting at ∼16.8 min, and also two new larger molecular weight species (arrowheads) appear which were not detectable in the control.

that are noncovalently bound through large hydrophobic interfaces and coordination of three free thiols of Cys230 of each subunit to a common Zn atom.[24] This trimeric form binds two known death receptors, DR4 (also called TRAIL-R1) and DR5 (also called TRAIL-R2), which subsequently engage the apoptotic caspase machinery leading to cell death[25] and *in vivo* antitumor activity in mouse models.[26,27]

5.1. Size Heterogeneity in Apo2L/TRAIL Species

Recombinant trimeric Apo2L/TRAIL (amino acids 114-281) formed higher molecular weight species upon storage at high concentrations in solution. These oligomers reverted to trimer upon dilution. Size-exclusion HPLC was used to determine the molecular weight of Apo2L/TRAIL association states formed upon storage in solution. Figure 3 shows the SEC chromatogram of Apo2L/TRAIL when stored as a 100 mg/ml solution for 2 months at 2–8°C compared to a control frozen sample. The area percent of the peak at ∼16.8 min retention time increased concomitant with a decrease in the main peak area, and two new minor species with shorter retention times also appeared (Figure 3, arrowheads).

The apparent molecular weights of the four peaks seen in the incubated sample were estimated at 240 kDa, 180 kDa, 125 kDa, and 55 kDa based on comparison of retention times to those of molecular weight markers. Since these apparent molecular weights could be affected by the shape and other hydrodynamic properties of the aggregates, the actual molecular weights were confirmed by on-line multiangle laser light-scattering (LS) detection. The molecular weights determined by LS were 247 ± 12 kDa, 188 ± 7 kDa,

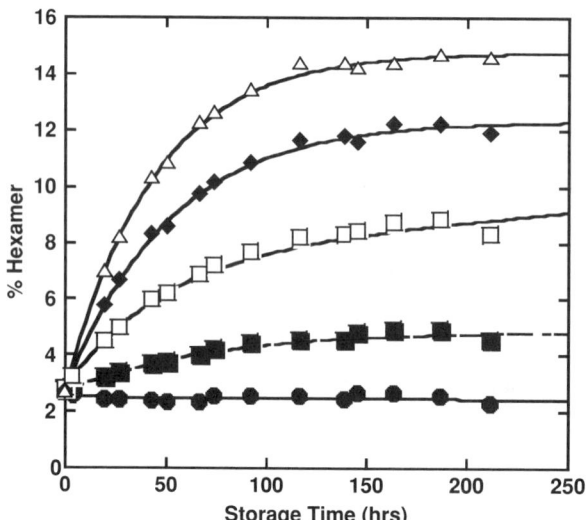

FIGURE 4. Kinetics of Apo2L/TRAIL hexamer formation in solution at 30°C. The concentrations of solutions tested were 10 mg/ml (●), 20 mg/ml (■), 40 mg/ml (□), 60 mg/ml (◆), and 80 mg/ml (○). Samples were analyzed by SEC at each time point to determine hexamer content.

122 ± 5 kDa, and 60 ± 3 kDa, corresponding to dodecamer, nonamer, hexamer, and trimer, respectively.

5.2. Kinetics of Association and Dissociation

The kinetics of Apo2L/TRAIL self-association were followed at various protein concentrations and temperatures. The rate of hexamer formation at 30°C for samples at 10 to 80 mg/ml and starting with 2.7 % hexamer is shown in Figure 4. The observed rate of aggregation in solution is rapid compared to the timescale of manufacturing operations. For example, at 80 mg/ml the solution contains over 10% hexamer after storage at 30°C for only 3 days and comes to an equilibrium value of 14% in less than 1 week. The equilibrium value of hexamer depends on the protein concentration.

The observed plateau in percent hexamer suggests that either an equilibrium exists between hexamer and trimer or that only a subpopulation of partially unfolded trimer is prone to association. Several findings point to the former rather than the latter explanation. First, the percent hexamer at equilibrium varied proportionally to the square of the protein concentration (data not shown). If the hexamer content had reached a plateau due to the depletion of an aggregation-prone subpopulation, the percent hexamer would have reached the same plateau value at all protein concentrations. Second, biophysical characterization of hexamer, as described above, demonstrated structural similarity with the trimer. Finally, hexamer was reversible upon dilution and converted back to trimer (Figure 5). That hexamer formation decreases with increasing ionic strength indicates that charge interactions play a major role in the self-association of Apo2L/TRAIL.

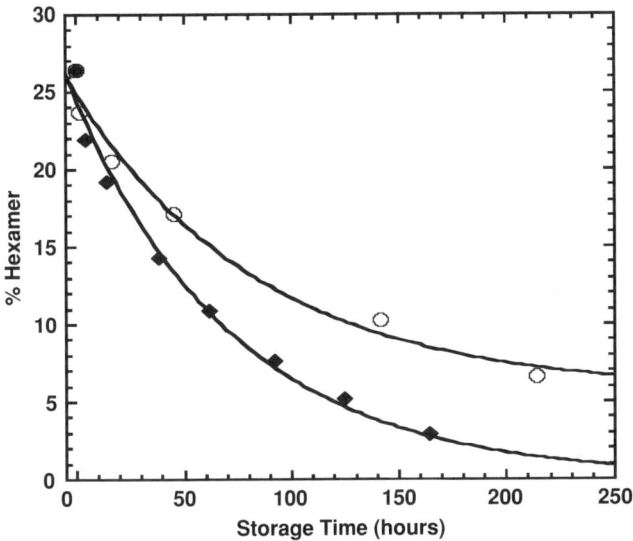

FIGURE 5. A solution containing 25% hexameric Apo2L/TRAIL was diluted to 1 mg/ml in a neutral pH buffer with 0.05M NaCl (o) and 0.5M NaCl (♦).

5.3. Implications for Product Development

There are several analytical and pharmaceutical implications of Apo2L/TRAIL reversible self-association. Fortunately, the observed rate of hexamer dissociation is relatively slow after dilution compared to the time required for the analytical SEC assay. Accurate quantitation of this species by SEC is made possible by controlling the dilution and handling time and temperature of samples for SEC analysis. The slow dissociation in isotonic buffers also means that dilution prior to delivery cannot be relied on to minimize the amount of aggregate delivered to the patient. For this reason, the formation of hexamer must be controlled throughout the manufacturing process.

The complexity of manufacture of Apo2L/TRAIL leads to a multistep process in which the protein may be held in solution after any of the steps prior to continuation of the process. Soluble human Apo2L/TRAIL containing amino acids 114–281[26] is expressed in the cytoplasm of *Escherichia coli*. The protein is purified by centrifugation of the lysed cells, followed by three chromatographic steps. The final purified protein is then concentrated and buffer exchanged into its final formulation by ultrafiltration/diafiltration (UF/DF). The well-defined kinetics and thermodynamics of Apo2L/TRAIL self-association described here enables control of the hexamer level during this purification process. The most effective method of minimizing the amount of hexamer formed in solution is to keep the solution at low protein concentration. However, this is not always feasible during large-scale manufacturing. The formation of hexamer can be slowed significantly, even at higher protein concentrations, by keeping the solution at low temperature and limiting the amount of time the solutions are stored. Further, as hexamer formation is slower in high ionic strength solutions, optimization of the chromatographic steps in the process can include the use of high ionic strength buffer systems, where appropriate, to shift the equilibrium toward the trimer state. This results in an increase in the time that the resulting column pool can be held without formation of significant quantities of aggregates.

The final formulation of Apo2L/TRAIL was carefully designed to hinder self-association to hexamers. Experiments were conducted utilizing a variety of pharmaceutically acceptable salts to determine if this primarily electrostatic interaction could be interfered with by the introduction of specific ions. The results of these experiments led to a formulation for Apo2L/TRAIL that included excipients that appear to interact with the protein in such a way as to block at least part of the interaction. Interference of this interaction led to a decreased observed rate of aggregation in the formulated bulk. As the self-association of Apo2L/TRAIL is not completely disrupted in the final formulation, limits are necessary on the handling times and temperatures of the concentrated protein solution to minimize the amount of aggregate in the final product. In particular, hexamer formation is completely halted by freezing of the solution to below $-20°C$ or by freeze-drying the preparation.

6. RECOMBINANT HUMANIZED ANTIBODY

Reversible protein self-association affects physical properties of a protein solution such as viscosity. As an example, physical properties of three recombinant humanized monoclonal antibodies with the same IgG1 constant region framework and in the same formulation were compared. The molecular mass as determined by SEC was similar for all three antibodies at the expected value of 150 kDa. After freeze-drying, one of the three monoclonal antibodies required a significantly longer time for reconstitution, and comparison of the viscosities after reconstitution to 85 mg/ml revealed that the viscosity of this antibody solution was ~three-fold higher than that of the other two antibodies (Table 3). This higher viscosity may be a prime reason why the reconstitution takes longer (Table 3) for this antibody since efficient mixing of the diluent with the lyophilized solid may not be as efficient in the higher viscosity solution. In addition, the viscosity of this antibody solution is highly concentration dependent, affecting the time required to remove formulation from a vial with a syringe.[28]

TABLE 3. Effect of viscosity on reconstitution time. In this experiment, the viscosity and reconstitution times were compared for three recombinant humanized monoclonal antibodies: Ab1, Ab2, and Ab3. The viscosity was measured at concentration of 85 mg/ml. The reconstitution times for three monoclonal antibodies were determined after reconstituting the lyophilized products with sterile water for injection to a final protein concentration of 100 mg/ml.

	Ab1	Ab2	Ab3
Viscosity (cP)	7.6	2.4	2.4
Reconstitution time (minutes)	8.3	5	4

The viscosity of a solution containing macromolecules depends on the concentration, shape, molecular weight, and interaction of the macromolecules. The viscosity of a protein solution η is related to the solvent viscosity η_0 and concentration in g/ml of the protein, c_p, by a power series[29]

$$\eta = \eta_0 \left(1 + k_1 c_p + k_2 c_p^2 + k_3 c_p^3 + \ldots\right), \tag{1}$$

where k_1 is associated with the contribution from individual solute molecules and k_2, k_3, and higher-order coefficients are related to interactions of protein molecules.

To determine if reversible protein self-association could account for the observed differences in viscosity between the three antibodies discussed above, the protein molecular weight in solution was determined at the high concentration of the formulations. Such an analysis is difficult to do with SEC because of the requirements for dilution before analysis. Indeed, in this study the SEC chromatograms show little difference between the three antibodies (data not shown). However, sedimentation equilibrium data can be collected at high concentration either with an analytical ultracentrifuge[30] or a preparative centrifuge and a microfractionator as described.[31–33] In the latter method, formulated protein at high concentration is centrifuged to equilibrium in a swinging bucket rotor and the contents are collected in 10 μl fractions using a Brandell microfractionator.[34] The collected fractions are placed into a 96-well plate for UV absorption spectroscopy analysis. Analysis of the concentration gradients to obtain actual weight average molecular weights is complicated because of the high protein concentrations, which results in a thermodynamically nonideal solution due to excluded volume and charge effects.[7,35,36] The data can be analyzed in a semiempirical fashion by assuming that the monoclonal antibodies that do not show a high concentration dependency on viscosity are not associating to any great extent in solution. This assumption appears to be valid since the viscosity dependence of these antibodies can be accounted for solely by the behavior of quasispherical particles at high concentration, i.e., the expected dependency due to excluded volume effects without any interaction between the antibodies.[37] The corrections due to excluded volume and protein charge can then be applied to obtain a corrected molecular weight as a function of concentration (Figure 6). The semiempirical corrections are estimates and probably do not account for the complete correction, but the analysis clearly demonstrates that the monoclonal antibody with the higher viscosity undergoes self-association as a function of concentration, and thus the impact of reversible self-association should not be ignored.

7. SUMMARY

The possibility of reversible self-association of therapeutic proteins must not be overlooked during product development. Currently, it is not possible to predict *a priori* which proteins may self-associate, and therefore care must be taken in selection of assays to monitor protein aggregation. The kinetics of dissociation play a critical factor in the selection of tools used for the assessment of aggregation for these proteins. The use of size exclusion chromatography has been demonstrated to be effective in assessing

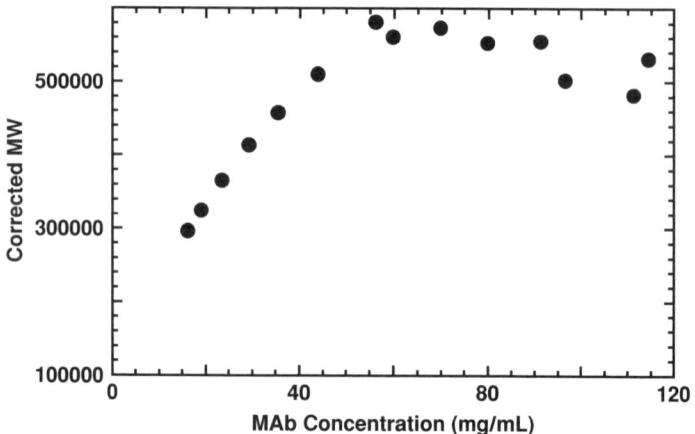

FIGURE 6. Corrected weight average molecular weight (MW) determined by sedimentation equilibrium analytical ultracentrifugation for MAb with high concentration dependency of viscosity.

the aggregate content in rhuMAb VEGF and Apo2L/TRAIL, proteins whose aggregate dissociation rates are sufficiently slow relative to the chromatography method used to study them. However, in the cases of human relaxin and a human IgG type 1 antibody, dissociation occurred much more rapidly than the chromatography, leading to an inaccurate assessment of aggregate content in those solutions. For these two proteins, it was necessary to use appropriate biophysical methods such as analytical ultracentrifugation to detect the presence of reversible aggregates in their formulations.

Understanding the factors governing the self-association (for example, pH, protein concentration) and the kinetics of the association/dissociation reaction are crucial to the development of a robust manufacturing process and pharmaceutical product. In the case of rhuMAb VEGF, it is vital to control the sample handling prior to assessment of aggregation. If this is not done, erroneous results are achieved that may cause one to come to inaccurate conclusions. With APO2L/TRAIL, the dissociation of the hexamer to the trimer form is slow and cannot be expected to occur to any significant extent when the drug is diluted for administration to patients. Therefore, it is important to control both the time and temperature during manufacture of this protein to minimize the formation of the hexamer. In the case of the antibody, these reversible multivalent aggregates led to the formation of a viscous solution. Minimization of the extent of self-association may result in a decrease in the viscosity for this protein.

REFERENCES

1. E. Y. Chi, S. Krishnan, T. W. Randolph, and J. F. Carpenter, Physical stability of proteins in aqueous solution: mechanism and driving forces in nonnative protein aggregation, *Pharm. Res.* **20**(9), 1325–1336 (2003).
2. J. L. Cleland, M. F. Powell, and S. J. Shire, The development of stable protein formulations: a close look at protein aggregation, deamidation and oxidation, *Crit. Rev. Ther. Drug Carrier Sys.* **10**(4), 307–377 (1993).

3. M. J. Treuheit, A. A. Kosky, and D. N. Brems, Inverse relationship of protein concentration and aggregation, *Pharm. Res.* **19**(4), 511–516 (2002).
4. A. Braun, L. Kwee, M. A. Labow, and J. Alsenz, Protein aggregates seem to play a key role among the parameters influencing the antigenicity of interferon alpha (IFN-alpha) in normal and transgenic mice, *Pharm. Res.* **14**(10), 1472–1478 (1997).
5. A. P. Minton, Confinement as a determinant of macromolecular structure and reactivity, *Biophys. J.* **63**(4), 1090–1100 (1992).
6. J. Wilf and A. P. Minton, Evidence for protein self-association induced by excluded volume. Myoglobin in the presence of globular proteins, *Biochim. Biophys. Acta* **670**(3), 316–322 (1981).
7. S. B. Zimmerman and A. P. Minton, Macromolecular crowding: biochemical, biophysical, and physiological consequences, *Annu. Rev. Biophys. Biomol. Struct.* **22**, 27–65 (1993).
8. J. Lebowitz, M. S. Lewis, and P. Schuck, Modern analytical ultracentrifugation in protein science: a tutorial review, *Prot. Sci.* **11**(9), 2067–2079 (2002).
9. S. J. Shire, Analytical ultracentrifugation and its use in biotechnology, in *Modern Analytical Ultracentrifugation*, eds. T. M. Schuster and T.M. Laue, (Boston: Birkhauser, 1992, 261–297).
10. J. Wen, T. Arakawa, and J. S. Philo, Size-exclusion chromatography with on-line light-scattering, absorbance, and refractive index detectors for studying proteins and their interactions, *Anal. Biochem.* **240**(2), 155–166 (1996).
11. J. E. Brennand, A. A. Calder, C. R. Leitch, I. A. Greer, M. M. Chou, and I. Z. MacKenzie, Recombinant human relaxin as a cervical ripening agent, *Br. J. Obstet. Gynaecol.* **104**(7), 775–778 (1997).
12. R. J. Bell, M. Permezel, A. MacLennan, C. Hughes, D. Healy, and S. Brennecke, A randomized, double-blind, placebo-controlled trial of the safety of vaginal recombinant human relaxin for cervical ripening, *Obstet. Gynecol.* **82**(3), 328–333 (1993).
13. A. H. MacLennan, R. C. Green, P. Grant, and R. Nicolson, Ripening of the human cervix and induction of labor with intracervical purified porcine relaxin, *Obstet. Gynecol.* **68**(5), 598–601 (1986).
14. M. I. Evans, M. B. Dougan, A. H. Moawad, W. J. Evans, G. D. Bryant-Greenwood, and F. C. Greenwood, Ripening of the human cervix with porcine ovarian relaxin, *Am. J. Obstet. Gynecol.* **147**(4), 410–414 (1983).
15. J. R. Seibold, P. J. Clements, D. E. Furst, M. D. Mayes, D. A. McCloskey, L. W. Moreland, B. White, F. M. Wigley, S. Rocco, M. Erikson, J. F. Hannigan, M. E. Sanders, and E. P. Amento, Safety and pharmacokinetics of recombinant human relaxin in systemic sclerosis, *J. Rheumatol.* **25**(2), 302–307 (1998).
16. J. R. Seibold, J. H. Korn, R. Simms, P. J. Clements, L. W. Moreland, M. D. Mayes, D. E. Furst, N. Rothfield, V. Steen, M. Weisman, D. Collier, F. M. Wigley, P. A. Merkel, M. E. Csuka, V. Hsu, S. Rocco, M. Erikson, J. Hannigan, W. S. Harkonen, and M. E. Sanders, Recombinant human relaxin in the treatment of scleroderma. A randomized, double-blind, placebo-controlled trial, *Ann. Intern. Med.* **132** (11), 871–879 (2000).
17. E. Canova-Davis, I. P. Baldonado, and G. M. Teshima, Characterization of chemically synthesized human relaxin by high-performance liquid chromatography, *J. Chromat. A* **508**(1), 81–96 (1990).
18. S. J. Shire, L. A. Holladay, and E. Rinderknecht, Self-association of human and porcine relaxin as assessed by analytical ultracentrifugation and circular dichroism, *Biochemistry* **30**(31), 7703–7711 (1991).
19. S. J. Shire, D. L. Foster, and E. Rinderknecht, Spectroscopic and hydrodynamic characterization of synthetic human and porcine relaxin, *Biophys. Chem.* **53**, 72a (1988).
20. T. W. Patapoff, R. J. Mrsny, and W. A. Lee, The application of size exclusion chromatography and computer simulation to study the thermodynamic and kinetic parameters for short-lived dissociable protein aggregates, *Anal. Biochem.* **212**(1), 71–78 (1993).
21. C. Eigenbrot, M. Randal, C. Quan, J. Burnier, L. O'Connell, E. Rinderknecht, and A. A. Kossiakoff, X-ray structure of human relaxin at 1.5 A. Comparison to insulin and implications for receptor binding determinants, *J. Mol. Biol.* **221**(1), 15–21 (1991).
22. N. H. Fernando and H. I. Hurwitz, Inhibition of vascular endothelial growth factor in the treatment of colorectal cancer, *Semin. Oncol.* **30** (3 Suppl 6), 39–50 (2003).

23. J. M. R. Moore, T. W. Patapoff, and M. E. M. Cromwell, Kinetics and thermodynamics of dimer formation and dissociation for a recombinant humanized monoclonal antibody to vascular endothelial growth factor, *Biochemistry* **38**(42), 13960–13967 (1999).
24. S. G. Hymowitz, M. P. O'Connell, M. A. Ultsch, A. Hurst, K. Totpal, A. Ashkenazi, B. de Vos, and R. F. Kelley, A unique zinc binding site revealed by a high-resolution x-ray structure of homotrimeric Apo2L/TRAIL, *Biochemistry* **39**(4), 633–640 (2000).
25. A. Ashkenazi and V. M. Dixit, Death receptors: signaling and modulation, *Science* **281**, 1305–1308 (1998).
26. A. Ashkenazi, R. C. Pai, S. Fong, S. Leung, D. A. Lawrence, S. A. Marsters, C. Blackie, L. Chang, A. E. McMurtrey, A. Hebert, L. DeForge, I. L. Koumenis, D. Lewis, J. Bussiere, H. Koeppen, Z. Shahrokh, and R. H. Schwall, Safety and anti-tumor activity of recombinant soluble Apo2 ligand, *J. Clin. Inv.* **104**(2), 155–162 (1999).
27. H. Walczak, R. E. Miller, K. Ariail, B. Gliniak, T. S. Griffith, M. Kubin, W. Chin, J. Jones, A. Woodward, T. Le, C. Smith, P. Smolak, R. G. Goodwin, C. T. Rauch, J. C. Schuh, and D. H. Lynch, Tumoricidal activity of tumor necrosis factor apoptosis-inducing ligand *in vivo*, *Nature Med.* **5**(2), 157–163 (1999).
28. S. J. Shire, Z. Shahrokh, and J. Liu, Challenges in the development of high protein concentration formulations, *J. Pharm. Sci.* **93**(6), 1390–1402 (2004).
29. C. R. Cantor and P. R. Schimmel, *Biophysical Chemistry Part II: Techniques for the Study of Biological Structure and Function*, ed. A.C. Bartlett (San Francisco: W. H. Freeman & Company, 1980, 503).
30. A. P. Minton and M. S. Lewis, Self-association in highly concentrated solutions of myoglobin: a novel analysis of sedimentation equilibrium of highly nonideal solutions, *Biophys. Chem.* **14**(4), 317–324 (1981).
31. A. P. Minton, Analytical centrifugation with preparative ultracentrifuges, *Anal. Biochem.* **176**, 209–216 (1989).
32. S. Darawshe and A. P. Minton, Quantitative characterization of macromolecular associations in solution via real-time and postcentrifugation measurements of sedimentation equilibrium: a comparison, *Anal. Biochem.* **220**, 1–4 (1994).
33. S. Darawshe, G. Rivas, and A. P. Minton, Rapid and accurate microfractionation of the contents of small centrifuge tubes: application in the measurement of molecular weight of proteins via sedimentation equilibrium, *Anal. Biochem.* **209**, 130–135 (1993).
34. A. K. Attri and A. P. Minton, Technique and apparatus for automated fractionation of the contents of small centrifuge tubes: application to analytical ultracentrifugation, *Anal. Biochem.* **152**(2), 319–328 (1986).
35. D. E. Roark and D. A. Yphantis, Equilibrium centrifugation of nonideal systems. The Donnan effect in self-associating systems, *Biochemistry* **10**(17), 3241–3249 (1971).
36. A. P. Minton, The effect of volume occupancy upon the thermodynamic activity of proteins: some biochemical consequences, *Mol. Cell. Biochem.* **55**, 119–140 (1983).
37. P. D. Ross and A. P. Minton, Hard quasispherical model for the viscosity of hemoglobin solutions, *Biochem. Biophys. Res. Commun.* **76**(4), 971–976 (1977).

Mutational Approach to Improve Physical Stability of Protein Therapeutics Susceptible to Aggregation

Role of Altered Conformation in Irreversible Precipitation

Margaret Speed Ricci,[1,4] Monica M. Pallitto,[1] Linda Owers Narhi,[2] Thomas Boone,[3] and David N. Brems[1]

Aggregation is a major challenge in the development of high dosage protein formulations. Administration of therapeutically effective doses of leptin is limited by its solubility at neutral pH. To achieve higher therapeutic doses, an acidic pH was utilized for the formulation. However, the propensity to form soluble aggregates in the acid formulation led to rapid precipitation at neutral pH. Biophysical characterization of leptin has led to insights into the mechanism of self-association. Thermal-, chemical-, and pH-induced denaturation was explored. In conditions that partially denature leptin (e.g., urea or acid pH), folding intermediates were populated and had the propensity to self-associate. These self-associated intermediates were soluble in acid but less soluble than the native or unfolded conformations and readily precipitated in nondenaturing aqueous solutions at neutral pH. Biophysical characterization was used to develop leptin analogs with fewer self-associated intermediates and a higher solubility at neutral pH. Equilibrium denaturation studies demonstrated that analogs with increased solubility had decreased secondary structural stability

---

[1] Department of Pharmaceutics, Amgen, Inc., Thousand Oaks, CA 91320
[2] Department of Analytical Sciences, Amgen, Inc., Thousand Oaks, CA 91320
[3] Department of Protein Science, Amgen, Inc., Thousand Oaks, CA 91320
[4] To whom correspondence should be addressed: Margaret Speed Ricci, Amgen Inc., Pharmaceutics Department, Thousand Oaks, CA 91320; email: mspeed@amgen.com

compared to leptin. Acid denaturation of these analogs, probed by pH titration effects on 1-anilino-8-naphthalene sulfonate (ANS) binding, followed a two-state model, whereas human leptin showed multistate behavior. These results suggest that folding intermediates play a role in the formation of soluble multimers in acid and in the precipitation observed at neutral pH. The irreversibility of neutral pH precipitation and its dependence on the initial conditions in acid suggest that solubility is controlled by some folding or aggregation event in acid, rather than being a simple-phase equilibrium phenomenon. Therefore, key mutations confer increased solubility at neutral pH by reducing the population and/or the hydrophobicity of structurally altered conformers, thereby reducing their tendency to precipitate at neutral pH.

1. INTRODUCTION

With the advent of high concentration protein therapeutics, physical stability is often a limiting factor in the manufacture, formulation, and delivery of these therapeutics. Formation of soluble aggregates and insoluble precipitates may be detrimental to the development of a therapeutically active protein or even lead to immunogenicity. Immunogenic responses can vary in severity from reduced efficacy to cross-reactivity with endogenous proteins. The formation of neutralizing antibodies to endogenous erythropoietin causing pure red cell aplasia has been proposed to result from aggregation or structural perturbation of Eprex® (S. Deechongkit et al., unpublished data).[1] The mechanisms for physical degradation have been investigated and reviewed to provide general formulation strategies to maximize stability.[2-4] Aggregation due to physical stresses such as agitation during processing or shipping may be ameliorated by the addition of polysorbates.[5] Self-association during the product's shelf life can be mitigated by adding appropriate stabilizing excipients (sugars, salts, detergents), optimizing the pH, or using a lyophilized formulation.

The first recombinant proteins to be used therapeutically were naturally occurring proteins, such as insulin and human growth hormone.[6] Acceptance of engineered proteins as protein therapeutics with improved pharmaceutical properties has increased more recently.[7] Recombinant therapeutic antibodies have progressed from murine to chimeric to fully humanized. Nonnative modifications to naturally occurring proteins may affect stability, immunogenicity, or pharmacokinetics. Nonnative modifications in approved protein therapeutics include site-specific mutations that reduce aggregation (Proleukin® and Betaseron®) or alter the injection site absorption kinetics (Humalog®). Likewise, pegylation (PEGasys®, Neulasta® and Oncaspar®) and increased glycosylation (Aranesp®), can improve serum half-life.[6] Some fusion protein therapeutics also have been developed and commercialized. These include fusion of an active protein to a carrier moiety such as Fc (Enbrel®, Amevive®), which increases serum half-life, and fusion of a targeting protein to a toxin (Ontak®), which leads to site-specific toxicity.[6,7]

Site-directed mutagenesis has been used to improve the solubility and stability of proteins by affecting properties such as hydrophobicity and pI. Surface-exposed hydrophobic residues were replaced by hydrophilic counterparts, leading to more soluble mutants of methyltransferase,[8] ankyrin repeat proteins,[9] and cholera toxin A1 subunit.[10] Proteins are expected to have a minimum solubility near their isoelectric point,[11] as was demonstrated for ribonuclease Sa and its lysine variants.[12] The strategy to increase solubility by changing the pI was used to raise the solubility of a single-chain Fv antibody at physiological pH by two orders of magnitude with the addition of five glutamic acids to the C-terminus, lowering the pI from 7.5 to 6.1.[13]

In this review, we describe the mutational approach to improving the solubility of leptin. Leptin is the protein product of the *ob* (obese) gene, which is expressed in adipose tissue.[14] Leptin is a 16-kDa member of the four-helical bundle family of cytokines. Other therapeutics that are members of this cytokine structural family include human growth hormone, erythropoietin, and granulocyte-colony stimulating factor.[15] The crystal structure of leptin was determined using a mutant with a single amino-acid substitution (W100E) that more readily crystallized.[16] Previous investigations of the structure-function relationship of leptin have focused on the necessity of the Cys96-Cys146 disulfide bond for stability and therapeutic activity[17,18] or the effect of methionine oxidation on structural stability and biological activity.[19] The secondary structure of murine leptin was evaluated at various pH's (5.35 to 8.45) using attenuated total reflection/Fourier transform infrared spectrometry.[20] Structural and stability studies of human leptin by spectroscopic (fluorescence and circular dichroism) and hydrodynamic methods (size-exclusion HPLC, dynamic light scattering, and small-angle X-ray scattering) indicated that leptin formed oligomers of different sizes depending on factors such as pH, concentration, and ionic strength (Z. Lakos and A.L. Fink, unpublished data). Folding and stability have been studied by equilibrium denaturation as a function of pH for leptin and other related cytokines.[21-23]

After the identification of the *ob* gene product, there was immense commercial interest in developing leptin as a therapeutic agent for weight loss or appetite control.[24] Clinical research into the use of leptin as a weight loss agent in general obesity has been discontinued owing to the modest weight loss in a subset of the obese population. However, clinically significant results have been observed in the treatment of leptin-deficient metabolic disease states.[25] For clinical trials, the formulation for recombinant human leptin was optimized for maximum solubility of the protein (S. Roy and A. McAuley, unpublished data). Several formulations were developed at protein concentrations of 5–20 mg/ml at pH 4.00–4.25. However, clinical dosing was limited by injection site reactions associated with the subcutaneous administration of higher concentrations of protein.[24] Limited solubility of leptin at physiological pH may contribute to precipitation at the injection site. Therefore, efforts to understand the mechanism of limited solubility and to reengineer the molecule for increased solubility at neutral pH were undertaken.

2. DESIGN AND CHARACTERIZATION OF SOLUBILITY ANALOGS

Due to the limited solubility of human leptin at neutral pH, extensive efforts were made to understand the aggregation mechanism and to design analogs that have a higher solubility at neutral pH to achieve therapeutic dosing levels. The basis for designing these analogs was the observation that murine leptin was significantly more soluble

(A)

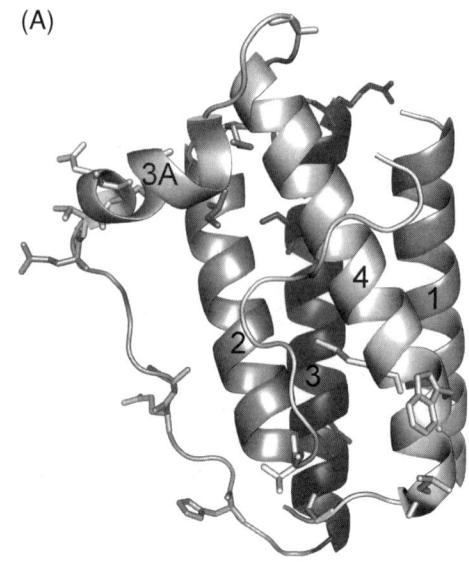

(B)

	0	10	20	30	40
HUMAN:	MVPIQKVQDD	TKTLIKTIVT	RINDISHTQS	VSSKQKVTGL	DFIPGLHPIL
MURINE:	MVPIQKVQDD	TKTLIKTIVT	RINDISHTQS	VSAKQRVTGL	DFIPGLHPIL

	50	60	70	80	90
HUMAN:	TLSKMDQTLA	VYQQILTSMP	SRNVIQISND	LENLRDLLHV	LAFSKSCHLP
MURINE:	SLSKMDQTLA	VYQQVLTSLP	SQNVLQIAND	LENLRDLLHL	LAFSKSCSLP

	100	110	120	130	140
HUMAN:	WASGLETLDS	LGGVLEASGY	STEVVALSRL	QGSLQDMLWQ	LDLSPGC
MURINE:	QTSGLQKPES	LDGVLEASLY	STEVVALSRL	QGSLQDILQQ	LDVSPEC

FIGURE 1. A) Ribbon diagram of human leptin structure. Helices are numbered, and side chains are shown for residues that differ between the human and murine sequences. B) Human and murine leptin sequences are provided, and differences between the sequences are highlighted. Helices are outlined as follows: helix 1 (Pro 2-His 26), helix 2 (Leu 51-Ser 67), helix 3 (Arg 71-Lys 94), helix 3A (Thr 106-Glu 115), and helix 4 (Ser 120-Ser 143). Note that the first methionine residue associated with *E. coli* production of the leptin protein is not counted.

than human leptin at neutral pH (43 mg/ml for murine vs 2–3 mg/ml for human leptin). Therefore, the strategy to improve neutral pH solubility of human leptin involved introducing various combinations of mutations based on the murine leptin sequence, which differs from the human leptin sequence at 22 sites. In examining the three-dimensional structure, the murine-like residues that were introduced into human leptin were located within helices 2, 3, 3A, and 4, and within the flexible loop regions connecting the helices (Figure 1). Extensive studies have focused on understanding the limited solubility of human leptin at neutral pH, identifying subtle conformational changes that may lead to self-association, and determining whether soluble multimers in acid induce precipitation at neutral pH. Human leptin and an array of leptin analogs were characterized using biophysical tools to try to elucidate several key aspects of the aggregation mechanism: (1) the identity of sites involved in self-association, (2) the relationship, if any, between structural stability and solubility, and (3) the role of acid soluble multimers in neutral pH precipitation.

2.1. Solubility of Leptin and Analogs

The method for determining solubility at neutral pH was carefully chosen to avoid any acid-state conformational contributions. The procedure involved dialyzing the protein into buffered solution at low protein concentrations under refrigerated conditions. Leptin is more soluble at low temperatures. To obtain supersaturation, the protein was concentrated at 2–8°C, using membrane/filter centrifuge concentration devices, to levels exceeding the solubility at room temperature. The protein was introduced to conditions of limited solubility in the gentlest manner possible by simply subjecting the sample to higher temperature. Upon exposure to room temperature, precipitation occurred. The time course of the soluble protein in the supernatant was determined following centrifugation. Final solubility was achieved when no further changes were observed, which took hours or days depending on the particular analog.

The pH-dependence and reversibility of precipitation are shown in Figure 2. As indicated, the solubility of leptin varied widely with pH. Acidic pH provided greater solubility

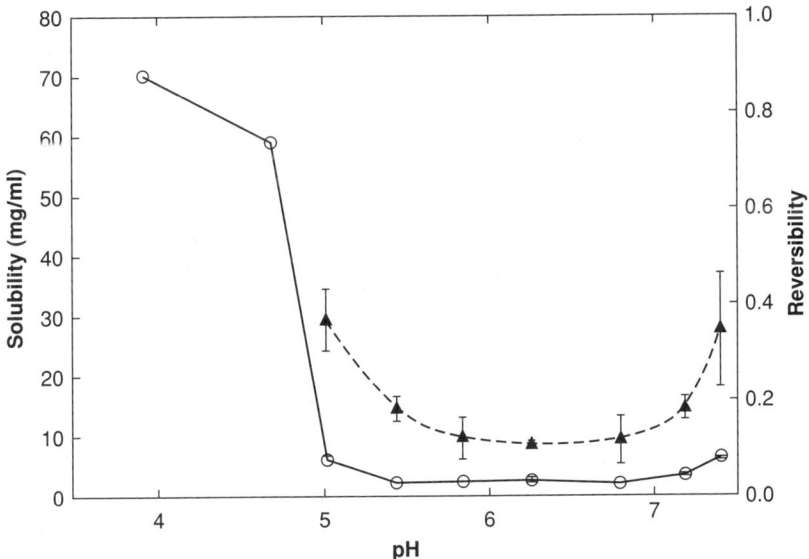

FIGURE 2. Effect of pH on leptin solubility (○) and reversibility of precipitation (▲) at room temperature in 25 mM Tris, 10 mM sodium acetate, 10 mM MES at the indicated pH ($n = 3$). At pH 4 and 4.7, no precipitation occurred at room temperature; therefore the solubility may actually be in excess of the measured value. Briefly, solubility was measured by preparing dilute solutions at various pH's and concentrating using membrane/filter centrifuge devices under refrigerated conditions. Each solution was then allowed to equilibrate at room temperature and the concentration (solubility) determined. The supernatant was carefully removed from the precipitate. The precipitate was then diluted with corresponding buffer to ∼20x the initial volume and allowed to equilibrate at room temperature. Reversibility of the precipitate was determined by dividing the mass of solubilized precipitate by the total mass of the initial precipitate.

than physiological/neutral pH. The solubility of leptin at pH 4 was in excess of 70 mg/ml, but the solubility at neutral pH was significantly lower (2 mg/ml). The difference in solubility may lead to protein precipitation upon a jump in pH from acidic to neutral pH that is amplified at higher starting concentrations. The implication of low solubility at neutral pH is that upon injection of leptin from an acidic formulation into physiological conditions, protein precipitation may occur, leading to injection site reactions.

The reversibility of precipitation was also evaluated as a function of pH. To test this, the supernatant was removed from the precipitate, and the pellet then was resolubilized by excess dilution with the corresponding buffer and allowed to equilibrate at room temperature. The amount of leptin that could be resolubilized was normalized to the amount of precipitate initially formed, giving the reversibility of precipitation. The data indicate that precipitation was partially reversible under acidic conditions (near pH 5) and became irreversible as the pH increased. Typical pI precipitation of proteins is reversible by dilution. Since the isoelectric point of leptin is 6.2, one would expect the reversibility of precipitation to be relatively high within the pH range of pH 6–7. However, contrary to typical pI-induced precipitation phenomena, a minimal fraction of the precipitated material at pH 6-7 could be resolubilized upon excess dilution with buffer. The pH dependence of the reversibility of precipitation indicates that the mechanism of limited solubility is more complicated than a simple equilibrium between soluble and associated forms. Therefore, although electrostatic repulsion may keep leptin in solution at low pH, the significant reduction in solubility at neutral pH is not due merely to the loss of the electrostatic repulsion effect near the pI but rather an irreversible conformational phenomenon.

Neutral pH solubility in phosphate-buffered saline solution was studied in depth for human leptin and the leptin analogs. Over 100 leptin analogs were produced and purified and the results of the solubility studies are shown in Figure 3. The particular subset of data shown demonstrates the solubility effect of murine-like substitutions in each region of the molecule and the combinatorial effects for the regions identified as domains 3 and 4. Certain murine-like mutations were found to be critical in modulating solubility; in particular, those residues comprising the third and fourth helical domain and the loop connecting these two domains. Not all murine-like mutations improved solubility; in fact, some mutations resulted in decreased solubility. Eliminating the residues that decrease solubility and optimizing combinations of mutations that included the key substitutions of W100Q and W138Q resulted in solubility that exceeded that of murine leptin by three-fold. There was no strong correlation between solubility and pI or hydrophobicity. Therefore, limited solubility at neutral pH is not due simply to pI precipitation or net hydrophobicity, but rather results from irreversible self-association involving specific residues.

In addition to improved solubility, the murine-like leptin mutants exhibited lower amounts of soluble multimers detected by size-exclusion HPLC. The molecular weights of the main peak and several prepeak species were confirmed by on-line multiangle light scattering.[26] The main peak represented a mixture of monomer and dimer in rapid equilibrium, and the first prepeak represented a dimer species in slow equilibrium with the monomer, consistent with the analytical ultracentrifugation data. An earlier eluting prepeak represented higher-order aggregates of tetramer molecular weight. Concomitant with lower amounts of soluble multimers, the murine-like leptin mutants exhibited less

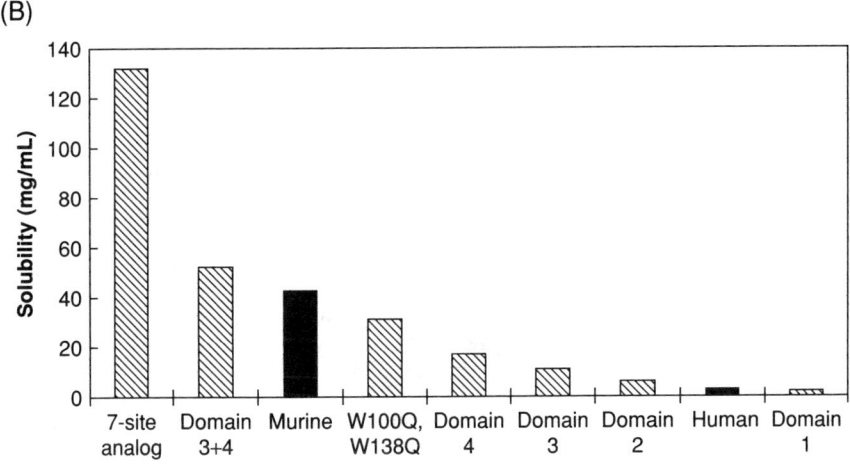

FIGURE 3. Neutral pH solubility of human leptin, murine leptin, and leptin analogs. Solubilities were measured in a phosphate buffered saline solution at neutral pH at room temperature, as described. From threading model predictions of the loop regions, four domains of human leptin are defined as the following regions. Domain 1: His 26-Ala 59; domain 2: Met 68-Ser 77; domain 3: Cys 96-Tyr 119 (with subdomain 3A: His 97-Ala 101); and domain 4: Met 136-Cys 146. The leptin analogs depicted have murine-like substitutions within the specified domains, in addition to the non-murine substitution G112E at a nonconserved site, as defined in the table.

precipitation in a two-step dilution of denaturant assay. Furthermore, when the ability of leptin to self-associate was reduced by lowering the initial protein concentration in acid or by the addition of the detergent CHAPS, the solubility of leptin was increased at neutral pH. This suggests that the acid-soluble multimers may play a role in precipitation at neutral pH.

The mechanistic basis for the increased solubility of the leptin analogs is uncertain. Both tryptophan residues, W100 and W138, are located in solvent-exposed domains in the loop region and in the outside face of the fourth helix, respectively. The dramatic effects of each single-site mutation W100Q and W138Q on solubility indicate the molecular specificity of self-association. In addition to the significant effects of each point mutation on aggregation, the combinatorial effects of specific murine-like mutations on improving solubility indicate the complexity of the self-association mechanism. Various quantitative hydrophobicity scales have been used to predict the miscibility of proteins in aqueous solutions.[27–29] However, net hydrophobicity of the complete polypeptide chain does not reflect the local solvent-exposed structural elements of the metastable folding

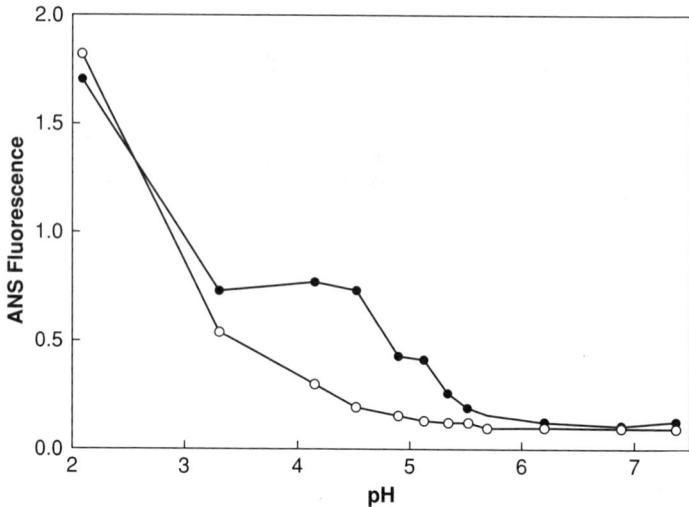

FIGURE 4. Structural perturbations detected by ANS fluorescence as a function of pH for human (●) and murine (○) leptin. Solution conditions were 0.05 mg/ml leptin in 2 mM citrate buffer, pH 3-8, in the presence of 50 μM 1,8-ANS. Fluorescence was measured at ex 390 nm, em 485 nm using a PTI fluorometer.

intermediate state, and therefore global hydrophobicity does not accurately predict aggregation propensity. The pronounced difference in leptin aggregation due to each point mutation W100Q and W138Q suggests that specific aggregation sites or hydrophobic patches rather than net hydrophobicity control self-association. However, using the crystal structure of the native state, structural analysis of the effects of each point mutation and combinatorial effects of the dozens of mutations screened provides limited insight into the aggregation mechanism. The precursor to soluble multimers leading to precipitation may not be the native state but rather a partially unfolded species.

2.2. pH-Dependent Conformational Stability

To elucidate the structural basis of aggregation, ANS binding studies were performed to assess conformational perturbations resulting in changes in hydrophobicity. ANS serves as a hydrophobic probe that produces a strong fluorescence signal upon binding to hydrophobic patches.[30,31] Figure 4 shows the increased ANS fluorescence for human leptin at pH 4-5, followed by a second transition at lower pH. Note that murine leptin and the leptin analogs did not display the increased ANS fluorescence at pH 4-5. The nature and shape of this transition suggest that a folding intermediate with increased hydrophobicity is populated at pH 4-5. The increased hydrophobicity may result in the formation of soluble multimeric species observed under acidic conditions. In addition, the absence of a hydrophobic intermediate species for the analogs may be related to their increased solubility.

Near-UV CD and fluorescence data also confirmed the presence of a conformationally perturbed intermediate species in the pH range of 4.5 to 5.5. The near-UV CD signal

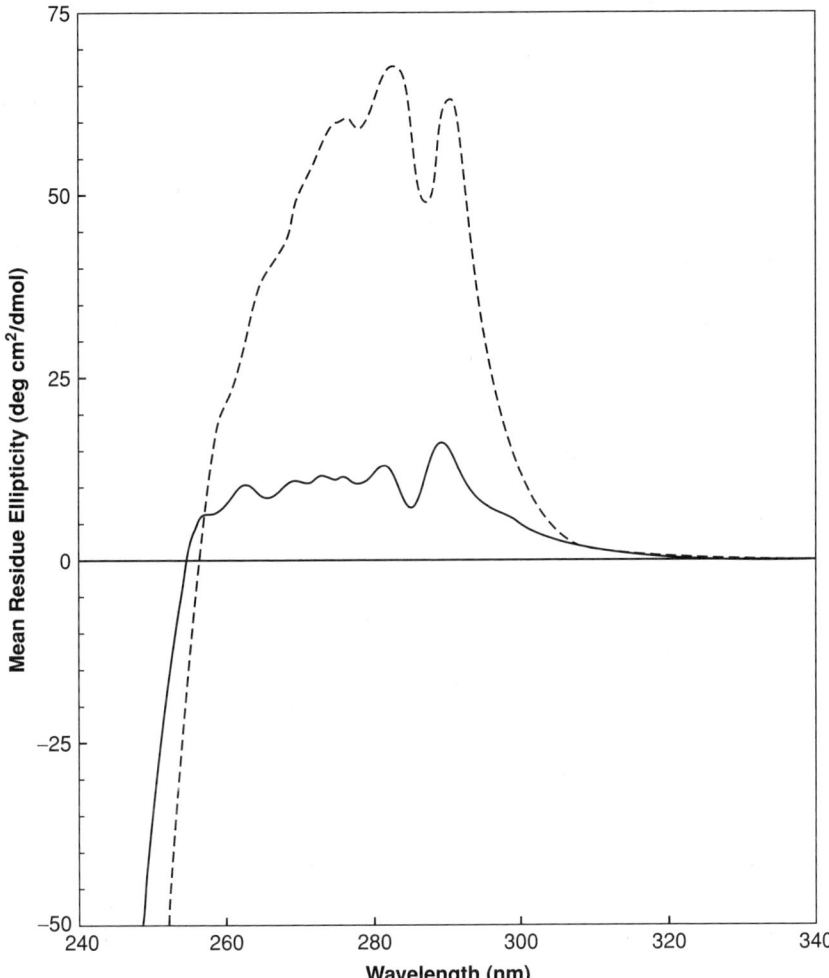

FIGURE 5. Near-UV CD spectrum of human leptin at pH 7 (solid line) and pH 5 (dashed line). Solution conditions were 0.05 mg/ml leptin in 2 mM citrate buffer. CD spectra were measured using a Jasco J-720 spectropolarimeter.

in the range of 250–300 nm reflects the environment of the aromatic residues, which is indicative of tertiary structure.[32] For leptin, the near-UV CD spectrum at pH 5 had double maxima at 282 and 290 nm that were increased about six-fold compared to the signal at pH 7, with a shoulder at 275 nm that was also greatly increased in intensity (Figure 5). Similar to the ANS results, the folding intermediate characterized by this unique near-UV CD signal was observed for human leptin but not murine leptin. The double mutant W100Q/W138Q also did not populate the pH 5 intermediate; this suggests that the presence of one or more tryptophan residues is critical for the formation of this folding intermediate.

The mechanism by which partial unfolding occurs near pH 5 is uncertain. Generally electrostatic repulsion of charged residues at low pH may induce acid unfolding, and this phenomenon likely is what occurs below pH 3, as seen by increased ANS binding for both murine and human leptin. However, since the two solvent-exposed tryptophan residues are critical for the partial unfolding transition near pH 5, the folding intermediate likely is due to local structural perturbation of hydrophobic sites rather than electrostatic repulsion unfolding the molecule.

The substitution of W138Q did not affect the near-UV CD peak at 290 nm, which suggests that this point mutation does not affect the folding intermediate state. However, this mutation did remove most of the fluorescence at 353 nm and resulted in the resolution of the tyrosine peak at 304 nm. These results imply that W138 is in a solvent-exposed, flexible region of the molecule and is an energy transfer partner for one or more tyrosines. Removing W100 changed the near-UV CD spectrum in the 290-nm region and decreased the intensity of the intrinsic fluorescence, but did not result in the resolution of the tyrosine peak. Therefore, W100 is in a more rigid and less solvent-exposed environment than W138, but is also not an energy transfer partner of a tyrosine.

2.3. Equilibrium Denaturation

Equilibrium denaturation of human leptin and the leptin analogs was conducted to determine the relationship between conformational stability and solubility. Figure 6 shows that the denaturation midpoint of the more soluble analogs was lower than that of human leptin. Equilibrium denaturation was monitored by far-UV CD at 222 nm, indicative of secondary structure. This implies that the phenomenon of precipitation at neutral pH is not governed by the stability of the secondary structure. Other properties that potentially may govern stability include the stability of the tertiary structure and the population of folding intermediates. Interestingly, urea-induced unfolding data demonstrated that the secondary structural stability of human leptin was greater under acidic conditions.[23] This suggested that tertiary structural perturbations near pH 5 were concomitant with increased secondary structural stability. The widening of the noncoincidence of secondary and tertiary structural transitions is consistent with increasing the population of folding intermediates. Therefore, folding intermediates with nativelike secondary structure and disrupted tertiary structure are involved in leptin precipitation. Whether in pH- or denaturant-induced folding studies, both the pH jump precipitation studies and equilibrium denaturation experiments provide evidence of the general importance of partially unfolded states to the overall solubility and precipitation of proteins. Furthermore, based on the solubility and equilibrium denaturation data for the single-site mutants, specific residues are involved in populating the folding intermediates that are responsible for aggregation.

The mutagenesis approach has been used extensively for increased solubility and/or improved pharmaceutical properties of protein therapeutics, including insulin (Humalog),[33,34] leptin,[23] granulocyte macrophage colony-stimulating factor (Leukine®), interleukin-2 (Proleukin®), interferon β-1B (Betaseron®), IFN-α (Intron-A®), granulocyte colony-stimulating factor,[35–37] and human growth hormone.[38] In most cases, the strategy for improved physical stability involves increasing the conformational

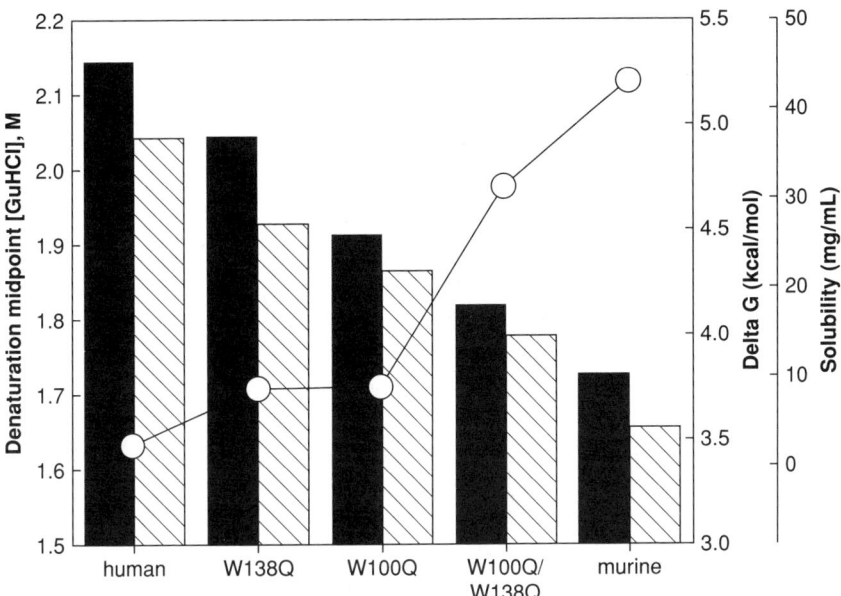

FIGURE 6. Equilibrium denaturation and solubility data for human leptin, murine leptin, and analogs. The equilibrium denaturation studies were done by monitoring the far-UV CD signal at 222 nm for leptin in a tri-buffer solution (5 mM Tris, 2 mM sodium acetate, 2 mM MES, pH 7) as a function of GuHCl concentration using an Aviv CD with an automated titrator. The fraction unfolded as a function of denaturant was calculated from the change in spectroscopic signal relative to the signal baselines for the folded and unfolded forms. From the unfolding transition, the free energy of unfolding (ΔG) was calculated as a function of denaturant using the equation $\Delta G = RT \ln(K_u)$, where K_u is the equilibrium constant for unfolding. The denaturation midpoint was the denaturant concentration for which $\Delta G = 0$, and the extrapolated values of ΔG in the absence of denaturant (ΔG_0) were reported in units of kcal/mol. Secondary structural transitions were analyzed by calculating the denaturation midpoint (solid bars) and ΔG of unfolding (hatched bars). Solubility (○) at neutral pH in phosphate-buffered saline solution was determined as described previously.

stability of the native state or decreasing the population of folding intermediates.[23] Although the inverse correlation of solubility and stability seems counterintuitive, there are several examples of therapeutic proteins formulated under acidic conditions that induce minor perturbations in tertiary structure but allow for increased solubility. This includes leptin, G-CSF, IFN-α, megakaryocyte growth and development factor, consensus interferon, and other small therapeutic cytokines formulated at pH 4-5.

2.4. Thermal Denaturation

Thermal unfolding studies were conducted to determine the relationship between conformational stability and solubility. The thermal unfolding was irreversible for human leptin and many of the analogs, perhaps due to precipitation of the unfolded or partially unfolded protein at neutral pH. Thermal denaturation of murine leptin was reversible under conditions where human leptin melted irreversibly (in 5 mM sodium acetate, pH 6.5). This result implies that the unfolded or partially unfolded conformer

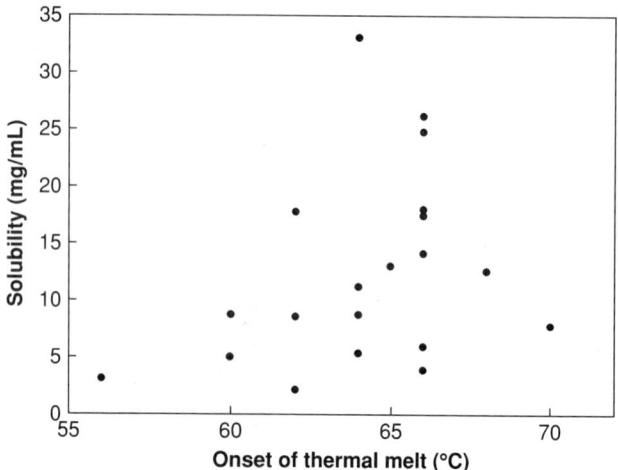

FIGURE 7. Relation between onset of thermal denaturation and solubility at neutral pH. The thermal denaturation studies were performed on a Jasco J-720 spectropolarimeter with a Peltier thermal unit and rectangular-shaped cuvettes with a pathlength of 0.1 cm. The protein was heated form 20°C to 80°C at a rate of 60°C/hr, recording the ellipticity at 222 nm every 0.5°C, The onset of thermal melt was determined for human leptin and several leptin solubility analogs. Thermal data were plotted versus neutral pH solubility (●), which was determined as described previously.

of human leptin is less soluble, resulting in irreversible precipitation. The substitution G112E/W138Q/M136I/G145E was the minimum motif that resulted in reversible thermal denaturation, suggesting that these specific sites are critical.

The onset of thermal unfolding was pH-dependent. At pH 7, human leptin began to melt and irreversibly aggregate at about 56°C, compared to 62°C at pH 6.5. The lower thermal stability at neutral pH and irreversibility of thermal unfolding may be related to the low solubility. However, there was no strong correlation between solubility and thermal stability (Figure 7). Due to the irreversibility, no thermodynamic parameters are reported here.

The lack of correlation between thermal stability and solubility is unexpected. There are many examples of utilizing thermal stability data by differential scanning calorimetry as a tool for selecting mutants and guiding formulation development for increased stability and solubility.[39,40] Likewise, recent computational approaches have been successful in designing cytokine analogs with increased thermal stability and improved pharmaceutical properties.[41,42] Although significant effects of mutations on solubility, thermal stability, *in vitro* refolding yields, and *in vivo* inclusion body formation have been observed for other proteins, these mutational effects do not necessarily correlate with the thermal stability of the native state. For example, a set of temperature-sensitive folding mutations have been identified in the phage P22 tailspike endorhamnosidase that predominantly altered the thermal stability of the folding intermediate state rather the native state, resulting in dramatic differences in *in vitro* refolding yields and *in vivo* inclusion body formation.[43–46] For interleukin-1, a series of mutants displayed no correlation between inclusion body propensity and thermal stability or thermodynamic stability of the native state by equilibrium denaturation.[47] Similar temperature-sensitive folding mutants were generated and identified for interferon-γ. Although thermal denaturation of wild-type interferon-γ was irreversible at neutral pH, the reversibility at low pH did not reflect increased stability of the native state in acid but rather increased solubility of the unfolded or partially unfolded intermediates under acidic conditions.[48]

3. NEUTRAL pH PRECIPITATION

To measure the aggregation propensity, the protein was taken from favorable solution conditions to those in which the protein is less soluble, by altering the pH, denaturant level, or temperature. Precipitation of soluble multimers and/or structurally perturbed conformers may require a second dilution to less favorable solution conditions. The first step in such a two-step dilution protocol was used to populate soluble multimers and/or folding intermediates. The second dilution to lower denaturant concentration and/or jump to neutral pH specifically induced precipitation of those intermediates.

3.1. Implication of Folding and Aggregation Intermediates

Using the two-step dilution protocol described above, the extent of leptin precipitation was measured upon jumping from varying initial solution conditions to constant final conditions at neutral pH. As Figure 8 depicts, the amount of precipitation was a function of the initial denaturant level rather than the final solution conditions. Denatured leptin diluted from 10 M urea to low denaturant levels at neutral pH did not precipitate. Likewise, leptin with intact nativelike structure diluted from <4 M urea did not precipitate. However, leptin initially incubated under partially denaturing conditions in 6.1 M urea did precipitate upon dilution. This level of denaturant corresponds with the equilibrium denaturation transition by far-UV CD, representing a loss in secondary structure. The precipitation data suggest that an unfolding intermediate is responsible for precipitation at neutral pH.

Furthermore, the extent of precipitation at this denaturant concentration also involved a time-dependent event. Rapidly adjusting the denaturant level from fully denaturing (10 M urea) to 6.1 M and then immediately diluting to < 4M urea did not induce precipitation. However, incubating at 6.1 M for greater equilibration times (on the order of several hours) resulted in complete precipitation. The time dependence of precipitation suggests that a folding or self-association phenomenon occurs as a function of the time spent under partially denaturing conditions. The slow time dependence of hours versus milliseconds is inconsistent with typical folding events and likely reflects self-association of a sparsely populated species, such as a transient folding intermediate.

3.2. Role of Acid Multimers

In the absence of denaturant, self-association of leptin and its analogs was explored in an acidic pH where solubility was high. Using size-exclusion chromatography to

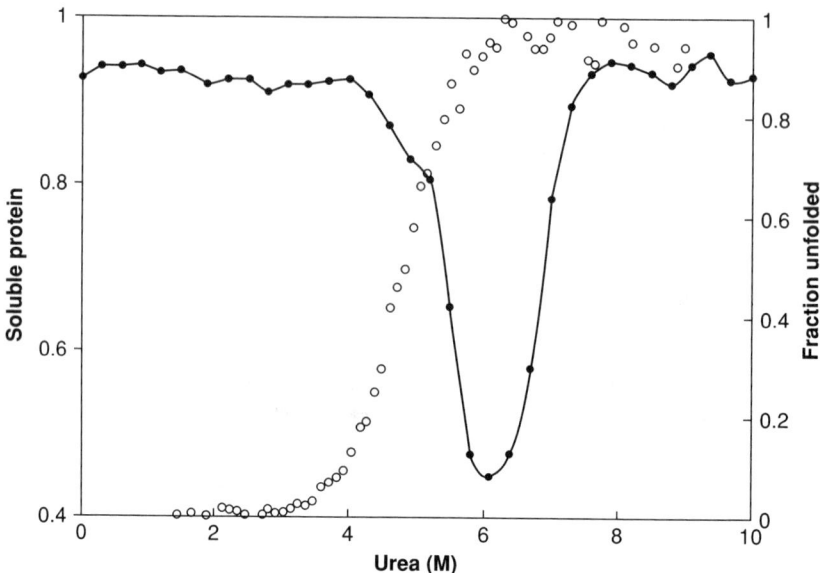

FIGURE 8. Aggregation of a folding intermediate of leptin, as depicted in an overlay of urea-induced equilibrium denaturation by circular dichroism (○) and precipitation following a urea jump from different initial urea concentrations (●). Equilibrium denaturation was done in urea at a protein concentration of 0.1 mg/ml in 20 mM sodium acetate, 20 mM MES, 50 mM Tris buffer and monitored by CD at 222 nm. The precipitation study utilized the two-step procedure, as follows. Only the initial urea concentration in the first step was varied, as depicted on the x-axis. In the first step, leptin at 37.4 mg/ml was equilibrated (>18 hr) in varying concentrations of urea, 1 mM acetate, and pH 4.0. To induce precipitation, the samples were diluted in the second step to 2.5 M urea, 0.93 mg/ml leptin, 20 mM sodium phosphate, 25 μM acetate, pH 7.0. Reprinted from Ricci and Brems.[23]

measure soluble aggregate formation, human leptin self-associated as a function of protein concentration under acidic solution conditions (Figure 9). In contrast, murine leptin did not form soluble aggregates in acid. Comparison with the ANS binding data indicated the more hydrophobic the leptin analog, the more populated the soluble multimer state by size-exclusion chromatography.

The extent of precipitation in the presence of denaturant was measured for human and murine leptin as a function of initial protein concentration after rapidly adjusting the solution from acidic to neutral pH (Figure 9). A two-step procedure was utilized and only the initial protein concentration under acid conditions was varied. In the first step, protein at varying concentrations was equilibrated in 6.1 M urea at pH 4.0 to populate soluble multimers, as described in Figure 7. Precipitation was induced in the second step by diluting to constant final solution conditions at low protein concentration in 2.5 M urea at neutral pH. Human leptin precipitated upon jumping from acidic to neutral pH. In contrast, murine leptin did not precipitate after the pH jump.

Kinetic studies were performed to determine whether the population of soluble multimers in acid contributed directly to precipitation at neutral pH or whether the soluble multimers were a secondary effect of the physical properties of leptin (Figure 9

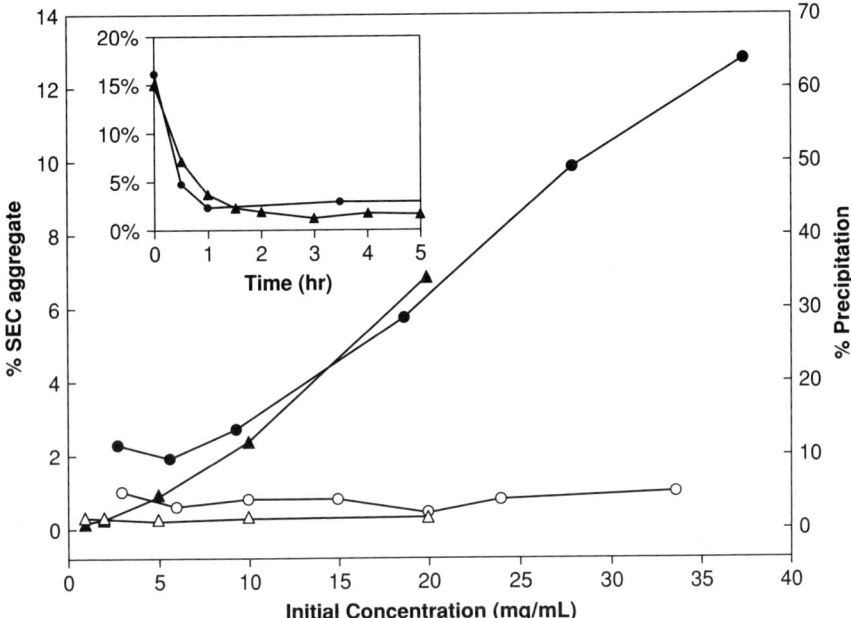

FIGURE 9. Concentration dependence of soluble multimer formation in acid (▲,△) and precipitation at neutral pH (●,○). Solid symbols represent human leptin and open symbols represent murine leptin. Soluble multimers were determined by quantifying the aggregate prepeak on size-exclusion chromatography using a TosoHaas TSK-GEL G2000SW$_{XL}$ analytical column (7.8 mm × 30 cm) with a mobile phase of 100 mM sodium phosphate, 0.5 M NaCl, pH 6.9. For the precipitation assay, a two-step procedure was utilized and only the initial protein concentration under acid conditions was varied. In the first step, protein at varying concentrations was equilibrated (>18 hr) in 6.1 M urea, 1 mM acetate, and pH 4.0. Precipitation was induced in the second step by diluting the samples to 0.93 mg/ml leptin, 2.5 M urea, 20 mM sodium phosphate, 0.33 mM acetate, and pH 7.0. Precipitation was quantified by absorbance of the supernatant at 280 nm, as described. *Inset.* Kinetic effect of predilution in acid on neutral pH precipitation. Concentrated stock at 50 mg/ml human leptin at pH 4 was diluted to 5 mg/ml at pH 4 to induce dissociation of soluble multimers. SEC-HPLC analysis was performed to quantify the level of soluble aggregates (▲), as described above. After varying equilibration times, the pH was rapidly adjusted to neutral pH with a 1:1 dilution with phosphate buffer at pH 7 to a final concentration of 2.5 mg/ml. Precipitation (●) was quantified by absorbance of the supernatant at 280 nm, as described.

inset). In the absence of denaturant, concentrated leptin at 50 mg/ml and pH 4 was diluted in acid to 5 mg/ml and incubated for varying times to induce dissociation of the soluble multimers. The time course of multimer dissociation was followed by size-exclusion HPLC. In the second step, a pH jump to neutral conditions was performed after different lengths of incubation under dilute conditions in acid to test the effects on solubility. The longer the equilibration times after initial dilution in acid, the less the precipitation after final dilution to neutral pH. The relationship between precipitation and the equilibration time after leptin dilution in acid suggests that the self-associated species in acid leads directly to precipitation as a result of the rapid pH adjustment.

To further assess the role of acid multimers in neutral pH precipitation, soluble multimer-enriched and monomer-enriched fractions were prepared using size-exclusion

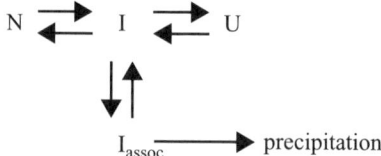

FIGURE 10. Proposed aggregation mechanism. The folding pathway involves a multistate mechanism with the native state *(N)* in equilibrium with the folding intermediate state *(I)* and unfolded conformer *(U)*. Along the aggregation pathway, the folding intermediate can self-associate to form soluble multimers (I_{assoc}), which can lead to irreversible precipitation.

chromatography. Once both samples had been collected, concentrated, and adjusted to equivalent concentrations, the dimer-enriched sample had begun to revert to a monomeric state. In the absence of denaturant, the monomer- and multimer-enriched fractions were induced to precipitate by rapidly adjusting to neutral pH. Despite the partial reequilibration of soluble multimer into monomer, there was a significant difference in the solubility of the dimer-enriched and monomer-enriched samples. During the pH jump, the multimer-enriched sample was observed to precipitate significantly more than the monomer-enriched sample; this was confirmed by absorbance measurement of the supernatants. A precipitation factor was calculated for monomeric and multimeric leptin from the precipitation data for the monomer-enriched and multimer-enriched samples using two equations and two unknowns. The precipitation factor represents the intrinsic solubility of the monomer or multimer, as well as the ability of each species to seed the precipitation of other protein. From the data, the precipitation factor was 17% for monomeric leptin and 260% for multimeric leptin. The value above 100% suggested that the multimeric leptin induced soluble monomeric leptin to precipitate. This is strong evidence for the theory that a small fraction of soluble multimers in acid can induce substantial precipitation at neutral pH and severely limit the overall solubility of leptin at neutral pH.

4. MECHANISM OF AGGREGATION

Based on the leptin precipitation studies, we hypothesize that the mechanism of leptin solubility involves structural perturbation of the native conformation, which in acid leads to the formation of soluble multimers that are insoluble at neutral pH (Figure 10). Specific residues, including W100 and W138, are involved in populating the folding intermediates that are prone to aggregation. Murine leptin or human leptin analogs with key murine-like mutations are more soluble because their structurally altered conformers are less populated or are less hydrophobic and less prone to precipitation. Such species may be difficult to detect in low amounts or to distinguish from the native conformer.

The experimental evidence for this hypothesis of solubility differences of the intermediate forms is as follows: (1) higher levels of ANS binding to human leptin versus the leptin analogs under acidic conditions suggest hydrophobicity differences that may govern self-association; (2) the two-step denaturant dilution studies identify a time-dependent event under partially denaturing conditions (at 6.1M urea) that controls the extent of precipitation which may involve structural perturbations and/or formation of aggregation intermediates; (3) the population of soluble multimers that can be separated by size-exclusion chromatography correlates with neutral pH precipitation and isolation of the multimer-enriched fractions confirms the relationship to neutral pH precipitation.

The ability of a small amount of multimers to induce the precipitation of otherwise soluble monomers upon jumping to neutral pH offers an explanation for the mechanism of the solubility limitation of leptin. For example, low concentrations of an altered conformation at neutral pH could recruit native leptin into an irreversible precipitate and limit solubility. The pH jump studies in the presence of 6.1 M urea, which populated a precipitation-prone species, may have identified conditions in which the altered conformer can effectively seed the monomeric protein to precipitate at neutral pH. Therefore, efforts to improve solubility need to decrease the population of the altered conformation and/or interfere with the ability of such a species to seed precipitation of monomeric leptin.

The mutational approach can be implemented to improve the solubility of protein therapeutics. The rational approach to reengineering cytokines for improved solubility requires knowing the molecular basis for structural instability of the native state and/or mechanism of aggregation of folding intermediates. Increasing the structural stability of the native state or increasing the free energy difference between the partially unfolded intermediate and the unfolded state are viable strategies to improve solubility. To elucidate the structural basis for limited solubility, insights on the mechanism of aggregation can be gained from structural analysis, the pH-dependence of stability, and the potential link between structural stability and solubility.

ACKNOWLEDGMENTS

We thank Karen Sitney and Janice Perez for cloning leptin analogs; Randy Hecht and Dwight Winters for purification of analogs; Sean Gallaher for soluble acid multimer studies; Byeong Chang, Arnold McAuley, Carl Kolvenbach, and Brian Peterson for leptin formulation and characterization efforts; Tsutomu Arakawa, John Philo, and Jennifer Liu for helpful discussions; and Gaya Ratnaswamy for assistance with structural graphics.

REFERENCES

1. A. Haselbeck, Epoetins: differences and their relevance to immunogenicity, *Curr. Med. Res. Opin.* **19**(5), 430–432 (2003).
2. J. L. Cleland, M. F. Powell, and S. J. Shire, The development of stable protein formulations: a close look at protein aggregation, deamidation, and oxidation, *Crit. Rev. Ther. Drug Carrier. Syst.* **10**(4), 307–377 (1993).
3. E. Y. Chi, S. Krishnan, T. W. Randolph, and J. F. Carpenter, Physical stability of proteins in aqueous solution: Mechanism and driving forces in nonnative protein aggregation, *Pharm. Res.* **20**(9), 1325–1336 (2003).

4. R. Krishnamurthy and M. C. Manning, The stability factor: importance in formulation development, *Cur. Pharm. Biotech.* **3**(4), 361–371 (2002).
5. M. J. Treuheit, A. A. Kosky, and D. N. Brems, Inverse relationship of protein concentration and aggregation, *Pharm. Res.* **19**(4), 511–516 (2002).
6. G. Walsh, Biopharmaceutical benchmarks—2003, *Nat. Biotech.* **21**(8), 865–870 (2003).
7. S. A. Marshall, G. A. Lazar, A. J. Chirin, and J. R. Desjarlais, Rational design and engineering of therapeutic proteins, *Drug Discov. Today* **8**(5), 212–221 (2003).
8. D. Daujotyt, G. Vilkaitis, L. Manelyt, J. Skalicky, T. Szyperski, and S. Klimaauskas, Solubility engineering of the HhaI methyltransferase, *Protein Eng.* **16**(4), 295–301 (2003).
9. L. K. Mosavi and Z. Peng, Structure-based substitutions for increased solubility of a designed protein, *Protein Eng.* **16**(10), 739–745 (2003).
10. L. Ågren, M. Norin, N. Lycke, and B. Löwenadler, Hydrophobicity engineering of cholera toxin A1 subunit in the strong adjuvant fusion protein CTA1-DD, *Protein Eng.* **12**(2), 173–178 (1999).
11. C. Tanford, *Physical Chemistry of Macromolecules* (New York: John Wiley and Sons, 1961).
12. K. L. Shaw, G. R. Grimsley, G. I. Yakovlev, A. A. Makarov, and C. N. Pace, The effect of net charge on the solubility, activity, and stability of ribonuclease Sa, *Protein Sci.* **10**(6), 1206–1215 (2001).
13. P. H. Tan, V. Chu, J. E. Stray, D. K. Hamlin, D. Pettit, D. S. Wilbur, R. L. Vessella, and P. S. Stayton, Engineering the isoelectric point of a renal cell carcinoma targeting antibody greatly enhances scFv solubility, *Immunotechnology* **4**(2), 107–114 (1998).
14. Y. Zhang, R. Proenca, M. Maffei, M. Barone, L. Leopold, and J. M. Friedman, Positional cloning of the mouse obese gene and its human homologue, *Nature* **372**(6505), 425–432 (1994).
15. C. P. Hill, T. D. Osslund, and D. Eisenberg, The structure of granulocyte-colony-stimulating factor and its relationship to other growth factors, *Proc. Nat. Acad. Sci. USA* **90**(11), 5167–5171 (1993).
16. F. Zhang, M. B. Basinski, J. M. Beals, S. L. Briggs, L. M. Churgay, D. K. Clawson, R. D. DiMarchi, T. C. Furman, J. E. Hale, H. M. Hsiung, B. E. Schoner, D. P. Smith, X. Y. Zhang, J.-P. Wery, and R. W. Schevitz, Crystal structure of the obese protein leptin-E100, *Nature* **387**(6629), 206–209 (1997).
17. K. Imagawa, Y. Numata, G. Katsuura, I. Sakaguchi, A. Morita, S. Kikuoka, Y. Matumoto, T. Tsuji, M. Tamaki, K. Sasakura, H. Teraoka, K. Hosoda, Y. Ogawa, and K. Nakao, Structure-function studies of human leptin, *J. Biol. Chem.* **273**(52), 35245–35249 (1998).
18. F. L. Rock, S. W. Altmann, M. van Heek, R. A. Kastelein, and J. F. Bazan, The leptin haemopoietic cytokine fold is stabilized by an intrachain disulfide bond, *Horm. Metab. Res.* **28**(12), 649–652 (1996).
19. J. L. Liu, K. V. Lu, T. Eris, V. Katta, K. R. Westcott, L. O. Narhi, and H. S. Lu, In vitro methionine oxidation of recombinant human leptin, *Pharm. Res.* **15**(4), 632–640 (1998).
20. L.-C. Au, S.-Y. Lin, M.-J. Li, and C.-J. Ho, pH-dependent secondary conformation of the peptide hormone leptin in different buffer solutions, *Artif. Cells Blood Substit. Immobil. Biotechol.* **27**(2), 119–134 (1999).
21. M. S. Ricci, C. A. Sarkar, E. M. Fallon, D. A. Lauffenburger, and D. N. Brems. pH dependence of structural stability of interleukin-2 and granulocyte colony-stimulating factor, *Protein Sci.* **12**(5), 1030–1038 (2003).
22. L. D. Ward, J. G. Zhang, G. Checkley, B. Preston, and R. J. Simpson, Effect of pH and denaturants on the folding and stability of murine interleukin-6, *Protein Sci.* **2**(8), 1291–1300 (1993).
23. M. S. Ricci and D. N. Brems, Common structural stability properties of 4-helical bundle cytokines: Possible physiological and pharmaceutical consequences, *Curr. Pharm. Design*, **10**:3901–3911 (2004).
24. S. B. Heymsfield, A. S. Greenberg, K. Fujioka, R. M. Dixon, R. Kushner, T. Hunt, J. A. Lubina, J. Patane, B. Self, P. Hunt, and M. McCamish, Recombinant leptin for weight loss in obese and lean adults: A randomized, controlled, dose-escalation trial, *J. Am. Med. Assoc.* **282**(16), 1568–1575 (1999).
25. I. S. Farooqi, G. Matarese, G. M. Lord, J. M. Keogh, E. Lawrence, C. Agwu, V. Sanna, S. A. Jebb, F. Perna, S. Fontana, R. I. Lechler, A. M. DePaoli, and S. O'Rahilly, Beneficial effects of leptin on obesity, T cell hyporesponsiveness, and neuroendocrine/metabolic dysfunction of human congenital leptin deficiency, *J. Clin. Invest.* **110**(8), 1093–1103 (2002).

26. G.-M. Wu, D. Hummel, and A. Herman, Analysis of the solution behavior of protein pharmaceuticals by laser light scattering photometry, in *Therapeutic Protein and Peptide Formulation and Delivery* (Washington, DC: American Chemical Society Symposium Series 675, 1997).
27. G. D. Rose, A. R. Geselowitz, G. J. Lesser, R. H. Lee, and M. H. Zehfus, Hydrophobicity of amino acid residues in globular proteins, *Science* **229**(4716), 834–838 (1985).
28. J. L. Fauchère and V. Pliska, Hydrophobic parameters pi of amino-acid side chains from the partitioning of N-acetyl-amino acid amides, *Eur. J. Med. Chem.* **18**, 369–375 (1983).
29. D. Eisenberg, R. M. Weiss, T. C. Terwilliger, and W. Wilcox, Hydrophobic moments and protein structure, *Faraday Symp. Chem. Soc.* **17**, 109–120 (1982).
30. P. Horowitz and N. L. Criscimagna, Low concentrations of guanidinium chloride expose apolar surfaces and cause differential perturbation in catalytic intermediates of rhodanese, *J. Biol. Chem.* **261**(33), 15652–15658 (1986).
31. N. A. Rodionova, G. V. Semisotnov, V. P. Kutyshenko, V. N. Uverskii, and I. A. Bolotina, Staged equilibrium of carbonic anhydrase unfolding in strong denaturants, *Mol. Biol.* (Mosk) **23**(3), 683–692 (1989).
32. E. H. Strickland, Aromatic contributions to circular dichroism spectra of proteins, *CRC Crit. Rev. Biochem.* **2**(1), 113–175 (1974).
33. D. C. Howey, R. R. Bowsher, R. L. Brunelle, and J. R. Woodworth, [Lys(B28), Pro(B29)]-Human insulin: a rapidly absorbed analog of human insulin, *Diabetes* **43**(3), 396–402 (1994).
34. N. C. Kaarsholm, K. Norris, R. J. Jorgensen, J. Mikkelsen, S. Ludvigsen, O. H. Olsen, A. R. Sorensen, and S. Havelund, Engineering stability of the insulin monomer fold with application to structure-activity relationships, *Biochem.* **32**(40), 10773–10778 (1993).
35. M. Ishikawa, H. Iijima, R. Satake-Ishikawa, H. Tsumura, A. Iwamatsu, T. Kadoya, Y. Shimada, H. Fukamachi, K. Kobayashi, S. Matsuki, and K. Asano, The substitution of cysteine 17 of recombinant human G-CSF with alanine greatly enhanced its stability, *Cell Struct. Funct.* **17**(1), 61–65 (1992).
36. T. Arakawa, S. J. Prestrelski, L. O. Narhi, T. C. Boone, and W. C. Kenney, Cysteine 17 of recombinant human granulocyte-colony stimulating factor is partially solvent-exposed, *J. Protein Chem.* **12**(5), 525–531 (1993).
37. D. Bishop, D. C. Koay, A. C. Sartorelli, and L. Regan, Reengineering granulocyte colony-stimulating factor for enhanced stability, *J. Biol. Chem.* **276**(36), 33465–33470 (2001).
38. S. R. Lehrman, J. L. Tuls, H. A. Havel, R. J. Haskell, S. D. Putnam, and C. S. Tomich, Site-directed mutagenesis to probe protein folding: Evidence that the formation and aggregation of a bovine growth hormone folding intermediate are dissociable processes, *Biochem.* **30**(23), 5777–5784 (1991).
39. P. K. Tsai, D. B. Volkin, J. M. Dabora, K. C. Thompson, M. W. Bruner, J. O. Gress, B. Matuszewska, M. Keogan, J. V. Bondi, and C. R. Middaugh, Formulation design of acidic fibroblast growth factor, *Pharm. Res.* **10**(5), 649–659 (1993).
40. R. L. Remmele Jr, N. S. Nightlinger, S. Srinivasan, and W. R. Gombotz, Interleukin-1 receptor (IL-1R) liquid formulation development using differential scanning calorimetry, *Pharm. Res.* **15**(2), 200–208 (1998).
41. A. V. Filikov, R. J. Hayes, P. Luo, D. M. Stark, C. Chan, A. Kundu, and B. I. Dahiyat, Computational stabilization of human growth hormone, *Protein Sci.* **11**(6), 1452–1461 (2002).
42. P. Luo, R. J. Hayes, C. Chan, D. M. Stark, M. Y. Hwang, J. M. Jacinto, P. Juvvadi, H. S. Chung, A. Kundu, M. L. Ary, and B. I. Dahiyat, Development of a cytokine analog with enhanced stability using computational ultrahigh throughput screening, *Protein Sci.* **11**(5), 1218–1226 (2002).
43. J. M. Sturtevant, M. H. Yu, C. Haase-Pettingell, and J. King, Thermostability of temperature-sensitive folding mutants of the P22 tailspike protein, *J. Biol. Chem.* **264**(18), 10693–10698 (1989).
44. B. Fane, R. Villafane, A. Mitraki, and J. King, Identification of global suppressors for temperature-sensitive folding mutations of the P22 tailspike protein, *J. Biol. Chem.* **266**(18), 11640–11648 (1991).
45. M. Danner and R. Seckler, Mechanism of phage P22 tailspike protein folding mutations, *Protein Sci.* **2**(11), 1869–1881 (1993).

46. C. Haase-Pettingell and J. King, Prevalence of temperature sensitive folding mutations in the parallel beta coil domain of the phage P22 tailspike endorhamnosidase, *J. Mol. Biol.* **267**(1), 88–102 (1997).
47. B. A. Chrunyk, J. Evans, J. Lillquist, P. Young, and R. Wetzel, Inclusion body formation and protein stability in sequence variants of interleukin-1β, *J. Biol. Chem.* **268**(24), 18053–18061 (1993).
48. R. Wetzel, L. J. Perry, and C. Veilleux, Mutations in human interferon-gamma affecting inclusion body formation identified by a general immunochemical screen, *Bio/Tech.* **9**(8), 731–737 (1991).

Index

Absorbance spectroscopy, 299
Actin, 143
Activation energy, 43
Adsorption, 115
Agitation, 269
Alcohol, 277
α-Chymotrypsinogen, 39
α-Crystallin, 163
α-Synuclein, 257, 265–281
Amino acids, 4, 6
Ammonium sulfate, 87
Amorphous aggregates, 61, 63
Amyloid (see also β-amyloid, fibrils), 134, 168, 176, 183, 226, 267
Analytical ultracentrifugation, 315, 316–317, 327–328, 335
Apo2L / TRAIL, 322–326
Apomyoglobin, 220, 221, 225, 231–233, 236
Arrhenius, 40
Atomic force microscopy, 257–258
Avastin™, 321

Barnase, 220, 221, 223, 231–233, 236, 248
Bead-string, 48, 51, 54
Benzyl alcohol, 94, 118
β-2-Microglobulin, 59, 248
β-Amyloid peptide, 58, 133, 156, 184, 186, 227, 257
β-Barrel, 219, 222
Betabellin, 183–184
β-Hairpin, 227
β-Helical peptides, 182
Beta-lactoglobulin, 143, 161
β-Strand peptides, 56
Bovine granulocyte colony stimulating factor (bovine G-CSF), 39, 290, 295, 296, 297, 304
Bovine pancreatic trypsin inhibitor (BPTI), 50
Bragg reflection, 172

Calorimetry, differential scanning (DSC), 102–104, 109, 110, 342
Calorimetry, isothermal titration (ITC), 104–105, 109
Capillary electrophoresis, 314, 317
Carbonic anhydrase, 256
Chaotrope, 8, 9, 88, 89

Chaperone, 163, 240, 301, 305
CHARMM, 58, 59
Chymotrypsin inhibitor, 2, 220, 221, 224, 231–233, 236
Circular diichroism (CD), 105–107, 109, 112–113, 115, 117, 134, 158, 221, 223, 225, 228, 250, 275, 319, 388
Coagulation, 173–174
Confocal microscopy, 254, 256, 258
Conformational heterogeneity, 224
Contact order, 238
Covalent aggregates, 101
Cryoelectron microscopy, 141
Cryoprotectant, 115
Crystallite, 167
Cytochrome C, 112

Debye-Hückel effect, 270
Debye-Waller factor, 199, 202, 203
Disease, protein misfolding, 4, 5
Disulfide bond, 8, 249
DLVO theory, 170, 173
Domain swapping, 228
Dye release assay, 292, 293–295, 300, 306
Dynamics, 195, 208, 209

Electrostatic interaction, 8, 170–171, 187, 228, 273, 314, 326, 340
Elongation, 156
Enthalpy, 102, 230–231
Entropy, 230, 236, 237
Equilibrium constant, 106, 320, 321, 341
Ethanol, 134, 163
Excipient, 109

Ferguson plot, 255
Fibril, 62, 134, 156, 175, 176, 180, 187, 260, 266–277
Fibrin, 143, 157, 170
Fibroblast growth factor 1 (FGF-1), 290, 295, 297, 301–302, 304
Flagellin, 143
Fluorescence spectroscopy, 250, 333, 338
Folding additive, 254

Folding intermediate, 52, 82, 162, 223, 255, 266, 268, 270, 276, 337, 339, 340, 341, 343
FoldX, 9
Formulation, 103, 105, 107, 108, 109, 114, 119, 326, 332, 333
Free energy (*see* Gibbs energy)
Freeze-drying (*also see* lyophilization), 109, 114, 200, 209, 326
Freezing, 109, 114

Gel electrophoresis, 251, 314, 317
Gel permeation chromatography (*see* size exclusion chromatography)
Gibbs energy, 9, 21, 84, 106, 218, 231, 341
Glass, 200, 203, 208
Glass transition temperature, 200, 204
Glassy solvent, 135
Glycerol, 135–140, 200, 201, 205, 207, 208
Glycosaminoglycans, 271–272
Glycosylation, 332
Go potential, 50
Guanidinium hydrochloride, 89, 91, 107, 112, 158, 162
Guinier plot, 130, 132, 181

Heat capacity, 43, 228, 230
Helical aggregates, 90
Helix bundle, 60
Heparin, 272
Herpes virus protein, 290, 295, 296, 297
Histones, 274
Hofmeister series, 8, 275
Horseradish peroxidase, 200
Hydration repulsion, 171, 187
Hydrodynamic diameter, 153, 154, 156, 288
Hydrogen bonding, 7, 51, 60, 143, 211, 228
Hydrogen-deuterium exchange, 82, 83, 85, 91, 108, 112, 126, 223
Hydrophobic core, 223, 236
Hydrophobic effect, 229, 230
Hydrophobic interaction, 7, 51, 60, 101, 184, 225, 228, 249, 266, 276
Hydrophobic interaction chromatography, 223
Hydrophobic interaction strength parameter, 61
Hydrophobicity, 7, 102, 267, 332, 337, 344

Immunogenicity, 100, 314, 332
In vivo folding, 218, 219, 248
Infrared (IR) spectroscopy, 107–109, 112–113, 114, 115, 158, 225, 333
Interferon-γ, 89, 91, 94, 95, 117
Interleukin-1 (IL-1), 342
Interleukin-2 (IL-2), 115

Intermolecular native contact parameter, 56
Irreversible aggregation, 18, 101, 314

Keratinocyte growth factor-2 (KGF-2), 290, 295–297, 304
Kinetics of aggregation, 11, 19–23, 35, 101, 158, 161, 162, 167, 268, 281, 314, 324
Kinetics of aggregation, association-limited, 26–30, 40, 41, 42
Kinetics of aggregation, rearrangement-limited, 30–32, 38, 40, 41, 42
Kinetics of aggregation, unfolding-limited, 23–26, 37, 40, 41
Kinetics of fibril formation, 65
Kinetics of refolding, 257
Kinetics of unfolding, 89, 95
Kosmotrope, 8, 9, 88
Kratky plot, 156, 159
Kyte-Doolittle hydropathy index, 231, 234

Lattice model, 50
Lennard-Jones potential, 48
Leptin, 333–347
Light scattering, 36, 156, 296–297
Light scattering, dynamic, 95, 118, 148, 152–154, 157, 158, 160, 161, 162, 163, 317, 333
Light scattering, static, 127, 148, 149–152, 158, 161, 162, 256, 315, 317, 323
Lipids, 276, 288, 289, 290, 295, 301
Liposome, 291, 296, 301–302, 305
Lorentz factor, 178
Lumry-Eyring model, 20, 24, 101
Lyophilization, 116
Lyoprotective, 200, 211
Lysozyme, 87, 134, 135–140, 163, 208, 209, 247

Mass spectroscopy, 82, 85, 314
Metals, 269
Methionine oxidation, 278
Micelle, 133
Microtubules, 141–143
Model, high resolution folding, 49, 58
Model, low resolution model, 49
Molecular crowding, 240, 272, 314
Molecular dynamics, 48, 56
Molten globule, 58, 88, 228, 235, 288–289, 295, 296, 299, 304
Monoclonal antibody (MAb), 116, 319–321, 326
Monte Carlo methods, 48, 54, 141
Morphology, 257, 267
Multimeric addition, 256
Multistate transition, 10, 32–34

INDEX

Mutagenesis, 332, 333, 340
Myelin, 168, 171, 174

Neutron scattering, inelastic, 194, 195–199, 206, 209
Neutron scattering, small angle (SANS), 125–133, 134, 140, 143, 160, 201, 205
NMR spectroscopy, 82, 85, 222, 223, 224, 228
Nonclassic transport, 289, 306
Noncovalent aggregates, 101
Nuclear magnetic resonance spectroscopy (see NMR spectroscopy)
Nucleation, 18, 135, 156, 167, 268

Off-lattice model, 50, 55, 58, 59
Osmolytes, 277

Patatin, 160
Pegylation, 332
Pesticides, 270–271
Phase diagram, 53, 56, 57, 67, 134
pKa, 7
Polar zipper, 158
Polyalanine, 59, 62, 66, 83
Polyamines, 273
Polydiacetylene, 291, 297, 300
Polydispersity, 132, 152, 154
Polyglutamine, 156
Potassium thioyanate, 87, 89, 91
Precipitation, 89, 335, 343–346
Preferential exclusion, 112
Preservation, 199, 209
Preservatives, 115
Pressure, hydrostatic, 161
Prion protein, 51, 59, 158, 180–182, 184, 227
Protofilaments, 134, 141, 143, 160, 167, 184, 187, 267
Pseudoequilibrium, 21, 26

Quasi-elastic light scattering (see also light scattering, dynamic), 195

Radius of gyration, 131, 137, 141, 150, 154
Raman scattering, 211
Rate constants, 35, 83
Rate-determining step, 37
Relaxin, 315, 319
Reversible aggregation, 101, 110, 111, 314, 319, 324
Rhodanese, 161
Ribbons, 176
Ribonuclease, 112, 227

Scattering form factor (particle scattering factor), 130, 133, 137, 150, 151, 153, 154
Scattering structure factor, 133, 135, 137, 169, 181
Sequential addition, 256
Serum amyloid A (SAA), 186
Size exclusion chromatography, 36, 95, 252, 315, 316, 319, 320, 323, 333, 343
Small angle neutron scattering (see neutron scattering)
Smoluchowski-Fuchs, 160
Solubility, 333, 335
Solvent accessible, 84, 93, 288
Solvent accessible surface area, 7, 227, 230, 233, 235, 340
Somatotropin, 290, 296, 297, 304
Stability, 84, 110, 111, 112, 200, 231, 249, 332, 340
Stability, accelerated, 42, 43, 100
Staphylococcal nuclease, 219, 221, 223, 231–233, 236
Stokes-Einstein equation, 153
Sucrose, 112, 116
Sugar, 200
Surface Racer, 234

Tailspike protein P22, 248–251, 253–262, 342
TANGO, 9
Temperature effects, 39
Temperature-senstive folding mutants, 253, 343
Thermal unfolding, 341
Thioflavin T fluorescence, 268
Tobacco mosaic virus, 143
Transthyretin, 183
Trehalose, 115, 200, 201, 202, 204, 205, 206, 211
Tubulin, 141–143
Two-state transition, 10, 19, 96, 102

Urea, 257, 340, 344

van der Waals, 187, 228, 314
Virial coefficient, 150, 169, 174
Viscosity, 326–327

X-ray crystallography, 224
X-ray diffraction (XRD), 141, 168–170, 172, 174, 175, 179, 180, 183, 184, 207
X-ray scattering, inelastic, 195
X-ray scattering, small angle (SAXS), 127, 134, 333

Yeast alcohol dehydrogenase, 200

Zimm plot, 150

THE UNIVERSITY OF MICHIGAN

DATE DUE

DEC 2 1 2006